A Textbook of Quantum Mechanics

McGraw-Hill Offices:

New Delhi
New York
St Louis
San Francisco
Auckland
Bogotá
Guatemala
Hamburg
Lisbon
London
Madrid
Mexico
Montreal
Panama
Paris
San Juan
São Paulo
Singapore
Sydney
Tokyo
Toronto

A Textbook of QUANTUM MECHANICS

P M Mathews
Professor and Head of the Department of Theoretical Physics
University of Madras

K Venkatesan
Department of Mathematics
Indian Institute of Technology, Madras

Tata McGraw-Hill Publishing Company Limited
New Delhi

1st Reprint 1977
7th Reprint 1984
8th Reprint 1984

Sponsoring Editor : Rajiv Beri

Published by Tata McGraw-Hill Publishing Company Limited
12/4 Asaf Ali Road, New Delhi 110 002 and
printed at Pearl Offset Press Private Limited, 5/33, Kirti Nagar Industrial Area
New Delhi-110015.

Preface

The aim of this book is to give a reasonably comprehensive introduction to the fundamental concepts, mathematical formalism and methodology of quantum mechanics, without assuming any previous acquaintance with the subject.

Quantum mechanics provides a characterization of microscopic physical systems in terms of essentially mathematical objects (wave functions, or at a deeper level, vectors in a linear vector space), and a set of rules enabling the information contained in the mathematical representation to be translated into physical terms. It is possible (and currently not uncommon) to develop the subject starting with these 'rules of the game' as postulates. This approach is economical and adequate if the reader's interest is limited to the use of quantum mechanics as a readymade and trustworthy tool for exploring the properties of specific systems. However, to one who is encountering the subject for the first time, the conceptual picture of physical objects (on which the mathematical base rests) would appear strange indeed, and we believe that an account of the experimental developments which compelled the adoption of such a picture, superseding the classical picture, is an essential part of an introduction to quantum mechanics. Accordingly we have devoted Chapter 1 to these developments and to a discussion of the compatibility of the quantum picture with our experience in the macroscopic domain. Chapter 2 is designed to exhibit the main characteristic features of quantum systems with the aid of simple examples and to show how these features arise from the conditions on the Schrödinger wave function. It also provides the motivating background to the basic postulates of quantum mechanics, which are formally stated in Chapter 3. In developing the general formalism based on these postulates in this Chapter, and in presenting exact solutions (Chapter 4) and approximate methods of solution of a variety of eigenvalue problems (Chapter 5) and scattering problems (Chapter 6), we have adhered to Schrödinger's wave mechanical language for the most part. However, we have sought, from an early stage, to bring out the correspondence between quantum states and vectors in a Hilbert space, and to make it clear that the Schrödinger language is just one way of describing such vectors and operations on them. To reinforce this point we have presented solutions to certain problems (e.g. the harmonic oscillator) employing algebraic methods which do not use the Schrödinger representation at all.

It is our hope that the student would find the idea of an abstract vector space underlying the Schrödinger description to be natural if not self-evident by the time the Dirac notation and representation theory are formally introduced in Chapter 7. Transformations generated by changes of coordinate frame, and the question of symmetries under such transformations, are also dealt with in this chapter. These developments are directly utilized in Chapter 8 in the treatment of angular momentum theory: the representation of spin, the coupling of two angular momenta, tensor operators, etc. The time evolution of quantum systems is dealt with in Chapter 9. Parts A and B of this Chapter are elementary and may be read before Chapter 7. In the context of the Heisenberg picture of time evolution (which, along with other pictures, is discussed in Part C), we give an elementary introduction to the concept of the *quantized* electromagnetic field and use it to calculate the rates of emission and absorption of radiation by atoms. A brief account of the representation of states of ensembles by density matrices, and of their time evolution, makes up the last part of Chapter 9. The final chapter is devoted to relativistic quantum mechanics, with emphasis on the Dirac equation and the natural way in which spin and its manifestations (magnetic moment, spin-orbit interaction) as well as the concept of the antiparticle emerge from it.

We have included a large number of examples distributed through the text, especially in the earlier chapters, to facilitate a quick grasp of the principal ideas and methods, and in some cases, to indicate their extensions or applications. Some supplementary material, as well as background material for ready reference, is given in the Appendices.

In the preparation of the book for publication, we have received cheerful cooperation and assistance from the members of the Department of Theoretical Physics, University of Madras, for which we are grateful. It is a special pleasure to thank Dr. M. Seetharaman who has read the manuscript critically and rendered valuable help in many other ways.

The idea of writing a book of this kind grew out of courses given by the first author for over ten years at Madras University. The work on the book has been supported by the University Grants Commission through a Fellowship awarded to the second author and financial assistance towards preparation of the manuscript. For this encouragement by the Commission, the authors are deeply grateful.

October 1975 P. M. Mathews
Madras K. Venkatesan

Contents

1

Towards Quantum Mechanics

A. CONCEPTS OF CLASSICAL MECHANICS

The formulation of quantum mechanics in 1925 was the culmination of the search, that began around 1900, for a rational basis for the understanding of the submicroscopic world of the atom and its constituents. At the end of the nineteenth century, physicists had every reason to regard the Newtonian laws governing the motion of material bodies, and Maxwell's theory of electromagnetism, as fundamental laws of physics. There was little or no reason to suspect the existence of any limitation on the validity of these theories which constitute what we now call *classical* mechanics. However, the discovery of the phenomenon of radioactivity and of X-rays and the electron, in the 1890s, set in motion a series of experiments yielding results which could not be reconciled with classical mechanics. For the resolution of the apparent paradoxes posed by these observations and certain other experimental facts, it became necessary to introduce new ideas quite foreign to commonsense concepts regarding the nature of matter and radiation—concepts which were implicit in classical mechanics and had an essential role in determining its consequences. It was this revolution in concepts, which led to the mathematical formulation of quantum mechanics, that had an immediate and spectacular success in the explanation of the experimental observations. Thus, to appreciate the part played by various discoveries in bringing about this revolution, it is necessary, from the very outset, to have a clear idea of the classical concepts. We, therefore, begin our studies with a brief discussion of these concepts. The rest of this chapter reviews the developments during the first quarter of this century which culminated in the establishment of the authority of quantum mechanics over the domain of microscopic phenomena.

1.1 Mechanics of Material Systems

There are two broad categories of entities which physicists have to deal with: material bodies, whose essential attribute is mass, and electromagnetic (and gravitational) fields which are fundamentally distinct from matter.

The most basic concept in the mechanics of material bodies is that of the *particle*, a point object endowed with mass. This concept emerged from Newton's observation that while the mass of a body has a central role in determining its motions, there is a wide variety of circumstances in which the size of the body is quite immaterial (e.g. in the motion of the planets around the sun). The idealization to point size was a convenient abstraction under such circumstances. Even when such an idealization of a body as a whole is obviously impossible, as when considering the internal motions of an extended object, one could imagine the object to be made up of myriads of minuscule parts, each little part being then visualized as an idealized particle. In this manner, the mechanics of any material system could be reduced to the mechanics of a system of particles. It was taken for granted that the motion of such particles, however large or small their intrinsic mass and size may be, can be pictured in just the same way as the motion of projectiles or other macroscopic objects of everyday experience. An essential part of this picture is the idea that any material object has a definite position at any instant of time. The particle idealization makes it possible to specify the position precisely, it being intuitively obvious that the position of a point object is perfectly well defined. The *trajectory* or path followed by the particle is then pictured by a sharply defined line, and the instantaneous position of the particle on the trajectory, its velocity, and its acceleration are represented by vectors $\mathbf{x}(t)$, $\dot{\mathbf{x}}(t)$ and $\ddot{\mathbf{x}}(t)$ with definite numerical values for their components. The manner in which these vary with time is governed by Newton's famous equation of motion. Newton's equation for the ith member of a system of N particles, is given by

$$m_i \ddot{\mathbf{x}}_i = \mathbf{F}_i \quad (i = 1,2,\ldots N) \tag{1.1}$$

The force $\mathbf{F}_i$ acting on the ith particle (of mass m_i) is a function of the positions of all the particles (and possibly also of the velocities). The *form* of the function is determined by the *nature* of the interactions of the particles among themselves and with external agencies. Once these are specified, Eqs. 1.1 which form a coupled system of second order differential equations can, in principle, be solved to obtain all the $\mathbf{x}_i$ as functions of t. Since the general solution of a second order ordinary differential equation contains two arbitrary constants, the general solution of the system of N such equations for vectors $\mathbf{x}_i$ depends on $2N$ arbitrary constant vectors. These may be chosen so as to satisfy specified initial conditions. In particular, given the $2N$ vectors $\{\mathbf{x}_i(t_0)\}$ and $\{\dot{\mathbf{x}}_i(t_0)\}$, i.e. the positions and velocities of all the particles at some instant t_0, as initial conditions, the $\mathbf{x}_i(t)$ are completely determined as functions of t for all t. The velocities $\dot{\mathbf{x}}_i(t)$ are then obtained by differentiation of $\mathbf{x}_i(t)$, thus determining completely the state of the system at an arbitrary time t.

The meaning ascribed to the word *state* is a crucial aspect of the difference between classical and quantum mechanics. In the classical context, knowledge of the state of a system of particles means knowing the instantaneous values of all the dynamical variables (like position, momentum, angular momentum, energy, etc.). Since these are all functions of the position coordinates and velocities (or momenta) of the constituent particles, complete information about the state is implicit in the knowledge of just these quantities. One may, therefore, say that *in classical mechanics, the state of a system of particles at any time t is represented by the instantaneous values* $\{\mathbf{x}_i(t)\}$ *and* $\{\dot{\mathbf{x}}_i(t)\}$ *of the position coordinates and velocities of all the particles.* The concluding statements of the last paragraph may now be re-expressed as follows: Given the state at any instant of time t_0, the state at any other time t is uniquely determined by the equations of motion (1.1). It is the system of equations 1.1 taken *together with* the concept of the state as outlined above which constitutes the classical mechanics of particle systems.

We have based our discussion of classical mechanics above on the Newtonian form of the equations of motion. As is well known, there are other equivalent formulations of classical mechanics, notably the Lagrangian and Hamiltonian formulations. We present a brief account of these in Appendix A. As we shall see later, the Hamiltonian formalism has a special role to play in the process of setting up the quantum mechanical equations of motion. But for the present we will content ourselves with the following general observations: In the Hamiltonian formalism, corresponding to each of the $3N$ independent coordinate variables of an N-particle system a canonically conjugate momentum variable is defined. The $3N$ coordinates and the $3N$ momenta conjugate to them are independent variables at any instant of time. The state of the system is completely specified by giving precise numerical values to these *momenta* (instead of velocities as in the Newtonian formulation) *and coordinates*. The Hamiltonian equations of motion, which are a system of $6N$ first order differential equations for the coordinates and momenta, determine the manner in which the state changes with time. These $6N$ equations are of course equivalent to the $3N$ second order equations 1.1.

It is worthwhile, at this point, to recall Einstein's discovery that the Newtonian and equivalent forms of mechanics (with the conventional definitions for forces, momenta, energy, etc.) hold good only if the speeds of all the particles are very small compared to c, the velocity of light in vacuo. Whenever velocities v of the order of c are involved, the more general equations following from Einstein's *special theory of relativity* (1904) have to be used. The mechanics of classical particles based on these generalized (*relativistic*) equations is called the *relativistic classical mechanics*. The Newtonian mechanics, being not valid when relativistic effects are present, is said to be *non-relativistic*. We will need to use the concepts and results of relativity theory at several points in our study. A summary of those aspects of the theory which are most relevant for our purposes is presented in Appendix B.

While relativity theory made possible a consistent description of physical phenomena involving fast-moving objects or observers, it is quantum mechanics which provided the key to the understanding of the behaviour of very small objects like the atom and its constituents. The failure of classical mechanics when applied to the submicroscopic world of the atom arose not from any inadequacy of the *form* of the equations of motion as such but from investing the symbols occurring in them with the conventional meanings idealized and extrapolated from the mechanics of macroscopic objects. This will be particularly evident when we consider the Heisenberg version of quantum mechanics. The essential step in the formulation of the new (quantum) mechanics was the abandonment of the underlying *concepts* of classical mechanics, like the mental picture of the path of a particle as a sharply defined line and the possibility of characterizing the instantaneous state by precise positions and velocities. The new concepts which replaced these are necessarily divorced from such intuitive pictures and therefore appear rather strange initially, but they have their own beauty (especially in the simplicity and elegance of the associated mathematical structures) which becomes evident with a little familiarity.

1.2 Electromagnetic Fields and Light

The discovery of the phenomena of interference and polarization of light early in the nineteenth century provided convincing evidence that light is a wave phenomenon. The *nature* of light waves was identified some decades later, following Maxwell's formulation of the electromagnetic theory. It was then recognized that light consists of electromagnetic waves and presents just one manifestation of the general phenomena of electromagnetism, governed by the fundamental equations of Maxwell:

$$\left.\begin{aligned} \frac{1}{c}\frac{\partial \mathcal{E}}{\partial t} &= \text{curl}\,\mathcal{H} - \frac{4\pi}{c}\mathbf{j}, \\ \text{div}\,\mathcal{E} &= 4\pi\rho, \\ \frac{1}{c}\frac{\partial \mathcal{H}}{\partial t} &= -\,\text{curl}\,\mathcal{E}, \\ \text{div}\,\mathcal{H} &= 0 \end{aligned}\right\} \tag{1.2}$$

Here $\mathcal{E} \equiv \mathcal{E}(\mathbf{x}, t)$ and $\mathcal{H} \equiv \mathcal{H}(\mathbf{x}, t)$ stand for the electric and magnetic fields at $\mathbf{x}$ at time t, and $\rho(\mathbf{x},t)$ and $\mathbf{j}(\mathbf{x},t)$ are the electric charge and current densities, respectively. At any time t, the vector fields $\mathcal{E}(\mathbf{x},t)$ and $\mathcal{H}(\mathbf{x},t)$ are implicitly assumed to have definite numerical values for their components at any point $\mathbf{x}$. This seemingly self-evident supposition is the essential feature of *classical* electromagnetic theory. Since all properties like the energy density, momentum density, etc., of the electromagnetic field are functions of the instantaneous values of $\mathcal{E}$ and $\mathcal{H}$, specification of $\mathcal{E}$ and $\mathcal{H}$ for all $\mathbf{x}$ at t_0 amounts to a complete description of the *state* of the classical electromagnetic field at that instant.

Further, such a specification provides the initial conditions necessary for identifying a particular solution of the first order differential equations (1.2). Therefore, if the state of the field at some instant t_0 is given, the state at any other time can be determined uniquely (at least in principle), provided of course that the charge and current densities ρ and $\mathbf{j}$ are known as functions of $\mathbf{x}$ and t. It is the set of equations (1.2) *together with* the concept of $\mathcal{E}$ and $\mathcal{H}$ as ordinary vector fields amenable to simultaneous specification with arbitrary precision at any given time, that constitutes what is known as the classical electromagnetic theory.

It is worthwhile at this point to notice the obvious but important fact that the classical pictures of material particles on the one hand, and of light (or more generally, electromagnetic fields) on the other, are mutually exclusive. While the ideal particle is a point object, a field necessarily exists over a region of space. In particular, the purest form of light, which is monochromatic, is a simple harmonic wave with a definite wavelength, existing throughout space with a uniform energy density everywhere. It would be impossible within the framework of this classical picture to conceive of a particle possessing wave-like properties or of light exhibiting particle-like behaviour. Yet, experiments in the early 1900s gave firm evidence for the existence of both these kinds of phenomena. Let us now go on to a discussion of these and other experimental results which revealed the limitations on the validity of classical concepts.

B. INADEQUACY OF CLASSICAL CONCEPTS[1]

(i) Macroscopic Statistical Phenomena

1.3 Black Body Radiation; Planck's Quantum Hypothesis

One of the very few things for which classical theory had been unable to offer an explanation till the end of the last century was the nature of the distribution of energy in the spectrum of radiation from a black body. By definition, a black body is one which absorbs all the radiation it receives. As is well known, the best practical realization is an isothermal cavity with a small aperture through which radiation from outside may be admitted. The cavity always contains radiation emitted by the walls, the spectrum of radiation being characterized by a function $u(\nu)$ where $u(\nu)d\nu$ is the energy (per unit volume of the cavity) contributed by radiation with frequencies between ν and $\nu + d\nu$. It had been deduced from very general thermodynamical arguments that the form of the function $u(\nu)$ depends only on the temperature T of the cavity. Efforts to deduce the actual functional form

[1] For a fuller account, especially of experimental details, see for example, Max Born, *Atomic Physics*, 5th ed., Blackie and Sons, London, 1952; F. K. Richtmyer, E. H. Kennard and T. Lauritsen, *Introduction to Modern Physics*, McGraw-Hill, New York, 1955.

from classical theory led to the Rayleigh-Jeans formula, $u(\nu) = \text{const.}\ \nu^2$. Except at low frequencies this law was in violent disagreement with experimental observations which showed $u(\nu)$ falling off after reaching a maximum as ν was increased. That was how matters stood until 1900, when Planck announced the discovery of a law which reproduced perfectly the experimental curve for $u(\nu)$:

$$u(\nu) = \frac{8\pi\nu^2}{c^3} \cdot \frac{h\nu}{e^{h\nu/kT} - 1} \tag{1.3}$$

This formula contains, besides the Boltzmann constant[2] k, a new fundamental constant h, with the value

$$h = 6{\cdot}626 \times 10^{-27} \text{ erg-sec.} \tag{1.4}$$

It is called *Planck's constant.*

The essential new ingredient in the derivation of the law was the following *ad hoc* hypothesis:

The emission and absorption of radiation by matter takes place, not as a continuous process, but in *indivisible discrete units* or *quanta* of energy. The magnitude ε of the quantum is determined solely by the frequency of the radiation concerned, and is given by

$$\varepsilon = h\nu \tag{1.5}$$

where h is Planck's constant.

This hypothesis was a revolutionary break from classical radiation theory based on Maxwell's equations (1.2). According to the classical theory, oscillating charges are responsible for the emission (or absorption) of electromagnetic radiation with frequency equal to that of the charge oscillations. Emission or absorption takes place *continuously* at a rate determined by the parameters of the oscillating system. The success of Planck's hypothesis was the first indication that one might have to look beyond classical theories for the understanding of at least some areas of physics. Fig. 1 shows the various regions of the electromagnetic spectrum viewed from both the wave and quantum points of view.

Let us now examine briefly how Planck's law (1.3) follows from his quantum hypothesis. We make use of the fact that the electromagnetic waves, which

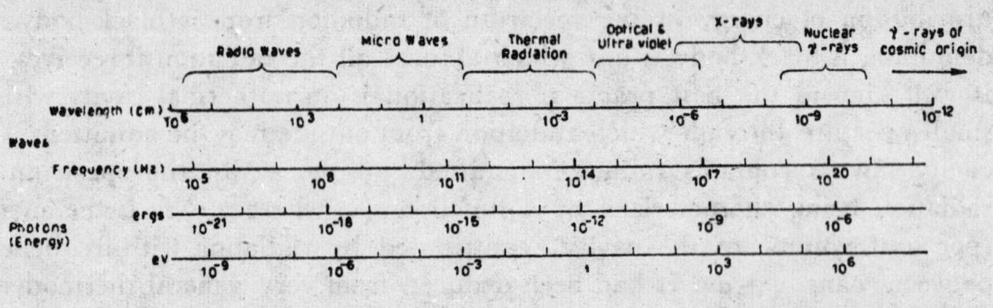

Fig. 1.1 Characterization of electromagnetic waves and quanta in various regions of the spectrum.

[2] A table of fundamental constants and other data of interest is provided at the end of the book. As each physical problem is considered, the student is urged to familiarize himself with the orders of magnitude of the numbers involved.

constitute the radiation in the cavity, can be analyzed into a superposition of normal modes characteristic of the cavity. In each normal mode, the fields vary with time in simple harmonic fashion, in unison throughout the cavity. Thus each normal mode is equivalent to a simple harmonic oscillator, and the radiation field forms an assembly or *ensemble* of such oscillators. The absorption (or emission) of radiation by the walls of the cavity is equivalent to a transfer of energy to (or from) the walls by (or to) the oscillators. As a result of such energy exchanges, which are continually taking place, the ensemble of radiation oscillators comes into thermal equilibrium at the temperature T of the walls of the cavity. Under these conditions, different oscillators having a given frequency ν have different energies at any given time, but their *average* energy $\bar{E}(\nu)$ has a definite value determined by the temperature T. The energy of radiation in the frequency range ν to $\nu + d\nu$ is then simply the number of normal mode oscillators $n(\nu)d\nu$ having frequencies within this range, multiplied by the average energy $\bar{E}(\nu)$ per oscillator. Thus if V is the volume of the cavity,

$$V.u(\nu)d\nu = n(\nu)d\nu.\,\bar{E}(\nu) \tag{1.6}$$

The counting of the oscillators is simply a geometrical problem, and it is easily shown (see end of Sec. 2.5) that

$$n(\nu) = \frac{8\pi\nu^2 V}{c^3} \tag{1.7}$$

The determination of $\bar{E}(\nu)$ is done by applying the standard results of statistical mechanics to the ensemble of oscillators. Statistical mechanics tells us that an individual member of an ensemble in thermal equilibrium at temperature T has energy E with probability

$$P_E = \frac{e^{-E/kT}}{\sum\limits_E e^{-E/kT}} \tag{1.8a}$$

so that its average energy is

$$\bar{E} = \sum_E EP_E \tag{1.8b}$$

The summations are to be taken over all values which the energy E (of any member of the ensemble) may take. The spectrum of permissible values of E is thus of crucial importance in determining $\bar{E}$. It is here that Planck's hypothesis comes into play. It implies that normal mode oscillators of the radiation field with the frequency ν can have only the energy values

$$E(\nu) = nh\nu, \quad n = 0,1,2,\ldots \tag{1.9}$$

This is because the field oscillator gets its energy by emission from the cavity walls (and loses energy through absorption by the walls) only in packets or quanta of magnitude $h\nu$. On substituting the values (1.9) for $E(\nu)$ in Eqs. 1.8, we immediately get[3]

[3] The denominator is a geometric series whose sum is $D = (1 - e^{-\beta h\nu})^{-1}$, where $\beta = (1/kT)$. The numerator is observed to be nothing but $-\,\partial D/\partial\beta = h\nu e^{-\beta h\nu}\,(1 - e^{-\beta h\nu})^{-2}$.

$$\left.\begin{aligned}\bar{E}(\nu) &= \frac{\sum_{n=0}^{\infty} nh\nu . e^{-nh\nu/kT}}{\sum_{n=0}^{\infty} e^{-nh\nu/kT}} \\ &= \frac{h\nu}{e^{h\nu/kT}-1}\end{aligned}\right\} \tag{1.10}$$

for the mean energy of a field oscillator. When the expressions (1.10) for $\bar{E}(\nu)$ and (1.7) for $n(\nu)$ are employed in Eq. 1.6 we obtain the Planck distribution law (1.3).

According to the classical theory, $E(\nu)$ could have *any* value from 0 to ∞, and the same thing would effectively happen if the quantum $h\nu$ in the above treatment had a vanishingly small magnitude. Therefore, passage to the limit $h \rightarrow 0$ in Eq. 1.10 should lead to the value kT predicted by the equipartition theorem of classical statistical mechanics, and indeed it does. With this value [$\bar{E}(\nu) = kT$] substituted in Eq. 1.6, one gets the Rayleigh-Jeans law for $u(\nu)$ which we have already seen to be incorrect. It appears, therefore, that the very small but *nonzero* value of the constant h is a measure of the failure of classical mechanics. This surmise is indeed confirmed by the mathematical formulation of quantum mechanics.

EXAMPLE 1.1 — If quantum effects are to be manifested through a departure of $\bar{E}(\nu)$ from its classical value kT, the frequency should be high enough, so that $(h\nu/kT)$ becomes comparable to unity. For room temperatures ($T \approx 300°$ K), $(h\nu/kT) \approx 1/6$ for $\nu = 10^{12}$ Hz. It is only when 'oscillators' of at least this frequency are involved, that quantum statistical effects become noticeable at room temperature.

1.4 Specific Heats of Solids

That the success of Planck's hypothesis was no mere accident became evident when precisely the same kind of ideas provided the solution for another puzzling problem of classical physics. It is well known that atoms in solids execute oscillations about their mean positions due to thermal agitation. Each atom may be thought of as a three-dimensional harmonic oscillator and its mean thermal energy should be three times that of a simple (one-dimensional) harmonic oscillator. As we saw in the last paragraph, classical theory predicts the latter to be kT. Therefore, the thermal energy of a solid should be $3kT$ per atom, or $3RT = 3NkT$ per gram-atom (containing N atoms where N is the Avogadro number, $6{\cdot}022 \times 10^{23}$). The atomic heat (i.e., the rate of increase of thermal energy with temperature, per gram-atom) then becomes $3R$, a universal constant. Many solids do conform to this expectation (at least approximately) at ordinary temperatures, as observed by Dulong and Petit empirically. But when the temperature is lowered sufficiently, the specific heat decreases instead of remaining constant, and indeed goes down to zero as T approaches 0 K. Einstein[4] observed that this behaviour can be simply

[4] A. Einstein, *Ann. d. Physik*, **22**, 180, 1907.

explained if it is postulated that the energy of oscillation of any atom in a solid can take only a *discrete* set of values—just like the energy of Planck's field oscillators. More precisely, it was proposed that the energy associated with each component (in the x, y, z directions) of the oscillation of an atom be constrained to take only one of the values $nh\nu$ ($n = 0, 1, 2, \ldots$), where ν is now the frequency of oscillation of the *atom*. The mean energy per atom then has exactly the form (1.10), except for an extra factor 3 coming from the three directions of motion. If it is assumed that all atoms have the same frequency of oscillation, one immediately obtains the atomic heat as

$$C = \frac{d}{dT}\left(\frac{3Nh\nu}{e^{h\nu/kT} - 1}\right) = 3R\left[\frac{e^{h\nu/kT}}{(e^{h\nu/kT} - 1)^2}\cdot\left(\frac{h\nu}{kT}\right)^2\right] \tag{1.11}$$

It is evident that this formula has the desired property of a gradual decrease in C as the temperature is lowered. It was found in fact that the behaviour of specific heats of solids is rather well accounted for by this formula, with a suitable choice of ν in each case.

Einstein's derivation is by no means the last word on the theory of specific heats of solids. But it suffices for the purpose of displaying one of the early manifestations of the inadequacy of classical concepts, namely the need for the supposition that harmonic oscillators—whether radiation oscillators as in Planck's theory, or material oscillators as in Einstein's theory—can take only discrete energy values.

(ii) Electromagnetic Radiation—Wave-Particle Duality

1.5 The Photoelectric Effect

We have already mentioned the success of Planck's hypothesis, which makes it appear that in the process of emission or absorption, light behaves as if it were a particle-like bundle of energy. However, the evidence here for the quantum nature of energy exchange between radiation and matter is indirect. It was Einstein[5] who drew attention to the fact that the phenomenon of *photoelectric emission* could be understood in terms of Planck's hypothesis and provides a direct verification of the hypothesis. This phenomenon is the emission of electrons by many metals (especially the alkali metals) when irradiated by light. Applying Planck's hypothesis, Einstein proposed that absorption of light of frequency ν by electrons in the metal takes place as discrete quanta of energy $h\nu$. If $h\nu$ exceeds the amount of energy W needed for an electron to escape from the surface of the metal, the electrons absorbing such quanta may escape with energies upto a maximum value

$$E_{max} = h\nu - W \tag{1.12}$$

Available experimental data were in conformity with this equation, which was accurately confirmed by later experiments. On the basis of the classical theory of absorption of light, it would be practically impossible to understand the existence of a maximum electron energy related to the frequency ν of the

[5] A. Einstein, *Ann. d. Physik*, **17**, 132, 1905.

incident radiation, as well as the absence of any photoelectric emission (i.e. the inability of electrons to acquire the amount of energy required for escape) when ν is below a definite value $\nu_0 \equiv W/h$. In the classical theory there is no reason to expect any sensitive frequency dependence for the amount of energy which an electron in the metal could ultimately accumulate by gradual absorption from the incident light wave; nor can one see why there should be a definite upper limit on the energy so absorbed. Other properties of the photoelectric emission are also equally difficult to understand on the classical picture, but are almost self-evident when this phenomenon is viewed as the instantaneous absorption of *light quanta* by the electrons with which they collide. For example, photoelectric emission starts instantly when light falls on the emitter, however weak the light intensity may be. (Classically, the electron would need some time to absorb enough energy to escape.) Under irradiation with monochromatic light, the rate of emission of electrons is directly proportional to the light intensity—which is exactly what would be expected on the quantum picture since the number of quanta (and hence, of the collisions with electrons) is evidently proportional to the intensity.

We conclude, therefore, without further discussion that in the photoelectric effect, light behaves as a collection of corpuscles and not as a wave. At the same time, we know only too well that the phenomena of diffraction, etc, require light to be waves. How are we to escape the paradox created by the existence of two quite irreconcilable manifestations for one and the same physical entity? One possibility is to suppose that light *propagates* in the form of waves and therefore undergoes diffraction etc., but assumes corpuscular character (in some unexplained manner) at the instant of absorption (or emission) by material objects. However, even this supposition, far-fetched as it is, was made untenable by the discovery of the *Compton effect*[6] in the scattering of X-rays.

EXAMPLE 1.2 — It is an experimental fact that if at all there is any delay between the commencement of irradiation and the emission of photoelectrons, it is less than 10^{-9} sec. An electron requires, let us say, 5×10^{-12} ergs (about 3 eV) to escape from the irradiated metal. If this much energy is to be absorbed classically (in a continuous fashion), the rate of absorption must be at least 5×10^{-3} ergs/sec. If the light energy is continuously distributed over the wave front, the electron can only absorb the light incident within a small area near it, say 10^{-15} cm^2 (i.e. of the order of the square of the interatomic distance). Therefore, the intensity of illumination required would be at least $(5 \times 10^{-3}/10^{-15}) = 5 \times 10^{12}$ ergs/sec/cm^2, that is half a million watts/cm^2! Clearly, explanation of the photo-effect in classical terms is not feasible.

1.6 The Compton Effect

That X-rays are electromagnetic waves (differing from light only in the considerably higher values of frequency) had become clear fairly soon after

[6] A. H. Compton, *Phys. Rev.*, 21, 483, 1923; 22, 409, 1923.

their discovery. Their wave nature was amply confirmed by the Laue photographs (1913) showing the diffraction of X-rays by crystals. Yet, barely ten years later, Compton had to invoke the extreme quantum picture to explain the fact that when monochromatic X-rays are scattered, part of the radiation scattered in any given direction has a definite wavelength *higher* than that of the primary beam. Assuming that X-rays of wavelength λ consist of a stream of corpuscles or quanta of energy $E = h\nu = hc/\lambda$, Compton theorized that when one of these quanta hits any free or loosely bound electron in the scatterer, the electron (being quite a light particle) would recoil. Its kinetic energy has to come from the energy of the incident quantum, and the latter would be left with an energy $E' < E$ after the collision (in which it gets scattered). The frequency $\nu' = E'/h$ of the X-rays so scattered would therefore be less than ν and the corresponding wave length $\lambda' > \lambda$. On this picture, quantitative calculation of $\Delta\lambda = \lambda' - \lambda$ can be made from considerations of energy and momentum conservation in the collision. Since the energy and momentum transported by electromagnetic radiation are known to be related by a factor c, the momenta of the X-ray quantum before and after scattering are given by

$$p = \frac{E}{c} = \frac{h\nu}{c} = \frac{h}{\lambda}, \text{ and } p' = \frac{E'}{c} = \frac{h\nu'}{c} = \frac{h}{\lambda'} \tag{1.13}$$

The electron recoiling from the collision may have a velocity comparable to c, and therefore the relation between its energy W and its momentum P has to be taken as the relativistic one (see Appendix B):

$$W = (m^2c^4 + c^2P^2)^{1/2} \tag{1.14}$$

where m is the rest mass of the electron. The energy of the electron before collision may be taken to be the rest energy $W_0 = mc^2$ since the initial kinetic energy is relatively very small. The configuration of the scattering event is shown in Fig. 1.2 where, following current practice, the quantum of radiation is depicted by a wavy line and the electron by a straight line. The equations of conservation of the momentum components perpendicular and parallel to the direction of the incident quantum are

$$p' \sin\theta = P \sin\varphi \tag{1.15a}$$

$$p - p' \cos\theta = P \cos\varphi. \tag{1.15b}$$

The energy conservation equation is $E + W_0 = E' + W$, or in view of Eqs. 1.13 and 1.14

$$cp + mc^2 = cp' + (m^2c^4 + c^2P^2)^{1/2} \tag{1.15c}$$

These equations are to be solved for p' which is related to λ'. We can eliminate φ first by squaring Eqs. 1.15a and 1.15b and adding. Then we get $P^2 = p^2 + p'^2 - 2pp' \cos\theta$. On substituting this expression for P^2 in Eq. 1.15c and eliminating the square root, we obtain

$$[c\,(p - p') + mc^2]^2 = m^2c^4 + c^2\,(p^2 + p'^2 - 2pp' \cos\theta)$$

This simplifies to $2mc^2.c\,(p - p') - 2c^2\,pp' = -\,2c^2pp'\cos\theta$, whence

$$mc\left(\frac{1}{p'} - \frac{1}{p}\right) = 1 - \cos\theta.$$

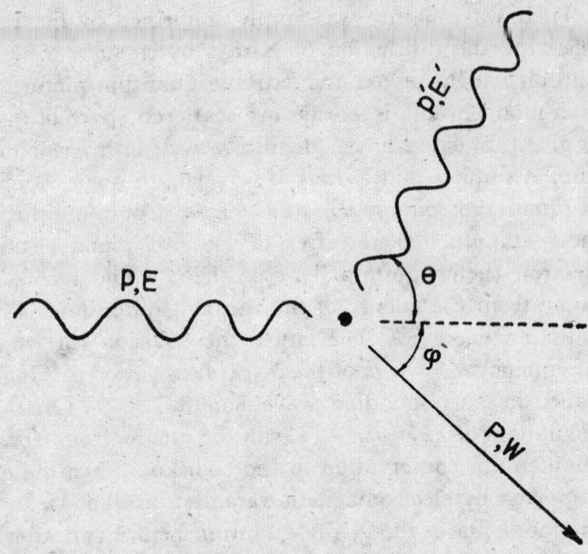

Fig. 1.2 Compton scattering

The use of Eqs. 1.13 enables us to write this finally in terms of λ, λ′ as

$$\Delta\lambda \equiv \lambda' - \lambda = \frac{h}{mc}(1 - \cos\theta) \tag{1.16}$$

This result was verified by Compton and his coworkers immediately after its derivation.

The wave length shift (h/mc) when the scattering is at 90° is called the *Compton wavelength* associated with the electron:

$$\lambda_C = (h/mc) = 2{\cdot}426 \times 10^{-10} \text{ cm} \tag{1.17}$$

This constant, which is characteristic of the X-ray scattering with wave length shift (Compton effect) could not be reproduced by applying classical electromagnetic theory to the scattering process. The success of the theory presented above thus implies unequivocally that in the Compton effect, electromagnetic radiation manifests itself as a stream of corpuscles or quanta *during propagation*. The occurrence of corpuscular behaviour in emission or absorption processes (as in photoelectric effect) has been already referred to. It seems therefore that electromagnetic quanta are not merely some shadowy concept but have real physical existence. Such quanta are now known under the name of *photons*.

The ability of radiation to manifest itself either as waves or as photons is referred to as the *wave-particle* dualism. Side by side with the developments leading to the recognition of the dual character of electromagnetic radiation, other equally startling discoveries were being made regarding the fundamental nature of *material* systems. We now turn our attention to these developments.

(iii) Atomic Structure and Atomic Spectra

1.7. The Rutherford Atom Model

The first outlines of the structure of atoms, of which all matter is constituted, began to be discernible soon after Thomson's discovery of the electron (1897). It became known then that an atom consists of a number of negatively charged electrons plus a positive residue which carries almost the entire mass of the atom. But it was only in 1911, with Rutherford's[7] analysis of the data on scattering of alpha particles by thin foils, that the picture of the atom became clearly defined. Rutherford came to the conclusion that the observed high proportion of alpha particles suffering large-angle scattering required that the heavy positive part of the atom be concentrated in a *nucleus*, whose size is extremely small compared to the dimension of the atom itself. This immediately suggested a structure for the atom resembling that of the solar system, with the electrons revolving in orbits around the nucleus (like planets around the sun). The Coulomb (electrostatic) attraction between each electron and the oppositely charged nucleus provides the force which holds the atom together. It was realized immediately that this picture of the atom encounters serious difficulties of principle in the context of classical theory. In fact, such a structure should not be stable at all, for the orbital motion of the electrons (which are charged particles) should cause them to emit radiation continuously. The consequent loss of energy should make the paths go spiralling inwards until the electrons 'collapsed' into the nucleus. During this process the frequency of the emitted radiation, which coincides with that of orbital motion, should be continually increasing. Obviously none of these things happens. The collapse of the kind envisaged does not take place; in fact, atoms have tremendous stability. Nor does the light actually emitted by atoms have the continuum character demanded by the above picture. As is well known the most important feature of atomic spectra is the presence of very sharp, discrete, spectral lines which are characteristic of the emitting atom. In brief, acceptance of Rutherford's nuclear model of the atom meant also recognition of a complete breakdown of the classical mechanism of radiation in the case of the atom.

1.8 Bohr's Postulates

This situation was tackled by Niels Bohr[8] who adopted the Rutherford model of the atom, overcoming its unacceptable consequences by *postulating* that the classical theory of radiation does not apply to the atom. He enunciated the following further postulates concerning the dynamical behaviour of atoms:

(i) The system of electrons and nucleus which constitute the atom cannot exist in any arbitrary state of motion allowed by the classical mechanics. The system can exist only in certain special states characterized by definite

[7] E. Rutherford, *Phil. Mag.*, **21**, 669, 1911.
[8] N. Bohr, *Phil. Mag.*, **26**, 1, 1913.

discrete values of the total energy. These are *stationary states*, in any of which the atom can remain indefinitely without radiating.

(ii) Emission of electromagnetic radiation takes place when (and only when) the atom 'jumps' from one of the stationary states with energy E_i to another with energy $E_f < E_i$. The frequency of the radiation emitted is given by

$$\nu = \frac{E_i - E_f}{h} \tag{1.18}$$

The first postulate extends the idea of discreteness of energy values, which originated with Planck, to the individual atom. The energy values associated with stationary states are called the *energy levels* of the atom. The second postulate incorporates the idea that emission of radiation takes place in discrete quanta, and adopts the Einstein relation between the energy of a quantum and the frequency of the associated radiation (first employed in the explanation of the photoelectric effect). It is an immediate consequence of these postulates that atomic spectra should consist of discrete lines, as observed. Eq. 1.18 states that the frequencies of spectral lines of any atom are differences between 'spectral terms' (stationary state energies, divided by h) which are characteristic of the atom. This general property of atomic spectra had been already observed empirically and was known as the Rydberg-Ritz combination principle (1905). Thus it seemed certain that Bohr's ideas were essentially sound. In fact, the supposition that the atom can have only discrete energy levels was directly verified from experiments by Franck and Hertz[9] on the scattering of monoenergetic electrons by atoms. They found that as long as the electron energy was below a certain minimum value, the scattering was purely elastic, indicating that the atom is incapable of accepting energies less than this amount. This behaviour is exactly what is demanded by the Bohr picture: an atom in its lowest energy level E_0 must get an amount of energy $(E_1 - E_0)$ in order to go to the next permissible level E_1. If energy exceeding this minimum is supplied to the atom, it can still take up only the exact amount $(E_1 - E_0)$, or $(E_2 - E_0)$ etc. The results of the Franck-Hertz experiment were indeed in accordance with this expectation. When the energy of the incident electrons was increased sufficiently, inelastic scattering with the absorption of discrete amounts of energy was found to take place.

While the postulates stated above provide a frame-work for the understanding of the stability of atoms and the general features of atomic spectra, they do not indicate how the stationary states are to be identified or how the energy levels E_i are to be determined for a particular system. But by supplementing these postulates by a *quantum condition*, Bohr was able to calculate the energy levels of the hydrogen atom and thus determine its spectral frequencies. His results were in agreement with the empirically deduced Balmer formula,

$$\frac{\nu}{c} = R\left(\frac{1}{n^2} - \frac{1}{m^2}\right), \quad n,m = 1,2,\ldots; \; m > n \tag{1.19}$$

[9] J. Franck and G. Hertz, *Verhandl. deut. phys. Ges.*, **16**, 457, 512, 1914.

where R is the so-called Rydberg constant. This was a spectacular triumph for the Bohr theory and inspired much of the later work which provided the guidance towards a more fundamental theory. Let us therefore consider it briefly before discussing the implications and limitations of Bohr's theory in general.

1.9 Bohr's Theory of the Hydrogen Spectrum

The hydrogen atom consists of a proton (which is its nucleus) and a single electron. According to the Rutherford model, the electron moves in an orbit around the nucleus; the latter, being relatively very heavy, remains practically at rest. Since the force of attraction between the two is electrostatic and therefore obeys the inverse square law (just like the gravitational force in the problem of planetary motion) the possible orbits are circular or elliptical. Suppose the electron is moving in a circular orbit of radius a with speed v (which is constant). In this orbit, the electrostatic attractive force e^2/a^2 is balanced by the centrifugal force mv^2/a, e being the charge of the electron. This fact gives us the relation

$$mv^2 = \frac{e^2}{a} \tag{1.20}$$

which can be used to eliminate v from the expression for the energy E:

$$E = \tfrac{1}{2}\, mv^2 - \frac{e^2}{a} = -\tfrac{1}{2}\frac{e^2}{a} \tag{1.21}$$

Thus the total energy is half the potential energy. Classically, the orbital radius a can take any positive value, and therefore E can be anything from $-\infty$ to 0. However, Bohr's first postulate asserts that only a special set from among these orbits is available to the electron. Bohr proposed that these special orbits which characterize the stationary states are those in which the angular momentum l of the electron about the centre of the orbit (i.e. the position of the nucleus) is an *integral* multiple of $\hbar \equiv h/2\pi$. The angular momentum in a *circular* orbit is of course just the product of the linear momentum mv and the orbital radius a. Thus Bohr's quantum condition is given by

$$mva = n\hbar \tag{1.22}$$

The integer n whose values identify the various stationary states is called a *quantum number*. The radii of the allowed ('quantized') orbits are now obtained by eliminating v between the Eqs. 1.20 and 1.22. For the nth orbit one has

$$a = \frac{n^2\hbar^2}{me^2} \tag{1.23}$$

Substitution of this expression for a into Eq. 1.21 gives us the quantized energy levels of the hydrogen atom as:

$$E_n = -\frac{me^4}{2n^2\hbar^2}, \quad n = 1,2,\ldots \tag{1.24}$$

The state of the lowest energy (the *ground state*) corresponds to $n = 1$. The energies of the other 'excited' states increase with n, tending to 0 as $n \to \infty$.

Knowing the energy levels, we can immediately obtain the frequencies of the hydrogen spectrum, using Eq. 1.18. If the atom jumps from an initial state with the quantum numbers $n = n_i$ to another with $n = n_f < n_i$ we find by substitution of the corresponding energies from Eq. 1.24 into 1.18, that the radiation emitted has the frequency

$$\nu = \frac{me^4}{4\pi\hbar^3}\left(\frac{1}{n_f^2} - \frac{1}{n_i^2}\right) \tag{1.25}$$

As already noted, this result agrees with the Balmer formula (1.19) and provides a theoretical value $R = (me^4/4\pi c\hbar^3)$ for the Rydberg constant.[10] We will see later that exactly the same formula follows from the quantum mechanical theory also, though the meaning of the quantum number n there is quite different.

1.10 Bohr-Sommerfeld Quantum Rules; Degeneracy

While this first-ever theoretical derivation of an atomic spectrum was indeed an exciting event, it was clear even then that the quantum condition (1.22) used in the derivation could not be applied, in that form, in more complicated cases. However, it was soon realized that Eq. 1.22 is a special case of the condition

$$\oint p\, dq = nh \tag{1.26}$$

which had been employed already by Planck in connection with his theory of black body radiation.[11] Here q is some generalized coordinate and p, the corresponding canonically conjugate momentum. The condition is applicable in the case of periodic motion only, and the integral is to be taken over one period, treating p as a function of the position q of the particle on the actual trajectory. Eq. 1.22 corresponds to choosing q to be the angular position φ (varying from 0 to 2π for one period) and p as its conjugate, the angular momentum (which is independent of φ, being a constant of the motion.) Bohr postulated that the quantum condition (1.26) is applicable to any system with one degree of freedom. This quantum rule was further generalized by Sommerfeld to *multiply-periodic* systems, with many degrees of freedom, i.e. systems which can be described by pairs of coordinate and momentum variables (q_1,p_1), (q_2,p_2), ..., (q_N,p_N), each of which is periodic, with possibly different periods for different pairs. The generalized *Bohr-Sommerfeld quantum rule*[12] is given by:

[10] This value is strictly correct only for an infinitely heavy nucleus which remains perfectly static. To indicate this fact explicitly, the notation R_∞ is often used for the constant. To take into account the finiteness of the mass (m_N) of the nucleus (and the consequent motion of the nucleus) we have to replace the electronic mass m in the expression for R by the reduced mass $m\, m_N/(m + m_N)$. The Rydberg constant for a finite nucleus is thus $R_N = R_\infty\ (1 + m/m_N)^{-1}$.

[11] Planck assumed that the emission and absorption of radiation in quanta $h\nu$ was done by hypothetical harmonic oscillators capable of having only discrete energy values, $nh\nu$ ($n = 0$, 1, 2,...). He showed that this quantum condition on the energy levels of the oscillator was equivalent to quantizing the 'action integral' as in Eq. 1.26. For this reason, the name 'quantum of *action*' has been applied to Planck's constant. Note that the value $n = 0$ does not make any sense in Bohr's quantum condition, and had to be dropped.

[12] W. Wilson, *Phil. Mag.*, **29**, 795, 1915; A. Sommerfeld, *Ann. d. Physik*, **51**, 1, 1916.

$$\oint p_r\, dq_r = n_r h, \quad r = 1,2,\ldots \mathcal{N} \qquad (1.27)$$

where the integration in the case of each pair of conjugate variables is to be taken over one period of that particular pair. The quantum numbers n_r take integral values.

Bohr's general postulates together with the quantum rule (1.27) constitute what is now known as the *Old Quantum Theory*.[13] Details of the applications of the theory are of no particular interest now, since the theory itself has been superseded by the *new* quantum theory or *Quantum Mechanics*. But the quantization of the elliptical orbits of the hydrogen atom deserves mention because it gave the first example of two general properties which persist in the quantum mechanical theory. One is the property of degeneracy of energy levels of systems possessing symmetries. Sommerfeld observed that the quantized elliptical orbits in a given plane are identified by two quantum numbers n' and k, characterizing the radial and angular parts of the motion in the orbit. He found however that the energies associated with such quantum states depend only on the sum $n = n' + k$ of these quantum numbers, and are given by the Bohr formula (1.24). Thus for a given value of the total or 'principal' quantum number n, there are n different states (elliptical orbits of various eccentricities, corresponding to[14] $k = 1, 2, \ldots n$) all of which have the same energy E_n. We say that the energy level E_n is *n-fold degenerate* (when the quantum orbits in one plane alone are considered). It is now recognized that the degeneracy in the case of the hydrogen atom is due to the special nature or 'symmetry' of the distance-dependence of the electrostatic potential, and does not occur for other potentials (unless they have other symmetries, of course). The second general property exemplified in the Sommerfeld treatment is the removal of degeneracy (i.e., *departure from equality* of the energy values of the previously degenerate quantum states) when the symmetry is 'broken'. In the case of the hydrogen atom there is indeed a slight departure from symmetry. It is caused by the fact that the variation of the speed of the electron as it moves along an elliptical orbit induces corresponding changes in its mass as given by the theory of relativity. Sommerfeld showed that this mass variation gives rise to a slow precession of the orbit in its own plane,[15] and that because of this, the energy associated with each orbit is changed by a very small amount depending on k. Consequently, the nth Bohr level gets split into n closely-spaced levels—there is no more degeneracy. The spectral lines also then get split, developing what is called a *fine structure*. Sommerfeld's calculation gave complete agreement with the observed fine structure.

[13] Discussion of the old quantum theory and many of its applications may be found in L. Pauling and E. B. Wilson, *Introduction to Quantum Mechanics*, McGraw-Hill, New York, 1935; A. Sommerfeld, *Atomic Structure and Spectral Lines*, 3rd ed., Methuen and Co., London, 1934.

[14] In the Bohr-Sommerfeld theory the angular momentum in a stationary state was identified as $k\hbar$ but quantum mechanics shows that it is given by $l\hbar$, where l takes the same values as $(k-1)$ for given n, namely $l = 0, 1, \ldots, n-1$. In the discussion of space quantization below, we use this quantum number l in preference to k.

[15] More precisely, the orbit no longer closes on itself, but remains very nearly elliptical for each revolution, with the direction of the major axis changing slightly (in the plane of the ellipse) from one revolution to the next. This change of orientation of the ellipse at a steady rate is called precession.

1.11 Space Quantization

In giving the *degree of degeneracy* of the nth Bohr level (i.e. the number of quantum states belonging to this level) as n, and in asserting that the relativistic mass variation removes this degeneracy, we have taken account of orbits in any one plane only. To put this in another way, only orbits with a specific direction for the angular momentum vector (which is normal to the plane of the orbit) were considered. Actually there is a further degeneracy associated with the possibility of various *orientations* for the angular momentum vector with respect to any fixed axis. Such an axis may be defined, for instance, by the direction of some externally applied field. From the consideration of the quantum condition (1.27) in axially symmetric situations it was inferred that the *direction* of angular momentum should be quantized. To be more specific, the *component* of angular momentum parallel to the axis has to be $m\hbar$ with the quantum number m taking integer values only. This is called *quantization of direction* or *space quantization*. When the quantum number characterizing the magnitude of the angular momentum has a value l, the values which m can take are limited to $m = l, l-1, \ldots, -l+1, -l$.

The existence of space quantization was experimentally demonstrated in a very direct and beautiful fashion by Stern and Gerlach.[16] They exploited the fact that an atom with nonzero angular momentum has a magnetic moment $\boldsymbol{\mu}$ in the same direction as the angular momentum vector $\mathbf{L}$; for, the orbital motion of the electron (with which the angular momentum is associated in the Rutherford-Bohr picture) also produces a magnetic moment since the electron is a charged particle. If such an atom is placed in a magnetic field $\mathcal{H}$, the field exerts a torque $\boldsymbol{\mu} \times \mathcal{H}$ tending to turn the direction of $\boldsymbol{\mu}$ and hence that of $\mathbf{L}$ too into alignment with the field $\mathcal{H}$. Now, it is well known that any torque acting on an angular momentum vector has a gyroscopic effect. Therefore $\mathbf{L}$ precesses around the direction of $\mathcal{H}$, keeping a constant angle θ to $\mathcal{H}$ all the time. This is all that happens if $\mathcal{H}$ is a uniform field. If it is not uniform, there is also a net force $(\boldsymbol{\mu}.\nabla)\mathcal{H}$ on the atom. Thus, if a coordinate system is chosen with z-axis in the direction (Fig. 1.3) of $\mathcal{H}$ and if the field

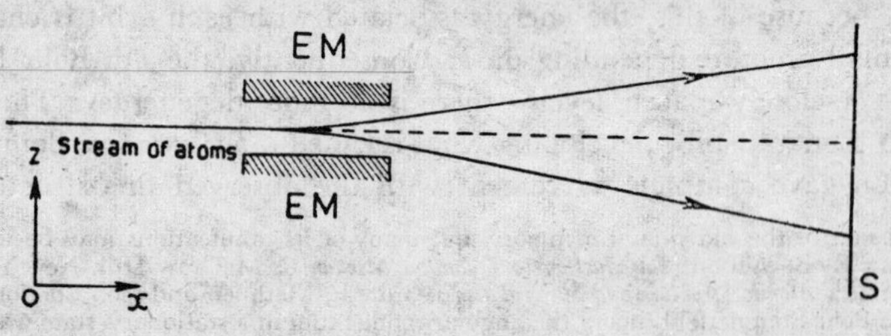

Fig. 1.3 Schematic diagram of the Stern-Gerlach experiment showing splitting of the beam of atoms between the pole-pieces (EM) of electromagnet.

[16] O. Stern and W. Gerlach, *Z. Physik*, **8**, 110, 1922.

strength $\mathcal{H}$ increases in the z-direction, we have a situation where μ_z and L_z are constants for the atom, and there is also a force μ_z $(\partial\mathcal{H}/\partial z)$ on the atom, acting in the z-direction. If a beam of atoms is shot through the field in a direction perpendicular to the field, say along the x-direction, the above force causes the individual atoms to be deflected up or down (i.e., in the positive or negative z-directions) by amounts proportional to their respective values of μ_z. Therefore, if the values of μ_z form a continuous range (in accordance with classical concepts) the beam would widen out into an expanding strip in the x-z plane. On the other hand, if there is space quantization, so that L_z (and hence μ_z) can take only a discrete set of values, the beam of atoms would split into a number of distinct diverging beams, each of which is characterized by a specific value of μ_z. Impinging on a plane perpendicular to the x-axis, these beams would leave spots spaced out along the z-direction (instead of a continuous line which would appear if there were no space quantization). It is the appearance of such distinct spots in the Stern-Gerlach experiment which gave direct confirmation of the idea of space quantization. In this experiment a fine, well-collimated beam of silver atoms was passed through an inhomogeneous field created by an electromagnet with specially shaped pole pieces. One of these had a ridge along the middle, and facing this was a hollowed-out channel in the other pole piece. The net result was a concentration of field lines (high intensity) near the former and dispersal (low intensity) near the latter. It was found that when the atomic beam was made to pass between the pole pieces, traversing their whole length, the beam split into two—one part deflected upwards, and the other downwards.

While this confirmation of space quantization was a success for the *Old Quantum Theory*, it must be mentioned that the appearance of just *two* values for μ_z was correctly explained only after the discovery of the *spin* of the electron a couple of years later. Unlike l, which can have only integral values, the quantum number s characterizing the angular momentum associated with the spinning motion has the value $\frac{1}{2}$; and the component of spin in any direction can have just the *two* values $+\frac{1}{2}\hbar$ or $-\frac{1}{2}\hbar$. Deferring further discussion of this subject, we return now to our main theme: the progression of concepts from the classical to the quantum mechanical.

1.12 Limitations of the Old Quantum Theory

We have seen above how the Old Quantum Theory has been able to provide the explanation of the spectrum of the hydrogen atom, including its fine structure. Recognition of the quantum character of the magnitude and direction of angular momentum remains as one of its finest achievements. However, despite these and other very considerable successes of the Old Quantum Theory, it is quite obvious that it is not really a fundamental theory and is, in any case, only of limited applicability. The scope of the Bohr-Sommerfeld quantum rules is restricted to periodic or multiply-periodic motions; they have nothing to say about situations where other kinds of motion are involved. Even in the Franck-Hertz experiment which gave direct support to Bohr's concept, the behaviour of the electrons scattered by the atoms are outside the purview of the Old Quantum Theory. The limitations of the theory were greatly mitigated by skilful exploitation of the idea that the results of quantum theory should tend to those of the classical theory under circumstances where the quantum discontinuities are negligibly small. This idea, which places powerful constraints on the quantum theory, played a considerable role in the developments of the decade preceding the birth of quantum mechanics. A formal enunciation of the idea, under the name *Correspondence Principle*, was given by Bohr.[17] In considering the quantum mechanics in later chapters we will have occasion to discuss its correspondence with classical mechanics in certain aspects of their mathematical structures as well as in the sense of a passage to the limit $h \to 0$.

Looking back on the essentials of the Old Quantum Theory, we see that its fundamental shortcoming is that it is a peculiar hybrid of quantum concepts grafted on to classical mechanics. The existence of discrete stationary states is experimentally well substantiated. But as long as the classical picture of well defined particle orbits is retained, it remains incomprehensible why certain orbits should be completely stable and others not allowed to exist at all. This perplexing question was responsible in part for the ultimate realization that particle states at the microscopic level are not describable in terms of well defined orbits, but must be pictured in terms of some kind of waves.

(iv) Matter Waves

1.13 De Broglie's Hypothesis

The suggestion that matter may have wave-like properties was first put forward in 1924-25 by Louis de Broglie.[18] He argued that if light (which consists of waves according to the classical picture) can sometimes behave like particles, then it should be possible for matter (which consists of particles, classically) to exhibit wave-like behaviour under suitable circumstances. He made the hypothesis that the relation between the energy E of a particle and the frequency ν of the associated wave is exactly the same as that between the energy of a photon (the particle-like quantum of light) and the frequency of the light radiation:

$$E = h\nu = \hbar\omega \tag{1.28a}$$

where $\omega = 2\pi\nu$ is the angular frequency. He noted then that according to the theory of relativity, the energy E and the momentum $\mathbf{p}$ of a particle form the components of a single four-vector, and so do the angular frequency ω and the propagation vector $\mathbf{k}$ of a wave. Since the components of both the four-vectors have to transform in an identical manner (according to the Lorentz transformation) for a given change of reference frame, any proportionality relation like (1.28a) between E and ω must necessarily hold also between $\mathbf{p}$ and $\mathbf{k}$. It was concluded, therefore, that

[17] N. Bohr, *Nature*, **121**, 580, 1923.

[18] L. de Broglie, *Phil. Mag.*, **47**, 446, 1924; *Annales de Physique*, **3**, 22, 1925.

$$\mathbf{p} = \hbar \mathbf{k} \tag{1.28b}$$

In terms of the magnitudes of the vectors, this gives

$$p = \hbar k = \frac{h}{\lambda}, \text{ or } \lambda = \frac{h}{p} \tag{1.28c}$$

De Broglie's hypothesis attributing a dual particle-wave character to matter appears very strange at first sight, it being very much against the direct evidence of one's senses at the macroscopic level. But at the level of the atom, the behaviour of matter had been already found to be unconventional as we have seen in the last few sections. It was such evidence of non-classical behaviour which gave scope for de Broglie's proposal. The precise form of the hypothesis was determined however by the speculative postulate that at the most fundamental level, matter and radiation (which form the basic constituents of the physical world) should be similar in nature. Confirmation that Nature does exhibit such an aesthetically pleasing symmetry between matter and radiation came from experiments by Davisson and Germer,[19] Kikuchi[20] and G. P. Thomson[21]. These experiments demonstrated the existence of 'de Broglie waves' associated with electrons in a very direct fashion. This was done by showing that crystal diffraction patterns just like those produced by X-rays are obtained even when a beam of electrons of appropriate momentum[22] is used instead of X-rays. Imparting of the desired momentum to the electrons (obtained by thermionic emission from a heated filament) was done by electrostatic acceleration. An electron accelerated through a potential difference of V volts (or $V/300$ electrostatic units) acquires a kinetic energy $E_{kin} = (eV/300)$ ergs. The momentum p of the electron is then obtained from the relativistic relation $p^2 = 2m\, E_{kin} + (E^2_{kin}/c^2)$, which is equation (B.16) of Appendix B with E identified as $mc^2 + E_{kin}$. Thus

$$\lambda = \frac{h}{p} = \frac{h}{\sqrt{2mE_{kin}}} \left(1 + \frac{E_{kin}}{2mc^2}\right)^{-1/2} \tag{1.29}$$

When E_{kin} is much less than the rest energy mc^2, the last factor is negligible; and then, on substituting $E_{kin} = eV/300$, we get

$$\lambda = h\sqrt{\frac{150}{meV}} = \frac{12{\cdot}25}{\sqrt{V}} \times 10^{-8} \text{ cm},$$

where V is to be expressed in volts.

EXAMPLE 1.3 — For an electron accelerated through 100 volts, the wavelength is $(12{\cdot}25 \times 10^{-8}/10)$ cm $= 1{\cdot}225$ Å. Thermal neutrons (with mean kinetic energy $E_{kin} = \frac{3}{2}\, kT \approx 6{\cdot}2 \times 10^{-14}$ ergs at 300°K) also have wavelengths of the same order. Substituting the neutron mass and the above value of E_{kin} in Eq. 1.29 we get $\lambda \approx 1{\cdot}5$ Å. In contrast the wavelength of a one gram mass with the same thermal energy, has the fantastically small value of about

[19] C. Davisson and L. H. Germer, *Nature*, **119**, 558, 1927; *Phys. Rev.*, **30**, 705, 1927.
[20] Kikuchi, *Japan. J. Phys.*, **5**, 83, 1928.
[21] G. P. Thomson, *Nature*, **120**, 802, 1927; *Proc. Roy. Soc., London*, **A117**, 600, 1928.
[22] Appropriate momenta are those for which the de Broglie wavelengths as given by Eq. 1.28c are in the range of X-ray wavelengths used in diffraction studies—from a few angstroms down to a tenth of an angstrom or less.

1.2×10^{-20} cm. There is no wonder then that macroscopic objects show no noticeable diffraction effects or other wave-like behaviour.

The Davisson-Germer experiment involved diffraction by a single crystal; Kikuchi reproduced the Laue type of diffraction pattern by transmission through very thin mica crystals, and Thomson obtained the analogue of the Debye-Scherrer rings by passage of electrons through thin (poly-crystalline) metal foils. In each case, the wavelength λ of the electron waves, as determined from the diffraction pattern, was found to be equal to (h/p) in agreement with de Broglie's theoretical prediction. Thus the dual nature of matter, like that of radiation, was firmly established.

Actually the theoretical exploitation of de Broglie's idea did not await its experimental verification. Immediately after the publication of the hypothesis, Erwin Schrödinger[23] proposed that the behaviour of matter waves associated with material particles (whether free or subject to forces) is governed by a certain differential equation for the *wave function* ψ (i.e., the function which represents the matter wave). He showed that this *wave equation*, together with physically-motivated conditions on the wave function leads in a very natural way to stationary states characterized by discrete energy values. Thus Schrodinger was able to explain the basic fact of quantization as a consequence of the wave nature of matter, and to replace the *ad hoc* quantization rules of the Old Quantum Theory by mathematical conditions of a very general nature on the wave functions. The Schrödinger theory, called *wave mechanics*, is one of the alternative formulations of the general theory of quantum mechanics which, as far as we know at present, gives the correct fundamental theory of the physical world at the microscopic level. Much of this book will deal with the development of quantum mechanics following Schrödinger's approach. We defer the introduction of the wave equation to the next chapter, and devote the rest of this chapter to a discussion of the conceptual problems raised by the developments described in the preceding sections.

1.14 The Motion of a Free Wave Packet; Classical Approximation and the Uncertainty Principle

The most serious problem raised by the discovery of the wave nature of matter concerns the very definition of the word 'particle'. We have already noted that classically, the mental picture of a particle was that of a point object endowed with a precise position and momentum at every instant of time. The discovery of the quantum properties (discreteness of energy values, etc.) of the atom made it clear that such a picture cannot hold, at least in its entirety, in the atomic domain. However, the success of the de Broglie hypothesis, that a particle of momentum p is to be associated with a wave of definite wavelength $\lambda = h/p$, leaves one in a very uncomfortable situation. For, a pure harmonic wave necessarily extends over all space, and this fact makes it impossible to get any idea of the position of a particle described by such a wave. On the other hand, if we have a *wave packet*, i.e., a wave which is confined to

[23] E. Schrödinger, *Ann. d. Physik*, **79**, 361, 489, 1926; **80**, 437, 1926; **81**, 109, 1926.

a small region of space, it would seem reasonable to suppose that the particle is within the region of the packet. The position of the particle would then be *approximately* determined by that of the wave packet, there being an uncertainty in the position which is of the order of the dimension or size of the packet. If this 'fuzzy' wave packet picture of a particle is to be taken seriously, it is necessary that such a wave packet should move like a classical particle, to a good approximation. Let us now test whether this is the case. For simplicity we consider first a wave packet in one-dimensional space.

Consider a wave packet represented by a wave function $\psi(x,t)$ which, at the instant t, has a maximum at the point $X(t)$. We will suppose for convenience that the function ψ falls off monotonically as in Fig. 1.4 as x moves to either

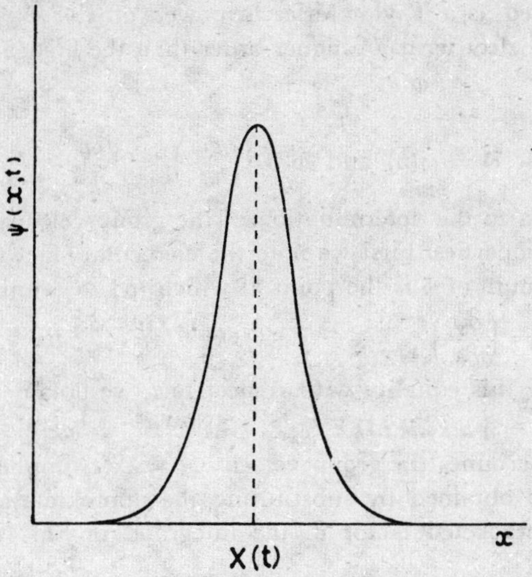

Fig. 1.4 A wave packet pictured at a particular instant

side from the maximum position. If the position of the wave packet changes with time, the rate at which the maximum point moves is clearly a good measure of the velocity v_g of the packet:

$$v_g = \frac{dX(t)}{dt} \tag{1.30}$$

The subscript g on v stands for group velocity[24] and is a reminder that the wave packet is composed of a group of harmonic waves having a certain range of wavelengths:

$$\psi(x,t) = \int_{-\infty}^{\infty} a(k)\ e^{i(kx-\omega t)}\ dk \tag{1.31}$$

[24] The velocity of propagation of a harmonic wave exp $[i\ (kx-\omega t)]$ is the rate at which surfaces of constant *phase* ($kx-\omega t$ = const.) move. It is therefore called the *phase velocity*, and is given by (ω/k). The velocity of a wave packet formed by a group of harmonic waves is the *group velocity*. Its value is $(d\omega/dk)$.

where $a(k)$ is the amplitude of the harmonic component with propagation constant k (wavelength $\lambda = 2\pi/k$) and angular frequency ω, in the wave function ψ. Equation 1.31 simply expresses the well known mathematical fact that any reasonable function can be analyzed into a Fourier integral. We will assume that the harmonic waves present in $\psi\,(x,t)$ have values of k lying within a small range centred about some value $\bar{k}$. In other words, $a(k)$ in Eq. 1.31 is taken to be non-vanishing only when k is very close to $\bar{k}$. This assumption is made in order to ensure that the momentum of the particle described by the wave packet is reasonably well defined: When k is restricted to a narrow range, the momentum associated with it according to the de Broglie relation is also restricted likewise. We note further that the frequencies ω of these harmonic waves will also, for the same reason, remain very close to $\bar{\omega} = \omega(\bar{k})$. Therefore, if $\omega(k)$ is expanded as a Taylor series in powers of $(k - \bar{k})$ we can, to a good approximation, neglect terms of higher order than the first, and write $\omega(k)$ as

$$\omega(k) = \bar{\omega} + \bar{\omega}'.\,(k - \bar{k}) \tag{1.32}$$

where

$$\bar{\omega} = \omega(\bar{k}) \text{ and } \bar{\omega}' = \left(\frac{d\omega}{dk}\right)_{\bar{k}} = \frac{d\bar{\omega}}{d\bar{k}}.$$

Let us now turn to the determination of the group velocity v_g for a packet with the above properties. First we note the elementary fact that the position $X(t)$ of the maximum of ψ is the point at which $(\partial\psi/\partial x)$ vanishes. Thus

$$0 = \left(\frac{\partial\psi}{\partial x}\right)_{x=X(t)} = \int a(k).ik\, e^{i[kX(t)-\omega t]}\,dk \tag{1.33}$$

On differentiating this equality with respect to t, we obtain

$$0 = \int a(k).ik.i\,[kX'(t) - \omega].e^{i[kX(t)-\omega t]}\,dk \tag{1.34}$$

This equation determines the group velocity $v_g = X'(t)$ implicitly. An explicit expression can be obtained by substituting the approximation (1.32) for ω into the square bracketed factor in the integrand in Eq. 1.34 which then becomes

$$0 = [X'(t) - \bar{\omega}'] \int a(k) i^2 k^2\, e^{i[kX(t)-\omega t]}\,dk$$
$$- (\bar{\omega} - \bar{\omega}'\bar{k}) \int a(k) i^2 k\, e^{i[kX(t)-\omega t]}\,dk$$

The second term on the right hand side of this equation vanishes on account of Eq. 1.33. The integral in the first term is nonzero because it is just the value of the second derivative $(\partial^2\psi/\partial x^2)$ at the *maximum* point $x = X(t)$ We conclude, therefore, that $X'(t) = \bar{\omega}'$, or more explicitly,

$$v_g \equiv X'(t) = \frac{d\bar{\omega}}{d\bar{k}} \tag{1.35}$$

This is the velocity of the *wave packet*. On the other hand, it is easily verified that the velocity v of a *particle* (whether relativistic or non-relativistic) is given by

$$v = \frac{dE}{dp} \tag{1.36}$$

The similarity of the two forms (1.35) and (1.36) is obvious. The two velocities become identical if the particle parameters E, p are related to the wave parameters $\bar{\omega}$, $\bar{k}$ through the de Broglie relations $E = \hbar\bar{\omega}$, $p = \hbar\bar{k}$.

We conclude, therefore, that a small wave packet composed of a small band of de Broglie waves does move like a classical particle at least in the case of free particles. This result is reassuring (despite the lack of rigour in the derivation leading to it) since it enables us to view classical particle mechanics as an approximation to the wave (quantum) mechanics. The approximation is good only to the extent that one can ignore the fundamental limitations placed by the wave packet picture on the accuracy with which *both* the position and momentum of a particle can be specified. That such limitations exist is obvious. We have already noted that if the momentum is precisely specified we have no wave packet at all, but only a pure harmonic wave, so that the particle position is completely unknown. On the other hand, if the position is to be absolutely precise, the wave packet must be of infinitely small extension, and to construct such a packet one needs waves of all wave numbers k from $-\infty$ to $+\infty$. In terms of the de Broglie relation $p = \hbar k$ this means that the momentum of the particle is completely indeterminate. It is apparent therefore that there is an inverse relationship between the accuracies to which the position and momentum can be simultaneously specified. In fact, it can be shown quite generally from the theory of Fourier transforms that if a packet represented by $\psi(x,t)$ of Eq. 1.31 has a size Δx, then the Fourier transform $a(k)$ must be nonvanishing over a range of width Δk with $\Delta k \gtrsim (1/\Delta x)$. This means that the spread in momenta (Δp) of the de Broglie waves contributing to the packet has to be such that

$$(\Delta p)\ (\Delta x) \gtrsim \hbar \tag{1.37}$$

This inequality is the essential content of the *Heisenberg Uncertainty Principle*, which will be derived later (Sec. 3.11) in a more precise form from quantum mechanics. It is easy to convince oneself that in the case of a three-dimensional wave packet[25]

$$\psi(\mathbf{x},t) = \int a(\mathbf{k})\ e^{i(\mathbf{k}.\mathbf{x}-\omega t)}\ d^3k \tag{1.38}$$

there are three uncertainty relations similar to (1.37):

$$(\Delta x)(\Delta p_x) \gtrsim \hbar, \quad (\Delta y)(\Delta p_y) \gtrsim \hbar, \quad (\Delta z)(\Delta p_z) \gtrsim \hbar \tag{1.39}$$

EXAMPLE 1.4 — The uncertainty principle shows immediately that the motion of the electron in a hydrogen atom cannot be described even approximately in classical terms. There is direct experimental evidence to indicate that the size of the hydrogen atom is of the order of 1Å. Thus the uncertainty in the position of the electron must be of this order: $\Delta x \sim 10^{-8}$ cm. Eqation 1.37 then implies that $\Delta p \gtrsim 10^8 \hbar \approx 10^{-19}$ gm.cm/sec. Since the mass of the electron is approximasely 10^{-27} gm, the *uncertainty* in the velocity of the electron has a tremendously high value, $\gtrsim 10^8$ cm/sec. It is evident that under these circumstances it would make no sense to talk of a trajectory.

EXAMPLE 1.5.—GAUSSIAN WAVE PACKET. To get a clearer idea as to how a wave packet evolves with time, we consider the example of a one-dimensional

[25] To save writing, we shall use only a single integral sign even in the case of multiple integrals. Wherever the number of integrals involved is relevant, it will be indicated by the notation for the 'volume element' such as d^3k (meaning $dk_x\, dk_y\, dk_z$) or $d^3x \equiv dx\, dy\, dz$. The integration is always from $-\infty$ to $+\infty$ for each variable, unless otherwise specified.

packet characterized by $a(k) = a_0\sigma^{1/2} \exp[-\frac{1}{2}\sigma^2 (k - \bar{k})^2]$ in Eq. 1.31. In this case, the propagation constant k is said to have a Gaussian spectrum, with a width $\Delta k \sim (1/\sigma)$ about the mean value $\bar{k}$. We introduce this form of $a(k)$ in Eq. 1.31, and also substitute $\omega = \hbar k^2/2m$. The latter relation is obtained from the energy-momentum relation $E = p^2/2m$ for a classical particle on using Eqs. 1.28. Then we have

$$\psi(x,t) = a_0\sigma^{1/2} \int_{-\infty}^{+\infty} e^{-\frac{1}{2}\sigma^2(k-\bar{k})^2}\, e^{i(kx-\hbar k^2 t/2m)}\, dk$$

$$= a_0\sigma^{1/2} \exp(-\tfrac{1}{2}\sigma^2\bar{k}^2) \int_{-\infty}^{+\infty} e^{-\frac{1}{2}(\sigma^2+i\hbar t/m)k^2}\, e^{ik(x-i\sigma^2\bar{k})}\, dk$$

$$= a_0 \left(\frac{2\pi\sigma}{\sigma^2 + i\hbar t/m}\right)^{1/2} \exp\left\{-\frac{(x - i\sigma^2\bar{k})^2}{2(\sigma^2 + i\hbar t/m)} - \frac{\sigma^2\bar{k}^2}{2}\right\}$$

and hence

$$|\psi(x,t)|^2 = \frac{2\pi |a_0|^2}{[\sigma^2 + \hbar^2 t^2/m^2\sigma^2]^{1/2}} \exp\left\{\frac{-(x - t\hbar\bar{k}/m)^2}{\sigma^2 + (\hbar^2 t^2/m^2\sigma^2)}\right\}$$

(We omit details of evaluation of the integral above. Its value may be obtained from the last equation under Example 3.12 (p. 90), by making the replacements $\sigma^2 \to \sigma^2 + i\hbar t/m$, $x \to x - i\sigma^2\bar{k}$).

Observe that the absolute value of the wave function, $|\psi(x,t)|$, has a Gaussian form. Its peak is at $x = t\hbar\bar{k}/m$ i.e. it moves with the velocity $(\hbar\bar{k}/m)$. This is just what we expect of a particle of momentum $\bar{p} = \hbar\bar{k}$ and is in conformity with Eq. 1.35. However, it is to be noted that the size or width of the wave packet at time t is $[\sigma^2 + \hbar^2 t^2/m^2\sigma^2]^{1/2}$, and it increases indefinitely with the passage of time. On account of this spreading, any possibility of identifying a wave packet directly as a particle is ruled out.

1.15 Uncertainties Introduced in the Process of Measurement

We have deduced the uncertainty principle above as a consequence of the essential wave nature of physical entities. This principle is of fundamental importance in that it makes precise the limitations on the validity of classical concepts. It would be of interest therefore to see how it manifests itself as an obstacle in the way of measurements aimed at determining the state of a particle with precision in classical terms.

Suppose we wish to determine the position of a small object with precision. Let us imagine we use a microscope for the purpose. The absolute limit to the accuracy Δx with which a position determination can be made by the microscope is given by its resolving power

$$\Delta x = \frac{\lambda}{\sin\alpha}.$$

Here λ is the wavelength of light used for illuminating the object, and α is the half-angle subtended by the objective lens of the microscope at the position of the object being examined. The higher the accuracy needed, the smaller the wavelength λ has to be. But light consists of photons of momentum $p = h/\lambda$ and the x-component of the momentum of a photon scattered by the object into the microscope is uncertain by an amount $\pm (h/\lambda) \sin\alpha$. In scattering the light, the object itself recoils, and the x-component of the recoil momentum is also evidently uncertain to the same extent:

$$\Delta p_x \sim \frac{h}{\lambda} \sin\alpha = \frac{h}{\Delta x}.$$

It follows, therefore, that in the very process of trying to determine position precisely, the momentum is made uncertain, the extent of the uncertainty being what is demanded by the Uncertainty Principle.

It is worth observing that in this discussion, the wave characteristics of the object whose position-momentum uncertainty is under consideration have played no role. (In this sense the uncertainty we are talking of here is quite a different thing from what was considered in the last section). It is the quantum properties of the agent (light) used in the measuring process which brought about the uncertainty. As long as all possible agents (matter and light) have quantum properties, no measurement can, *even in principle*, lead to absolutely precise determinations of positions and momenta (even if there were no theoretical objection to the object itself having a precise position and momentum).

EXAMPLE 1.6 — Uncertainties arising from the disturbance due to measurements is extremely small on the macroscopic scale. If, for instance, the position of an object is determined to an accuracy of 10^{-6} cm using an electron microscope, the resulting uncertainty in momentum is only of the order of $(\hbar/10^{-6}) \sim 10^{-21}$ gm.cm/sec. Thus, unless the mass of the object is of the order of an atomic mass or less the uncertainty in velocity is negligibly small.

1.16 Approximate Classical Motion in Slowly Varying Fields

Let us now return to Sec. 1.14 and summarize the main conclusions. The wave nature of matter has the consequence that particles in the strict classical sense (with precise positions and momenta) are unrealizable in nature. The closest one can get to a classical particle is through a wave packet with Δx and Δp made as small as possible subject to the constraint (1.37). Such a wave packet does move approximately like a classical particle. Our verification of this last statement is actually incomplete. To justify it fully we must show that the correspondence (1.28) between the corpuscular and wave aspects holds good also in the case of particles moving under the influence of forces. Such a proof can be given on the basis of the perfect analogy which exists between Maupertuis' principle of least action in classical particle mechanics, and Fermat's principle of least time in the classical theory of wave propagation. *Fermat's principle* concerns the passage of *light* through an inhomogeneous medium, in which the refractive index and hence the velocity of light v varies from point to point. To travel from a point $\mathbf{x}_1$ to $\mathbf{x}_2$ along some hypothetical path, the time taken by the light ray would be given by the line integral

$$\int_{\mathbf{x}_1}^{\mathbf{x}_2} \frac{ds}{v} = \frac{1}{\nu} \int_{\mathbf{x}_1}^{\mathbf{x}_2} \frac{ds}{\lambda} \tag{1.40}$$

where ds is the length of an infinitesimal segment of the path. Fermat's principle states that the actual path taken by the light ray is one which minimizes the integral. *Maupertuis' principle* concerns the motion of a *particle* in a varying force field. It states that the actual path taken by the particle is one which minimizes the action integral

$$\int_{x_1}^{x_2} p \, ds \tag{1.41}$$

A comparison of the two principles shows that the path of a ray associated with a wave of variable wavelength λ would be identical with that of a particle in a force field if $p \propto (1/\lambda)$. But this is just what we have if we assume the de Broglie relation to be applicable also to particles subject to forces. We note that, if a particle moving nonrelativistically in a force field has potential energy $V(\mathbf{x})$ at the position $\mathbf{x}$, its momentum (for a given total energy E) is given by

$$p = [2m \ (E - V)]^{1/2}$$

It varies with the position of the particle. Correspondingly, the wavelength λ attributed to the particle is also a function of $\mathbf{x}$.

$$\lambda = \frac{h}{p} = \frac{h}{[2m\,(E - V(\mathbf{x}))]^{1/2}} \tag{1.42}$$

Now, it makes sense to talk of a position-dependent wave-length only if $\lambda(\mathbf{x})$ has an appreciably constant value over many waves in the neighbourhood of $\mathbf{x}$, i.e., from one wave to the next, the fractional change in wavelength should be very small. In symbols,

$$|\nabla\lambda| \ll 1 \tag{1.43}$$

This means in turn that the potential energy $V(\mathbf{x})$ should be a slowly-varying function of $\mathbf{x}$. Thus, the motion of a matter-wave packet in a force field will coincide approximately with that of a classical particle provided the variation of the field is slow.

1.17 Diffraction Phenomena: Interpretation of the Wave-Particle Dualism

Our efforts in the last section have been to convince ourselves that despite the wave nature of matter, classical mechanics holds in an approximate sense under suitable circumstances, and that the approximation is in fact perfectly adequate for the description of the macroscopic world. Under such circumstances, the wave aspect of matter remains unobtrusive and does not create any serious problems of visualization. Difficulties of interpretation arising from the wave-particle dualism appear in an acute form when phenomena (e.g., diffraction) in which the wave aspect plays an essential role are considered. The fundamental problem then is how to reconcile the *discrete* nature of material entities with the wave-like behaviour exhibited by them. The same problem occurs in reverse in the case of radiation, whenever it exhibits the discrete photon character in an essential way.

Let us now analyse a diffraction experiment to see how the mutual association of the two (seemingly incompatible) aspects, whether of matter or of radiation, is to be understood. To be specific, let us think of the experiments of G. P. Thomson or Kikuchi. We recall that electron diffraction photographs identical with the Laue and Debye-Scherrer X-ray diffraction patterns were obtained in these experiments. More specifically, in electron as well as X-ray diffraction, the observed pattern is precisely what one would expect on the basis of the wave theory, assuming the incident beam to be a harmonic wave and *the intensity at any point on the pattern to be proportional to the absolute square of the amplitude of the wave at that point.* It is also an empirical fact that the *nature* of the diffraction pattern is quite independent of the intensity of the incident beam, i.e. on taking a long-exposure diffraction photograph with a weak incident beam one gets exactly the same pattern as with a beam many times more intense and a correspondingly shorter exposure. This is again just what one would expect on the wave theory, but it acquires profound significance when we recall that the beam actually consists of discrete particles (electrons or photons)[26]. For it implies that the diffraction process is independent of the number of particles simultaneously present in the beam, and hence that the wave property manifested through diffraction is not the result of some conspiracy among the particles present. Instead, *the wave nature has to be an inherent property of each particle.* This inference is confirmed by experiments in which the beam intensity is made so low that there is effectively only one particle at a time going through the apparatus; the diffraction pattern still emerges if recording is made for a sufficiently long period of time. We have to conclude, therefore, that associated with *each* particle there is a wave which undergoes diffraction in the crystal. If the wave is represented by a (possibly time dependent) function $\psi(\mathbf{x},t)$, diffraction manifests itself in a variation of $|\psi|^2$ with $\mathbf{x}$ at points along the surface of the photographic film. The form of $|\psi|^2$ as a function of $\mathbf{x}$ is the same for all the particles, and it is the accumulated effect of all of them which makes up the actually observed diffraction pattern. The intensity distribution $I(\mathbf{x})$ in the diffraction photograph is then proportional to $|\psi(\mathbf{x})|^2$.

This conceptual picture of the diffraction phenomenon still leaves us with the following question. Does the electron (or photon) smear itself out over the photographic plate in a manner determined by its own wave function (or rather, by $|\psi|^2$)? The answer is no. It remains discrete and may be detected as a discrete particle at some point of the diffraction pattern. In that case, how can the photographic plate possibly know of the existence of peaks and zeroes of $|\psi|^2$? The only possible answer seems to be a *statistical* one. Each individual particle is recorded at some point or other of the photographic plate, at random, with a *probability* proportional to the value of $|\psi|^2$ at that point. When a large number of particles get recorded in this fashion, the *average number* arriving in the neighbourhood of a particular point $\mathbf{x}$ of the

[26] One needs no persuasion to accept that an electron beam consists of electrons as discrete particles. In the case of X-ray diffraction, one could keep a Geiger counter in the path of the incident or diffracted X-rays and actually count the individual photons. The counting rate is proportional to the X-ray intensity (for monochromatic X-rays). In fact, intensity measurements are often made using Geiger counters.

photographic plate is proportional to $|\psi(\mathbf{x})|^2$; on the other hand, the intensity $I(\mathbf{x})$ of the diffraction pattern (represented for instance by the degree of blackening on the photograph) is also proportional to this average number. Hence $I(\mathbf{x}) \propto |\psi(\mathbf{x})|^2$, and thus the formation of the diffraction pattern through the diffraction of individual particles is explained.

We will see later that the above statistical interpretation is a fundamental feature of wave mechanics. Every particle (or system of particles) has a wave function associated with it. The wave function determines the *corpuscular* characteristics like position, momentum, angular momentum, etc. in a statistical sense. For instance a particle with a wave function $\psi(\mathbf{x},t)$ has a probability proportional to $|\psi(\mathbf{x},t)|^2$ for being found in the neighbourhood of the point $\mathbf{x}$, and there are other formulae which give probability distributions for momentum, etc., in terms of ψ. *Nature does not permit any more precise specification of the state* of a particle than what is provided by the wave function and the probability distributions obtainable from it.

1.18 Complementarity

There is one final question we would like to consider. To make its import clear, we pose it in the context of a simple situation like Young's two-slit experiment showing the interference of light, or a hypothetical analogue involving electrons instead of light. Can we say that the photon (or electron) which reaches some point of the interference pattern has followed a definite trajectory passing through a particular one of the two slits? The probability interpretation, which we were forced to adopt above, makes it pretty clear that we cannot. For, the particle has a definite probability of appearing at any point where the wave function is non-vanishing, and the wave function has to have non-vanishing values in a region which includes the locations of *both* the slits if interference is to take place. Thus the particle has a chance of being found at either of the slits and we cannot say that it has passed through a particular one of them. Another slightly different view point which also leads to the same conclusion, is the following: Let us assume that the particle has a trajectory. We have seen already that sharp trajectories in the classical sense are not possible; so let us suppose the trajectory is a moving wave packet, with a lateral extension less than half the distance a between the two slits. By this condition we make the trajectory sufficiently well defined, so that we can be sure that it can go through only one of the two slits, if at all. Suppose all particle trajectories are defined with this degree of precision (though the individual trajectories may be shifted sideways with respect to each other). Can we have an interference pattern under these conditions? The answer is clearly no, from the wave point of view, since any wave which does not cover both the slits cannot cause interference, as already noted. But it is instructive to note that the same result can be deduced from the wave-particle dualism. Since we have assumed that the extension of the wave packet laterally is less than $\frac{1}{2}a$, the uncertainty in any component of the position vector transverse to the direction of motion of the particle is $\Delta y < (\frac{1}{2}a)$. Therefore, according to the uncertainty principle [Eq. 1.39], the corresponding (transverse) momentum component has an uncertainty Δp_y of at least $(h/\Delta y) = (2h/a)$. The direction of motion of the particle is therefore uncertain by an angle of at least

$$\theta \approx \tan\theta = (\Delta p_y/p) = 2h/pa = 2\lambda/a \qquad (1.44)$$

where λ is the de Broglie wavelength of the particle. Now the angular separation between successive maxima of the interference pattern due to two slits separated by a distance a is (λ/a). If the (uncontrollable) uncertainty in the direction of motion of the particle exceeds this value, as it does according to Eq. 1.44, the interference pattern will obviously get washed out. It may be noted that for the purposes of the above argument it does not matter whether the 'particle' we refer to is a material particle or a photon.

The principal result of the above discussion may be stated in a general form as follows: *In any experimental situation in which a physical entity (matter or radiation) exhibits its wave properties, it is impossible to attribute corpuscular characteristics to it.* The entity does behave like a particle if its wave packet is made compact enough, as seen in Sec. 1.14, but when this is done, it loses the ability to display any wave properties. In other words, the particle and wave aspects of a physical entity are *complementary* and cannot be exhibited at the same time. This is the *Complementarity Principle* of Bohr.

1.19 The Formulation of Quantum Mechanics

The classification of the relation between the particle and wave aspects of physical entities, in the manner presented above, came about only some time after quantum mechanics had been formulated as a mathematical theory. As we have already observed in Sec. 1.13, Schrödinger's formulation (the so-called wave mechanics which is based on the de Broglie hypothesis of matter waves), preceded even the direct experimental verification of the existence of matter waves. Schrödinger (1926) proposed that the wave function ψ describing the matter waves satisfies a partial differential equation, and gave a prescription for writing down the equation for any particular system of particles.[27] He showed that there exist solutions of the equation which correspond to a stationary wave pattern, rather like the familiar standing waves or normal modes of a string or membrane. And just as boundary conditions on the latter restrict the possible normal mode frequencies to a discrete set of values, the natural boundary conditions on matter waves restrict the energies associated with the stationary patterns to discrete values (in the case of particles which are 'bound', i.e. confined to a finite region of space by the action of a potential). In this manner, a natural explanation was found for the existence of stationary states, originally postulated *ad hoc* by Bohr. This, and the correct quantitative prediction of the stationary state energy levels of some important systems, constituted the first spectacular success of Schrödinger's

[27] The Schrödinger equation is applicable only to non-relativistic particles. A new wave equation, which meets also the requirements of relativity theory, was formulated by Dirac in 1928. We will take up the Dirac equation towards the end of the book.

wave mechanics. We will not pursue here any further the successes of this theory since these form the subject matter of the next several chapters.

At just about the same time as the announcement of Schrödinger's theory came that of another theory (now called Matrix Mechanics), due to Heisenberg[28]. Despite the vastly different appearance of the two theories it was very soon recognized that they are completely equivalent. While the Schrödinger formulation is based on explicit use of a mental picture of matter as waves (replacing the classical point-particle picture), Matrix Mechanics was born out of a deliberate effort to exclude from the theory any reference to conceptual pictures which are not amenable to direct experimental verification. Working against the background of the Old Quantum Theory of atomic spectra, Heisenberg noted that the inevitable disturbance of the motion of a particle which is caused by any attempt to observe it (*cf.* Sec. 1.15) is, in the case of electrons in atoms, of sufficient magnitude to make their orbits unobservable. He set out then to reformulate the quantum theory in terms of observable quantities alone (like intensities and frequencies of spectral lines). The resulting formulation was found to be equivalent to replacing the position, momentum and other dynamical variables of classical mechanics by *matrices*, while keeping the form of the equations of motion superficially the same as in the classical mechanics. As is well known, the product of two matrices depends, in general, on the order in which they are multiplied, i.e. matrix multiplication is not *commutative*. It is through the non-commutativity of the product of the position and momentum variables (contrasted with their perfect commutativity, as ordinary numbers, in classical mechanics) that the quantum nature of the Heisenberg theory is manifested.

Of the two alternative forms of quantum mechanics — wave mechanics and matrix mechanics — it is the former which lends itself more easily to the solution of a wide variety of practical problems. For this reason, as well as because of the insight provided by the underlying wave picture, we will be dealing, for the most part, with the Schrödinger version of quantum mechanics. However, the Heisenberg form has its advantages, especially in considering formal questions. And in the quantum theory of electromagnetic and other fields, the representation of dynamical variables [e.g., the field quantities $\mathcal{E}(\mathbf{x},t)$ and $\mathcal{H}(\mathbf{x},t)$] by matrix operators as in Heisenberg's theory is almost an inescapable necessity.

1.20 Photons: The Quantization of Fields

In closing this chapter we would like to make a few remarks on the photon aspect of light or more generally the discreteness properties manifested under suitable conditions by electromagnetic and other *fields*. We have seen in the preceding paragraphs how, by passing from the classical mechanics of particles to the quantum mechanics, one is led to the existence of discrete stationary

[28] W. Heisenberg, *Z. Phys.*, **33**, 879, 1925. M. Born, W. Heisenberg and P. Jordan, *Z. Phys.*, **35**, 557, 1925. The equivalence of wave and matrix mechanics was shown by E. Schrödinger, *Ann. d. Physik*, **79**, 734, 1926.

states (e.g., of atoms) in a natural way. The existence of the discrete photon characteristics of light is understood in a similar way, as a manifestation of the fact that the electromagnetic field is really not a classical field but a quantum field. There are two ways of picturing the process of quantizing fields, analogous to the Schrödinger and Heisenberg approaches to the quantization of material systems. In the Schrödinger approach, the position $\mathbf{x}_i$ of some member (identified by the label i) of a system of material particles is supposed to be defined only to within a certain region where its wave function does not vanish. Correspondingly, in quantum field theory, the field quantity [e.g., the electric field $\mathcal{E}(\mathbf{x})$] at some point of space — identified by the label $\mathbf{x}$ — is taken to have *no precise value* but to be defined only within a certain range determined by a *functional*[29] $\Psi\,[\mathcal{E}(\mathbf{x})]$. This is illustrated in Fig. 1.5, where the

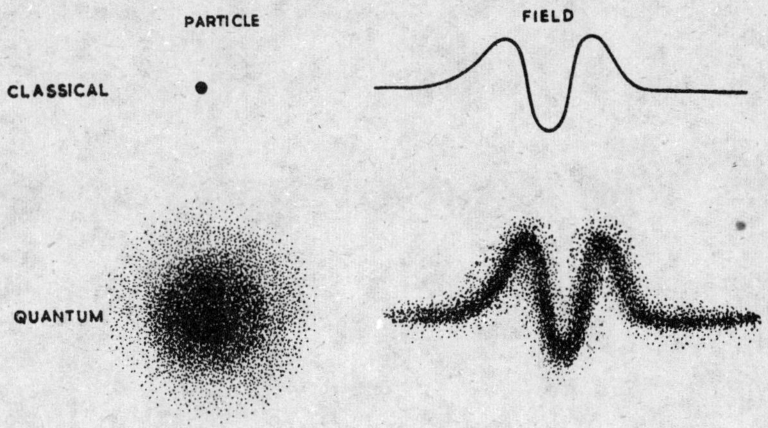

Fig. 1.5 Classical and quantum pictures of particles and fields. In the quantum picture, the particle position (or the value $\phi\,(x)$ of the field ϕ at x) has a probability distribution. The value of the probability density is indicated by darkness of shading in the figure.

classical and quantum concepts of material particles and fields are illustrated pictorially. Pursuing this analogy, one would be led to a functional differential equation for Ψ as the counterpart of the Schrödinger equation, and 'boundary' conditions on Ψ (e.g., that Ψ should vanish as $|\mathcal{E}(\mathbf{x})| \to \infty$ for any $\mathbf{x}$) would then result in the energy, etc., for the field being restricted to discrete values. In other words, photons emerge as a consequence of field quantization. Actually, the theory of functional differential equations is not well developed and therefore in dealing with quantum fields, it is the Heisenberg approach which is invariably employed. As we have indicated above, the quantum equations of motion in this approach remain formally the same as in the classical version. But the field variables (e.g., the components of the

[29] A functional is a function of a continuous infinity of variables. In the present example, $\mathcal{E}(\mathbf{x})$ at *each point* $\mathbf{x}$ is a variable, and since Ψ depends on the continuous infinity of such variables, defined at the continuous infinity of points $\mathbf{x}$ in space, Ψ is a functional.

electric and magnetic fields $\mathscr{E}(\mathbf{x},t)$ and $\mathscr{H}(\mathbf{x},t)$ for each $\mathbf{x}$ and t) become matrices which do not commute with each other. (Equivalently, the field functions may be Fourier analysed and the Fourier coefficients treated as matrices). Clearly, the basic ideas of field quantization are the same as those underlying the process of quantizing particle mechanics. But the fact that a field is a system with an infinite number of degrees of freedom creates special problems, and these are outside the scope of this book.

2

The Schrödinger Equation and Stationary States

As we have noted in the last chapter, the idea of matter waves, introduced as a speculative concept by de Broglie, was made the basis of a systematic and quantitative theory by Schrödinger. The crucial step was the setting up of a *wave equation* governing the behaviour of matter waves. We will show in this chapter how, by rather simple and straight-forward reasoning, the form of the Schrödinger equation for a free non-relativistic particle may be arrived at. Abstracting the essential elements in the derivation, we then deduce the rules whereby the Schrödinger equation for a particle subject to forces may also be written down. Following this we consider in some detail the physical interpretation of the wave function, and the constraints which this interpretation places on the possible wave functions of a particle. One of the most important and immediate applications of the wave equation is to show that the existence of stationary states with discrete energies, which was postulated *ad hoc* by Bohr, actually comes out as a natural consequence of wave mechanics. By solving the Schrödinger equation for a simple model system (a particle in a 'square well' potential) we illustrate the manner in which the energy levels and the wave functions associated with stationary states may be determined. We also demonstrate the existence of other non-classical phenomena like reflection of particles at the potential well and 'tunnelling' by particles through potential barriers.

We restrict the treatment in this chapter to the wave mechanics of a *single* particle, so that the essential features of the theory may not be obscured by the complexities arising from the presence of several particles. The examples to be considered here will serve to illuminate the general formalism of wave mechanics, which will be presented in the next chapter.

A. THE SCHRÖDINGER EQUATION

2.1 A Free Particle in One Dimension

Consider a free particle of velocity v, momentum p and energy E moving in one dimension. We know that for a nonrelativistic particle, $p = mv$ and $E = \frac{1}{2} mv^2$. (Since the particle is free, i.e., there are no forces acting on it, there is no potential energy and E is entirely kinetic). The energy-momentum relation is

$$E = \frac{p^2}{2m} \tag{2.1}$$

According to the de Broglie hypothesis, associated with a free particle there is a harmonic wave with propagation constant k and angular frequency ω, where

$$\hbar k = p \text{ and } \hbar\omega = E \tag{2.2a}$$

so that by virtue of Eq. 2.1,

$$\hbar\omega = \frac{\hbar^2 k^2}{2m} \tag{2.2b}$$

The possible forms for such a harmonic wave are cos $(kx - \omega t)$ or sin $(kx - \omega t)$, or more generally any linear combination of the two, namely

$$\psi\,(x,t) = a \cos\,(kx - \omega t) + b \sin\,(kx - \omega t) \tag{2.3}$$

where a and b are arbitrary constants. Let us try to construct a differential equation for ψ, making use of the elementary observation that

$$\left.\begin{aligned}
\frac{\partial\psi}{\partial x} &= k\,[-a \sin\,(kx - \omega t) + b \cos\,(kx - \omega t)],\\
\frac{\partial^2\psi}{\partial x^2} &= -k^2\,[a \cos\,(kx - \omega t) + b \sin\,(kx - \omega t)] = -k^2\psi,\\
\frac{\partial\psi}{\partial t} &= -\omega\,[-a \sin\,(kx - \omega t) + b \cos\,(kx - \omega t)]\\
\frac{\partial^2\psi}{\partial t^2} &= -\omega^2\psi, \text{ etc.}
\end{aligned}\right\} \tag{2.4}$$

Of the possible differential equations for ψ, two obvious ones are

$$\frac{\partial\psi}{\partial t} = -\frac{\omega}{k}\frac{\partial\psi}{\partial x}, \quad \frac{\partial^2\psi}{\partial t^2} = \frac{\omega^2}{k^2}\frac{\partial^2\psi}{\partial x^2} \tag{2.5}$$

But these are unsuitable for the description of matter waves because the quantity $(\omega/k) = (E/p) = (p/2m)$ depends on the particular state of motion of the particle, i.e. the parameter (ω/k) appearing in these equations has different values for different values of the particle momentum. What we need is a single equation whose solutions encompass *all* motions of a free particle, i.e. one which is satisfied by matter waves for all momenta.

The trouble with Eqs. 2.5 arose because (ω/k) is *not* independent of p. But we know that (ω/k^2) is; in fact its value, according to Eq. 2.2b, is $(\hbar/2m)$

What we have to look for, therefore, is an equation in which the first power of ω is divided by the second power of k. Inspection of Eq. 2.4 reveals that such a balancing would be accomplished if we could equate $\partial\psi/\partial t$ to $\partial^2\psi/\partial x^2$ apart from some proportionality constant. But this can be done only if the square bracketed factors in the expressions for $\partial\psi/\partial t$ and $\partial^2\psi/\partial x^2$ are themselves proportional, i.e. the ratio of the coefficients of the cosine functions in the two cases should be the same as that of the coefficients of the sine functions:

$$\frac{a}{b} = -\frac{b}{a}, \text{ or } b^2 = -a^2, \text{ i.e., } b = \pm ia \qquad (2.6)$$

With $b = ia$, ψ reduces to[1]

$$\psi(x,t) = a\,[\cos(kx - \omega t) + i\sin(kx - \omega t)] = ae^{i(kx-\omega t)} \qquad (2.7)$$

Thus matter waves for a particle of momentum p are to be represented by a *complex* harmonic function of x and t. By direct differentiation of ψ in Eq. 2.7, we find that

$$\frac{\partial\psi}{\partial t} = -i\omega\psi, \quad \frac{\partial^2\psi}{\partial x^2} = -k^2\psi = -\frac{2m\omega}{\hbar}\psi$$

where Eq. 2.2b has been used in the last step. Hence

$$i\hbar\frac{\partial\psi(x,t)}{\partial t} = -\frac{\hbar^2}{2m}\frac{\partial^2\psi(x,t)}{\partial x^2} \qquad (2.8)$$

This is the Schrödinger equation for a free particle in one dimension.

It is important to note that though we consider only harmonic waves of the form (2.7) in the derivation of Eq. 2.8, actually the solutions of this equation are not restricted to such waves only. Evidently any linear combination or *superposition* of harmonic waves, such as

$$\psi(x,t) = \int a(k)\, e^{i(kx-\omega t)}\, dk \qquad (2.9)$$

will also be a solution. This property is a simple consequence of the fact that ψ and its derivatives occur only *linearly* in the Schrödinger equation, i.e. second or higher powers of these quantities do not appear in the equation. It is an important characteristic of quantum mechanics that superpositions of wave functions of a system are also possible wave functions. We leave further consideration of this property (called the superposition principle) to a later stage. We content ourselves here with the observation that the expression (2.9) is of the same form as the 'wave packet' Eq. 1.31 of Sec. 1.14. Thus the free-particle Schrödinger equation has perfectly good wave-packet solutions besides simple harmonic ones.

A striking feature of the Schrödinger equation for any general system, which is exhibited in Eq. 2.8, is the explicit appearance of the imaginary unit i. This feature sets it apart from the partial differential equations of classical physics, all of which contain only real parameters. It goes without

[1] This is the conventional choice. We could equally well take $b = -ia$. Though this would necessitate reversal of the sign of i in all expressions, it would make no difference to any physical results.

saying that the function ψ representing matter waves must be necessarily a *complex* function of x and t.

2.2 Generalization to Three Dimensions

The energy-momentum relation for a free particle moving in three-dimensional space is given by

$$E = \frac{\mathbf{p}^2}{2m} = \frac{1}{2m}(p_x^2 + p_y^2 + p_z^2), \tag{2.10}$$

where p_x, p_y, p_z are the components of the momentum vector $\mathbf{p}$. The harmonic wave to be associated with the particle must have the form

$$\psi(\mathbf{x},t) = a \exp[i(\mathbf{k}.\mathbf{x} - \omega t)] \tag{2.11}$$

This is the generalization of Eq. 2.7 of the one-dimensional case. It is a simple matter to verify with the aid of the de Broglie relations $\hbar\mathbf{k} = \mathbf{p}$ and $\hbar\omega = E$, together with Eq. 2.10, that ψ of (2.11) satisfies

$$i\hbar\frac{\partial\psi(\mathbf{x},t)}{\partial t} = -\frac{\hbar^2}{2m}\nabla^2\psi(\mathbf{x},t) \tag{2.12}$$

Here ∇^2 is the Laplacian operator viz., $(\partial^2/\partial x^2) + (\partial^2/\partial y^2) + (\partial^2/\partial z^2)$. This is the Schrödinger equation for a free particle in three dimensions.

As in the one-dimensional case, any superposition of harmonic waves of the form (2.11), such as

$$\psi(\mathbf{x},t) = \int a(\mathbf{k})\, e^{i(\mathbf{k}.\mathbf{x}-\omega t)}\, d^3k \tag{2.13}$$

satisfies the Schrödinger equation.

2.3 The Operator Correspondence and the Schrödinger Equation for a Particle Subject to Forces

A comparison of Eqs. 2.10 and 2.12 shows that at least for a free particle, the Schrödinger equation can be obtained from the classical energy-momentum relation by the following procedure: replace the energy E and the momentum $\mathbf{p}$ by the following *differential* operators:

$$E \to i\hbar\frac{\partial}{\partial t}, \quad \mathbf{p} \to -i\hbar\nabla \tag{2.14}$$

Let the operators which result from making this replacement on the two sides of the energy-momentum relation (2.10) be supposed to act on the wave function $\psi(\mathbf{x},t)$. The equation so obtained is just the Schrödinger equation 2.12.

Let us now suppose that the same procedure remains applicable even when the particle is not free, and see what equation results therefrom. Consider the case where the total force $\mathbf{F}$ acting on the particle can be written as

$$\mathbf{F} = -\nabla V(\mathbf{x},t) \tag{2.15}$$

Then $V(\mathbf{x},t)$ is the potential energy the particle would have if it were at the point $\mathbf{x}$ at time t. The total energy E of the particle can be expressed as

$$E = \frac{\mathbf{p}^2}{2m} + V(\mathbf{x},t) \tag{2.16}$$

On employing the above-mentioned procedure utilizing the operator correspondence (2.14), we obtain the equation

$$i\hbar \frac{\partial \psi(\mathbf{x},t)}{\partial t} = -\frac{\hbar^2}{2m} \nabla^2 \psi(\mathbf{x},t) + V(\mathbf{x},t)\, \psi(\mathbf{x},t) \tag{2.17}$$

This is the Schrödinger equation for a particle moving in a potential. The method by which we have arrived at this equation is admittedly heuristic. Its validity rests upon the correctness of the results following from it, as evidenced by quantitative agreement with experimental observations on a vast variety of phenomena in microscopic physics.

EXAMPLE 2.1—If a particle carrying a charge Q is in a uniform but possibly time-dependent electric field $\mathcal{E}(t)$, the force $Q\mathcal{E}(t)$ experienced by it can be written as $-\nabla[-Q\mathcal{E}(t).\mathbf{x}]$. Therefore, the potential energy function in this case is $V(\mathbf{x},t) = [-Q\mathcal{E}(t).\mathbf{x}]$. The student is urged to write down for himself the Schrödinger equation for this and other simple cases, such as the simple harmonic oscillator in one dimension, a charged particle moving under the influence of another fixed charge, etc.

It is useful to remember at this point that in general, the energy of a particle is given by

$$E = H(\mathbf{x}, \mathbf{p}, t) \tag{2.18}$$

wherein the Hamiltonian function H need not always have the simple form (2.16). But the procedure which was followed in obtaining Eq. 2.17 can still be applied. On making the operator replacement (2.14) in Eq. 2.18, and making both sides operate on ψ, we get the general Schrödinger equation

$$i\hbar \frac{\partial \psi}{\partial t} = H(\mathbf{x}, -i\hbar\nabla, t)\psi \tag{2.19}$$

The *operator* $H(\mathbf{x}, -i\hbar\nabla, t)$ is the *quantum mechanical Hamiltonian* for the particle. Schrödinger equations for many-particle systems can be set up in a similar fashion, but we will consider these only after gaining some familiarity with the basic ideas of quantum mechanics from the relatively simple one-particle case. To illustrate the general form (2.19), let us consider the example of a charged particle moving in an electromagnetic field. Such motion is basic to all problems involving the interaction between matter and radiation, and is therefore of great practical importance. This example is also instructive in that it brings us face to face with one of the main consequences of replacing classical dynamical variables by operators, namely that $\mathbf{x}$ and $\mathbf{p}$ in quantum mechanics do not commute, i.e. the order in which they occur in any product is important.

It is known that the classical Hamiltonian of a charged particle moving in electric and magnetic fields is given by

$$H(\mathbf{x},\mathbf{p},t) = \frac{1}{2m}\left(\mathbf{p} - \frac{Q}{c}\mathbf{A}(\mathbf{x},t)\right)^2 + Q\phi(\mathbf{x},t) \tag{2.20}$$

where Q is the charge on the particle and $\mathbf{A}(\mathbf{x}, t)$ and $\phi(\mathbf{x},t)$ are the vector and scalar potentials of the electromagnetic field. (See Appendix A). The Schrödinger equation for the particle is obtained by introducing this form of H in Eq. 2.19:

$$i\hbar \frac{\partial \psi}{\partial t} = \frac{1}{2m}\left(-i\hbar\nabla - \frac{Q}{c}\mathbf{A}(\mathbf{x},t)\right)^2 \psi + Q\phi(\mathbf{x},t)\psi \tag{2.21}$$

The expression $(-i\hbar\nabla - Q\mathbf{A}/c)^2\psi$ which appears on the right hand side can be written in expanded form, but in making the expansion we have to be careful to preserve the order in which ∇ and $\mathbf{A}$ occur in each term of the product. This is because ∇ and $\mathbf{A}$ *do not commute:* $\nabla.\mathbf{A}\psi$ is not equal to $\mathbf{A}.\nabla\psi$, because in the first form, ∇ acts on the $\mathbf{x}$-dependent quantity $\mathbf{A}$ too.

Actually, $$\nabla.\mathbf{A}\,\psi \equiv \nabla.(\mathbf{A}\psi) = (\text{div }\mathbf{A})\,\psi + \mathbf{A}.\nabla\psi \tag{2.22}$$

Keeping this in mind, we write

$$\left(-i\hbar\,\nabla - \frac{Q}{c}\mathbf{A}\right)^2\psi = \left(-i\hbar\,\nabla - \frac{Q}{c}\mathbf{A}\right).\left(-i\hbar\,\nabla - \frac{Q}{c}\mathbf{A}\right)\psi$$
$$= -\hbar^2\nabla^2\psi + i\hbar\frac{Q}{c}\nabla.\mathbf{A}\,\psi + i\hbar\frac{Q}{c}\mathbf{A}.\nabla\,\psi + \frac{Q^2}{c^2}\mathbf{A}^2\,\psi \tag{2.23}$$

On inserting Eq. 2.23 along with Eqs. 2.22 into 2.21, the latter takes the more explicit form

$$i\hbar\frac{\partial\psi}{\partial t} = -\frac{\hbar^2}{2m}\nabla^2\psi + \frac{i\hbar Q}{mc}\mathbf{A}.\nabla\,\psi + \left(Q\phi + \frac{i\hbar Q}{2mc}\text{div }\mathbf{A} + \frac{Q^2}{2mc^2}\mathbf{A}^2\right)\psi \tag{2.24}$$

The factor $(Q\phi + i\hbar Q \text{ div } \mathbf{A}/2mc + Q^2\mathbf{A}^2/2mc^2)$ in the last term, which depends only on $\mathbf{x}$ and t, here plays the part of the potential energy function $V(\mathbf{x},t)$ of Eq. 2.17. But the second term on the right hand side of Eq. 2.24 has no counterpart in Eq. 2.17; it may be treated as being due to a *momentum dependent* potential (since ∇ operating on ψ is proportional to the momentum operator).

The above example illustrates the fact that whenever any product of position and momentum variables occurs in the classical Hamiltonian for any system, an ambiguity arises regarding the order in which the corresponding operators should be written when making the transition to the quantum mechanical version through the operator replacement as in Eq. 2.19. Fortunately we will meet only cases where the momentum appears at most linearly in any term containing $\mathbf{x}$, as in the problem just considered. In such cases all we have to do is to make a simple symmetrization of the classical expression before making the operator replacement. Thus

$$\mathbf{p}f(\mathbf{x}) = \tfrac{1}{2}[\mathbf{p}f(\mathbf{x}) + f(\mathbf{x})\,\mathbf{p}] \rightarrow -\tfrac{1}{2}i\hbar\,[\nabla f(\mathbf{x}) + f(\mathbf{x})\nabla] \tag{2.25}$$

Incidentally, it is important to keep in mind that all such operators are to act on a wave function to the right, and that ∇ operates on all factors to its right including the wave function. Thus ∇f operating on ψ does not mean $(\nabla f)\,\psi$ but $\nabla(f\psi)$.

B. PHYSICAL INTERPRETATION AND CONDITIONS ON ψ

Suppose we have set up the Schrödinger equation for a particle, following the prescription given in the last section. Can we say that each and every solution of the equation describes a possible matter wave associated with the particle? The answer is 'no'. Only those solutions which satisfy certain general conditions are admissible for the description of any physical system. The conditions arise partly from the physical interpretation of ψ and partly from the nature of the wave equation itself. We now turn to consider these.

2.4 Normalization and Probability Interpretation

We have already noted that the Schrödinger equation is *linear* and *homogeneous* in ψ and its derivatives, i.e. every term contains ψ or one of its derivatives in the first power only. Consequently if we multiply any solution of the equation by a constant, the resulting function is still a solution. We can take advantage of this arbitrariness in the definition of the wave function to *normalize* it in a convenient way. Suppose $\psi'(\mathbf{x},t)$ is a solution of the Schrödinger equation 2.19. Let

$$\int |\psi'(\mathbf{x},t)|^2 d^3x = \mathcal{N}^2 \tag{2.26}$$

Since $|\psi'|^2$, which is the absolute value squared of a complex number, is by definition a positive real number, its integral is also a positive real number as indicated by the notation $\mathcal{N}^2$. The number $\mathcal{N}^2$ is called the *norm* of the wave function ψ'. Let us now define

$$\psi(\mathbf{x},t) = \frac{1}{\mathcal{N}}\psi'(\mathbf{x},t) \tag{2.27}$$

As explained above, ψ, which differs from ψ' only by a constant factor, will also be a solution of Eq. 2.19. Further,

$$\int |\psi(\mathbf{x},t)|^2 d^3x = 1 \tag{2.28}$$

Any wave function ψ which has unit norm, i.e. which has the property (2.28), is said to be *normalized*. Evidently the step (2.27) through which the normalized wave function is defined makes sense only if $\mathcal{N}$ is finite. Thus *normalizable* wave functions are those with finite norm. Since the norm is defined as the integral of $|\psi|^2$ over *all* of space, its finiteness implies that $|\psi(\mathbf{x},t)|^2$ vanishes at infinity, and hence

$$\psi(\mathbf{x},t) \to 0 \quad \text{as } r \to \infty \text{ where } r = |\mathbf{x}|. \tag{2.29}$$

This is a *boundary condition* which holds for all normalizable wave functions.

It should be noted that even after normalization, the wave function ψ remains undetermined to the extent of an *arbitrary phase factor* (of the form $e^{i\alpha}$, α = real constant). In other words, ψ may be multiplied by $e^{i\alpha}$ without affecting Eq. 2.28.

Incidentally, if Eq. 2.28 is to be valid for all times t, the integral on the left hand side has to be time-independent. We shall verify in Sec. 2.6, with the aid of the Schrödinger equation, that this is indeed the case.

The motivation for normalizing wave functions in the above fashion comes from the probability interpretation[2] of ψ. We recall our conclusions in Chapter 1 that the position (or any other dynamical variable) of a particle cannot, in general, be precisely defined. One can only say that there is a certain probability that the particle is within any specified volume element.

[2] M. Born, *Z. Phys.*, **37**, 863, 1926.

It has been suggested that the need for a statistical interpretation arises from a lack of knowledge regarding the values of some hidden variables pertaining to the system (A. Einstein, B. Podolsky and N. Rosen, *Phys. Rev.*, **47**, 777, 1935). However, an experimental test of hidden variable theories based on an analysis by J. S. Bell (*Rev. Mod. Phys*, **38**, 447, 1966) has given evidence against this suggestion (S. J. Freedman and J. F. Clauser, *Phys. Rev. Lett.*, **28**, 938, 1972).

According to the discussion of Sec. 1.17, this probability has to be assumed to be proportional to the value of $|\psi|^2$ within the volume element. The normalization (2.28) enables us to sharpen this statement and say that $|\psi(\mathbf{x},t)|^2 d^3x$ *is equal to* (rather than just proportional to) *the probability that the particle is within the volume element* d^3x *around the point* $\mathbf{x}$. The total probability that the particle is somewhere in space is then $\int |\psi|^2 d^3x$. Since the particle certainly has to be somewhere in space,[3] this total probability must be unity. This is precisely what Eq. 2.28 says.

EXAMPLE 2.2.—To normalize the function $\psi' = e^{-|x|} \sin \alpha x$. Note that ψ' has the explicit forms, $\psi_{\mathrm{I}}' = e^x \sin \alpha x$ for $x < 0$ and $\psi_{\mathrm{II}}' = e^{-x} \sin \alpha x$ for $x > 0$. Therefore

$$\int_{-\infty}^{\infty} |\psi'|^2 dx = \int_{-\infty}^{0} |\psi_{\mathrm{I}}'|^2 dx + \int_{0}^{\infty} |\psi_{\mathrm{II}}'(x)|^2 dx = \frac{\alpha^2}{2(1+\alpha^2)},$$

on actual evaluation of the integrals. The normalized wave function is therefore

$$\psi(x) = \sqrt{\frac{2(1+\alpha^2)}{\alpha^2}}\, e^{-|x|} \sin \alpha x.$$

EXAMPLE 2.3.—If a particle has the wave function $\psi(x)$ of the last example, what is the probability that its position is to the right of the point $x = 1$? Clearly, the answer is

$$\int_{1}^{\infty} |\psi|^2 dx = \frac{e^{-2}}{2\alpha^2}[1 + \alpha^2 - \cos 2\alpha + \alpha \sin 2\alpha]$$

2.5 Non-normalizable Wave Functions and Box Normalization

We have already noted above that normalization will be possible only if the integral (2.26) is itself not infinite in the first place. Actually there exist wave functions of physical interest for which the integral *is* infinite. An outstanding example is the wave function (2.11) describing a particle of definite momentum. For this function, $\int |\psi|^2 d^3x = |a|^2 \int d^3x = \infty$, since a is a constant and $|\exp[i(\mathbf{k}.\mathbf{x} - \omega t)]| = 1$. The fact that $|\psi|^2$ has the same value $|a|^2$ everywhere implies that the probability of finding the particle in any given region of space is directly proportional to its volume. Consequently the chance of finding the particle in any *finite* region, however large, is vanishingly small compared to the chance of its being in the infinite volume of space outside of this region. This must be considered *unphysical*; thus it should not be possible rigorously for a particle to have an unnormalizable wave function. Nevertheless the use of such wave functions in quantum mechanics is very widespread, on account of the considerable mathematical advantages resulting therefrom. A convenient way of interpreting such functions is to say that the infinite value of $\int |\psi|^2 d^3x$ represents an infinite number of *noninteracting particles* each of which has a wave function propor-

[3] More precisely stated, the assertion is that it is possible to say *with certainty* that any physical particle will be found in a *finite* region of space, provided the region is chosen sufficiently large. It is hard to conceive of a particle which cannot be localized even in this broad sense.

tional to ψ. Then $|\psi|^2\, d^3x$ is the number of particles in the volume d^3x at any instant.

An alternative method of handling non-normalizable wave functions is by regarding them as an idealization—some kind of limit—of physically realizable (normalizable) wave functions. One particularly useful way of visualizing this limit is the following. Imagine the particle to be confined within a large box, and define the norm to be the integral of $|\psi|^2$ taken over the interior of the box only, entirely disregarding the (infinite) region of space outside. This makes the norm finite; it can be made unity by multiplying the wave function by a suitable constant. Use this normalized wave function for calculations; after completion of the calculations, pass to the limit of infinite volume for the box. This procedure is called *box normalization.* It leads to the correct results because physical systems are essentially of finite extent, and therefore as long as the wave functions in a sufficiently large finite region of space are given correctly, it does not really matter what one assumes about the infinitely distant reaches of space.

It is to the case of de Broglie waves (2.11) describing particles of definite momentum ($p = \hbar k$) that the box normalization procedure is most frequently applied. Let us consider these functions at some instant of time, say $t = 0$. If the box is taken to be a cube with edges of length L parallel to the x,y,z axes, the wave functions

$$\psi_{\mathbf{k}}(\mathbf{x}) = L^{-3/2}.\ e^{i\mathbf{k}.\mathbf{x}} \tag{2.30}$$

are evidently normalized within the box, since

$$\iiint_{-\frac{1}{2}L}^{\frac{1}{2}L} \left|\psi_{\mathbf{k}}(\mathbf{x})\right|^2 dx\,dy\,dz = \iiint_{-\frac{1}{2}L}^{\frac{1}{2}L} \frac{dx\,dy\,dz}{L^3} = 1 \tag{2.31}$$

The boundary condition (2.29) is not applicable in this case. Instead, we have to prescribe conditions on the 'walls' of the box. For reasons which will become clear in the next section (Example 2.4, p. 45), we require that the functions (2.30) be *periodic* with respect to the size of the box. This means that if x (or y or z) is increased by L, the wave function should remain unchanged. Hence

$$e^{ik_xL} = e^{ik_yL} = e^{ik_zL} = 1 \tag{2.32a}$$

so that k_x, k_y, k_z must be integer multiples of $(2\pi/L)$:

$$\mathbf{k} = \frac{2\pi}{L}\,\mathbf{n},\ (n_x,\ n_y,\ n_z = 0, \pm 1, \pm 2,\ldots) \tag{2.32b}$$

Thus the admissible momentum vectors get *quantized* for the 'particle in a box'.

In a variety of problems one needs to know the *number* of periodic waves in a box, which have wave vectors within some specified range. This is easily computed from the fact that the tips of the wave vectors (2.32b) form a regular 'lattice' of points with spacing $(2\pi/L)$ along each of the directions in a '**k**-space'. There is one such point per volume $(2\pi/L)^3$ of **k**-space. So the number of vectors **k** whose tips lie within an infinitesimal volume element $d\tau_{\mathbf{k}}$ is given by $(L/2\pi)^3$. $d\tau_{\mathbf{k}}$. Usually one is interested in the number of waves for which the length of **k** is between k and $k + dk$ and the orientations are within a solid angle $d\Omega = \sin\theta\, d\theta\, d\varphi$ around some direction with polar angles (θ, φ). The corresponding volume element is $k^2\, dk\, d\Omega$, and so the required number is $(L/2\pi)^3 k^2\, dk\, d\Omega$. If all directions are considered, the total angle 4π replaces $d\Omega$, and the number is $(L/2\pi)^3 k^2\, dk.4\pi$.

Incidentally it may be noted that in the case of electromagnetic waves the above numbers are to be doubled because for each k, two polarizations are possible. Since in this case $k = 2\pi\nu/c$, the number of light waves (periodic in a box) with frequencies between ν and $\nu + d\nu$ is $n(\nu)d\nu = (8\pi\nu^2 L^3/c^3)d\nu$.

2.6 Conservation of Probability

We have seen in Sec. 2.4 that according to the probability interpretation of the normalized wave function, Eq. 2.28 says simply that the particle is sure to be found somewhere in space. This statement has to be true at all times as long as the particle is a stable one which cannot decay or disappear in some other way. Therefore, the total probability must be *conserved*, i.e. $|\psi|^2$ must be time-independent, and

$$\frac{\partial}{\partial t}\int |\psi|^2\, d^3x \equiv \int \frac{\partial}{\partial t}|\psi|^2\, d^3x = 0 \tag{2.33}$$

Let us verify that this requirement is actually satisfied. We note first that

$$\begin{aligned}\frac{\partial}{\partial t}|\psi|^2 &= \frac{\partial}{\partial t}(\psi^*\psi) = \psi^*\frac{\partial\psi}{\partial t} + \frac{\partial\psi^*}{\partial t}\psi \\ &= (i\hbar)^{-1}[\psi^*(H\psi) - (H\psi)^*\psi]\end{aligned} \tag{2.34}$$

In the last step we have made use of the Schrödinger equation 2.19, namely $i\hbar\, \partial\psi/\partial t = H\psi$, together with its complex conjugate, $-i\hbar\, \partial\psi^*/\partial t = (H\psi)^*$. In the case of a single particle in a potential $V(\mathbf{x},t)$ the explicit form of the quantum Hamiltonian operator is given by

$$H = -\frac{\hbar^2}{2m}\nabla^2 + V(\mathbf{x}, t) \tag{2.35}$$

Eq. 2.34 then becomes

$$\begin{aligned}\frac{\partial}{\partial t}|\psi|^2 &= \frac{i\hbar}{2m}[\psi^*\nabla^2\psi - (\nabla^2\psi^*)\psi] \\ &= \frac{i\hbar}{2m}\nabla.[\psi^*\nabla\psi - (\nabla\psi^*)\psi]\end{aligned}$$

or

$$\frac{\partial}{\partial t}P(\mathbf{x}, t) + \operatorname{div} \mathbf{S}(\mathbf{x}, t) = 0 \tag{2.36}$$

where

$$P(\mathbf{x},t) = \psi^*\psi \tag{2.37a}$$

and

$$\mathbf{S}(\mathbf{x}, t) = -\frac{i\hbar}{2m}[\psi^*\nabla\psi - (\nabla\psi^*)\psi] \tag{2.37b}$$

On integrating Eq. 2.36 over all space we get

$$\begin{aligned}\frac{\partial}{\partial t}\int P\, d^3x &= -\int \operatorname{div} \mathbf{S}\, d^3x \\ &= -\int_\sigma \mathbf{S.n}\, d\sigma\end{aligned} \tag{2.38}$$

where we have used Gauss' theorem to reduce the volume integral of div **S** to the integral of the normal component of **S** over the surface σ bounding the volume. Since the volume integral in Eq. 2.38 is over all of space, the surface σ is at infinity, where ψ and $\nabla\psi$ vanish (for normalizable wave functions). Consequently **S** also vanishes, and in fact it vanishes sufficiently fast as $|\mathbf{x}| \to \infty$, that the surface integral (2.38) itself vanishes. In the case of box-normalized functions, ψ and $\nabla\psi$ do not necessarily vanish on the surface σ of the box. But the boundary conditions specifying their values on σ are chosen in such a way that once again, the integral (2.38) is zero. (See Example 2.4 below).

Thus we have verified that the condition (2.33) for conservation of total probability is indeed satisfied, and hence the normalization (2.28) is time-independent.

EXAMPLE 2.4.—Let us verify that the boundary conditions (2.32) on the wave functions of a particle in a box are necessary for probability conservation. Consider a particle in one dimension, having a wave function which is a simple superposition of two de Broglie waves:

$$\psi = a_1 \exp\,[i\,(k_1x - \omega_1t - \alpha_1)] + a_2 \exp\,[i\,(k_2x - \omega_2t - \alpha_2)]$$

where a_1, a_2 are the (real) amplitudes and α_1, α_2 the phases of the waves. The position probability density is $P = a_1^2 + a_2^2 + 2a_1a_2 \cos\,[(k_1 - k_2)x - (\omega_1 - \omega_2)t - (\alpha_1 - \alpha_2)]$. If the particle is confined to a 'box' (which, in one dimension, is simply the interval of x between $-\frac{1}{2}L$ and $+\frac{1}{2}L$), conservation of probability within the box requires that $\int_{-\frac{1}{2}L}^{\frac{1}{2}L} P\,dx$ should be time-independent. If this is to be true with the above expression for P, it is evidently necessary that

$$\int_{-L/2}^{L/2} \cos\,[(k_1 - k_2)\,x - (\omega_1 - \omega_2)t - (\alpha_1 - \alpha_2)]\,dx$$
$$\equiv [2/(k_1 - k_2)] \sin\,[\tfrac{1}{2}L\,(k_1 - k_2)] \cos\,[(\omega_1 - \omega_2)t + (\alpha_1 - \alpha_2)]$$

should be time-independent. The only way this can happen is by the vanishing of the sine factor, which means that $(k_1 - k_2)\,L = 2n\pi$ where n is any integer. Thus the admissible values for k are restricted to a set wherein the differences are integral multiples of $(2\pi/L)$. By taking $k = (2\pi n/L)$ we satisfy the above condition on the differences (and thus ensure conservation of probability), and further ensure that along with every admissible k, $-k$ also is admissible. Clearly there should be no discrimination between one direction of propagation and the opposite direction, since the 'box' itself has no asymmetry.

Eqs. 2.32 are a straightforward generalization of the above result to three dimensions, and can be obtained directly from probability conservation in a three-dimensional box.

It may be noted here that Eq. 2.36 has exactly the same form as the continuity equation, $\partial\rho/\partial t + \text{div}\,\mathbf{j} = 0$, of hydrodynamics. Instead of the matter density ρ and the matter current density $\mathbf{j}$, we have in Eq. 2.36 the *probability density P and the probability current density* **S**. Just as the continuity equation says that no sources or sinks of matter are present, Eq. 2.36 asserts that there is no creation or destruction of probability: any increase or decrease $(\partial P/\partial t)\,d\tau\,dt$ in the probability for finding the particle in a given volume element $d\tau \equiv d^3x$ is compensated by a corresponding decrease or increase elsewhere through an inflow or outflow of probability (div **S**) $d\tau\,dt$ across the boundaries of $d\tau$.

2.7 Expectation Values; Ehrenfest's Theorem[4]

The probability interpretation tells us what results to expect if a large number of observations are made of the positions of particles having a specified wave function ψ. We imagine we have an apparatus which can prepare particles with this particular wave function. Each observation causes the wave function inevitably to undergo some change. So before the next observation is made, the apparatus is to be used to restore the wave function to the original form.

Suppose we now make a large number of observations of the position of the particle. Though we take care to ensure that the particle has the same wave function ψ before each observation, we do not expect to get the same result each time. What is expected is that, of all the observations, a fraction equal to $|\psi|^2\, d^3x$ will show the particle as being within the volume element d^3x. Therefore if we take the *mean* or *average* of all the observed values of the position vectors, the result is expected to be

$$\langle \mathbf{x} \rangle = \int \mathbf{x}\,|\psi|^2\, d^3x = \int \psi^* \mathbf{x}\, \psi\, d^3x \tag{2.39}$$

i.e., the integral of possible results $\mathbf{x}$ weighted by the function $|\psi(\mathbf{x})|^2$. Remember, $|\psi(\mathbf{x})|^2$ represents the frequency with which (i.e. the fraction of the total number of times) the various values $\mathbf{x}$ occur. Eq. 2.39 gives the expected mean position of the particle; it is usually referred to simply as the *mean* position or *expectation value* of the position variable $\mathbf{x}$.

Now, how does the mean position change with time? By direct differentiation of Eq. 2.39 we obtain

$$\begin{aligned}\frac{d\langle \mathbf{x} \rangle}{dt} &= \int \left(\frac{\partial \psi^*}{\partial t}\,\mathbf{x}\,\psi + \psi^*\,\mathbf{x}\,\frac{\partial \psi}{\partial t}\right) d^3x \\ &= \frac{1}{i\hbar}\int \left[-\left(-\frac{\hbar^2}{2m}\nabla^2\psi^* + V\psi^*\right)\mathbf{x}\psi + \psi^*\mathbf{x}\left(-\frac{\hbar^2}{2m}\nabla^2\psi + V\psi\right)\right] d^3x \\ &= \frac{i\hbar}{2m}\int \left[\psi^*\,\mathbf{x}\,\nabla^2\psi - (\nabla^2\psi^*)\,\mathbf{x}\,\psi\right] d^3x\end{aligned}$$

Here we have assumed that the particle is moving in a potential V, and used the Schrödinger equation 2.17 as well as its complex conjugate. By carrying out two integrations by parts on the second term in the above expression and using the fact that ψ (and its derivatives) vanish at infinity according to Eq. 2.29, we obtain

$$\begin{aligned}\frac{d\langle \mathbf{x} \rangle}{dt} &= \frac{i\hbar}{2m}\int \left[\psi^*\,\mathbf{x}\,\nabla^2\psi - \psi^*\,\nabla^2(\mathbf{x}\,\psi)\right] d^3x \\ &= -\frac{i\hbar}{m}\int \psi^*\,\nabla\,\psi\, d^3x\end{aligned} \tag{2.40}$$

Now we know that in classical mechanics, the velocity is related to

[4] P. Ehrenfest, *Z. Physik.*, **45**, 455, 1927.

momentum through $\mathbf{p} = m\, d\mathbf{x}/dt$. We have seen in the last chapter that for the motion of a quantum mechanical wave packet *as a whole,* (which is what the rate of change of the *mean* position describes), the classical relations should apply at least approximately. Therefore Eq. 2.40 suggests that we define the *mean* momentum or *expectation value* of the momentum as

$$\langle \mathbf{p} \rangle = m \frac{d\langle \mathbf{x} \rangle}{dt} = \int \psi^*(-i\hbar\nabla)\psi\, d^3x \tag{2.41}$$

Observe that the integral has a very interesting form: it is the integral of the product of ψ^* and $\mathbf{p}_{op}\,\psi$ where $\mathbf{p}_{op}$ is the operator $(-i\hbar\nabla)$ representing momentum in quantum mechanics. This suggests that the expectation value of *any* dynamical variable, say $A(\mathbf{x},\mathbf{p})$ is to be identified as

$$\langle A \rangle = \int \psi^* A_{op}\, \psi\, d^3x \tag{2.42}$$

where A_{op} is the operator $A(\mathbf{x}, -i\hbar\nabla)$ which represents A in quantum mechanics. Note that A_{op} acts only on the wave function ψ standing to its right. More generally, when the wave function is *not* normalized as in Eq. 2.28,

$$\langle A \rangle = \frac{\int \psi^* A_{op}\, \psi\, d^3x}{\int \psi^* \psi\, d^3x} \tag{2.42a}$$

The definition of expectation values through the Eq. 2.42 is one of the basic postulates of quantum mechanics. The qualitative arguments of Sec. 1.16 regarding the approximately classical motion of small wave packets (in slowly varying fields) can be made precise on the basis of this definition. Indeed, by differentiating $\langle \mathbf{p} \rangle$, Eq. 2.41, and proceeding exactly as we did for determining $d\langle \mathbf{x} \rangle/dt$, we can easily show that

$$\frac{d\langle \mathbf{p} \rangle}{dt} = -\langle \nabla V(\mathbf{x}, t) \rangle \tag{2.43}$$

Since $\langle \mathbf{p} \rangle = m\, d\langle \mathbf{x} \rangle/dt$, this equation seems very similar to Newton's equation of motion. In fact it would have been identical to Newton's equation (except for the appearance of $\langle \mathbf{x} \rangle$ instead of the plain $\mathbf{x}$ of classical mechanics) if the right hand side of Eq. 2.40 had been

$$-[\nabla V(\mathbf{x},t)]_{\mathbf{x} = \langle \mathbf{x} \rangle} \tag{2.44}$$

The difference between this expression (∇V at $\langle \mathbf{x} \rangle$) and the average of ∇V which appears in Eq. 2.43 is small when the size of the wave packet is small, and in this limit the overall motion of the quantum wave packet is well approximated by classical mechanics. This important result, and more particularly Eq. 2.43, is known as Ehrenfest's Theorem.

Example 2.5.—Let us calculate $\langle x \rangle$ and $\langle p^2 \rangle$ for the wave function considered in Example 2.2 (p. 42), namely

$\psi = [2(1 + \alpha^{-2})]^{1/2} e^{x} \sin\alpha x,\ x < 0$, and $\psi = [2(1 + \alpha^{-2})]^{1/2} e^{-x} \sin\alpha x,\ x > 0$

Then

$$\langle x \rangle = \int \psi^* x\, \psi\, dx$$

$$= 2(1 + \alpha^{-2})\left[\int_{-\infty}^{0} x e^{2x} \sin^2 \alpha x\, dx + \int_{0}^{\infty} x e^{-2x} \sin^2 \alpha x\, dx\right] = 0,$$

since the two integrals cancel each other. This can be seen by changing the variable of integration in the first integral from x to $x' = -x$. Since $p^2_{op} =$

$-\hbar^2 d^2/dx^2$, we have $\langle p^2 \rangle = -\hbar^2 \int \psi^* (d^2\psi/dx^2)\, dx$. On substituting the above wave function, evaluating the integrals from $-\infty$ to 0 and 0 to ∞ and adding up, we get

$$\langle p^2 \rangle = \hbar^2 (1 + \alpha^2).$$

EXAMPLE 2.6.—As an example in three dimensions, let us evaluate $\langle r \rangle$ when the wave function (in spherical polar coordinates) is $\psi = (1/\pi a^3)^{1/2} e^{-r/a}$ where a is a constant. In this case it is simplest to express the integral in terms of the polar coordinates:

$$\int \psi^* r \psi d\tau = \int_0^\infty r^2 dr \int_0^\pi \sin\theta\, d\theta \int_0^{2\pi} d\varphi (\psi^* r \psi)$$

$$= \left(\frac{1}{\pi a^3}\right) 4\pi \int_0^\infty r^3 e^{-2r/a}\, dr = \frac{3a}{2}$$

Note that since the integral was independent of θ and φ, the integrations over these angles could be trivially evaluated, leading to the factor 4π.

2.8 Admissibility Conditions on the Wave Function

So far we have not raised any questions as to what kinds of mathematical functions can serve as wave functions for a particle in quantum mechanics. Actually, the probability interpretation of the wave function makes it necessary that ψ should satisfy certain general conditions. The conditions are intended to ensure that $|\psi|^2 d\tau$ does have the properties of a probability magnitude. We require that ψ *should be (a) finite and (b) single-valued*[5] *everywhere.*

Finiteness is needed in order that $|\psi|^2 d\tau$, for an infinitesimal volume element $d\tau$, should lie between 0 and 1, as any probability should. This argument does not rule out square integrable singularities (where ψ becomes infinite without making $\int |\psi|^2 d\tau$ infinite); but condition (c), stated below, does. Single-valuedness is the requirement that at any given physical point the wave function should have a unique value, in order that the probability density $|\psi|^2$ may be uniquely defined. This condition makes its appearance in non-trivial form whenever there exist several alternative values for the coordinates of a given physical point. This is the case, for example, when spherical polar coordinates (r,θ,φ) are employed: as is well known, changing φ by any integer multiple of 2π does not affect the point which is referred to. In this context, single-valuedness of the wave function means that we must have $\psi(r,\theta,\varphi) = \psi(r,\theta,\varphi + 2n\pi)$ for every integer n.

There is a further property which is required of the wave function, namely (c) that: ψ *and its first partial derivatives* $\frac{\partial\psi}{\partial x}, \frac{\partial\psi}{\partial y}$ *and* $\frac{\partial\psi}{\partial z}$ *should be continuous functions of* $\mathbf{x}$, *for all* $\mathbf{x}$.

This requirement arises from an implicit assumption regarding the nature of the potential function V, namely that V is a continuous function of $\mathbf{x}$

[5] For single-valuedness of $|\psi|^2$ it is actually not essential that ψ be single-valued. Nevertheless it has been traditional to require the latter, and we have followed the tradition as a matter of expediency. See the discussion at the end of Sec. 4.10.

except perhaps for a certain number of *finite* discontinuities. When V is of this kind, the condition (c) is necessary for consistency of the Schrödinger equation. To see this, let us consider the simplest case of a particle in one dimension, when the Schrödinger equation becomes

$$i\hbar \frac{\partial \psi}{\partial t} = -\frac{\hbar^2}{2m} \frac{\partial^2 \psi}{\partial x^2} + V\psi \tag{2.45}$$

If ψ is a continuous function of x at all times in accordance with (c), then $\partial\psi/\partial t$ is also evidently a continuous function of x. Therefore, the right hand side of Eq. 2.45 must be continuous, and any departure from continuity in one of the two terms must be cancelled by an opposite behaviour in the other term. For example if V (and hence $V\psi$) has a finite discontinuity at some point a, $\partial^2\psi/\partial x^2$ also must have a finite discontinuity at a, i.e. $(\partial\psi/\partial x)$ must be continuous at a but its slope (which is $\partial^2\psi/\partial x^2$) to the right of a must be unequal to that on the left of a. (See Fig. 2.1). Conversely, if $(\partial\psi/\partial x)$ has this behaviour, V must necessarily have a finite discontinuity. If $(\partial\psi/\partial x)$ behaves any worse, e.g., if it has a finite discontinuity of its own [violating the condition (c)], then the curve for $\partial\psi/\partial x$ becomes 'vertical' at that point, and its slope, $\partial^2\psi/\partial x^2$ at that point is infinite. This would force V to have an *infinite* discontinuity. Conversely, we have:

(*c'*) *at any point where the potential V makes a sudden jump of infinite magnitude, $\partial\psi/\partial x$ has a finite discontinuity but ψ remains continuous.*

This statement is for the one-dimensional case. The generalization to three dimensions is straightforward, but we will not need it.

Any function which meets all the above requirements is *admissible* as a wave function. The requirements themselves are so natural as to be almost self-evident. But their consequences for the theory are of profound importance, as we shall see immediately.

C. STATIONARY STATES AND ENERGY SPECTRA

We have already noted that the state of a quantum mechanical system is specified by giving its wave function ψ. We are now in a position to see that in the case of time-independent systems (such as a particle moving in a *static* or time-independent potential V) there exist a special category of solutions of the wave equation, which describe *stationary* states. In these states, the position probability density $|\psi|^2$ at every point $\mathbf{x}$ in space remains independent of time; so also do the expectation values of all dynamical variables. Further when a particle is described by such a wave function its energy has a perfectly definite value. The energy spectrum (i.e. the set of energy values associated with the various stationary states) is, in general, at least partly discrete. We will now see (in the context of a simple example) how all these results follow in a very natural way from the Schrödinger equation and the admissibility conditions on wave functions. The existence of stationary states with discrete energies, which was *postulated* by Bohr in order to account for the nature of

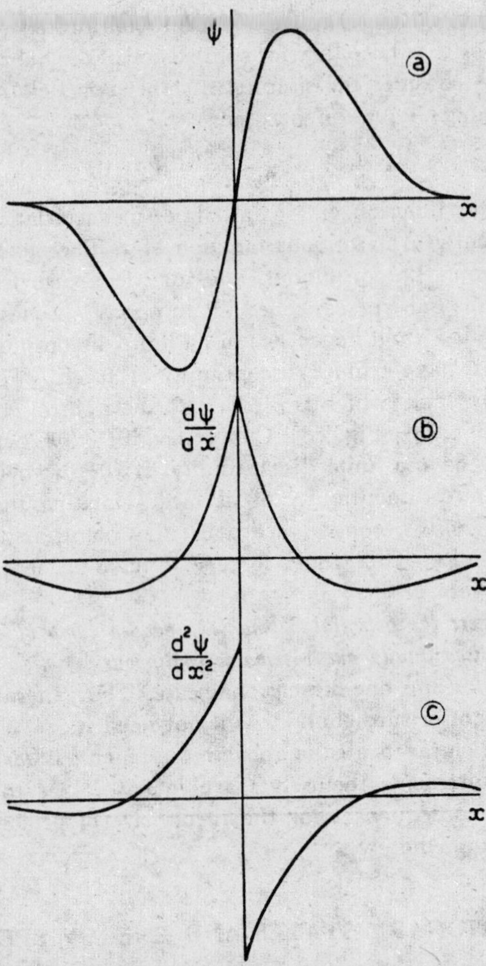

Fig. 2.1 Illustration of continuity properties (**a**) The wave function $\psi = e^{-|x|} \sin x$, continuous everywhere; (**b**) $d\psi/dx$: It has a 'cusp' at $x = 0$, but is still continuous; and (**c**) $d^2\psi/dx^2$: It has a finite discontinuity at $x = 0$.

atomic spectra, thus finds a rational explanation on the basis of quantum mechanics.

2.9 Stationary States; The Time-Independent Schrödinger Equation

Let us consider a particle moving in a *time-independent* potential $V(\mathbf{x})$. It is easy to verify that the Schrödinger equation in this case has solutions of the form

$$\psi(\mathbf{x}, t) = u(\mathbf{x}) f(t) \tag{2.46}$$

Substituting this assumed form in Eq. 2.17, and dividing throughout by $u(\mathbf{x}) f(t)$, we obtain

$$\frac{1}{f} i\hbar \frac{df}{dt} = -\frac{\hbar^2}{2m} \frac{\nabla^2 u}{u} + V(\mathbf{x}) \tag{2.47}$$

The right hand side of this equation is independent of t, and the left hand side is independent of $\mathbf{x}$. Their equality implies, therefore, that *both* sides must be independent of $\mathbf{x}$ and t, and hence must be equal to a constant, say E. Thus Eq. 2.47 separates into two equations:

$$i\hbar \frac{df(t)}{dt} = Ef(t) \tag{2.48}$$

and

$$\left[-\frac{\hbar^2}{2m}\nabla^2 + V(\mathbf{x})\right] u(\mathbf{x}) = Eu(\mathbf{x}) \tag{2.49}$$

The former is readily solved, and shows that $f(t)$ is proportional to $\exp[-iEt/\hbar]$. The solution of Eq. 2.49 depends on the value assumed for E, and so we write it as $u_E(\mathbf{x})$. Thus Eq. 2.46 reduces to

$$\psi(\mathbf{x},t) = u_E(\mathbf{x})\, e^{-iEt/\hbar} \tag{2.50}$$

The value of E has to be real. For if it had an imaginary part ε, the wave function ψ would vanish for all $\mathbf{x}$ as $t \to \infty$ or $-\infty$ according as the sign of ε is $-$ or $+$, and this is of course not admissible. It follows then that

$$|\psi(\mathbf{x},t)|^2 = |u_E(\mathbf{x})|^2 \tag{2.51}$$

i.e. the probability density is independent of t. Similarly expectation values, as defined by Eq. 2.42, are also evidently time-independent. In other words, ψ of Eq. 2.50 describes a *stationary state* in which none of the particle characteristics changes with time.

Let us return now to Eq. 2.49 and the interpretation of E. This equation (called the *time-independent Schrödinger equation*) states that the action of the Hamiltonian operator of the particle (the quantity in square brackets) on the wave function $u_E(\mathbf{x})$ is simply to reproduce the same wave function multiplied by the *constant* E. This property is reminiscent of characteristic equations or eigenvalue equations in matrix theory, where a (column) vector u is called an eigenvector belonging to the eigenvalue λ of a matrix M if it satisfies the equation $Mu = \lambda u$ (with λ = a constant number). By analogy, we say that since $u_E(\mathbf{x})$ satisfies Eq. (2.49), it is an *eigen-function* belonging to the *eigenvalue* E of the (differential) operator $H = -(\hbar^2/2m)\nabla^2 + V$. Remember that this operator represents the energy of the particle as a function of the position and momentum. So $u_E(\mathbf{x})$ may be called an energy eigenfunction. Anticipating the general interpretation (Sec. 3.9) of eigenvalues and eigenfunctions, we now assert that *when the state of a particle is described by an energy eigenfunction, the energy of the particle has a definite value, given by the eigenvalue E.* The reader is invited to verify for himself that this interpretation is consistent with what he already knows about the free-particle case, by substituting the wave function (2.11) into Eqs. 2.48 and 2.49 with $V = 0$.

The following question still remains: If we assign an *arbitrary* real value to E in Eq. 2.49, does there exist a corresponding eigenfunction $u_E(\mathbf{x})$? The answer is in the negative. This is because the solution of Eq. 2.49 does not satisfy the admissibility conditions of Sec. 2.8, unless E is restricted to certain specific values. Only these special values are to be considered as eigenvalues. The set of all such admissible values of E form what is called the eigenvalue spectrum of energy, or simply the *energy spectrum*. Following the terminology of old quantum theory, we shall frequently refer to the energy eigenvalues as *energy levels* of the system. We shall now determine the energy spectrum and energy eigenfunctions of a very simple system, in order to illustrate how the admissibility conditions on wave functions lead to restrictions on possible energy values.

2.10 A Particle in a Square Well Potential

The example we consider is that of a particle whose potential energy function has the shape of a 'well' with vertical sides. It is depicted in Fig. 2.2 and is defined by

$$\begin{aligned} V(x) &= 0 && \text{for} \quad x < -a && \text{(Region I)} \\ V(x) &= -V_0 && \text{for} \quad -a < x < a && \text{(Region II)} \\ V(x) &= 0 && \text{for} \quad x > a && \text{(Region III)} \end{aligned} \tag{2.52}$$

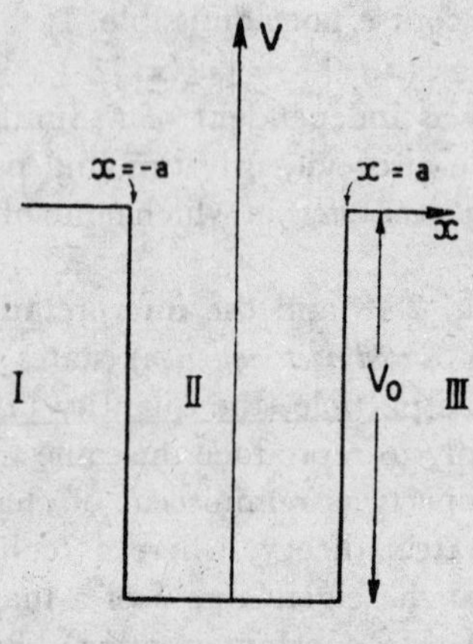

Fig. 2.2 The square well potential

If we were considering this problem in classical mechanics, we would have to keep in mind that the kinetic energy $(E - V)$ can never be negative. Since $V = 0$ for $|x| > a$, $(E - V)$ can be positive in this region only if $E > 0$. Therefore any particle with $E < 0$ cannot enter regions I and III, and will have to stay within the potential 'well', between $x = a$ and $x = -a$. We say in this case that the particle is *bound* by the potential. On the other hand if the particle has energy $E > 0$, it can go anywhere; it merely experiences momentary forces on crossing the points $x = -a$ and $x = +a$.

Let us now consider the quantum mechanical picture of this system. We confine our attention, for the present, to stationary states, which are described by solutions of Eq. 2.49. In the present case this equation, reduced to one dimension, takes distinct forms in the different regions:

$$-\frac{\hbar^2}{2m}\frac{d^2u}{dx^2} = Eu, \qquad |x| > a \tag{2.53a}$$

and

$$-\frac{\hbar^2}{2m}\frac{d^2u}{dx^2} - V_0 u = Eu, \; |x| < a \tag{2.53b}$$

The set of Eqs. 2.53 can be trivially solved. The nature of the solutions depends on whether $E < 0$ or $E > 0$. The former condition corresponds to bound states, as we shall see, and we take up this case first.

2.11 Bound States in a Square Well: (E < 0)

(*a*) *Admissible Solutions of Wave Equation:* For $E < 0$, we write

$$\frac{2mE}{\hbar^2} = -\alpha^2 \text{ and } \frac{2m(E + V_0)}{\hbar^2} = \beta^2; \; \alpha, \beta > 0 \tag{2.54}$$

Eqs. 2.53 can then be conveniently rewritten in terms of the positive constants α^2 and β^2 as

$$\frac{d^2u}{dx^2} - \alpha^2 u = 0, \quad |x| > a \tag{2.55a}$$

$$\frac{d^2u}{dx^2} + \beta^2 u = 0, \quad |x| < a \tag{2.55b}$$

The general solution of the second of these equations is obviously

$$u^{\text{II}}(x) = A \cos \beta x + B \sin \beta x \tag{2.56}$$

where A and B are any constants. The superscript II on $u(x)$ here indicates that it is the solution valid in Region II i.e. $(|x| < a)$. The equation which holds in Region I, $(-\infty < x < a)$, is (2.55*a*). Its general solution is a linear combination of its two independent solutions, $e^{\alpha x}$ and $e^{-\alpha x}$. However the latter *does not remain finite* everywhere in Region I. In fact, as $x \to -\infty$, it becomes infinitely large. Therefore the only admissible solution in Region I must be of the form

$$u^{\text{I}}(x) = C e^{\alpha x} \tag{2.57}$$

Region III, $(a < x < \infty)$, is also governed by Eq. 2.55*a* with the idependent solutions $e^{\alpha x}$ and $e^{-\alpha x}$. But now it is $e^{-\alpha x}$ which remains well behaved everywhere and $e^{\alpha x}$ which is not admissible (since it diverges as $x \to +\infty$). Therefore, we have

$$u^{\text{III}}(x) = D e^{-\alpha x} \tag{2.58}$$

where D, like A, B, and C, is an undetermined constant.

Thus the solution $u(x)$ has three different forms u^{I}, u^{II}, u^{III} in the three regions. We must now make sure that $u(x)$ and its first derivative (du/dx) are continuous everywhere,[6] as demanded by the condition (c) of Sec. 2.8. In particular, at the point $x = -a$ where Regions I and II meet, we should have

$$u^{\text{I}} = u^{\text{II}} \text{ and } \frac{du^{\text{I}}}{dx} = \frac{du^{\text{II}}}{dx}, \; (x = -a) \tag{2.59}$$

or explicitly,

$$Ce^{-\alpha a} = A \cos \beta a - B \sin \beta a, \quad C\alpha e^{-\alpha a} = A\beta \sin \beta a + B\beta \cos \beta a \tag{2.60}$$

[6] In view of the relation (2.50) between $\psi(\mathbf{x}, t)$ and $u(\mathbf{x})$, it is obvious that all the admissibility conditions on ψ must be equally satisfied by $u(\mathbf{x})$.

Similarly, at $x = a$, where Regions II and III meet, we must have

$$u^{\text{II}} = u^{\text{III}} \text{ and } \frac{du^{\text{II}}}{dx} = \frac{du^{\text{III}}}{dx}, \quad (x = a) \tag{2.61}$$

These lead to

$$De^{-\alpha a} = A\cos\beta a + B\sin\beta a, \quad -D\alpha e^{-\alpha a} = -A\beta\sin\beta a + B\beta\cos\beta a \tag{2.62}$$

From Eqs. 2.60 and 2.62 we readily find that

$$2A\cos\beta a = (C + D)\, e^{-\alpha a} \tag{2.63a}$$
$$2A\beta\sin\beta a = (C + D)\, \alpha e^{-\alpha a} \tag{2.63b}$$
$$2B\sin\beta a = -(C - D)\, e^{-\alpha a} \tag{2.64a}$$
$$2B\beta\cos\beta a = (C - D)\, \alpha e^{-\alpha a} \tag{2.64b}$$

Eqs. 2.63 show that if $(C + D) \neq 0$, then $A \neq 0$, and further,

$$\alpha = \beta\tan\beta a \tag{2.65a}$$

This relation implies that

$$C = D \text{ and } B = 0 \tag{2.65b}$$

and hence

$$D = Ae^{\alpha a}\cos\beta a \tag{2.65c}$$

To verify this, we substitute for the factor α in Eq. 2.64b the value determined above; then, multiplying the equation by $\sin\beta a$ and using Eq. 2.64a, we obtain $-(C - D)\cos^2\beta a = (C - D)\sin^2\beta a$. Since $\sin^2\beta a$ cannot be equal to the negative quantity $-\cos^2\beta a$, we must have $C = D$. Thus we obtain Eq. 2.65b, and on feeding this back into Eq. 2.63, Eq. 2.65*c* follows.

Eqs. 2.65 give one type of solution for our problem. Another type of solution exists for $C \neq D$ and $B \neq 0$, when we get from Eqs. 2.64

$$\alpha = -\beta\cot\beta a. \tag{2.66a}$$

Repetition of the kind of arguments used above will show that now

$$C = -D \text{ and } A = 0 \tag{2.66b}$$

and

$$D = Be^{\alpha a}\sin\beta a \tag{2.66c}$$

(*b*) *The Energy Eigenvalues—Discrete Spectrum:* We can now see that *both the types of solutions exist only for certain discrete values of the energy parameter E.* We observe first of all that by virtue of Eq. 2.54

$$(\alpha^2 + \beta^2)\, a^2 = \frac{2mV_0 a^2}{\hbar^2} = \frac{V_0}{\triangle} \tag{2.67}$$

where

$$\triangle = \frac{\hbar^2}{2ma^2} \tag{2.68}$$

The parameter $\triangle$ has an interesting interpretation. The half-width a of the potential well indicates the uncertainty in the position of a particle confined to the well, and associated with this, there is an uncertainty of the order of $(\hbar/a)$ in the momentum. The corresponding energy $\triangle = (\hbar/a)^2/2m$ is a natural unit in terms of which the depth of the potential may be measured. Thus the non-dimensional parameter $(V_0/\triangle)$ of Eq. 2.67 is a measure of the *strength* of the potential.

Now, in the case of solutions of the first type, we have, besides Eq. 2.67, the further relation (2.65a) between α and β, which has two consequences. One is that since α and β have been defined to be positive, $(\alpha/\beta) = \tan \beta a$ must be positive and hence only values of βa lying in the intervals

$$2r \frac{\pi}{2} \leqslant \beta a \leqslant (2r+1) \frac{\pi}{2} \quad (r = 0,1,2,\ldots) \tag{2.69a}$$

are admissible. Secondly, the substitution of $\alpha = \beta \tan \beta a$ into Eq. 2.67 leads to the requirement

$$\frac{V_0}{\Delta} = \beta^2 a^2 \sec^2 \beta a, \text{ or } \left(\frac{\Delta}{V_0}\right)^{1/2} \beta a = |\cos \beta a| \tag{2.69b}$$

The modulus sign arises because the left hand side of the equation is known to be positive.

Similarly, for the second type of solutions of the wave equation, given by (2.66), we find from Eqs. 2.66*a* and 2.67 that

$$(2r-1) \frac{\pi}{2} \leqslant \beta a \leqslant 2r \frac{\pi}{2}, \quad (r = 1,2,\ldots) \tag{2.70a}$$

and

$$\frac{V_0}{\Delta} = \beta^2 a^2 \operatorname{cosec}^2 \beta a, \text{ or } \left(\frac{\Delta}{V_0}\right)^{1/2} \beta a = |\sin \beta a| \tag{2.70b}$$

Eqs. 2.69 and 2.70 can be satisfied only by certain specific discrete values of β, which can be found graphically. These special values β_n are determined by the intersections of the straight line $(\Delta/V_0)^{1/2}\beta a$ with the curves for $|\cos \beta a|$ and $|\sin \beta a|$. The parts of $|\cos \beta a|$ and $|\sin \beta a|$ which lie within the respective allowed intervals — conditions (2.69a) and (2.70a) — are shown as solid lines and dashed lines respectively in Fig. 2.3. The parts to be ignored are indicated by dotted lines.

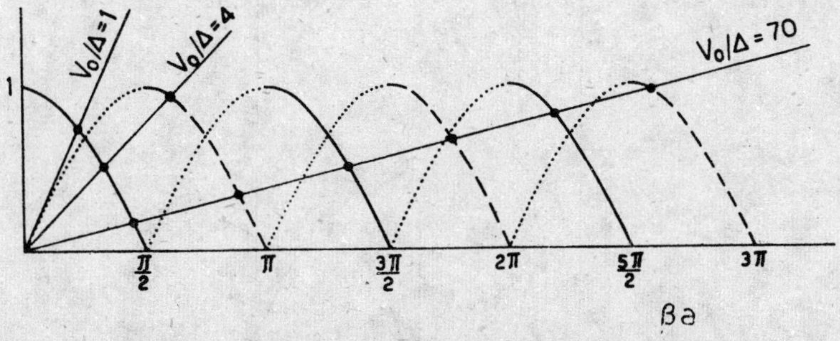

Fig. 2.3 Graphical solution for allowed values of β

If the intersections occur at $\beta = \beta_n$ $(n = 0, 1, 2, \ldots)$ the corresponding allowed values of the energy are obtained from Eq. (2.54) as

$$E_n = \frac{\hbar^2}{2m}\beta_n^2 - V_0 = \left[(\beta_n a)^2 \frac{\Delta}{V_0} - 1\right] V_0 \tag{2.71}$$

It may be noted that the value of the combination of parameters appearing in the square bracket may be read off directly from the figure. It is evident that the number of these energy levels is finite. Inspection of Fig. 2.3 shows that if $(\Delta/V_0)^{1/2}\,\beta a$ reaches the value unity for a value of βa in the interval $\frac{1}{2}\pi\mathcal{N} \leqslant \beta a < \frac{1}{2}\pi\,(\mathcal{N}+1)$, then there are $(\mathcal{N}+1)$ intersections. In other words, the number of discrete energy levels is $(\mathcal{N}+1)$ if $\frac{1}{2}\pi\mathcal{N}(\Delta/V_0)^{1/2} \leqslant 1 < \frac{1}{2}\pi\,(\mathcal{N}+1)\,(\Delta/V_0)^{1/2}$, that is, if

$$\mathcal{N} \leqslant \frac{2}{\pi}\left(\frac{V_0}{\Delta}\right)^{1/2} < (\mathcal{N}+1) \tag{2.72}$$

It is noteworthy that there exists at least one bound state, however weak the potential may be.

(*c*) *The Energy Eigenfunctions; Parity:* We observe that the energy levels E_n with $n = 0, 2, 4, \ldots$ correspond to solutions characterized by Eqs. 2.65. In this case any solution $u_n(x)$ has the following explicit forms in the three regions:

$$\left.\begin{aligned} u_n^{\mathrm{I}}(x) &= (A\,e^{\alpha_n a}\cos\beta_n a)\,e^{\alpha_n x} && (x < -a)\\ u_n^{\mathrm{II}}(x) &= A\cos\beta_n x && (-a < x < a)\\ u_n^{\mathrm{III}}(x) &= (A\,e^{\alpha_n a}\cos\beta_n a)\,e^{-\alpha_n x} && (x > a)\\ & && (n = 0,2,\ldots) \end{aligned}\right\} \tag{2.73}$$

The nature of such functions is illustrated graphically in Fig. 2.4(a). It is evident that $u_n(x)$ is symmetric about the origin:

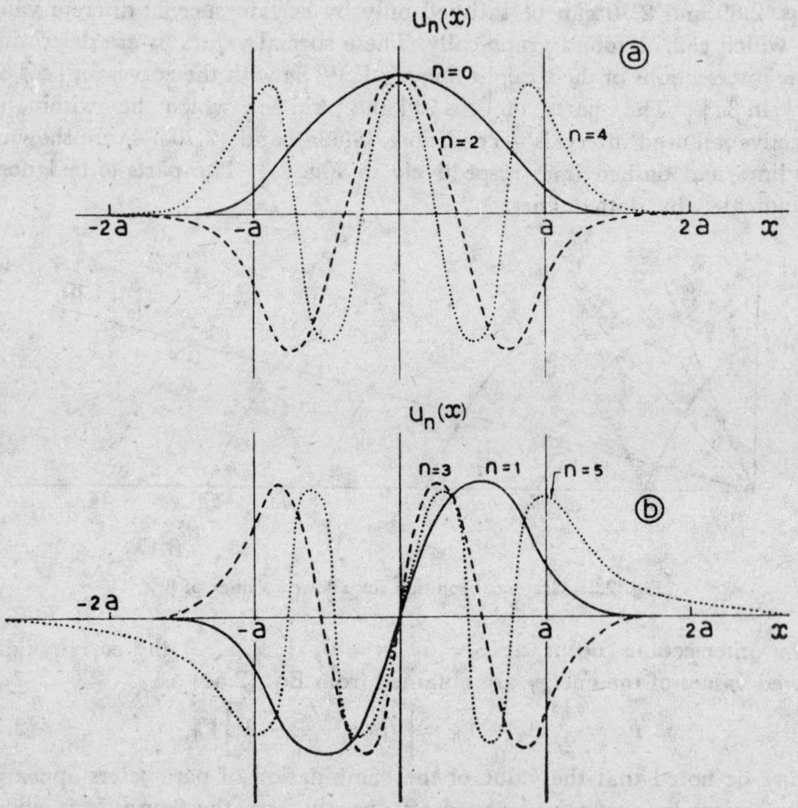

Fig. 2.4 (**a**) Even parity eigenfunctions; and (**b**) Odd parity eigenfunctions

$$u_n(x) = u_n(-x) \tag{2.74}$$

In general, any wave function which has this symmetry property is said to be of *even parity*.

The eigenfunctions corresponding to $n = 1,3,\ldots$ are characterized by Eqs. 2.66. For these we have, explicitly,

$$\left.\begin{aligned} u_n^{\mathrm{I}}(x) &= -(B\,e^{\alpha_n a}\sin\beta_n a)\,e^{\alpha_n x} && (x < -a) \\ u_n^{\mathrm{II}}(x) &= B\sin\beta_n x && (-a < x < a) \\ u_n^{\mathrm{III}}(x) &= (B\,e^{\alpha_n a}\sin\beta_n a)\,e^{-\alpha_n x} && (x > a) \\ & && (n = 1,3,\ldots). \end{aligned}\right\} \tag{2.75}$$

These functions are illustrated in Fig. 2.4(*b*). They are antisymmetric with respect to the origin, i.e.,

$$u_n(x) = -u_n(-x) \tag{2.76}$$

Any wave function which has this property of antisymmetry is said to be of *odd parity*.

Thus the eigenfunctions describing the stationary states of a particle in a square well potential, when considered in order of increasing energy, are alternately of even and odd parity. The fact that the eigenfunctions have even or odd parity is a consequence of the symmetry of the potential V itself with respect to the origin. The proof of this statement, as well as the definition of parity in the case of more general systems (many particles, in three dimensions) will be given in Sec. 4.11.

(*d*) *Penetration Into Classically Forbidden Regions:* The wave functions (2.74) and (2.75) provide an illustration of a feature of quantum mechanics which is of fundamental importance. We recall that as discussed at the beginning of this section, a classical particle of energy $E < 0$ can stay only in Region II and cannot at all enter Regions I and III. However, the quantum mechanical wave functions $u(x)$ have nonvanishing values in both these classically forbidden regions. Therefore, according to the probability interpretation, there exists a nonvanishing probability that the particle is somewhere within these regions. However as one goes from the boundary point ($x = -a$ or $x = +a$) deeper into the forbidden region, the probability density $|u(x)|^2$ decreases rapidly, (proportional to $e^{-2\alpha|x|}$) to zero. Therefore, the particle cannot escape to infinitely large distances; it stays bound to the potential well. Thus all the states which we have so far considered ($E < 0$) are *bound states*.

EXAMPLE 2.7.—The eigenfunctions (2.73) are normalized if we take $A = (a + \alpha^{-1})^{-1/2}$. Verify this. The probability of finding the particle in the classically forbidden regions is

$$\int_{-\infty}^{-a} (u^{\mathrm{I}})^2\,dx + \int_a^{\infty} (u^{\mathrm{III}})^2\,dx = 2\int_a^{\infty} (u^{\mathrm{III}})^2\,dx = (\cos^2\beta a)/(1 + \alpha a)$$

EXAMPLE 2.8.—Eq. 2.71 gives the positions of the energy levels, as measured from the bottom of the potential well, as $E_n + V_0 = (\beta_n a)^2\Delta$. For a very deep potential well ($V_0 \to \infty$) we see from Fig. 2.3 that $\beta a \to (n + 1)\pi/2$. Hence the energy levels in this case are given by $E_n + V_0 = \frac{1}{4}\pi^2\Delta(n + 1)^2$.

Further, in this limit the wave function in the classically forbidden regions tends to zero, as can be seen from Eqs. 2.73 and 2.75.

2.12 The Square Well: Non-localized States ($E > 0$)

When $E > 0$, $(2mE/\hbar^2)$ is positive and, therefore, instead of Eq. 2.54 we write

$$\frac{2mE}{\hbar^2} = k^2 \text{ and } \frac{2m(E + V_0)}{\hbar^2} = \beta^2, \quad (k, \beta > 0) \tag{2.77}$$

Clearly, the only change in the Schrödinger equations 2.55 is that $-\alpha^2$ is to be replaced by $+k^2$. The possible independent solutions in Regions I and III now become e^{ikx} and e^{-ikx} instead of $e^{\alpha x}$ and $e^{-\alpha x}$. But unlike the latter pair, of which one becomes infinite as $x \to +\infty$ or $x \to -\infty$, both e^{ikx} and e^{-ikx} remain finite. Therefore, both have to be retained in the general solution, and instead of Eqs. 2.57 and 2.58 we now have

$$u^{\mathrm{I}}(x) = C_+ \, e^{ikx} + C_- \, e^{-ikx} \tag{2.78a}$$

$$u^{\mathrm{III}}(x) = D_+ \, e^{ikx} + D_- \, e^{-ikx} \tag{2.78b}$$

where C_+, C_-, D_+, D_- are undetermined constants. In Region II the same Eq. 2.55a holds as before, but we will find it convenient to rewrite its general solution (2.56) in a form similar to u^{I} and u^{III}:

$$u^{\mathrm{II}}(x) = A_+ e^{i\beta x} + A_- e^{-i\beta x} \tag{2.78c}$$

where $A_+ = \frac{1}{2}(A - iB)$ and $A_- = \frac{1}{2}(A + iB)$. A particle with the wave function (2.78) is evidently *not localized*: it is not confined to any finite region of space, since $|u(x)|^2$ remains nonzero even when $x \to \pm\infty$. Such wave functions are of course not normalizable.

The forms u^{I}, u^{II}, u^{III} which define $u(x)$ in the different regions must be subjected to the continuity conditions (2.59) and (2.61) at $x = -a$ and $x = +a$ respectively. However now, unlike in the case of bound states, this procedure *does not lead to any restrictions on k or β. Hence any energy $E > 0$ is an eigenvalue.* The reason for the dramatic difference between the cases $E > 0$ and $E < 0$ can be seen very simply. For $E < 0$ we had *four* constants A,B,C,D and an equal number of *linear homogeneous* equations for them, arising from the continuity conditions. Under such circumstances it is well known that nontrivial solutions exist only if the determinant formed by the coefficients appearing in the equations is zero. Eqs. 2.65*a* and 2.66*a* merely express this restriction on the coefficients, which in turn forces E to take only a discrete set of values. On the other hand, when $E > 0$, the continuity conditions still give four linear homogeneous equations but now they involve *six* unknowns $A_\pm C_\pm$, $D_\pm$. Since the number of equations is less than the number of unknowns, an infinite number of solutions exist, whatever the coefficients in the equations may be.

As the coefficients depend on E, this amounts to saying that admissible solutions exist for *every* $E > 0$. Thus the energy eigenvalues form a *continuous* (not a discrete) set. We say that the energy spectrum (for $E > 0$) is a *continuum*.

To complete the determination of the eigenfunctions belonging to any $E > 0$, we have to obtain explicitly the relations among the coefficients

$A_\pm$, $C_\pm$, $D_\pm$ appearing in Eqs. 2.78, which result from the continuity conditions. This is a straightforward matter. But since we will be presenting in detail a very similar calculation in the next section, we will content ourselves here with only stating the most important result. It is that when a particle is incident on the potential well there is, in general, a non-zero probability of its being 'reflected' by the well. This is a typically quantum mechanical phenomenon.[7] It turns out that the probability of reflection is given by

$$\mathcal{R} = \left[1 + \frac{4E\,(E + V_0)}{V_0^2 \sin^2 \{2\sqrt{(E + V_0)/\Delta}\}}\right]^{-1} \tag{2.79}$$

This expression is shown graphically in Fig. 2.5. Note that for very low energies ($E \to 0$) the reflection is almost total. As (E/V_0) increases, $\mathcal{R}$ oscillates between zero and a steadily decreasing upper bound defined by the function $[1 + 4E\,(E + V_0)/V_0^2]^{-1}$. This bound depends only on (E/V_0) and not on

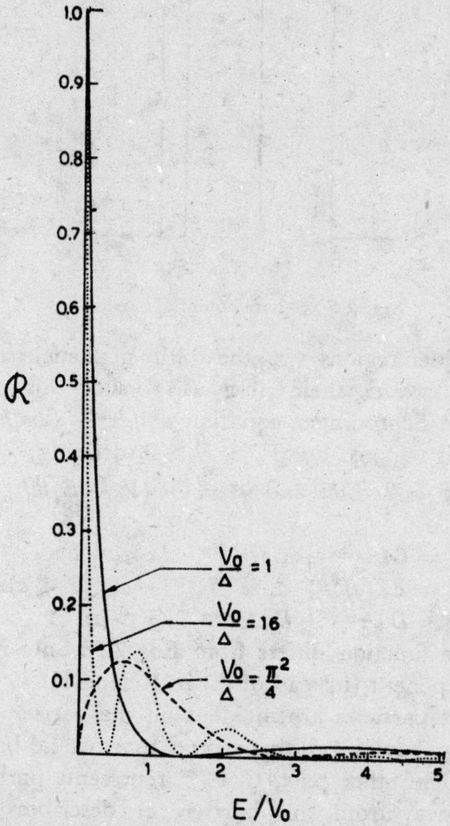

Fig. 2.5 $\mathcal{R}$, as a function of E/V_0 for various values of V_0/Δ.

[7] Classically, any particle approaching the well undergoes an instantaneous acceleration on reaching one edge of the well, and equal retardation at the opposite edge. But it would keep on going; there is no possibility of reflection.

the width of the potential well, but the frequency of the oscillatory changes in $\mathcal{R}$ does depend on a through the parameter Δ. Rigorously complete transmission occurs ($\mathcal{R} = 0$) only when the energy is such that $\sin \{2\sqrt{(E+V_0)/\Delta}\} \equiv \sin 2\beta a = 0$, i.e. when there is an integral number of half waves ($2\beta a = N\pi$) within the potential region where $V = -V_0$.

2.13 Square Potential Barrier

(*a*) *Quantum Mechanical Tunnelling:* Let us now consider a potential *barrier* (Fig. 2.6) instead of a potential well. This problem is of special interest since it exhibits dramatically one of the effects of the penetration of the wave function

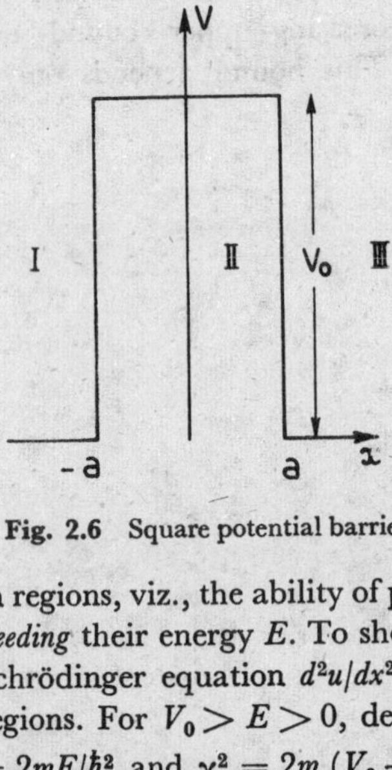

Fig. 2.6 Square potential barrier

into classically forbidden regions, viz., the ability of particles to 'tunnel' through barriers of height V_0 *exceeding* their energy E. To show this we start once again from solutions of the Schrödinger equation $d^2u/dx^2 + (2m/\hbar^2)\ (E - V)\ u = 0$ in the three different regions. For $V_0 > E > 0$, defining

$$k^2 = 2mE/\hbar^2 \text{ and } \gamma^2 = 2m\,(V_0 - E)/\hbar^2 \tag{2.80}$$

we have

$$u^{\text{I}} = C_+\ e^{ikx} + C_-\ e^{-ikx} \quad (x < -a) \tag{2.81a}$$

$$u^{\text{II}} = A_+\ e^{\gamma x} + A_-\ e^{-\gamma x} \quad (-a < x < a) \tag{2.81b}$$

$$u^{\text{III}} = D_+\ e^{ikx} + D_-\ e^{-ikx} \quad (x > a) \tag{2.81c}$$

Note that this wave function differs from Eqs 2.78 only in having the real constant γ in the exponent (instead of $i\beta$) in Region II.

Suppose now that particles are incident on the barrier only from the left side. These particles, having positive momentum, would be described by the part $C_+\ e^{ikx}$ of u^{I}. The other part, $C_-\ e^{-ikx}$ represents particles moving with momentum $-\hbar k$, away from the barrier; it describes particles reflected by the barrier. On the other side of the barrier (Region III), there are, by hypothesis, no particles moving to the left (i.e., approaching the barrier). Therefore, we must take

$$D_- = 0 \tag{2.82}$$

Then u^{III} reduces to $D_+\ e^{ikx}$, representing particles moving to the right, which could come only by tunnelling through the barrier from Region I. The ratio (D_+/C_+) gives the *amplitude for the tunnelling* to take place and (C_-/C_+) is the *amplitude for reflection.* The corresponding probabilities are $|D_+/C_+|^2$ and $|C_-/C_+|^2$ respectively. We will now determine these by applying the continuity conditions (2.59) and (2.61), at $x = -a$ and $x = +a$ respectively, to the wave function (2.81) subject to the condition $D_- = 0$.

It is simplest to proceed as follows: First determine the ratio (A_-/A_+) and hence (C_+/C_-) by solving successively the equations

$$\left(\frac{1}{u^{\text{II}}}\cdot\frac{du^{\text{II}}}{dx}\right)_{x=a} = \left(\frac{1}{u^{\text{III}}}\cdot\frac{du^{\text{III}}}{dx}\right)_{x=a} \text{ and } \left(\frac{1}{u^{\text{I}}}\frac{du^{\text{I}}}{dx}\right)_{x=-a} = \left(\frac{1}{u^{\text{II}}}\frac{du^{\text{II}}}{dx}\right)_{x=-a} \tag{2.83}$$

We get

$$\frac{A_-}{A_+} = \left(\frac{\gamma - ik}{\gamma + ik}\right)e^{2\gamma a} \tag{2.84}$$

$$\frac{C_-}{C_+} = \frac{-i(k^2+\gamma^2)\,e^{-2ika}\,.\sinh 2\gamma a}{-i(k^2-\gamma^2)\sinh 2\gamma a + 2\gamma k\cosh 2\gamma a} \tag{2.85}$$

We can then express (D_+/A_+) and (A_+/C_+) in terms of the quantities (2.84) and (2.85) by solving $u^{\text{II}} = u^{\text{III}}$ (at $x = a$) and $u^{\text{I}} = u^{\text{II}}$ (at $x = -a$). The product of the two gives the quantity of interest, viz., the transmission amplitude (D_+/C_+):

$$\frac{D_+}{C_+} = \frac{2\gamma k\, e^{-2ika}}{-i(k^2-\gamma^2)\sinh 2\gamma a + 2\gamma k\cosh 2\gamma a} \tag{2.86}$$

The transmission probability (i.e., the probability for tunnelling) is therefore given by

$$\mathcal{T} = \left|\frac{D_+}{C_+}\right|^2 = \left[1 + \frac{(k^2+\gamma^2)^2}{4\gamma^2k^2}\cdot\sinh^2 2\gamma a\right]^{-1} \tag{2.87}$$

It may be easily verified from Eq. 2.85 that the reflection probability $\mathcal{R} = |C_-/C_+|^2$ is just $1 - \mathcal{T}$, as it should be.

The expression (2.87) for $\mathcal{T}$ can be reduced to a more instructive form, directly in terms of the energies involved, if we use the relations

$$(\gamma a)^2 = (V_0 - E)/\Delta \text{ and } (ka)^2 = E/\Delta \tag{2.88}$$

which follow from the definitions in Eqs. 2.80 and 2.68. We have

$$\mathcal{T} = \left[1 + \frac{V_0^2}{4(V_0-E)E}\cdot\sinh^2\{2\sqrt{(V_0-E)/\Delta}\}\right]^{-1} \tag{2.89}$$

Observe that $\sinh y \equiv \frac{1}{2}(e^y - e^{-y})$ is a rapidly increasing function of y, which is closely approximated by $\frac{1}{2}e^y$ when $y \gg 1$. Therefore except when $2\sqrt{(V_0-E)/\Delta} \lessdot 1$, the second term within square brackets in Eq. 2.89 dominates, and using the above approximation we have

$$\mathcal{T} \approx \frac{16(V_0-E)E}{V_0^2}\cdot\exp\left[-4\sqrt{(V_0-E)/\Delta}\right] \tag{2.90}$$

Thus, as soon as the energy 'deficit' $(V_0 - E)$ exceeds the 'uncertainty energy' Δ, the tunnelling probability starts falling off exponentially. On the other hand if the particle energy is very close to the top of the potential barrier, $(V_0 - E)/\Delta \ll 1$, then the approximation $\sinh y \approx y$ is appropriate and Eq. 2.89 reduces to

$$\mathcal{R} \approx \left(1 + \frac{V_0^2}{E\Delta}\right)^{-1} \tag{2.91}$$

Fig. 2.7 shows the variation of the transmission probability with E. The part for $E < V_0$ (tunnelling) is taken from Eq. 2.87, while for $E > V_0$, $\mathcal{T} = 1 - \mathcal{R}$ is obtained from Eq. 2.95.

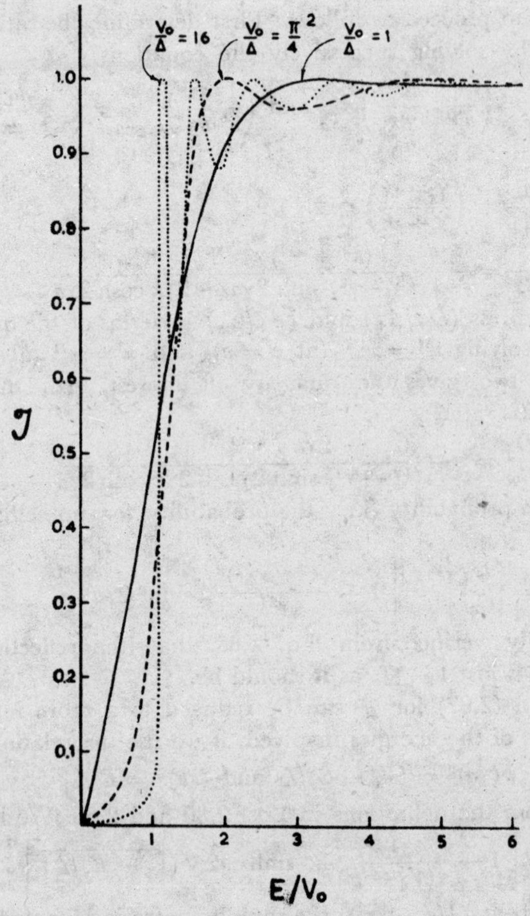

Fig. 2.7 $\mathcal{T}$ as a function of E/V_0 for various values of V_0/Δ.

(*b*) *Reflection at potential barriers and wells:* We have referred in the last section to the quantum mechanical phenomenon of reflection of particles by potential wells. A similar non-classical phenomenon is the reflection, at potential barriers, of particles having energy E exceeding the barrier height V_0. The reflection probability $\mathcal{R}$ can be obtained by making a simple modification of Eq. 2.85. Since $E > V_0$ now, we write

$$2m(V_0 - E)/\hbar^2 = -\gamma'^2 \tag{2.92}$$

instead of γ^2 as in Eq. 2.80. The consequent modification in subsequent equations is simply to replace γ everywhere by $i\gamma'$. The amplitude for reflection, given by Eq. 2.85, then becomes

$$\frac{C_-}{C_+} = \frac{(k^2 - \gamma'^2)\, e^{-2ika} \sin 2\gamma' a}{(k^2 + \gamma'^2) \sin 2\gamma' a + 2i\, \gamma' k \cos 2\gamma' a} \tag{2.93}$$

The reflection probability is therefore

$$\mathcal{R} = \left| \frac{C_-}{C_+} \right|^2 = \left[1 + \frac{4\, k^2\, \gamma'^2}{(k^2 - \gamma'^2)^2 \sin^2 2\, \gamma' a} \right]^{1} \tag{2.94}$$

Substituting the value of γ'^2 from Eq. 2.92 and $k^2 = 2m\, E/\hbar^2$, and writing $\Delta = \hbar^2/(2ma^2)$ as before, we finally have

$$\mathcal{R} = \left[1 + \frac{4E\,(E - V_0)}{V_0^2 \sin^2 \{2\sqrt{(E - V_0)/\Delta}\}} \right]^{-1} \tag{2.95}$$

Exactly the same treatment goes through for the nonlocalized states of a potential *well*. This is because the wave function (2.78) of that case differs from Eqs. 2.81 only in having $i\beta$ in the place of γ, or what is the same thing, β instead of γ'. With this replacement in Eq. 2.94, and the corresponding replacement of $(E - V_0)$ by $(E + V_0)$ in Eq. 2.95, we recover the formula (2.79) for the probability of reflection by a potential well. The variation of $\mathcal{T} = 1 - \mathcal{R}$ with energy is shown in Fig. 2.7.

2.14 Multiple Potential Wells: Splitting of Energy Levels; Energy Bands

Let us now consider a potential consisting of a repeating pattern of identical square wells (each of width w and depth V_0) separated by identical barriers (each of width w'). Such a potential (Fig. 2.8) with a very large number of repeating units has been used as an idealized one-dimensional model to represent the effect of the regular arrays of atoms on electrons in crystalline media.

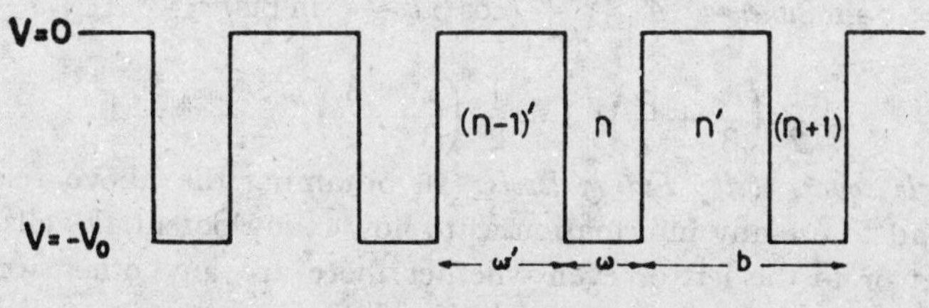

Fig. 2.8 An array of identical square wells

(*a*) *The wave function; Transfer across a potential well:* The key to the solution of the problem is the determination of the relation between the wave functions in two regions separated by a single potential well. The presence or absence of other wells does not affect this relation. But in determining it we shall use, for future convenience, a notation adapted to the case when the well is part of a regular array. To be specific, we fix our attention on the well labelled n in Fig. 2.8, which we show separately in Fig. 2.9.

We know already that in the 'primed' regions where $V = 0$ [marked $(n-1)'$ and n' in Fig. 2.9], the wave function for $E < 0$ must be linear combinations

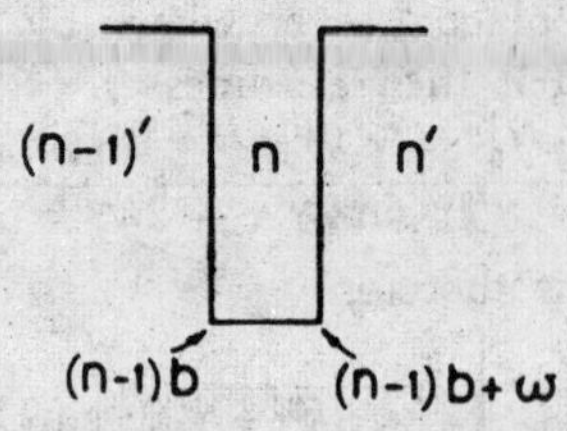

Fig. 2.9 A typical well from an array

of decreasing and increasing exponential functions. Let us write the combination, for the region n', as

$$u(x) = A_n^+ e^{\alpha(x-nb)} + A_n^- e^{-\alpha(x-nb)}, \; b = w + w' \tag{2.96}$$

(The constant factors $e^{\pm\alpha nb}$ have been pulled out of the coefficients for future convenience). Similarly the wave function within the well (region n, where $V = -V_0 < E$) will be written as

$$u(x) = B_n^+ e^{i\beta(x-nb)} + B_n^- e^{-i\beta(x-nb)} \tag{2.97}$$

Here, as in Sec. 2.10, we have denoted $2m(E-V)/\hbar^2$ in the two kinds of regions by

$$\frac{2mE}{\hbar^2} = -\alpha^2 \text{ and } \frac{2m}{\hbar^2}(E+V_0) = \frac{2m}{\hbar^2}(V_0 - |E|) = \beta^2 \tag{2.98}$$

The condition that $u(x)$ and du/dx be continuous at the boundary $x = (n-1)b + w$ between n and n' gives rise to two equations relating A_n^+ and A_n^- to B_n^+ and B_n^-. Similarly the continuity conditions at $x = (n-1)b$ between the regions $(n-1)'$ and n leads to two equations relating B_n^+ and B_n^- to A_{n-1}^+ and A_{n-1}^-. By eliminating B_n^+ and B_n^- from these four equations we get A_n^+ and A_n^- directly in terms of A_{n-1}^+ and A_{n-1}^-:

$$\begin{aligned} A_n^+ &= (\cos\beta w + s.\sin\beta w)\, e^{\alpha w'} A_{n-1}^+ - c\sin\beta w . e^{\alpha w'} A_{n-1}^- \\ A_n^- &= c\sin\beta w . e^{-\alpha w'} . A_{n-1}^+ + (\cos\beta w - s\sin\beta w)\, e^{-\alpha w'} A_{n-1}^- \end{aligned} \tag{2.99}$$

where

$$c = \frac{1}{2}\left(\frac{\alpha}{\beta} + \frac{\beta}{\alpha}\right); \; s = \frac{1}{2}\left(\frac{\alpha}{\beta} - \frac{\beta}{\alpha}\right); \; c^2 - s^2 = 1 \tag{2.100}$$

(*b*) *A single square well; Energy levels:* In obtaining the above relations we have not had to use any information as to how many potential wells there are to the right or to the left or even whether there are any other wells at all. In fact we could get back to the problem of the single square well by simply assuming that the $V = 0$ region which we have labelled by n' in Fig. 2.9 extends all the way to $x = +\infty$, and $(n-1)'$ extends to $-\infty$. If this were done, we would of course have to set

$$A_n^+ = A_{n-1}^- = 0 \tag{2.101}$$

to prevent the wave function from becoming infinite at $x = \pm\infty$. Then the first of Eqs. 2.99 reduces to

$$\cos\beta w + s.\sin\beta w = 0 \tag{2.102}$$

which can be immediately solved by substituting

$$\cos\beta w = \frac{1-\tau^2}{1+\tau^2}, \; \sin\beta w = \frac{2\tau}{1+\tau^2}, \; (\tau = \tan\tfrac{1}{2}\beta w) \tag{2.103}$$

We get

$$\tau^2 - 2\tau s - 1 = 0 \tag{2.104}$$

and hence

$$\tau = s \pm \sqrt{s^2 + 1} = s \pm c \tag{2.105}$$

Thus

$$\tan\tfrac{1}{2}\beta w = \frac{\alpha}{\beta} \text{ or } -\frac{\beta}{\alpha} \tag{2.106}$$

When it is remembered that in the notation of Sec. 2.10, the width of the well is $w = 2a$, it becomes clear that Eqs. 2.106 are simply the pair of Eqs. 2.65*a* and 2.66*a*. Thus the results of Sec. 2.11 are recovered.

(*c*) *The wave function; Transfer across N square wells:* Returning now to Eq. 2.99, we write it in matrix form as

$$\begin{pmatrix} A_n^+ \\ A_n^- \end{pmatrix} = M \begin{pmatrix} A_{n-1}^+ \\ A_{n-1}^- \end{pmatrix} \tag{2.107}$$

where M is the 2×2 matrix

$$M = \begin{pmatrix} (\cos\beta w + s.\sin\beta w)\ e^{\alpha w'} & -c\ \sin\beta w\ e^{\alpha w'} \\ c\ \sin\beta w.\ e^{-\alpha w'} & (\cos\beta w - s.\sin\beta w)\ e^{-\alpha w'} \end{pmatrix} \tag{2.108}$$

These equations show how the coefficients in the wave function change on transferring from any one barrier to the next one across one well. To transfer across any number of wells, all we have to do is to iterate this equation. Thus,

$$\begin{pmatrix} A_N^+ \\ A_N^- \end{pmatrix} = M^N \begin{pmatrix} A_0^+ \\ A_0^- \end{pmatrix} \tag{2.109}$$

Now, M^N may be expressed very simply in terms of the diagonalized form of M, say $M_D = SMS^{-1}$, in which the eigenvalues of M occur along the diagonal, and all other elements are zero. The eigenvalues are the roots of the characteristic equation det $(M - \lambda I) = 0$. For the matrix M of (2.108) the characteristic equation reduces to

$$\lambda^2 - 2\lambda\,(\cosh \alpha w'.\cos \beta w + s.\sinh \alpha w'.\sin \beta w) + 1 = 0 \tag{2.110}$$

The product of the two roots is clearly unity. On denoting them by μ, μ^{-1}, we have

$$M = S^{-1} M_D S = S^{-1} \begin{pmatrix} \mu & 0 \\ 0 & \mu^{-1} \end{pmatrix} S \tag{2.111}$$

and hence

$$M^N = S^{-1} \begin{pmatrix} \mu^N & 0 \\ 0 & \mu^{-N} \end{pmatrix} S \tag{2.112}$$

Eq. 2.109 may therefore be rewritten as

$$S \begin{pmatrix} A_N^+ \\ A_N^- \end{pmatrix} = \begin{pmatrix} \mu^N & 0 \\ 0 & \mu^{-N} \end{pmatrix} .S. \begin{pmatrix} A_0^+ \\ A_0^- \end{pmatrix} \tag{2.113}$$

(*d*) *A regular array of N square wells; Energy levels:* Suppose now that there are just N wells, numbered $1, 2, \ldots, N$ as in Fig. 2.8, the region $0'$ to the left of the wells being taken to extend all the way to $-\infty$ and the region N' to the right, to $+\infty$. Then we must have

$$A_N^+ = A_0^- = 0 \tag{2.114}$$

in order that the wave functions in these regions, which are of the form (2.96), be not divergent as $x \to +\infty$. With this substitution we have from Eq. 2.113 that

$$S_{12} A_N^- = \mu^N S_{11} A_0^+$$
$$S_{22} A_N^- = \mu^{-N} S_{21} A_0^+ \qquad (2.115)$$

For compatibility of these equations it is clearly necessary that

$$\frac{\mu^N}{\mu^{-N}} = \frac{S_{12} S_{21}}{S_{11} S_{22}} \qquad (2.116)$$

This is an implicit equation for E, since all quantities in it depend on E through α and β. To solve the equation we have to make the dependence more explicit. We proceed as follows.

First we note that the ratios (S_{12}/S_{11}) and (S_{21}/S_{22}) can be obtained from (2.111) by rewriting it as $SM = M_D S$ and considering the equality of the (1,1) and (2,1) elements of the matrices on the two sides. Using the expression (2.108) for M we then get

$$\frac{S_{12}}{S_{11}} = \frac{\mu - (\cos\beta w + s.\ \sin\beta w)\ e^{\alpha w'}}{c\ \sin\beta w\ .\ e^{-\alpha w'}}$$
$$\frac{S_{21}}{S_{22}} = \frac{c\ \sin\beta w\ .\ e^{-\alpha w'}}{\mu^{-1} - (\cos\beta w + s.\ \sin\beta w)\ e^{\alpha w'}} \qquad (2.117)$$

Introducing these relations in (2.116) and rearranging, we reduce it to

$$(\cos\beta w + s.\ \sin\beta w)\ e^{\alpha w'} = \frac{\mu^{N-1} - \mu^{-(N-1)}}{\mu^N - \mu^{-N}} \qquad (2.118)$$

This can be rewritten, after cancelling a common factor $(\mu - \mu^{-1})$, as

$$[\mu^{N-1} + \mu^{N-3} + \ldots + \mu^{-(N-3)} + \mu^{-(N-1)}]\ (\cos\beta w + s.\ \sin\beta w)\ e^{\alpha w'}$$
$$= [\mu^{N-2} + \mu^{N-4} + \ldots + \mu^{-(N-4)} + \mu^{-(N-2)}] \qquad (2.119)$$

It is evident that the factors in square brackets can be written as polynomials in $(\mu + \mu^{-1})$; and this quantity, being the sum of the two roots of (2.110), is given by

$$(\mu + \mu^{-1}) = 2\ (\cosh\alpha w'.\ \cos\beta w + s.\sinh\alpha w'.\sin\beta w)$$
$$= \frac{2}{1+\tau^2}\ [(1-\tau^2)\ \cosh\alpha w' + 2\tau s.\sinh \alpha w'] \qquad (2.120)$$

where $\tau = \tan \frac{1}{2}\beta w$. In view of this it is easy to see that Eq. 2.119 is an algebraic equation of degree $2N$ in τ. This result is the generalization of Eq. 2.104 to the case of N square wells. Instead of the two solutions (2.105) for τ *we now have 2N solutions, each of which provides an implicit equation for the energy.* Thus the net effect is that the number of energy levels increases by a large factor (compared to that of a single potential well) when N is large. If the barriers between successive wells are wide ($\alpha w' \gg 1$), the energy levels appear as closely spaced multiplets about the positions of the individual levels of the single potential well. In the limit $w' \to \infty$, each of these multiplets becomes degenerate in energy. These facts can be deduced fairly easily from Eqs. 2.119 and 2.120: When $\alpha w' \gg 1$, $\cosh \alpha w' \approx \sinh \alpha w' \approx e^{\alpha w'}$ and hence $(\mu + \mu^{-1}) \approx 2(\cos \beta w + s.\sin \beta w)\ e^{\alpha w'}$. Because of the presence of the large factor $e^{\alpha w'}$ in $(\mu + \mu^{-1})$, the highest power of $(\mu + \mu^{-1})$, namely the $(N-1)$*th* power on the left hand side in relations (2.119), dominates over other terms. Therefore, Eq. 2.119 becomes

$$(\cos\beta w + s.\ \sin\beta w)^N \approx 0 \tag{2.121}$$

the neglected terms being smaller than the dominant term by at least a factor $\epsilon^{-\alpha w'}$. If the right hand side is strictly zero, this equation gives just the same solutions as Eq. 2.102, but each is reproduced N times. When the right hand side is small but not negligible, the N solutions differ from each other by small amounts, i.e. there is an N-fold splitting of each level, as already mentioned.

(*e*) *An infinite array of square wells; The Kronig-Penney model:* When the array of potential wells extends to infinity on both sides, we have to make sure that the wave function does not diverge as $x \to \pm\infty$. Inspection of Eq. 2.113 shows that this is possible only if $|\mu| = 1$. For, if $|\mu| < 1$, $\mu^N \to \infty$ as $N \to -\infty$ and $\mu^{-N} \to \infty$ as $N \to +\infty$; when $\mu > 1$ the situation is reversed. In either case, the wave function in the Nth well will diverge in one or both of the limits $N \to +\infty$ and $\to -\infty$. Hence we must have

$$\mu = e^{i\theta} \ (\theta \text{ real}) \tag{2.122}$$

Then by Eq. 2.120,

$$\cos\theta = \tfrac{1}{2}(\mu + \mu^{-1}) = \cosh\alpha w'.\cos\beta w + s.\sinh\alpha w'.\sin\beta w = f\cos(\beta w - \varphi) \tag{2.123}$$

where

$$f = (\cosh^2 \alpha w' + s^2.\ \sinh^2 \alpha w')^{1/2}$$
$$\tan\varphi = s.\ \tanh \alpha w' \tag{2.124}$$

Eq. 2.123 determines what values of the energy are allowed, since α and β (and hence f and φ also) depend on the parameter E. From the structure of this equation it is immediately evident that the allowed energies form continuous *bands* with *gaps* in between. For, since the value of the left hand side of Eq. 2.123 is restricted to $-1 \leqslant \cos\theta \leqslant 1$, the equation can be satisfied only when the right hand side also is within this range. As the parameter E is varied, the right hand side, $f\cos(\beta w - \varphi)$, oscillates between the bounds $+f(E)$ and $-f(E)$ which lie *outside* the limits $+1$ and -1 admissible for $\cos\theta$, since $|f| \geqslant \cosh\alpha w' > 1$. The intervals of E for which $|f\cos(\beta w - \varphi)| \leqslant 1$ constitute the 'allowed' energy bands, and those for which $|f\cos(\beta w - \varphi|) > 1$ are the 'forbidden' regions. These form what are called energy gaps, between allowed bands. The variation of $f\cos(\beta w - \varphi)$ with E, and the identification of energy bands and gaps are illustrated in Fig. 2.10.

The nature of the energy eigenfunctions $u(x)$ corresponding to any energy E within an allowed band can now be easily seen. We recall that we have already imposed the necessary continuity conditions on the wave function, which resulted in the 'transfer' equation 2.107, and also the condition of finiteness at infinity which gave us the restriction (2.122) on the eigenvalues of the transfer matrix M. There are no further conditions on the wave function, and in particular, there is nothing to prevent us from choosing the two coefficients (A^+, A^- or B^+, B^-) in any *one* of the regions in Fig. 2.8 as we please. Suppose then that we choose some energy E within an allowed band, and after finding α, β, c, s for this E using (2.98) and (2.100), determine θ from (2.123) and the matrix M itself from (2.108). Now let us avail ourselves of the freedom of choosing the coefficients, and take A_0^+, A_0^- to be such that

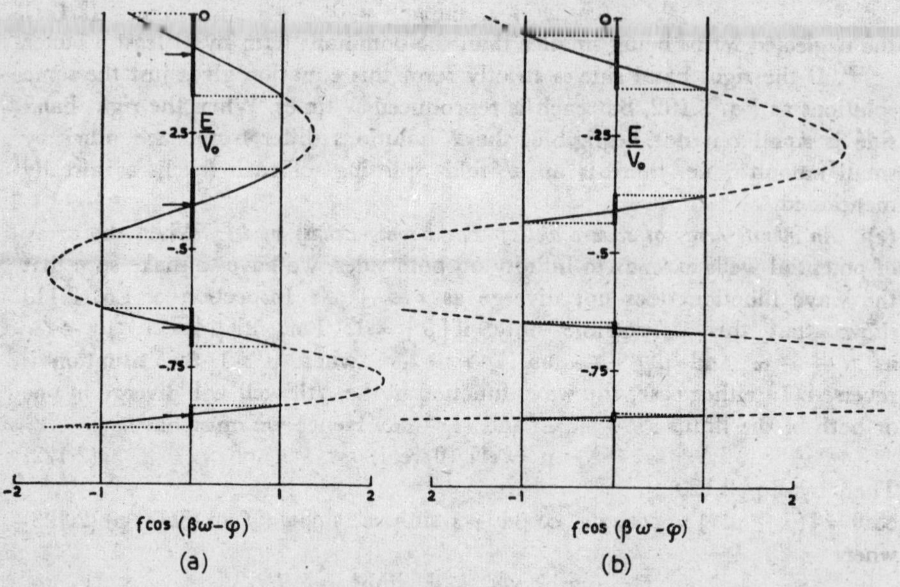

Fig. 2.10 Energy bands and gaps for $V_0w^2 = 200$ (**a**) $w' = 0{\cdot}1w$; and (**b**) $w' = 0{\cdot}2w$ Bands of allowed energies are shown as thickened sections on the energy axis. Note the narrowing of the bands as w' increases. In the limit $w' \to \infty$ they reduce to points at the positions of the energy levels of a single potential well, shown by arrows in the figure.

the column made up of these is the eigenvector belonging to the eigenvalue $e^{i\theta}$ of M, i.e.

$$M \begin{pmatrix} A_0^+ \\ A_0^- \end{pmatrix}_\theta = e^{i\theta} \begin{pmatrix} A_0^+ \\ A_0^- \end{pmatrix}_\theta \tag{2.125}$$

The θ-dependence of the coefficients so defined is explicitly indicated here. Eq. 2.109 now tells us that

$$\begin{pmatrix} A_n^+ \\ A_n^- \end{pmatrix}_\theta = e^{in\theta} \begin{pmatrix} A_0^+ \\ A_0^- \end{pmatrix}_\theta \tag{2.126}$$

i.e. the coefficients depend on n only through the simple factor $e^{in\theta}$. It is obvious that a similar result holds good for B_n^+, and B_n^- also. If we think of a well and the adjacent barrier together, as forming a 'unit cell' of the repeating pattern, the result just obtained may be stated in the following instructive form: The wave functions at corresponding points x and $x + nb$ of wells (or barriers) separated by n unit cells differ only by the constant factor $e^{in\theta}$:

$$u_\theta\,(x + nb) = e^{in\theta}\, u_\theta\,(x) \tag{2.127}$$

This property is often expressed in the form

$$u_\theta\,(x) = e^{i\theta x/b}\, v_\theta\,(x) \tag{2.128a}$$

$$v_\theta\,(x) = v_{\theta}\,(x + nb) \tag{2.128b}$$

Eq. 2.128a may be thought of as the definition of $v_\theta\,(x)$, and when it is introduced in Eq. 2.127, the periodicity of $v_\theta\,(x)$, [see 2.128b], follows. Besides $u_\theta(x)$ there is another independent solution $u_{-\theta}\,(x)$ obtained by using the eigenvalue $e^{-i\theta}$ of M instead of $e^{i\theta}$ in (2.125). The most general eigenfunction for a given energy E is a linear combination $c_1 u_\theta(x) + c_2 u_{-\theta}(x)$, with arbitrary complex coefficients c_1 and c_2.

The above results are of considerable physical importance in the study of crystalline solids[8] wherein the array of atoms produces a potential forming a regular repeating pattern. In the usual situation, the number of atoms is so large that the idealization to infinite number may be made to simplify matters. The Kronig-Penney model makes the further idealization of treating the potential due to each atom as a square well. But the fundamental results obtained above using this model are independent of this idealization. Specifically, the occurrence of allowed energy bands and forbidden gaps follows just from the periodicity property, $V(x) = V(x + nb)$, of the potential. So also does the general form (2.128) of the eigenfunctions, which are called *Bloch wave functions*. But the explicit structure of the function $v_\theta(x)$ in (2.128a) depends on the actual shape of the potential function.

PROBLEMS

1. Show that for a charged particle in the presence of electromagnetic potentials, the current density is
$$\mathbf{S} = -\frac{i\hbar}{2m}\left(\psi^*\nabla\psi - \psi\nabla\psi^* - 2i\frac{e}{\hbar c}\mathbf{A}\psi^*\psi\right)$$
2. Show that a complex potential $V = V_R + iV_I$ in the Schrödinger equation leads to a source or sink of probability according as V_I is positive or negative.
3. The solution of the eigenvalue problem for the potential: $V(x) = +\infty$, $x < 0$; $V(x) = -V_0$, $0 < x < a$; $V(x) = 0$, $x > a$ can be obtained directly from the known solution for the potential of Fig. 2.2. Explain why, and write down the eigenvalues and eigenfunctions.
4. Show that resonance effects occur (at certain energies) in the transmission of a particle through a pair of identical square potential barriers of width $2a$ separated by a distance b.
5. Given that $V = 0$ for $x < 0$ and $x > a$ and that a potential barrier of arbitrary shape $V(x)$ exists in $0 < x < a$, show that the coefficient of reflection of particles incident on the barrier is the same whether they approach from the right or from the left.
6. Consider a square well or barrier of width $2a$, centred at $x = 0$ with $V_0 = K/2a$. If the width is taken to be vanishingly small and V_0 infinitely large (as given by the limit $a \to 0$ with K held fixed), show that in the limit, ψ will be continuous across this potential spike while the values of $(\partial\psi/\partial x)$ on the two sides will differ by $(2mK/\hbar^2)\,\psi(0)$. (In the above limit $V(x) = \mp K\delta(x)$ where δ is the Dirac delta function, Sec. 3.6).
7. Consider an infinite periodic array of potential wells as in Fig. 2.8, but with energies measured from the bottom of the wells (i.e. $V = 0$ in the wells and $V = V_0$ in the barrier regions). Let $w' \to 0$ and $V_0 = K/w' \to \infty$ (K fixed), so that the barriers become infinite spikes. Determine the energy band structure in this case by direct solution of this problem, noting the behaviour of ψ and $\partial\psi/\partial x$ at the spikes as in Prob. 6.
8. Verify directly that if the potential function $V(x)$ is periodic, i.e. $V(x + a) = V(x)$ for constant a, the Schrödinger equation has solutions such that $u(x + a) = \lambda u(x)$ (Floquet's Theorem). Show that these solutions can be written as $u(x) = e^{ikx} v_k(x)$, (k being real) with $v_k(x + a) = v_k(x)$.

[8] See, for instance, S. Raimes, *The Wave Mechanics of Electrons in Metals*, North-Holland Publishing Co., Amsterdam, 1961.

3 General Formalism of Wave Mechanics

We have confined our attention so far to the wave mechanics of a single particle, so that the main features and typical results of the theory can be seen clearly without being clouded by the greater complexity of many-particle systems. We begin this chapter with a brief account of the generalization of the Schrödinger equation to any N-particle system and the interpretation of the wave function in this case. We then proceed to a systematic presentation of the general mathematical framework of wave mechanics and the fundamental postulates (and rules derived therefrom) for the interpretation of the mathematical structures in physical terms. These form the essential basis of quantum mechanics, which any serious student must strive to understand thoroughly.

3.1 The Schrödinger Equation and the Probability Interpretation for an N-Particle System

A system of N particles is characterized by the position and momentum variables $\mathbf{x}_1, \mathbf{x}_2, \ldots, \mathbf{x}_N$ and $\mathbf{p}_1, \mathbf{p}_2, \ldots, \mathbf{p}_N$. Its energy is given by the Hamiltonian function H depending in general on all the position and momentum variables:

$$E = H(\mathbf{x}_1, \mathbf{x}_2, \ldots \mathbf{x}_N;\ \mathbf{p}_1, \mathbf{p}_2 \ldots \mathbf{p}_N;\ t) \tag{3.1}$$

The explicit inclusion of the time parameter is intended to take care of the effect of time-dependent external forces acting on the system, if any. Knowing the Hamiltonian, one can write down the Schrödinger equation by a direct generalization of the method used in deriving Eq. 2.19. Firstly, the operator correspondence (2.14) is generalized to

$$E \to i\hbar \frac{\partial}{\partial t}, \quad \mathbf{p}_i \to -i\hbar\nabla_i \ (i = 1,2,\ldots, N) \tag{3.2}$$

Here ∇_i stands for the gradient operator taken with respect to the position variable $\mathbf{x}_i$ of the ith particle:

$$\nabla_i = \left(\frac{\partial}{\partial x_i}, \frac{\partial}{\partial y_i}, \frac{\partial}{\partial z_i}\right)$$

Since these operators have to act on the wave function ψ, the latter must necessarily depend on all the variables $\mathbf{x}_1,\mathbf{x}_2, \ldots,\mathbf{x}_N$. This can also be seen from a physical point of view. As we have noted in Sec. 1.17 the wave concept of matter does *not* envisage that the wave functions of different particles add up to a single wave into which they merge completely. The particles remain as separate entities, each with its own set of dynamical variables. This fact must be reflected in the nature of the wave function through a dependence on the position variables of all the particles. It is sometimes convenient to think of the $3N$ coordinates of the N particles as the coordinates of a single point in a $3N$-dimensional space. Such a space is called the *configuration space* of the system. *The wave function ψ is therefore a function defined in the configuration space.*

Using the prescription of Sec. 2.3 we can now write down the equation for the wave function ψ as

$$i\hbar \frac{\partial\psi(\mathbf{x}_1,\mathbf{x}_2,\ldots,\mathbf{x}_N;t)}{\partial t} = H(\mathbf{x}_1,\mathbf{x}_2,\ldots,\mathbf{x}_n, -i\hbar\nabla_1, -i\hbar\nabla_2,\ldots, -i\hbar\nabla_N)\psi(\mathbf{x}_1,\ldots,\mathbf{x}_N;t) \tag{3.3}$$

This is the general form of the Schrödinger equation.

It may be noted at this point that though the Hamiltonian can be expressed in terms of generalized coordinates q_α (in the sense of the Lagrangian form of mechanics) and their canonically conjugate momenta p_α ($\alpha = 1,2,\ldots,3N$), the replacement of p_α by $-i\hbar\partial/\partial q_\alpha$ is correct only if the q_α are *Cartesian* coordinates. This is why we have not used the generalized coordinate notation in Eqs. (3.1)—(3.3). If in a particular problem the use of some generalized coordinates is advantageous (e.g. spherical polar coordinates in systems with spherical symmetry), the operator replacement must be done first with the Hamiltonian expressed in Cartesian coordinates, and only thereafter should the transformation to generalized coordinates be made. A different result would follow if the steps were reversed, and the equation obtained would not be the correct Schrödinger equation.

Example 3.1 — The quantum Hamiltonian (operator) for a free particle in three dimensions is $-(\hbar^2/2m)\nabla^2$. When converted to spherical polar coordinates, it takes the form

$$-\frac{\hbar^2}{2m}\left[\frac{1}{r^2}\frac{\partial}{\partial r}\left(r^2\frac{\partial}{\partial r}\right) + \frac{1}{r^2\sin\theta}\frac{\partial}{\partial\theta}\left(\sin\theta\frac{\partial}{\partial\theta}\right) + \frac{1}{r^2\sin^2\theta}\frac{\partial^2}{\partial\varphi^2}\right].$$

On the other hand, if we had started with the form of $\mathbf{p}^2$ in spherical polar coordinates, viz.,

$$\mathbf{p}^2 = p_r^2 + \frac{1}{r^2}p_\theta^2 + \frac{1}{r^2\sin^2\theta}p_\varphi^2,$$

and *then* substituted $p_r = -i\hbar(\partial/\partial r)$, $p_\theta = -i\hbar(\partial/\partial\theta)$ and $p_\varphi = -i\hbar(\partial/\partial\varphi)$, we would get a different (and incorrect) expression for the Hamiltonian operator.

Example 3.2 — What is the Schrödinger equation of a system of two particles of masses m_1 and m_2 carrying charges e_1 and e_2 respectively? The Hamiltonian is obviously $(\mathbf{p}_1^2/2m_1) + (\mathbf{p}_2^2/2m_2) + (e_1e_2/r_{12})$, the last term being the electrostatic interaction energy of the two particles ($r_{12} = |\mathbf{x}_1 - \mathbf{x}_2|$). So the Schrödinger equation is

$$i\hbar\frac{\partial\psi(\mathbf{x}_1,\mathbf{x}_2;t)}{\partial t} = \left(-\frac{\hbar^2}{2m_1}\nabla_1^2 - \frac{\hbar^2}{2m_2}\nabla_2^2 + \frac{e_1e_2}{r_{12}}\right)\psi(\mathbf{x}_1,\mathbf{x}_2;t).$$

Examples of such systems are: Hydrogen-like atoms (hydrogen, singly ionized helium, doubly ionized lithium, etc.), mesic atoms (a μ^- or π^- meson together with a proton), positronium (electron-positron system). When considering atoms it is usual to consider the position of the nucleus (say $\mathbf{x}_2$) as fixed and the kinetic energy $(\mathbf{p}_2^2/2m_2)$ to be effectively zero, because of the large nuclear mass (some thousands of times the electron mass). Then the term in ∇_2^2 drops out, and the equation reduces to one involving $\mathbf{x}_1$ only, $\mathbf{x}_2$ being just a fixed vector.

The many-particle wave function appearing in Eq. 3.3, like that for a single particle, is interpreted in probabilistic terms. Specifically, it is postulated that

$$|\psi(\mathbf{x}_1,\mathbf{x}_2,\ldots,\mathbf{x}_N;t)|^2\, d^3x_1\, d^3x_2\ldots d^3x_N \tag{3.4}$$

is the joint probability that particle 1 is within the volume elment d^3x_1 around the point $\mathbf{x}_1$, particle 2 is within d^3x_2 around $\mathbf{x}_2,\ldots$, and particle N is within d^3x_N around $\mathbf{x}_N$ at time t. If this expression is integrated over the coordinates of some of the particles, the resulting expression gives the joint probability for finding the remaining particles within specified volume elements irrespective of the positions of the other particles. For example

$$[\textstyle\int d^3x_2\ldots\int d^3x_N\,|\psi(\mathbf{x}_1,\mathbf{x}_2,\ldots,\mathbf{x}_N;t)|^2]\, d^3x_1 \tag{3.5}$$

is the probability that particle 1 is within d^3x_1 when nothing is said about the positions of other particles. If one integrates the expression (3.4) over the coordinates of *all* the particles, one gets the probability that all the particles are to be found somewhere in space, and this should of course be unity:

$$\int d^3x_1 \int d^3x_2\ldots\int d^3x_N\,|\psi(\mathbf{x}_1,\mathbf{x}_2,\ldots,\mathbf{x}_N;t)|^2 = 1 \tag{3.6}$$

This *normalization* of the wave function has been assumed in making the statements above regarding the interpretation of ψ. The procedure used in Sec. 2.6 can be generalized in an obvious fashion to prove that the norm is conserved. A simpler proof depending only on a certain general property (self-adjointness) of the Hamiltonian operator will be given in Sec. 3.14. As regards non-normalizable wave functions, remarks similar to those made in Sec. 2.5 apply, and we will not consider them further here.

Associated with the probability interpretation is the postulate regarding the expectation values of dynamical variables. As a direct generalization of Eq. 2.42, we define

$$\langle A\rangle = \int\int\ldots\int\psi^*(\mathbf{x}_1,\mathbf{x}_2\ldots,\mathbf{x}_N)\, A_{op}\,\psi(\mathbf{x}_1,\mathbf{x}_2,\ldots,\mathbf{x}_N)\, d^3x_1\, d^3x_2\ldots d^3x_N \tag{3.7}$$

Here A_{op} is obtained by the operator replacement (3.2) from the function $A(\mathbf{x}_1,\ldots,\mathbf{x}_N;\, \mathbf{p}_1,\ldots,\mathbf{p}_N)$ defining the dynamical variable:

$$A_{op} = A\ (\mathbf{x}_1,\mathbf{x}_2,\ldots,\mathbf{x}_N;\ -i\hbar\nabla_1,\ldots,-i\hbar\nabla_N) \tag{3.8}$$

In trying to write down A_{op} explicitly, ambiguities can arise from the fact that the differential operators do not commute with the coordinate variables. As observed in Sec. 2.3, one must then use a symmetrized expression, but we will have no occasion to worry unduly about this problem.

Notation — To save writing, we will hereafter adopt the following convention: Whenever we are talking of a general system, we will write the wave function $\psi(\mathbf{x}_1,\mathbf{x}_2\ldots\mathbf{x}_N)$ in abbreviated form as $\psi(\mathbf{x})$, it being understood that $\mathbf{x}$ stands for the coordinates $\mathbf{x}_1,\mathbf{x}_2,\ldots,\mathbf{x}_N$ of all the particles. Similarly, we will write $\mathbf{p}$ for the set of momentum variables $\mathbf{p}_1,\mathbf{p}_2,\ldots,\mathbf{p}_N$. The individual volume elements d^3x_1, $d^3x_2\ldots$will be denoted by $d\tau_1,d\tau_2,\ldots$, and we will use $d\tau$ as an abbreviation for $d\tau_1 d\tau_2\ldots d\tau_N$. Thus Eqs. (3.6) and (3.7) would appear in this shorthand notation as

$$\int d\tau\ \psi^*(\mathbf{x})\psi(\mathbf{x}) = 1, \quad \langle A \rangle = \int d\tau\ \psi^*(\mathbf{x})A_{op}\ \psi(\mathbf{x}) \tag{3.9}$$

This convention will help us to avoid using cumbersome expressions in discussing general theorems. Of course in particular problems where the number of particles involved is important, the expressions will be written out in full.

3.2 The Fundamental Postulates of Wave Mechanics

What we have learned so far about the nature of the wave-mechanical description of physical systems and about the rules of interpretation, were deduced from observational facts by heuristic (non-rigorous) arguments. Beginning with a formal statement of these as basic postulates, we shall now study their consequences in some detail. In the course of this study, the mathematical structure of wave mechanics and its physical content will be made more explicit.

(*a*) *Representation of states*

POSTULATE 1 — The *state* of a quantum mechanical system *is described or represented by a wave function* $\psi\,(\mathbf{x}, t)$.

The wave function contains all the information which nature permits one to give about the state of the system at any instant of time. We will often speak of the 'state ψ' meaning thereby the physical concept (the state) described by the mathematical function $\psi(\mathbf{x},t)$. The wave function is not directly measurable. In fact, all *constant* multiples of a given ψ (namely $c\psi$, for all real or complex numbers c) describe one and the same state.

POSTULATE 1′ — THE SUPERPOSITION PRINCIPLE. If ψ_1 and ψ_2 are wave functions describing any two states of a given system, then corresponding to every linear combination[1] $(c_1\psi_1 + c_2\psi_2)$ of the two functions there exists a state of the system.

[1] The term 'linear combination' will appear repeatedly in the text. By a linear combination of $\psi_1, \psi_2, \ldots,\psi_r$ we mean any quantity of the form $c_1\psi_1 + c_2\psi_2 + \ldots c_r\psi_r$ where $c_1, c_2, \ldots, c_r$ are any complex numbers. We say that $\psi_1, \psi_2, \ldots, \psi_r$ form a ***linearly independent set*** if none of them can be expressed as a linear combination of the others; or equivalently, if the only way to make $c_1\psi_1 + c_2\psi_2 + \ldots + c_r\psi_r$ vanish is by taking ***all*** the constant coefficients to be zero.

This is a fundamental principle of quantum mechanics to which there is no correspondence in classical mechanics. It is the possibility of superposition which makes interference phenomena possible. We have already assumed it in constructing wave packet states of a free particle, as in Eq. 1.31, as superpositions of de Broglie waves corresponding to free particle states of various momenta.

The principle of superposition has deep mathematical significance too. This realization comes from the observation that this principle is a fundamental characteristic of *vectors* representing points in space. (If $\mathbf{v}_1,\mathbf{v}_2$ are the position vectors of any two points, $c_1\mathbf{v}_1 + c_2\mathbf{v}_2$ is again a vector which represents some point in space). Therefore, it is possible (and useful) to think of wave *functions* as vectors of some kind. The ordinary vectors which we are familiar with need only three real numbers each, for a complete specification; the components v_1,v_2,v_3 identify a vector $\mathbf{v}$ completely. But each wave function-vector requires an *infinite* set of *complex* numbers for its specification, namely the values of ψ *at all the points* $\mathbf{x}$ in (configuration) space. Therefore, we say that the function-vectors are *infinite-dimensional.* The set of all wave functions of a given system (which includes also all linear combinations of the wave functions) forms an infinite dimensional *complex linear vector space.*

In this 'geometrical' picture of wave functions, the quantities $\psi(\mathbf{x})$ for various $\mathbf{x}$ are thought of as different 'components' of the wave function-vector ψ. Here the continuous variable $\mathbf{x}$ serves as a label distinguishing the different components of ψ, and is the counterpart of the index i labelling the components v_i of an ordinary vector $\mathbf{v}$. With this identification, the norm $\int |\psi(\mathbf{x})|^2 d\tau$ becomes the analogue of $\Sigma_i v_i^2$, and it may therefore be viewed as the 'square of the length' of the function-vector. (Since $\psi(\mathbf{x})$ is complex, the absolute square $|\psi(\mathbf{x})|^2$ has to be taken to get a real 'length' and of course, it is 'summation' over $\mathbf{x}$ which is indicated by the integral sign). More generally, $\int\phi^*(\mathbf{x})\psi(\mathbf{x})d\tau$ is defined as the *scalar product* of ϕ and ψ, analogous to $\Sigma_i u_iv_i$ for ordinary vectors. In view of this, the notation (ϕ,ψ)—as the counterpart of $\mathbf{u}.\mathbf{v}$—is often used for such integrals. From the definition

$$(\phi,\psi) \equiv \int \phi^*(\mathbf{x})\psi(\mathbf{x})d\tau \qquad \text{(scalar product of } \phi,\psi\text{)},$$

it follows that

$$(\phi,\psi) = (\psi,\phi)^*,$$

$$(\phi,c\psi) = c(\phi,\psi), \; (c\phi,\psi) = c^*(\phi,\psi),$$

and that

$$\text{Norm of } \psi = (\psi,\psi) \geqslant 0$$

with the equality sign holding if and only if $\psi \equiv 0$.

The above notation is now in common usage. To help the student in gaining familiarity with it, we shall frequently write the scalar product notation (ϕ,ψ) side by side with its explicit integral expression.

(*b*) *Representation of dynamical variables; Expectation values, Observables*

Postulate 2 — Each *dynamical variable* $A(\mathbf{x};\mathbf{p})$ is represented in quantum mechanics by a *linear operator*[2] $A_{\text{op}} = A(\mathbf{x}_{\text{op}}, \mathbf{p}_{\text{op}}) \equiv A(\mathbf{x}, -i\hbar\nabla)$.

The operators act on the wave functions of the system. The effect of an operator A on a wave function ψ is to convert it into another wave function, denoted by $A\psi$. *Linearity* of the operator means that a linear combination of two (or more) wave functions, say ψ_1 and ψ_2 is converted into the *same* linear combination of $A\psi_1$ and $A\psi_2$:

$$A(c_1\psi_1 + c_2\psi_2) = c_1(A\psi_1) + c_2(A\psi_2) \tag{3.10}$$

Dynamical variables in quantum mechanics do not, in general, commute with each other: $AB \neq BA$. By this we mean that $A_{\text{op}}B_{\text{op}}\psi \neq B_{\text{op}}A_{\text{op}}\psi$ for *arbitrary* wave functions ψ (though there may be particular functions for which the two sides are equal). The difference $AB - BA$ is called the *commutator* of A and B. A bracket notation is used to denote the commutator

$$[A,B] = AB - BA \tag{3.11}$$

Commutation relations (relations giving the values of the commutators) of position and momentum variables may be easily deduced. For example, in one dimension,

$$(xp - px)\,\psi = -i\hbar\left[x\frac{\partial}{\partial x}\psi - \frac{\partial}{\partial x}(x\psi)\right] = i\hbar\psi$$

Thus the operator $[x,p]$ has the effect of simply multiplying any arbitrary wave function ψ by $i\hbar$ and we can therefore say that $[x,p] = i\hbar$. In three dimensions, if x_i and $p_j \equiv -i\hbar\,\partial/\partial x_j$ are the respective operators representing the ith and jth components of the position and momentum of a particle,[3] we have

$$x_i p_j \psi = -i\hbar\, x_i \frac{\partial\psi}{\partial x_j},$$

and

$$p_j x_i \psi = -i\hbar\frac{\partial}{\partial x_j}(x_i\psi) = -i\hbar\left[\delta_{ij}\psi + x_i\frac{\partial\psi}{\partial x_j}\right],$$

since $\partial x_i/\partial x_j$ is equal to the *Kronecker delta function* δ_{ij} *defined as:*

$$\delta_{ij} = 1 \text{ if } i = j, \text{ and } \delta_{ij} = 0 \text{ if } i \neq j \tag{3.12}$$

Thus $(x_i p_j - p_j x_i)\,\psi = i\hbar\,\delta_{ij}\psi$. Since this is true for *any* wave function ψ we can conclude that $x_i p_j - p_j x_i = i\hbar\delta_{ij}$ or

$$[x_i, p_j] = i\hbar\,\delta_{ij} \tag{3.13a}$$

It is a trivial matter to verify in a similar manner that

$$[x_i, x_j] = 0, \quad [p_i, p_j] = 0 \tag{3.13b}$$

The above equations assert that of all the commutators among the position and momentum components *the only non-vanishing ones* are

$$[x, p_x] = [y, p_y] = [z, p_z] = i\hbar \tag{3.13c}$$

[2] We will drop the subscript 'op' hereafter, and write, for example, $\mathbf{p} = -i\hbar\nabla$ which is to be understood as '$\mathbf{p}$ is represented by the operator $\mathbf{p}_{\text{op}} = -i\hbar\nabla$'. Therefore, the same symbol will hereafter be used for both the physical concept of the dynamical variable (as when we say 'the momentum $\mathbf{p}$') and for the quantum operator representing it (as when we write $\mathbf{p}\psi$ meaning $-i\hbar\nabla\psi$). Which of these is meant in a particular case will be clear from the context.

[3] We will use either of the notations (x, y, z) or (x_1, x_2, x_3) for the components of $\mathbf{x}$, as convenience dictates. Similarly the components of $\mathbf{p}$ will be written as (p_x, p_y, p_z) or (p_1, p_2, p_3).

These equations may be considered as definitions of *canonically conjugate pairs* of dynamical variables in quantum mechanics. More generally, if subscripts α,β denote the αth and βth particles of a many particle system, we have

$$[x_{\alpha i},p_{\beta j}] = i\hbar\delta_{\alpha\beta}\delta_{ij};$$
$$[x_{\alpha i},x_{\beta j}] = [p_{\alpha i},p_{\beta j}] = 0 \qquad (3.14)$$

These commutation relations are fundamental to all of quantum mechanics. Commutators involving products of components of position coordinates and momenta can be evaluated using these basic commutation relations and the identities

$$[AB,C] = A[B,C] + [A,C]B$$
$$[A,BC] = [A,B]C + B[A,C] \qquad (3.15)$$

which are valid for arbitrary operators A, B, C.

There are, in quantum mechanics, certain dynamical variables (e.g. parity) which do not have a classical counterpart, i.e. there is no classical function $A(\mathbf{x},\mathbf{p})$ which would yield the quantum operator for such a quantity through the operator replacement of postulate 3. In these cases the operators concerned are defined directly by stating what effect they produce on arbitrary states or wave functions of the system. (See, for example, Sec. 4.11 for parity). The commutation rules (3.13) have nothing to say about commutation relations involving such operators, which must be deduced directly from their definitions. We will consider these explicitly at the appropriate stage, but it may be kept in mind that the treatment of the remainder of this chapter holds good also for such nonclassical dynamical variables.

Finally, we remind the reader of the observation at the end of Sec. 1.11 that electrons (and other particles like the proton and neutron) carry *spin* angular momentum, i.e., angular momentum associated with what may be visualized as a spinning motion of the particle analogous to the spinning of a top or the rotation of the earth about its own axis. This spinning motion is of course quite independent of the position and momentum of linear motion of the particle. Therefore one needs new dynamical variables (which *cannot* be functions of $\mathbf{x}$ and $\mathbf{p}$) to represent this degree of freedom. Further, the concept of the wave function itself has to be generalized to include a dependence on the direction of spin also (besides the dependence on $\mathbf{x}$). Fortunately, in a wide variety of important phenomena, the presence of spin plays no significant role. So we postpone consideration of this complication to Sec. 8.4. However the general theorems of this chapter hold good for the spin operators too.

EXAMPLE 3.3 — Let us evaluate $[x,p^n]$ in the one-dimensional case from the basic relation $[x,p] = i\hbar$ using Eqs. (3.15). We have $[x,p^2] = p[x,p] + [x,p]p = 2i\hbar p$ and then $[x,p^3] = [x,p^2p] = p^2[x,p] + [x,p^2]p = 3i\hbar p^2$. Continuing this process, we get $[x,p^n] = i\hbar\, n\, p^{n-1}$. By a similar process we can show that $[p,x^n] = -i\hbar n x^{n-1}$. Actually this result is more easily obtained by direct use of the operator form $(-i\hbar d/dx)$ of p, as in the derivation of the commutation rules (3.13). The student may verify that for any function $f(x)$, $[p,f(x)] = -i\hbar df(x)/dx$. In three dimensions,

$$[\mathbf{p}, f(\mathbf{x})] = -i\hbar\nabla f(\mathbf{x})$$

and similarly,

$$[\mathbf{x}, f(\mathbf{p})] = i\hbar \nabla_p f(\mathbf{p}),$$

where ∇_p stands for $(\partial/\partial p_x, \partial/\partial p_y, \partial/\partial p_z)$.

EXAMPLE 3.4 — The angular momentum of a particle with respect to the origin is defined by $\mathbf{L} = \mathbf{x} \times \mathbf{p}$. To find the commutation relations of its components, consider

$$[L_x, L_y] \equiv [(yp_z - zp_y), (zp_x - xp_z)]$$
$$= [yp_z, zp_x] - [yp_z, xp_z] - [zp_y, zp_x] + [zp_y, xp_z].$$

Observe that in the two middle terms, none of the variables is accompanied by its canonically conjugate partner. So these terms vanish. In the first term y and p_x commute with everything else, so that they can be 'pulled out', reducing the commutator to $yp_x[p_z, z] = -i\hbar yp_x$. Similarly the last term reduces to $[z, p_z]xp_y = i\hbar xp_y$. Putting these together, we find $[L_x, L_y] = i\hbar(xp_y - yp_x) = i\hbar L_z$. The commutators of the other components can be evaluated similarly. We finally have

$$[L_x, L_y] = i\hbar L_z, \ [L_y, L_z] = i\hbar L_x, \ [L_z, L_x] = i\hbar L_y.$$

Note that we have not had to use the differential operator form for $\mathbf{p}$ to obtain these results; in fact it is usually much simpler to use the commutation rules (3.13) directly than to go back to $\mathbf{p} = -i\hbar\nabla$.

The above commutation rules hold also for the total angular momentum $\mathbf{L} = \Sigma \mathbf{L}_\alpha$ of a system of particles. The student is urged to verify this, keeping in mind that the angular momenta $\mathbf{L}_\alpha$ and $\mathbf{L}_\beta$ of two different particles ($\alpha \neq \beta$) commute in view of (3.14).

EXAMPLE 3.5 — In taking the square (or higher powers) of a product of operators, possible noncommutativity of the factors should not be overlooked. For example, $(xp_x)^2 \neq x^2p_x^2$. Actually $(xp_x)^2 = xp_xxp_x = x(xp_x - i\hbar) p_x = x^2p_x^2 - i\hbar xp_x$.

Similarly

$$(xp_x)\ (yp_y) = xp_x\ (p_y y + i\hbar) = (xp_y)\ (yp_x) + i\hbar\, xp_x.$$

Using results of this kind we can express the expansion of $(\mathbf{x}.\mathbf{p})^2$ as

$$(\mathbf{x}.\mathbf{p})^2 = (x^2 + y^2 + z^2)\ (p_x^2 + p_y^2 + p_z^2) - (xp_y - yp_x)^2$$
$$- (yp_z - zp_y)^2 - (zp_x - xp_z)^2 + i\hbar\ (xp_x + yp_y + zp_z)$$

or

$$(\mathbf{x}.\mathbf{p})^2 = \mathbf{x}^2\mathbf{p}^2 - \mathbf{L}^2 + i\hbar\ (\mathbf{x}.\mathbf{p}).$$

This result will be used in Sec. 4.6.

EXAMPLE 3.6 — When written explicitly as a differential operator, L_z can be written as equal to $-i\hbar\ (x\partial/\partial y - y\partial/\partial x)$. On transforming this operator to spherical polar coordinates, we get $L_z = -i\hbar\partial/\partial\varphi$. Hence we can formally write $[\varphi, L_z] = i\hbar$. This relation has the same form as Eq. 3.13c. Nevertheless φ and L_z should not be considered as canonically conjugate variables; indeed φ *itself cannot be regarded as a proper dynamical variable* in quantum mechanics. The reason is that it has the effect of converting a perfectly good wave function ψ into one with inadmissible properties. For, if φ is limited to the interval 0 to 2π, then after it has increased to 2π it jumps discontinuously down to zero, and the function $\varphi\psi$ will do the same. On the other hand, if φ is allowed an unlimited range of variation, $\varphi\psi(r,\theta,\varphi)$ will violate the single-valuedness condition. Thus $\varphi\psi$ is not an admissible wave function though ψ is, which is to say that φ is not a good operator. As a corollary, one cannot draw the same kind of conclusions from the formal relation $[\varphi,L_z] = i\hbar$ as can be done from Eqs. 3.13c. In particular it is not possible to infer an uncertainty relation $(\Delta\varphi)\ (\Delta L_z) \geqslant \frac{1}{2}\hbar$ analogous to Eq. 3.73.

POSTULATE 3 — If a large number of measurements of a dynamical variable A are made on a system which is prepared to be in one and the same state ψ before each measurement, the results of the individual measurements will in general be different, but the average of all the observed values is expected to be

$$\langle A\rangle_\psi = \int \psi^* A\psi d\tau \equiv (\psi,A\psi) \tag{3.16}$$

The quantity $\langle A\rangle_\psi$ is called the *expectation value* of the dynamical variable A. Note that inside the integral, A means A_{op}.

On inspection of the special case of Eq. 3.16 when A is the position variable of any of the particles, it becomes evident that the definition of the expectation value automatically implies the interpretation of $|\psi|^2$ as the probability density in configuration space, and the normalization of ψ according to Eq. 3.6. Unless otherwise stated, all wave functions will be taken to be so normalized. For the present we will assume that non-normalizable wave functions are also brought into this category through suitable limiting procedures like the technique of box normalization (Sec. 2.5). Later in this chapter we will consider the kind of normalization involving the Dirac delta function, which becomes necessary if unnormalizable functions are to be handled directly. Whenever a ψ having a different normalization from (3.6) is employed for any reason, the definition (3.16) of expectation values is to be replaced by

$$\langle A\rangle = \frac{\int \psi^* A\psi d\tau}{\int \psi^*\psi d\tau} \tag{3.16a}$$

The expectation value $\langle A\rangle$ may be real or complex, depending on the nature of A. However, from the experimental point of view, the result of any single observation or the measurement of a single quantity is necessarily a real number (e.g. the reading on a dial or scale, weights in a pan, etc). To measure any complex-valued variable, one has to separate it into two real variables (the real and the imaginary parts, or the modulus and the argument), and measure each separately. The measurement of each part always gives some real number as the result; and the average result of repeated measurements is also of course real. This fact implies, in the quantum mechanical context, that only those dynamical variables whose expectation values are real should be considered to be directly measurable, or *observable*. Thus if A is to be observable, it is necessary that $\langle A\rangle = \langle A\rangle^*$, or

$$\int \psi^* A\psi d\tau = \left(\int \psi^* A\psi d\tau\right)^* = \int (A\psi)^*\, \psi d\tau \tag{3.17}$$

If this is to be ensured for all wave functions ψ, the operator A must have a certain mathematical property called self-adjointness, which will be defined in the next section.

The three postulates considered so far concern the kinematics (description at a particular time) of a quantum system. The fourth basic postulate concerns the dynamics (development with time). This will be taken up in Sec. 3.14, after the discussion of kinematical questions is completed.

EXAMPLE 3.7 — To evaluate $\langle x^n \rangle$ when ψ is the Gaussian function, namely $\psi = (1/\sigma\sqrt{\pi})^{1/2}.\ \exp\ [-x^2/2\sigma^2]$:

$$(\sigma\sqrt{\pi})\ \langle x^n \rangle = \int_{-\infty}^{\infty} x^n\ e^{-x^2/\sigma^2}\ dx$$

$$= \int_{0}^{\infty} x^n\ e^{-x^2/\sigma^2}\ dx + \int_{-\infty}^{0} x^n\ e^{-x^2/\sigma^2}\ dx$$

$$= [1 + (-1)^n] \int_{0}^{\infty} x^n\ e^{-x^2/\sigma^2}\ dx.$$

The last expression has been obtained with the aid of the substitution $x = -x$ in the second integral in the previous step. We thus have immediately that

$$\langle x^n \rangle = 0 \text{ for odd values of } n.$$

This result is a consequence of the fact that the integrand, $x^n \exp\ [-x^2/\sigma^2]$ is of odd parity for odd n. It is true quite generally that the integral (from $-\infty$ to $+\infty$) of any function of odd parity is zero. This is a useful fact to remember for future calculations. From this fact we can directly conclude, for instance, that $\langle p \rangle = 0$ since in $\langle p \rangle = \int \psi^*(-i\hbar d/dx)\psi\ dx$, the Gaussian ψ is of even parity while $d\psi/dx$ is of odd parity.

To determine $\langle x^n \rangle$ when n is an even integer, say $n = 2s$, evaluation of the final integral is best done by converting it to the form of a gamma function, $\Gamma(m) = \int_0^\infty e^{-w}\ w^{m-1}\ dw$, by the substitution $(x^2/\sigma^2) = w$. Remembering $\Gamma(\frac{1}{2}) = \sqrt{\pi}$ and $\Gamma(m) = (m-1)\ \Gamma(m-1)$ we get

$$\langle x^{2s} \rangle = \frac{(2s)!}{s!} \cdot \left(\frac{\sigma}{2}\right)^{2s}, \quad (s = 0, 1, 2, \ldots).$$

Note that the value unity of $\langle x^n \rangle$ for $n = 0$ checks that the wave function is correctly normalized.

3·3 The Adjoint of an Operator and Self-Adjointness

Let us consider the integral

$$\int \phi^* A\psi\ d\tau \equiv (\phi,\ A\psi)$$

which involves two different functions ϕ and ψ and reduces to (3.16) in the special case when $\phi = \psi$. It can be shown that one can always find another operator, called the *adjoint* of A and denoted by $A^\dagger$ (read A dagger), such that

$$\int \phi^* A\psi d\tau = \int (A^\dagger\phi)^* d\tau, \text{ or } (\phi, A\psi) = (A^\dagger\phi, \psi). \tag{3.18}$$

In other words, as far as the value of the integral is concerned, it makes no difference whether A acts on ψ or its adjoint $A^\dagger$ acts on the other wave function ϕ. Eq. 3.18 serves to define the adjoint of any operator. From this definition it is easy to show

$$(A + B)^\dagger = A^\dagger + B^\dagger \tag{3.19}$$

and that if c is a complex number

$$(cA)^\dagger = c^* A^\dagger \tag{3.20}$$

i.e. in taking the adjoint, any complex number goes over into its complex conjugate. Further, since

$$\int \phi^*(A^\dagger\psi)\, d\tau = [\int (A^\dagger\psi)^*\phi d\tau]^* = [\int \psi^* A\phi d\tau]^* = \int (A\phi)^*\psi d\tau$$

we have

$$(A^\dagger)^\dagger = A \tag{3.21}$$

For the adjoint of the product of two operators A and B, by applying the definition given in Eq. 3.18 successively to A and B, one obtains

$$\int \phi^* AB\psi d\tau = \int (A^\dagger\phi)^* B\psi d\tau = \int (B^\dagger A^\dagger\phi)^*\psi d\tau,$$

or

$$(AB)^\dagger = B^\dagger A^\dagger \tag{3.22}$$

An operator A is said to be *self adjoint*[4] if its adjoint is equal to itself:

$$A^\dagger = A,$$

or

$$\int \phi^* A\psi d\tau = \int (A\phi)^*\psi d\tau, \text{ i.e., } (\phi, A\psi) = (A\phi, \psi). \tag{3.23}$$

Note that the product of two self-adjoint operators is *not* necessarily self-adjoint. For, if $A^\dagger = A$ and $B^\dagger = B$, then according to Eq. 3.20,

$$(AB)^\dagger = BA \tag{3.24}$$

Thus AB is self adjoint only if $BA = AB$, i.e., if A and B commute. However, the two combinations

$$(AB + BA) \text{ and } i(AB - BA) \tag{3.25}$$

are both self-adjoint, as the reader may verify.

It is now a trivial matter to see that the expectation value of a self-adjoint operator is real, for on setting $\phi = \psi$ in Eq. 3.23 it reduces to $\langle A\rangle = \langle A\rangle^*$. Thus self-adjoint operators are suitable for representing observable dynamical variables, and we proceed to study their most important properties. Incidentally, it may be observed that $A^\dagger A$ is always self-adjoint (even if A is not). Further, its expectation value is non-negative in *all* states. (Any operator with this property is said to be *positive*.)

$$\langle A^\dagger A \rangle = \int \psi^* A^\dagger A\, \psi\, d\tau = \int (A\psi)^*(A\psi)\, d\tau \geqslant 0 \tag{3.26}$$

since the integrand, being the absolute square of the function $A\psi$, is non-negative. Evidently if $\langle A^\dagger A \rangle$ is to vanish, the integrand must vanish identically. Thus,

$$\langle A^\dagger A \rangle = 0 \text{ implies } A\psi = 0 \tag{3.26a}$$

EXAMPLE 3.8 — The position operator is obviously self-adjoint. As for the momentum operator, we find by partial integration that

$$\int \phi^* p\psi dx \equiv \int \phi^*(-i\hbar)\frac{d\psi}{dx}\, dx$$

$$= -i\hbar\, \phi^*\psi + \int \left(-i\hbar\frac{d\phi}{dx}\right)^* \psi\, dx.$$

The first term drops out in the case of normalizable wave functions ϕ and ψ since they vanish at both the limits of integration ($x \to \pm\infty$). Then the expres-

[4] For reasons which will become clear later, in the context of quantum mechanics, the term *hermitian* operator is very often used for self-adjoint operators. We will also be using the words 'hermiticity' and 'self-adjointness' interchangeably.

sion reduces to $\int (p\phi)^*\psi dx$ showing that p is self-adjoint. In the case of box-normalized functions, the condition that all wave functions be periodic within the box (see Sec. 2.5) ensures that $(\phi^*\psi)$ cancels out between the upper and lower limits ($x = -L/2$ and $+L/2$). Therefore $\mathbf{p}$ is self-adjoint in this case too.

Knowing that $\mathbf{x}$ and $\mathbf{p}$ are self-adjoint, one can easily prove the same property for $\mathbf{L}$ and H (provided the potential function in the latter is a *real* function of $\mathbf{x}$).

3.4 The Eigenvalue Problem; Degeneracy

Further development of the formalism relies heavily on the concept of eigenfunctions and eigenvalues. We had occasion to consider these in the context of the energy or Hamiltonian operator in Sec. 2.9. In general, for any operator A one can set up the eigenvalue equation

$$A\phi_a = a\phi_a \tag{3.27}$$

Let us recall the relevant definitions. If a function ϕ_a is such that the action of the operator A on it has the simple effect of multiplying it by a constant factor a, then we say that ϕ_a is an *eigenfunction of A belonging to the eigenvalue a*. The set of all eigenvalues of A form what is called the *eigenvalue spectrum* (or simply the spectrum) of A. The spectrum may be continuous, or discrete, or partly continuous and partly discrete (as is the case with the energy spectrum for the square well potential). If there exists only one eigenfunction belonging to a given eigenvalue, the eigenvalue is said to be *non-degenerate*; otherwise it is *degenerate*.

Associated with any degenerate eigenvalue there is always an *infinite* number of eigenfunctions. For example, if ϕ_a, χ_a belong to the same eigenvalue a, then so do the combinations $c_1\phi_a + c_2\chi_a$ for *all* values of c_1 and c_2:

$$A(c_1\phi_a + c_2\chi_a) = c_1 A\phi_a + c_2 A\chi_a = a\,(c_1\phi_a + c_2\chi_a).$$

Since the set of all eigenfunctions belonging to a given degenerate eigenvalue a is *closed under linear combinations* (i.e. any linear combination of members of the set is again a member), this set forms a linear space. This space may be called the eigenspace belonging to the eigenvalue a of A. We can always choose from this function-space a linearly independent subset of functions, say $\phi_{a1}, \phi_{a2}, \ldots, \phi_{ar}$, such that *every* eigenfunction belonging to a can be expressed uniquely as a linear combination

$$c_1\phi_{a1} + c_2\phi_{a2} + \ldots + c_r\phi_{ar},$$

with suitable coefficients $c_1, c_2, \ldots c_r$. We say then that $\phi_{a1}, \phi_{a2} \ldots \phi_{ar}$ form a set of basis functions (or simply a *basis*) which *spans* the linear space. There is an infinite number of ways of choosing a basis, just as there is an infinite number of ways of choosing coordinate axes in three-dimensional space. But the number r is characteristic of the space. This means, in the present context, that among the infinite number of eigenfunctions belonging to a given degenerate eigen value, there is a definite number r of linearly independent ones. This number is called the *degree of degeneracy* of the eigenvalue; we say that the eigenvalue is *r-fold degenerate*.

The task of determining all the eigenvalues and eigenfunctions of any operator is referred to as the eigenvalue problem. We will be solving a number of important eigenvalue problems in the next chapter. Here we proceed with the consideration of general properties of eigenvalues and eigenfunctions.

EXAMPLE 3.9 — To see the significance of eigenvalues and eigenfunctions, consider an eigenstate ϕ_a of A. Then $A^n\phi_a = a^n\phi_a$, and hence $\langle A^n \rangle = a^n = \langle A \rangle^n$, i.e. in this state, the average value of the nth power of A is just the nth power of $\langle A \rangle$. This implies that when the system is in an *eigenstate* of A, the dynamical variable A has a *definite value equal to the eigenvalue* to which the state belongs. This result will be confirmed in a different way later.

EXAMPLE 3.10 — In one dimension, the eigenvalue equation for the energy of a free particle is $(-\hbar^2/2m)\ (d^2\phi/dx^2) = E\phi$, or $d^2\phi/dx^2 + \alpha^2\phi = 0$, with $\alpha^2 = (2mE/\hbar^2)$. For any given E, this equation has *two independent* solutions, $e^{i\alpha x}$ and $e^{-i\alpha x}$. So each eigenvalue is *doubly degenerate*. Of course all linear combinations of these are also solutions, e.g. $\sin\alpha x$, $\cos\alpha x$. The set $e^{i\alpha x}$, $e^{-i\alpha x}$ forms a basis for the space of these solutions. So also does $\sin\alpha x$ and $\cos\alpha x$ or $e^{i\alpha x}$ and $\cos\alpha x$, etc.

3.5 Eigenvalues and Eigenfunctions of Self-Adjoint Operators

Let A be a self-adjoint operator, and ϕ_a, $\phi_{a'}$ be two of its eigenfunctions:

$$A\phi_a = a\phi_a,\quad A\phi_{a'} = a'\phi_{a'} \tag{3.28}$$

Substituting $\phi = \phi_a$ and $\psi = \phi_{a'}$ in the self-adjointness condition (3.23), we get

$$\int \phi_a^* A\phi_{a'}\, d\tau = \int (A\phi_a)^*\phi_{a'}\, d\tau \tag{3.29}$$

Using Eq. 3.28 we replace the operator A in the integral by the relevant eigenvalues. Eq. 3.29 then reduces to

$$(a' - a^*) \int \phi_a^*\phi_{a'}\, d\tau = 0 \tag{3.30}$$

This is valid for any two eigenfunctions of A. Let us consider the special case $\phi_{a'} = \phi_a$, which automatically implies $a' = a$. Then

$$(a - a^*) \int \phi_a^*\, \phi_a\, d\tau = 0 \tag{3.31}$$

The integral here is the norm of ϕ_a which can never vanish. Therefore,

$$a = a^* \tag{3.32}$$

Thus the eigenvalues of a self-adjoint operator are real.

Feeding this information back into Eq. 3.30 we conclude that

$$\int \phi_a^*\phi_{a'} d\tau = 0 \text{ or } (\phi_a, \phi_{a'}) = 0 \text{ for } a \neq a' \tag{3.33}$$

This result may be stated in words as follows:

Any two eigenfunctions belonging to distinct (unequal) eigenvalues of a self-adjoint operator are mutually orthogonal[5].

If a and a' have the same value but ϕ_a is different from $\phi_{a'}$, (i.e. if ϕ_a and $\phi_{a'}$, belong to the same degenerate eigenvalue), we can draw no conclusions from Eq. 3.31. In fact, it is not to be expected that all the (infinite number of) eigenfunctions belonging to a degenerate eigenvalue will be mutually ortho-

[5] The term 'orthogonal' is taken from the language of vector spaces; it means that the scalar product (of the function-vectors ϕ_a, $\phi_{a'}$, in the present case) vanishes.

gonal. However, *it is always possible to choose* from among these *a linearly independent set of basis functions which are mutually orthogonal* (see Example 3.11 below).

In order to keep the essentials of the further development clear, in the discussions of the next few sections, we will assume that all eigenvalues of A are non-degenerate. The elaborations which are necessary when degenerate eigenvalues are present will be carried out in Sec. 3.13.

So far, we have assumed nothing about the norm of the eigenfunctions. Actually we know from the example of the square well potential problem considered in Chapter 2 that an operator may have both normalizable and non-normalizable eigenfunctions. Thus the norm of ϕ_a here may be either 1 or ∞. We, therefore, write

$$\int \phi_a{}^*\phi_{a'}d\tau = \delta(a,a') \tag{3.34a}$$

where $\delta\,(a,a') = 0$ for $a \neq a'$ and $\delta\,(a,a) = 1$ or ∞ according as ϕ_a is, or is not, normalizable. In the former case $\delta\,(a,a')$ is clearly nothing but the Kronecker delta function:

$$\delta(a,a') = \delta_{aa'} \tag{3.34b}$$

In the latter case (eigenfunctions of infinite norm) we write

$$\delta(a,a') = \delta(a-a') \tag{3.34c}$$

where $\delta\,(a-a')$ is the Dirac delta function, defined below. We will see shortly that Eq. 3.34b applies if a belongs to the discrete part of the eigenvalue spectrum, and (3.34c) in the case of eigenvalues belonging to the continuum.

EXAMPLE 3.11 — From an arbitrary linearly independent set of normalized functions $\chi_1, \chi_2, \ldots, \chi_r$ we can construct a new set of *orthogonal* functions $\psi_1, \psi_2, \ldots, \psi_r$ by the *Schmidt Orthogonalization Procedure*. Understanding of this procedure is facilitated by recalling that normalized wave functions can be thought of as unit vectors in a function space. To construct from an ordinary vector **V** another vector orthogonal to a given *unit* vector **U**, all we have to do is to remove from **V** its component parallel to **U**, namely **U** (**U.V**). The result is **V** − **U** (**U.V**). In analogy with this we have on taking $\psi_1 \equiv \chi_1$ in place of **U** and χ_2 instead of **V**,

$$\psi_2' = \chi_2 - \psi_1\,(\psi_1,\chi_2) \equiv \chi_2 - \psi_1 \int \psi_1{}^*\chi_2 d\tau$$

It is a trivial matter to satisfy oneself that ψ_2' is indeed orthogonal to ψ_1, i.e. $(\psi_1, \psi_2') = 0$. Normalization of ψ_2' may be carried out as usual: we have

$$\psi_2 = \psi_2'/[\int\psi_2'{}^*\psi_2'\,d\tau]^{1/2} \equiv \psi_2'/(\psi_2',\psi_2')^{1/2}.$$

This procedure may be continued by subtracting from χ_3 the parts 'parallel' to ψ_1 and ψ_2 and then normalizing:

$$\psi_3' = \chi_3 - \psi_1\,(\psi_1,\,\psi_3) - \psi_2\,(\psi_2,\,\psi_3),$$
$$\psi_3 = \psi_3'/(\psi_3',\psi_3')^{1/2},$$

and so on.

3.6 The Dirac Delta Function

The Kronecker delta function δ_{mn} is a function of two discrete parameters m,n. As an immediate consequence of its definition, Eq. 3.12, we have the following important property:

$$\sum_m c_m\delta_{mn} = c_n \tag{3.35}$$

where c_m is any function of the discrete variable m. That is, δ_{mn} annihilates all terms in the sum for which $m \neq n$, and the only surviving term is $c_n\delta_{nn} = c_n$. Eq. 3.35 may be used as the *definition* of the Kronecker delta, since its validity for all functions c_m implies the conditions of Eq. 3.12.

The Dirac delta function $\delta(x - x')$ plays the same role for functions of a continuous variable as the Kronecker delta does in the case of functions of a discrete variable. It is defined through an equation entirely analogous to (3.35):

$$\int_a^b f(x)\, \delta(x - x')\, dx = f(x'), \quad a < x' < b \tag{3.36}$$

Here the integration sign naturally replaces the summation sign of Eq. 3.35, since x is a continuous variable. According to the definition, whatever the function $f(x)$ may be, the delta function appearing in the integral picks out the value of $f(x)$ at the single point x'; the integral does not take account of the behaviour of $f(x)$ anywhere else. Evidently, this can be the case only if

$$\delta(x - x') = 0 \text{ for all } x \neq x'. \tag{3.37a}$$

At $x = x'$, the delta function *cannot be finite;* for a function which is finite at a single point and is zero everywhere else does not 'enclose any area' under it, and its integral must vanish. Actually, from Eq. 3.36, $\int \delta(x - x')\, dx = 1$. Therefore, we conclude that

$$\delta(x,x') = \infty \text{ when } x = x' \tag{3.37b}$$

The form of the Dirac delta function as given by Eqs. (3.37) is very peculiar indeed. Nevertheless, it is of great utility. For ease of visualization one may think of it as the limit of a suitable sequence of more ordinary functions $\delta_\varepsilon(x - x')$ as $\varepsilon \to 0$. One of the possibilities, illustrated in Fig. 3.1 is the following:

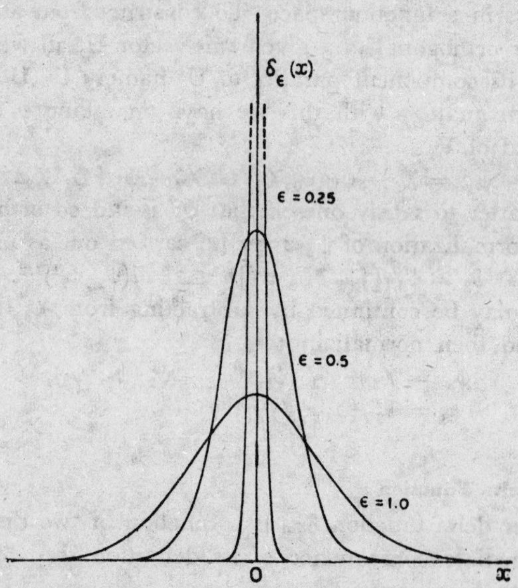

Fig. 3.1 The Dirac delta function as a limit of a sequence of functions $\delta_\varepsilon(x)$.

$$\delta_\varepsilon (x - x') = (2\pi\varepsilon^2)^{-1/2} \exp [-(x - x')^2/2\varepsilon^2]$$

In the limit $\varepsilon \to 0$ it satisfies Eqs. 3.37 and also the condition

$$\int \delta(x - x')\, dx = 1 \qquad (3.37c)$$

as required by Eq. 3.36. Other ways of representing the delta function, and its most important properties, are listed in Appendix C.

In our studies we will have to use frequently the three-dimensional Dirac delta function $\delta(\mathbf{x} - \mathbf{x}')$. It is defined by

$$\delta(\mathbf{x} - \mathbf{x}') = \delta(x - x')\ \delta(y - y')\ \delta(z - z') \qquad (3.38)$$

3.7 Observables: Completeness and Normalization of Eigenfunctions

Let us recall now the observation at the end of Sec. 3.2 that if a dynamical variable is to be considered as observable, the operator representing it must be self adjoint. A further requirement, the origin of which is more subtle, is that the eigenfunctions of the operator should form a complete set, in a sense now to be explained. Any dynamical variable represented by a self-adjoint operator having a complete set of eigenfunctions qualifies to be called an *observable*.

Let A be a self adjoint operator pertaining to some physical system. Its eigenfunctions $\{\phi_a\}$ are said to form a *complete set* if *any arbitrary wave function* ψ of the system can be 'expanded' into a linear combination,

$$\psi = \sum_{a \in D} c_a \phi_a + \int_{a \in C} c_a \phi_a \, da \qquad (3.39)$$

of the members of the set. The linear combination includes summation over the discrete part of the eigenvalue spectrum (indicated by $a \in D$, D for discrete), as well as integration over the continuous part. The assumption that a set $\{\phi_a\}$ is complete, i.e. the postulate that the expansion (3.39) is possible for arbitrary ψ, is often referred to as the *expansion postulate.*

Assuming the expansion (3.39), let us evaluate the norm of ψ (taken to be 1) in terms of the coefficients c_a. Consider first the case of an operator A whose eigenvalue spectrum is wholly discrete, so that the second (integral) term is not present in Eq. 3.39. We have, on using the orthonormality property, Eq. (3.34a), of the ϕ_a,

$$\begin{aligned} 1 &= \int \psi^* \psi \, d\tau = \int \Big(\sum_{a'} c_{a'}^* \phi_{a'}^*\Big) \Big(\sum_a c_a \phi_a\Big) d\tau \\ &= \sum_{a'} \sum_a c_{a'}^* c_a \int \phi_{a'}^* \phi_a \, d\tau \\ &= \sum_{a'} \sum_a c_{a'}^* c_a \, \delta(a', a) \\ &= \sum_a c_a^* c_a \, \delta(a, a) \end{aligned} \qquad (3.40)$$

In the last step, the sum over a' has been reduced to the single non-vanishing term, for which $a' = a$. Now, of the two possibilities corresponding to Eqs. (3.34b) and (3.34c), namely $\delta(a,a) = 1$ or ∞, we must obviously reject the latter, since that would make the above equation inconsistent. Hence, we conclude that: *Eigenfunctions belonging to discrete eigenvalues are normalizable.*

Setting $\delta(a,a') = \delta_{aa'}$ in the above equation, we obtain

$$\sum_a | c_a |^2 = 1 \tag{3.41}$$

If we had a *continuous* instead of a discrete spectrum for a, integrals would appear in the place of summations in Eq. 3.40 and the penultimate expression in Eq. 3.40 would become

$$1 = \int c_a da \int c_{a'}^* \, \delta(a,a') \, da' \tag{3.42}$$

The integral over a' here vanishes if $\delta(a', a)$ is taken to be the Kronecker delta, and of course this would make Eq. 3.42 inconsistent. Hence we have to take $\delta(a,a')$ as the Dirac delta function in this case, so that: *Eigenfunctions belonging to continuous eigenvalues are of infinite norm.*
Eq. 3.42 now simplifies to

$$\int | c_a |^2 \, da = 1 \tag{3.43}$$

More generally, if the spectrum of A has both discrete and continuous parts, we have

$$\sum_{a \in D} | c_a |^2 + \int_{a \in C} | c_a |^2 \, da = 1 \tag{3.44}$$

Hereafter, for economy of notation we will write out all delta functions, sums over eigenvalues, etc., as if the spectrum were discrete. For example, we abbreviate (3.39) and (3.34a) to

$$\psi = \sum_a c_a \phi_a \tag{3.45}$$

and

$$\int \phi_a^* \phi_{a'} \, d\tau = \delta_{aa'} \tag{3.46}$$

respectively.
Whenever the spectrum has a continuous part, the summation signs are to be understood as including integrations over the continuous part.

3.8 Closure

Any set of functions $\{\phi_a\}$ which is orthonormal and complete, i.e., for which Eq. (3.46) and the expansion postulate (3.45) are valid, has the important property of *closure:*

$$\sum_a \phi_a(\mathbf{x}) \, \phi_a^*(\mathbf{x}') = \delta(\mathbf{x} - \mathbf{x}') \tag{3.47}$$

The proof depends on the fact that the coefficients c_a in Eq. 3.45 can be expressed as

$$c_a = \int \phi_a^* \, \psi \, d\tau \tag{3.48}$$

This may be verified by evaluating the right hand side using Eqs. 3.45 and 3.46. On feeding this expression back into Eq. 3.45 we get

$$\psi(\mathbf{x}) = \sum_a c_a \phi_a(\mathbf{x}) = \sum_a \left[\int \phi_a^*(\mathbf{x}') \, \psi(\mathbf{x}') \, d\tau' \right] \phi_a(\mathbf{x})$$

$$= \int \left[\sum_a \phi_a(\mathbf{x}) \, \phi_a^*(\mathbf{x}') \right] \psi(\mathbf{x}') \, d\tau'$$

The last step is simply a matter of interchanging the order of integration and summation, to carry out the summation first. The above equation states that the value of the integral is just the value, at the particular point $\mathbf{x} = \mathbf{x}'$, of the function $\psi(\mathbf{x}')$ appearing in the integrand. Evidently, the function which multiplies $\psi(\mathbf{x}')$ must be the Dirac delta function. This is just what Eq. 3.47 asserts.

The converse of what we have just proved can also be shown easily. If the set $\{\phi_a\}$ obeys the closure relation (3.47) and is orthonormal, then it satisfies the expansion postulate. For, on writing

$$\psi(\mathbf{x}) = \int \psi(\mathbf{x}')\, \delta(\mathbf{x} - \mathbf{x}')\, d\tau'$$

and substituting from Eq. 3.47 for the delta function, we recover the expansion (3.45) with c_a given by Eq. 3.48.

3.9 Physical Interpretation of Eigenvalues, Eigenfunctions and Expansion Coefficients

To bring out the physical meaning of eigenvalues, etc., we consider the expectation value of an observable A in some arbitrary state ψ. On expanding ψ in terms of the eigenfunctions ϕ_a of the *same* observable A, we have

$$\begin{aligned} \langle A \rangle &= \int \psi^* A \psi d\tau \\ &= \int (\Sigma c_{a'}^* \phi_{a'}^*)\, A\, (\Sigma c_a \phi_a)\, d\tau \\ &= \Sigma\Sigma\, c_{a'}^* c_a \int \phi_{a'}^* A \phi_a\, d\tau \\ &= \Sigma\Sigma\, c_{a'}^* c_a\, a\, \delta_{a'a}. \end{aligned}$$

In the last step we have used the fact that $A\phi_a = a\phi_a$ and that the functions ϕ_a are orthonormal. Hence

$$\langle A \rangle = \sum_a |c_a|^2 a \tag{3.49}$$

This equation states that $\langle A \rangle$ is the *weighted average* of the eigenvalues a of A. The weight factors are the positive quantities $|c_a|^2$ whose sum, according to Eq. 3.41 is unity. It is evident that the meaning of these observations is the following:

The result of any measurement of A is one of its eigenvalues. The probability that a particular value a comes out as the answer, when the system is in the state ψ, is given by[6] $|c_a|^2$ with c_a given by Eq. 3.46; when repeated measurements of A are made on systems in the state ψ the number of times the answer a is obtained is expected to be proportional to $|c_a|^2$. We see thus that *the physical significance of the eigenvalues of any observable is that they are the possible results of measurement of the observable.* The significance of the eigenfunctions can also be seen now. Suppose ψ is itself chosen to be one of the eigenfunctions of A, say $\phi_{a'}$. Eq. 3.46 tells us that in this case $c_a = \int \phi_a^* \phi_{a'} d\tau = \delta_{aa'}$. Thus $c_a = 1$ for $a = a'$ and zero for all other a. Hence the probability $|c_a|^2$ for getting the answer a on measuring A is unity for $a = a'$. In other words, the answer is certain to be a'. Thus *the eigenfunction $\phi_{a'}$ of A represents a state in which the observable A has a definite value a'. Expressed differently, the uncertainty in the value of A is zero if the system is in one of the eigenstates of A.*

Returning to the interpretation of c_a as a probability amplitude and $|c_a|^2$ as a probability (or probability density if a is continuous), we observe that this is exactly the same kind of interpretation which was given to the wave

[6] If a is in the continuous part of the spectrum, then the probability that the value of A lies between a and $a + da$ is given by $|c_a|^2 da$. Then $|c_a|^2$ is the corresponding probability *density*. Whether a is in the continuous part or is a discrete eigenvalue, c_a itself is referred to as a *probability amplitude*.

function $\psi(\mathbf{x})$. The only difference is that instead of the position $\mathbf{x}$ we have the values a of the observable A. Therefore, considering c_a as a function of a, we can legitimately call it the 'A-space wave function' just as $\psi(\mathbf{x})$ was called the coordinate space or configuration space wave function. The function c_a contains just the same information about the system as $\psi(\mathbf{x})$ does, since we can get $\psi(\mathbf{x})$ from c_a through Eq. 3.45; so it may be used instead of $\psi(\mathbf{x})$ for describing the state. We say that c_a and $\psi(\mathbf{x})$ are different *representations* of the state. Representation theory will be considered in detail in Chapter 7.

3.10 Momentum Eigenfunctions; Wave Functions in Momentum Space

To illustrate the general properties proved in the last three sections we consider now the eigenvalue problem for the momentum operator. Consider first the one-dimensional case. The eigenvalue equation is

$$-i\hbar \frac{d\phi_{p'}(x)}{dx} = p'\phi_{p'}(x)$$

In view of the result proved in Example 3.9, p. 82, p' is the unique value of momentum which a particle has when it is in the eigenstate $\phi_{p'}(x)$. The solution of the equation is $\phi_{p'}(x) = c \exp{[ip'x/\hbar]}$, c being an arbitrary constant.

(*a*) *Self-adjointness and Reality of Eigenvalues:* We have seen in Example 3.8, p. 80, that $(-i\hbar\, d/dx)$ is self-adjoint only with respect to functions which either vanish as $x \to \pm\infty$ or obey periodic boundary conditions in a box. Neither of these would be satisfied by $\phi_{p'}(x)$ if p' had an imaginary part. (Verify). Thus reality of the eigenvalues p' is explicitly seen to follow as a consequence of self-adjointness.

(*b*) *Normalization and Closure:* When normalized within a (one-dimensional) box of length L, with periodic boundary conditions, the momentum eigenfunctions are given by

$$\phi_{p'}(x) = \frac{1}{\sqrt{L}}\, e^{ip'x/\hbar}, \quad p' = \frac{2\pi\hbar n}{L} \quad (n = 0, \pm 1, \pm 2 \ldots) \tag{3.50}$$

They are mutually orthogonal:

$$\int \phi_{p'}^* \phi_{p''}\, dx = \frac{1}{L}\int_{-L/2}^{L/2} e^{i(p''-p')x/\hbar}\, dx$$

$$= \frac{2\hbar \sin{[(p''-p')\, L/2\hbar]}}{(p''-p')\, L} \tag{3.51}$$

which vanishes when $p'' \neq p'$ since $(p''-p')\, L/2\hbar$ is an integral multiple of π. Further, the set of functions (3.50) possesses the closure property:

$$\sum_{p'} \phi_{p'}(x)\, \phi_{p'}^*(x') = \sum_{n=-\infty}^{\infty} \frac{1}{L} \exp\left[\frac{2\pi i n (x-x')}{L}\right]$$

$$= \lim_{N\to\infty} \sum_{n=-N}^{N} \frac{1}{L} \exp\left[\frac{2\pi i n (x-x')}{L}\right]$$

$$= \lim_{N\to\infty} \frac{1}{L} \cdot \frac{\sin [(2N+1)(x-x')\pi/L]}{\sin [(x-x')\pi/L]}$$

$$= \frac{\pi}{L} \delta\left(\frac{\pi}{L}(x-x')\right) = \delta(x-x') \qquad (3.52)$$

(The properties of delta functions which we have used here may be found in Appendix C.) Thus the eigenfunctions of the self-adjoint operator representing momentum form a complete set. This confirms that momentum is indeed an observable in quantum mechanics. Though we have considered above only the one-dimensional case explicitly, it is obvious that the same kind of arguments hold good in three dimensions, where the box-normalized eigenfunctions of momentum $\mathbf{p}'$ are just the functions (2.30) with $\mathbf{k} = (\mathbf{p}'/\hbar)$.

If we consider momentum eigenfunctions as defined in the whole of space rather than within a box, the eigenvalues no longer form a discrete set as in Eq. 3.50. Any p' from $-\infty$ to $+\infty$ (in three dimensions, any real vector $\mathbf{p}$) is an eigenvalue of momentum. Since the eigenvalue spectrum is thus continuous, the orthonormality relation of the eigenfunctions involves the Dirac (rather than Kronecker) delta function:

$$\int \phi_{\mathbf{p}'}^* (\mathbf{x})\, \phi_{\mathbf{p}''}(\mathbf{x})\, d\tau = \delta(\mathbf{p}' - \mathbf{p}'') \qquad (3.53)$$

The momentum eigenfunctions with this normalization are

$$\phi_{\mathbf{p}'}(\mathbf{x}) = \frac{1}{(2\pi\hbar)^{3/2}}\, e^{i\mathbf{p}'\cdot\mathbf{x}/\hbar} \qquad (3.54)$$

To verify this statement we substitute Eq. 3.54 in Eq. 3.53 and take the integrals with respect to x, y and z each from $-\frac{1}{2}L$ to $+\frac{1}{2}L$, and finally make $L \to \infty$. The triple integral evidently reduces to a product of three single integrals. A typical one is

$$\lim_{L\to\infty} \int_{-L/2}^{L/2} \left(\frac{1}{\sqrt{2\pi\hbar}} \cdot e^{-ip_x'x/\hbar} \cdot \frac{1}{\sqrt{2\pi\hbar}}\, e^{ip_x''x/\hbar}\right) dx$$

$$= \lim_{L\to\infty} \frac{1}{(2\pi\hbar)} \frac{2\hbar}{(p_x' - p_x'')} \cdot \sin\left(\frac{(p_x' - p_x'')L}{2\hbar}\right)$$

$$= \frac{1}{2\hbar}\, \delta\left(\frac{p_x' - p_x''}{2\hbar}\right) = \delta(p_x' - p_x'').$$

Similarly the y and z integrals yield $\delta(p_y' - p_y'')$ and $\delta(p_z' - p_z'')$. On taking the product of all these we obtain $\delta(\mathbf{p}' - \mathbf{p}'')$, thus confirming that the eigenfunctions of Eq. 3.54 are normalized according to Eq. 3.53. The closure property for these functions takes the form

$$\int \phi_{\mathbf{p}}(\mathbf{x})\, \phi_{\mathbf{p}'}^*(\mathbf{x}')\, d^3p' = \delta(\mathbf{x} - \mathbf{x}') \qquad (3.55)$$

It is usually the practice to write the momentum eigenfunctions in terms of the propagation vector $\mathbf{k}'$ as

$$\phi_{\mathbf{k}'}(\mathbf{x}) = \frac{1}{(2\pi)^{3/2}}\, e^{i\mathbf{k}'\cdot\mathbf{x}} \qquad (3.56)$$

to save writing factors $\hbar$. Here $\mathbf{k}' = \mathbf{p}'/\hbar$. Note that the normalization constant here differs from that in Eq. 3.54 by a factor $\hbar^{3/2}$. It may be easily verified that with this normalization

$$\int \phi_{\mathbf{k}'}^{*}(\mathbf{x})\, \phi_{\mathbf{k}''}(\mathbf{x})\, d\tau = \delta(\mathbf{k}' - \mathbf{k}'') \tag{3.57a}$$

$$\int \phi_{\mathbf{k}}(\mathbf{x})\, \phi_{\mathbf{k}}^{*}(\mathbf{x}')\, d^3k = \delta(\mathbf{x} - \mathbf{x}') \tag{3.57b}$$

(*c*) *The Wave Function and Operators in Momentum Space:* Since the momentum eigenfunctions form a complete orthonormal set, we can expand any arbitrary wave function $\psi(\mathbf{x})$ in terms of them as

$$\psi(\mathbf{x}) = \int c(\mathbf{p})\, \phi_{\mathbf{p}}(\mathbf{x})\, d^3p = \frac{1}{(2\pi\hbar)^{3/2}} \int c(\mathbf{p})\, e^{i\mathbf{p}.\mathbf{x}/\hbar}\, d^3p \tag{3.58a}$$

which is a special case of Eq. 3.45. Conversely, from Eq. 3.48 we get,

$$c(\mathbf{p}) = \int \phi_{\mathbf{p}}^{*}(\mathbf{x})\, \psi(\mathbf{x})\, d^3x = \frac{1}{(2\pi\hbar)^{3/2}} \int \psi(\mathbf{x})\, e^{-i\mathbf{p}.\mathbf{x}/\hbar}\, d^3x \tag{3.58b}$$

Equivalently, we have in terms of the propagation vector $\mathbf{k} = \mathbf{p}/\hbar$,

$$\psi(\mathbf{x}) = \frac{1}{(2\pi)^{3/2}} \int c(\mathbf{k})\, e^{i\mathbf{k}.\mathbf{x}}\, d^3k, \quad c(\mathbf{k}) = \frac{1}{(2\pi)^{3/2}} \int \psi(\mathbf{x})\, e^{-i\mathbf{k}.\mathbf{x}}\, d^3x \tag{3.59}$$

where

$$c(\mathbf{k}) = \hbar^{3/2}\, c(\mathbf{p}) \tag{3.60}$$

It is evident that $\psi(\mathbf{x})$ is the *Fourier transform* of $c(\mathbf{k})$.

The physical interpretation of $c(\mathbf{p})$ follows from the discussion of Sec. 3.9: $c(\mathbf{p})$ gives the probability amplitude, and $|c(\mathbf{p})|^2$ the probability density, for the momentum to be in the neighbourhood of $\mathbf{p}$. The expectation value of any function of momentum can, therefore, be expressed as

$$\langle f(\mathbf{p}) \rangle = \int |c(\mathbf{p})|^2 f(\mathbf{p})\, d^3p \tag{3.61}$$

As indicated at the end of Sec. 3.9, $c(\mathbf{p})$ may be called the *momentum space wave function;* it contains the same information as the configuration space wave function $\psi(\mathbf{x})$, in view of Eq. 3.58a, and is an equally valid representation of the state of the system. In momentum space, the dynamical variables would be represented by operators which act on $c(\mathbf{p})$. In particular, position and momentum are represented by

$$\mathbf{x}_{op} = i\hbar\, \nabla_p, \quad \mathbf{p}_{op} = \mathbf{p} \tag{3.62}$$

where $\nabla_p = (\partial/\partial p_x, \partial/\partial p_y, \partial/\partial p_z)$. In other words the momentum space wave functions corresponding to $\mathbf{x}\,\psi(\mathbf{x})$ and $-i\hbar\nabla\psi(\mathbf{x})$ are $i\hbar\nabla_{\mathbf{p}} c(\mathbf{p})$ and $\mathbf{p}\, c(\mathbf{p})$ respectively. This is easily verified by noting that on acting with $-i\hbar\nabla$ on Eq. 3.58a, $c(\mathbf{p})$ in the integral gets replaced by $\mathbf{p}\, c(\mathbf{p})$ and further that to replace $\psi(\mathbf{x})$ by $\mathbf{x}\psi(\mathbf{x})$ what one has to do is to operate on $c(\mathbf{p})$ with $i\hbar\nabla_p$ as is evident from Eq. 3.58b.

It is important to note that though the explicit forms of the operators in momentum space are necessarily different from those in configuration space, the commutation relations given by Eqs. 3.13 hold in both cases.

EXAMPLE 3.12 — What is the probability distribution of momentum of a particle with the Gaussian wave function $\psi(x) = (\sigma.\sqrt{\pi})^{-1/2}.\exp[-x^2/2\sigma^2]$? Specializing the above treatment to the one-dimensional case, we have

$$c(k) = \frac{1}{(2\pi)^{1/2}} \int \psi(x)\, e^{-ikx}\, dx.$$

This integral can be evaluated by expanding e^{-ikx} in series and using the fact that

$$\int_{-\infty}^{\infty} e^{-x^2/2\sigma^2} . x^{2s} \, dx = (2\pi)^{1/2} \frac{(2s)!}{s!} \frac{\sigma^{2s+1}}{2^s}, \quad (s = 0, 1, 2, \ldots)$$

(This may be inferred from the integral evaluated in Example 3.7, p. 79). Since integrals involving odd powers (x^{2s+1}) vanish, we have

$$\int_{-\infty}^{\infty} e^{-x^2/2\sigma^2} \, e^{-ikx} \, dx = (2\pi)^{1/2} \sum_{s=0}^{\infty} \frac{(2s)!}{s!} \frac{\sigma^{2s+1}}{2^s} \frac{(-ik)^{2s}}{(2s)!} = (2\pi)^{1/2} \sigma e^{-\frac{1}{2}\sigma^2 k^2},$$

and hence

$$c(k) = (\sigma/\sqrt{\pi})^{1/2} \exp\left[-\tfrac{1}{2}\sigma^2 k^2\right].$$

The probability density function for momentum is $|c(p)|^2 = \hbar^{-1} |c(k)|^2$. It is noteworthy that the Gaussian wave function has a Gaussian form in momentum space too. (Compare the widths of the wave packets $\psi(x)$ and $c(p)$). The relation $\psi(x) = (2\pi)^{-1/2} \int c(k) e^{ikx} \, dk$ now becomes

$$\left(\frac{1}{\sigma\sqrt{\pi}}\right)^{1/2} e^{-x^2/2\sigma^2} = \frac{1}{(2\pi)^{1/2}} \left(\frac{\sigma}{\sqrt{\pi}}\right)^{1/2} \int e^{-\frac{1}{2}\sigma^2 k^2} e^{ikx} \, dk$$

EXAMPLE 3.13 — As an example in three-dimensions, consider the ground state wave function of the hydrogen atom (See Sec. 4.16):

$$\psi(\mathbf{x}) = \left(\frac{1}{\pi a_0^3}\right)^{1/2} e^{-r/a_0}$$

It depends only on the radial distance r. The constant a_0 is the radius of the first Bohr orbit. In momentum space, we have

$$c(\mathbf{p}) = \left(\frac{1}{2\pi\hbar}\right)^{3/2} \left(\frac{1}{\pi a_0^3}\right)^{1/2} \int e^{-r/a_0} . e^{-i\mathbf{p}.\mathbf{x}/\hbar} \, d\tau.$$

It is advantageous to transform to spherical polar coordinates, choosing the polar axis along the direction of $\mathbf{p}$. Then $\mathbf{p}.\mathbf{x} = pr\cos\theta$, and the integrand is independent of the azimuthal angle φ. We thus have

$$c(\mathbf{p}) = (8\pi^4 \hbar^3 a_0^3)^{-1/2} \int_0^{\infty}\int_0^{\pi}\int_0^{2\pi} \exp\left[-\frac{r}{a_0} - \frac{ipr\cos\theta}{\hbar}\right] d\varphi \sin\theta \, d\theta . r^2 \, dr.$$

On carrying out the integrations we get

$$c(\mathbf{p}) = \frac{1}{\pi} \cdot \left(\frac{2a_0}{\hbar}\right)^{3/2} \cdot \frac{1}{[(p^2 a_0^2/\hbar^2) + 1]^2} \cdot$$

The probability that the magnitude and direction lie within $(p, p + dp)$, $(\alpha, \alpha + d\alpha)$, $(\beta, \beta + d\beta)$ is *not* $|c(\mathbf{p})|^2 \, dp \, d\alpha \, d\beta$ but $|c(\mathbf{p})|^2 \, p^2 dp \sin\alpha \, d\alpha \, d\beta$. If the probability distribution of only one of the variables is of interest, integrate over the others. For instance, in the present case where $c(\mathbf{p})$ is independent of α, β we get $P(p)\,dp = |c(\mathbf{p})|^2 \, 4\pi p^2 \, dp$.

3.11 The Uncertainty Principle

Whenever we have some quantity which can assume several values with various probabilities, it is customary to use the mean squared deviation as a measure of the 'width' of the probability distribution or in other words, the uncertainty in the value of the quantity. The uncertainty in the values of quantum mechanical observables also is defined in the same way. If A is an observable and $\langle A \rangle$ its expectation or mean value in the state ψ,

$$\mathcal{A} \equiv A - \langle A \rangle \tag{3.63}$$

is the self-adjoint operator representing the deviation of A from its mean. The expectation value of $\mathcal{A}^2$ is the mean squared deviation; and the uncertainty in A (usually denoted by ΔA) is defined by

$$(\Delta A)^2 = \langle \mathcal{A}^2 \rangle \tag{3.64a}$$

$$\equiv \langle (A - \langle A \rangle)^2 \rangle = \langle A^2 \rangle - \langle A \rangle^2 \tag{3.64b}$$

The uncertainty depends on the state ψ in which the system is.
We can always find states in which ΔA vanishes; in fact we know that *the eigenfunctions ϕ_a of A represent such states* (See Sec. 3.9).

Now we ask: Given *two* observables A, B, can we find some state in which *both* ΔA and ΔB are zero, so that both A and B have precise values? We already know that at least in one case (position and momentum) the answer is in the negative. To study this question in more general terms we make essential use of the positivity property (3.26). We apply it to the operator $\mathcal{A} - i\lambda\mathcal{B}$ where $\mathcal{A}$ is defined by Eq. 3.63; $\mathcal{B}$ is defined similarly as $\mathcal{B} = B - \langle B \rangle$ and λ is a real parameter (not an operator). Noting that the adjoint of $(\mathcal{A} - i\lambda\mathcal{B})$ is $(\mathcal{A} + i\lambda\mathcal{B})$ since $\mathcal{A}$ and $\mathcal{B}$ are self-adjoint, we have from Eq. 3.26a

$$\langle (\mathcal{A} - i\lambda\mathcal{B})(\mathcal{A} + i\lambda\mathcal{B}) \rangle \geqslant 0 \tag{3.65}$$

Expanding the product, keeping in mind that $\mathcal{A}$ and $\mathcal{B}$ may not commute, we simplify the expression in Eq. 3.65:

$$\langle \mathcal{A}^2 \rangle + \lambda^2 \langle \mathcal{B}^2 \rangle - \lambda \langle C \rangle \geqslant 0 \tag{3.66}$$

Here C is the self-adjoint operator defined by

$$iC = [\mathcal{A}, \mathcal{B}] = [A, B] \tag{3.67}$$

The left hand side of (3.66) has its minimum value when its derivative with respect to λ is zero. This happens when

$$\lambda = \frac{\langle C \rangle}{2 \langle \mathcal{B}^2 \rangle} \tag{3.68}$$

For this value of λ Eq. 3.66 becomes

$$\langle \mathcal{A}^2 \rangle - \frac{\langle C \rangle^2}{4 \langle \mathcal{B}^2 \rangle} \geqslant 0 \tag{3.69}$$

Hence we have $\langle \mathcal{A}^2 \rangle \langle \mathcal{B}^2 \rangle \geqslant \frac{1}{4} \langle C^2 \rangle$ or, in view of the definition (3.64a) of the uncertainty, and Eq. 3.67 for C,

$$(\Delta A)^2 (\Delta B)^2 \geqslant -\tfrac{1}{4} \langle [A, B] \rangle^2 \tag{3.70}$$

Thus a lower limit on the product of the uncertainties in A and B is set by the minimum value possible for the expectation value of the commutator $[A, B]$. Eq. 3.70 *gives the general statement of the uncertainty principle for any pair of observables A, B.*

In particular, if A, B are a *canonically conjugate* pair of operators, which are characterized by

$$[A, B] = i\hbar \tag{3.71}$$

then

$$(\Delta A)^2 (\Delta B)^2 \geqslant \tfrac{1}{4}\hbar^2, \text{ or } (\Delta A)(\Delta B) \geqslant \tfrac{1}{2}\hbar \tag{3.72}$$

Any position coordinate variable x and its canonically conjugate momentum variable p of course satisfy Eq. 3.71 and, therefore, we have

$$(\Delta x)(\Delta p) \geqslant \tfrac{1}{2}\hbar \tag{3.73}$$

This is the precise statement of the position-momentum uncertainty principle. In three dimensions,

$$(\Delta x)(\Delta p_x) \geqslant \tfrac{1}{2}\hbar, \quad (\Delta y)(\Delta p_y) \geqslant \tfrac{1}{2}\hbar, \quad (\Delta z)(\Delta p_z) \geqslant \tfrac{1}{2}\hbar \tag{3.74}$$

It may be recalled that a qualitative version of it was deduced from heuristic arguments in Sec. 1.14.

We have seen in Example 3.6 (p. 77) that the azimuthal angle φ is not a proper dynamical variable[7] in quantum mechanics. Hence, even though a commutation rule $[\varphi, L_z] = i\hbar$ of the form (3.71) can be formally written down, we cannot talk of an uncertainty relation $(\Delta\varphi)(\Delta L_z) \geqslant \frac{1}{2}\hbar$ since $\Delta\varphi$ itself cannot be properly defined.

As far as the energy and time variables are concerned, one cannot talk of a Δt at a single instant of time, nor is the operator $i\hbar\,(\partial/\partial t)$, associated with E by Eq. 2.14, definable on an instantaneous wave function. So, even though we do have $[i\hbar\,\partial/\partial t, t] = i\hbar$, it cannot be employed in the above analysis which is based on expectation values with respect to wave functions at a single instant. However, a relation $(\Delta E)(\Delta t) \geqslant \hbar$ does exist. Its interpretation, which is quite different from that of Eq. 3.74 will be given elsewhere.

EXAMPLE 3.14 — As an example involving a pair of observables which are *not* canonically conjugate, let us consider the following question: Does there exist a state ϕ in which $\Delta L_x = \Delta L_y = 0$, where **L** is the angular momentum vector? This question is equivalent to asking whether there is any state ϕ which is an eigenstate of both L_x and L_y: $L_x\phi = m_x\phi$, $L_y\phi = m_y\phi$ where m_x, m_y are constants. If such a state exists, $L_xL_y\phi = L_xm_y\phi = m_xm_y\phi = L_yL_x\phi$. Using this result and the commutation relation $[L_x, L_y] = i\hbar L_z$ proved in Example 3.4 (p. 77), one finds $L_z\phi = (i\hbar)^{-1}(L_xL_y\phi - L_yL_x\phi) = 0$. So ϕ must belong to the zero eigenvalue of L_z. Repeating the above arguments, now starting with $L_y\phi = m_y\phi$ and $L_z\phi = 0$, one finds $m_x = 0$ and similarly $m_y = 0$. Thus if more than one component of angular momentum is to have a definite value in a state ϕ, it must satisfy $L_x\phi = L_y\phi = L_z\phi = 0$. Such a state does exist, as will be seen in Sec. 4.10.

3.12 States with Minimum Value for Uncertainty Product

Whether the equality or inequality holds in (3.70) depends on the state ψ in which the system is. The minimum value of $(\Delta A)(\Delta B)$ occurs evidently

[7] For a review on phase variables and operators to be associated with them in quantum mechanics, *see* P. Carruthers and M. M. Nieto, *Rev. Mod. Phys.*, **40**, 411, 1968.

in those states for which the equality sign holds in (3.70) and hence in (3.65) with λ given by Eq. 3.68. But Eq. 3.26a applied to the present case shows that the left hand side of (3.65) can vanish only in states ψ for which $(\mathcal{A} + i\lambda\mathcal{B})\,\psi$ vanishes, i.e.

$$\left(\mathcal{A} + \frac{i\langle \mathcal{C}\rangle}{2\langle \mathcal{B}^2\rangle}\cdot \mathcal{B}\right)\psi = 0 \tag{3.75}$$

The case of greatest interest is when A and B are canonically conjugate position and momentum variables x and p. Then

$$\mathcal{A} = x - \langle x\rangle,\ \mathcal{B} = p - \langle p\rangle,\ \langle \mathcal{B}^2\rangle = (\Delta p)^2,$$
$$\mathcal{C} = -i\,[A, B] = \hbar \tag{3.76}$$

Using the explicit differential operator form for B, we rewrite Eq. 3.75 in this case as

$$(x - \langle x\rangle)\,\psi + \frac{i\hbar}{2(\Delta p)^2}\left(-i\hbar\frac{d}{dx} - \langle p\rangle\right)\psi = 0$$

or

$$\frac{d\psi}{dx} + \left\{\frac{2(\Delta p)^2}{\hbar^2}(x - \langle x\rangle) - \frac{i\langle p\rangle}{\hbar}\right\}\psi = 0 \tag{3.77}$$

It is a trivial matter to verify that the solution of this equation is

$$\psi = \mathcal{N}\exp\left[-\frac{(\Delta p)^2}{\hbar^2}(x - \langle x\rangle)^2 + \frac{i\langle p\rangle x}{\hbar}\right] \tag{3.78a}$$

The wave function is normalized, i.e. $\int |\psi|^2\,dx = 1$ if $\mathcal{N}$ is chosen as

$$\mathcal{N} = \left[\frac{2(\Delta p)^2}{\pi\hbar^2}\right]^{1/4} \tag{3.78b}$$

Eqs. 3.78 give the normalized wave functions for which $(\Delta x)\,(\Delta p)$ has the minimum value $\frac{1}{2}\hbar$. Note that ψ has the form of a Gaussian function 'modulated' by the oscillatory factor $\exp\,[i\langle p\rangle x/\hbar]$. The latter, being of unit modulus, does not affect the probability density $|\psi|^2$ which is Gaussian with maximum at $x = \langle x\rangle$:

$$|\psi|^2 = \left[2\pi(\Delta x)^2\right]^{-\frac{1}{2}}.\exp\left[-\frac{(x - \langle x\rangle)^2}{2(\Delta x)^2}\right] \tag{3.79}$$

(The normalization may be checked by comparison of this expression with Example 3.7 p. 79.) Since $|\psi|^2$ is negligibly small outside a region having dimensions of the order $\Delta x = (\frac{1}{2}\hbar/\Delta p)$, the wave function (3.78a) is said to describe a minimum uncertainty *wave 'packet'*.

3.13 Commuting Observables; Removal of Degeneracy

In the case of two commuting observables A, B, the right hand side of the general uncertainty relation (3.70) vanishes, so that states in which ΔA and ΔB are *both* zero can exist. In other words, there exist states in which both A and B have precise values and which are, therefore, eigenstates of A and B simultaneously.

Consider an eigenstate ϕ_a of A. When $AB = BA$, we have

$$AB\phi_a = BA\phi_a = B.a\phi_a,$$

i.e.

$$A\,(B\phi_a) = a\,(B\phi_a) \tag{3.80}$$

Thus not only ϕ_a but $B\phi_a$ also is an eigenstate of A belonging to the *same* eigenvalue a. If a happens to be a non-degenerate eigenvalue, there is (by definition) only one eigenfunction belonging to it and hence $B\phi_a$ must be a constant multiple of ϕ_a, say

$$B\phi_a = b\phi_a \tag{3.81}$$

But this means that ϕ_a is also an eigenfunction of B, belonging to the eigenvalue b. This information may be included in the notation for the eigenfunction by renaming it as ϕ_{ab}.

Thus any eigenfunction belonging to a *non-degenerate* eigenvalue of either of a pair of commuting operators A, B is necessarily an eigenfunction of the other operator too.

What if a happens to be a degenerate eigenvalue? Then if ϕ_a is an *arbitrary* one of the (infinite number of) eigenfunctions belonging to a, all we can say from Eq. 3.80 is that $B\phi_a$ is also *one of the* eigenfunctions belonging to a. But it is always possible to choose a basic set of r eigenfunctions (where r is the degree of degeneracy of the eigenvalue a) in such a way that each of them is an eigenfunction of B too.[8] In this manner one can obtain a *complete set of simultaneous eigenfunctions* ϕ_{ab} for any pair of commuting observables A, B.

It may be observed that the label b serves to distinguish between different eigenfunctions belonging to any degenerate eigenvalue a of A. If the r independent eigenfunctions belonging to a given degenerate eigenvalue a are characterized by r *distinct* values of b, then we say that the degeneracy of a is *completely removed*. This is because for that value of a, by specifying a particular value of b a particular eigenfunction is uniquely identified.

Suppose now that the degeneracy of every eigenvalue a is not removed completely by the introduction of the additional label b. Then one can introduce one or more further observables $C, \ldots$ which commute with one another and with A and B, till the set of eigenvalues $a, b, c, \ldots$ identify the state $\phi_{abc\ldots}$ unambiguously. That is to say, $\phi_{abc\ldots}$ and $\phi_{a'b'c'\ldots}$ are identical if and only if $a = a'$, $b = b'$, $c = c', \ldots$ When this is accomplished, we say that $A, B, C, \ldots$ form a *complete commuting set of observables*. It is obvious that if each $\phi_{abc\ldots}$ is individually normalized, the set of all such simultaneous eigenfunctions forms an orthonormal set:

$$\int \phi^*_{abc\ldots} \phi_{a'b'c'\ldots} d\tau = \delta_{aa'} \delta_{bb'} \delta_{cc'} \ldots \tag{3.82}$$

For, if any two similar indices, e.g. b, b' are unequal, it means that the two wave functions belong to *distinct* eigenvalues of the corresponding operator B, and must, therefore, be orthogonal as shown in Sec. 3.5.

Since no degeneracy exists when all the eigenvalues $a, b, c, \ldots$ are specified, the considerations of Sec. 3.4-3.8 can be made completely general by using the $\phi_{abc\ldots}$ in the place of the ϕ_a. Apart from the appearance of a number of indices and consequent changes in wording of some of the statements, no further complication arises. The completeness postulate, for instance, becomes

[8] The proof of this statement will not be given at this stage since it involves the use of matrix concepts to be introduced later (Sec. 7.6).

$\psi = \Sigma_{abc\ldots} c_{abc\ldots} \phi_{abc\ldots}$, and $|c_{abc\ldots}|^2$ is now the *joint* probability[9] that the observables $A, B, C, \ldots$ have the respective values $a, b, c, \ldots$ etc.

3.14 Evolution of System with Time; Constants of the Motion

The consideration of general properties of quantum systems upto this point have been concerned with various aspects of the description at any given instant of time. The basic picture of the changes which take place as time progresses is given, in the Schrödinger form of quantum mechanics, by the following postulate.

POSTULATE 4 — The state ψ varies with time in a manner determined by the Schrödinger equation

$$i\hbar \frac{\partial \psi}{\partial t} = H_{op}\, \psi \tag{3.83}$$

where H_{op} is the Hamiltonian operator.[10] The basic dynamical variables $\mathbf{x}$ and $\mathbf{p}$ (or more precisely, the operators representing them) do not change with time.

However, there is nothing to prevent us from considering operators $A_{op}(\mathbf{x}, \mathbf{p}; t)$ which are explicitly time dependent. In such cases

$$\frac{d}{dt} A_{op}\,(\mathbf{x}, \mathbf{p}; t) = \frac{\partial}{\partial t} A_{op}\,(\mathbf{x}, \mathbf{p}; t) \tag{3.84}$$

i.e. the change in A_{op} is solely due to the explicit time variable since $\mathbf{x}_{op}$ and $\mathbf{p}_{op}$ are time independent.

Let us now see how expectation values change, according to the above postulate. We have

$$\begin{aligned} \frac{d}{dt} \langle A\,(\mathbf{x}, \mathbf{p}; t) \rangle &= \frac{d}{dt} \int \psi^* A_{op}\psi d\tau \\ &= \int \left[\frac{\partial \psi^*}{\partial t} A_{op}\, \psi + \psi^* \left(\frac{\partial A_{op}}{\partial t} \right) \psi + \psi^* A_{op} \frac{\partial \psi}{\partial t} \right] d\tau \\ &= \int \left[-\frac{1}{i\hbar} (H_{op}\, \psi)^* A_{op}\psi + \psi^* \frac{\partial A_{op}}{\partial t} \cdot \psi + \frac{1}{i\hbar} \psi^* A_{op} H_{op}\psi \right] d\tau \end{aligned} \tag{3.85}$$

In the last step we have used the Schrödinger equation and its conjugate, $-i\hbar\, \partial\psi^*/\partial t = (H_{op}\psi)^*$. Now, since the energy is an observable, H_{op} is a self adjoint operator. This fact enables us to reduce the above equation to

$$\frac{d\langle A \rangle}{dt} = \int \psi^* \left\{ \frac{1}{i\hbar} [A_{op}, H_{op}] + \frac{\partial A_{op}}{\partial t} \right\} \psi\, d\tau \tag{3.86}$$

Thus the rate of change of the *expectation value* of any dynamical variable A may be obtained as the expectation value of $(i\hbar)^{-1} [A, H] + \partial A/\partial t$. This operator may be taken to represent the dynamical variable (dA/dt):

$$\left(\frac{dA}{dt} \right)_{op} = \frac{1}{i\hbar} [A_{op}, H_{op}] + \frac{\partial A_{op}}{\partial t} \tag{3.87}$$

[9] Or joint probability *density*, if any of the indices belongs to a continuum.
[10] The subscript 'op' is temporarily revived here, because the distinction between the dynamical variable as a physical concept on the one hand and the operator representing it on the other hand, is important for the following discussion.

It is important to note the distinction between this operator for (dA/dt), and the rate of change of A_{op} which is given by Eq. 3.84.

It may be observed in particular that Eq. 3.85 contains the statement that the norm of ψ is conserved. For, the norm is nothing but the expectation value of unity, and its time derivative is trivially seen to be zero by substituting $A = 1$ in Eq. 3.86. We have used only the hermiticity (self adjointness) of the Hamiltonian in the proof of this result. The hermiticity of H is thus linked to conservation of probability.

Finally we observe that if A is any dynamical variable which is not explicitly time-dependent $(dA_{op}/dt = 0)$ *and it commutes with* H, then $\langle A \rangle$ is independent of time, whatever ψ may be. Any such A is said to be a *conserved quantity* or *constant of the motion.* Whenever it is desired to choose a complete commuting set of observables (See Sec. 3.13) for any system, it is standard practice to take H as one of them, so that all the observables of the set will be constants of the motion because of their commutability with H. Thus if at some instant of time a system is in a state wherein these observables have specific values, i.e. in a simultaneous eigenstate of the observables, it remains in the same state for all time. The eigenvalues or related quantum numbers which label the state are *good quantum numbers* for the system, since the values of the labels of such a state remain good (unaltered) for all time.

Example 3.15—It has been noted in Example 3.6, (p. 77) that the operator for L_z in polar coordinates is $L_z = -i\hbar\partial/\partial\varphi$. Therefore, L_z commutes with any operator which does not depend on φ. In particular it commutes with ∇^2 which is independent of φ as may be seen from its polar coordinate form (See Example 3.1, p. 71). Therefore L_z is a constant of the motion for a free particle, whose Hamiltonian is $-(\hbar^2/2m)\nabla^2$. In fact, it is a constant of the motion even if a potential V is present, provided V is independent of φ, i.e. symmetric with respect to the z axis. Similarly L_x, L_y are constants of the motion if V is symmetric about the x and y axes respectively. If the potential is spherically symmetric, depending on r only and not on θ or φ then all components of the angular momentum are conserved.

3.15 Non-interacting and Interacting Systems

Suppose we have two systems, each with its own set of basic dynamical variables (coordinates and conjugate momenta). The two 'systems' may, for example, be two individual particles, or a particle and an atom, or two atoms etc. Let us use the symbol '1' to denote collectively the dynamical variables of the first system, and '2' for the other. Any of the variables 1 will commute with the variable 2, because they pertain to different systems. The Hamiltonian of the *combined* system will depend on both the sets of variables, and we denote it by $H(1, 2)$. If it can be expressed as a sum of two parts

$$H(1, 2) = H_1(1) + H_2(2), \tag{3.88}$$

where either part depends on the variables of one system only, as indicated, then the two systems are *non-interacting;* they do not influence each other.

This is evident from the fact that if system 1 is in an eigenstate $u(1)$ of $H_1(1)$ and system 2 is in an eigenstate $v(2)$ of $H_2(2)$ at some instant of time, this state of affairs persists for all time. For $H_1(1)$ and $H_2(2)$ commute with each other and hence with $H(1, 2)$, and so their eigenvalues are constants of the motion. Note that if

$$H_1(1)\, u(1) = E'u(1) \text{ and } H_2(2)\, v(2) = E''v(2), \tag{3.89}$$

then

$$\begin{aligned} H(1,2)\, u(1)\, v(2) &= [H_1(1)\, u(1)]\, v(2) + u(1)\, [H_2(2)\, v(2)] \\ &= E\, u(1)\, v(2) \end{aligned} \tag{3.90}$$

with $E = E' + E''$.

Thus the eigenfunctions of the combined system consisting of two non-interacting subsystems are *products* of the eigenfunctions of the individual subsystems, while the energy eigenvalue E is a *sum* of E' and E''. It may be observed that operators of one system have no effect on operators or wave functions of the other. Therefore in products, we can freely change the order of occurrence of the quantities of one system relative to the other.

If the systems 1 and 2 interact mutually, then $H(1,2)$ cannot be separated as before; the presence of the interaction is usually manifested as an additional term $H'(1,2)$ depending on the variables of both:

$$H(1, 2) = H_1(1) + H_2(2) + H'(1, 2) \tag{3.91}$$

Then one refers to $H'(1, 2)$ as the *interaction part* of the Hamiltonian, and $H_1(1) + H_2(2)$ as the '*free*' part. When H' is present, one no longer has eigenstates of $H(1, 2)$ as simple products $u(1)\, v(2)$. Viewed another way, if the system at some instant of time has the wave function $u(1)\, v(2)$ which is an eigenfunction of the 'free' part of $H(1, 2)$, at a later time the wave function will not be an eigenfunction of the 'free' part but a superposition of several such eigenfunctions. The determination of the manner in which the superposition develops as a consequence of interaction is one of the major problems in the application of quantum mechanics.

3.16 Systems of Identical Particles

Finally, we establish certain important properties of systems consisting of identical particles, and incidentally, provide an illustration of operators which are not definable as functions of $\mathbf{x}$ and $\mathbf{p}$.

(*a*) *Interchange of particles; Symmetric and antisymmetric wave functions:* By definition, if two particles are *identical*, no observable effects whatever can arise from interchanging them. More precisely, *all observable quantities* must remain unaltered if the position, momentum and other dynamical variables such as spin of the first particle (which we collectively denote by 1) are interchanged with those of the second particle (collectively denoted by 2). In particular, this should be true of the Hamiltonian:

$$H(1, 2) = H(2, 1) \tag{3.92}$$

Now let us consider the *exchange operator* $\mathcal{P}$ which is *defined* by saying that it has the following effect on an arbitrary wave function $\psi(1, 2)$:

$$\mathcal{P}\psi(1, 2) \equiv \psi(2, 1) \tag{3.93}$$

By operating with $\mathcal{P}$ on this equation, we obtain

$$\mathcal{P}^2\psi(1,2) = \mathcal{P}.\mathcal{P}\psi(1,2) = \mathcal{P}\psi(2,1) = \psi(1,2) \tag{3.94}$$

so that

$$\mathcal{P}^2 = 1 \tag{3.95}$$

Taking ψ in particular to be an eigenfunction, ϕ, of $\mathcal{P}$ we have $\mathcal{P}^2\phi = \phi$ from Eq. 3.94 and further, from $\mathcal{P}\phi = \lambda\phi$, $\mathcal{P}^2\phi = \lambda^2\phi$. Hence $\lambda^2 = 1$. Thus the eigenvalues of $\mathcal{P}$ are $+1$ and -1; denoting the types of eigenfunction belonging to these eigenvalues by ϕ_+ and ϕ_- respectively, we have

$$\mathcal{P}\phi_+(1,2) = \phi_+(1,2); \; \mathcal{P}\phi_-(1,2) = -\phi_-(1,2) \tag{3.96}$$

In view of the definition of $\mathcal{P}$, these lead to

$$\phi_+(2,1) = \phi_+(1,2); \; \phi_-(2,1) = -\phi_-(1,2) \tag{3.97}$$

so that ϕ_+ and ϕ_- are symmetric and antisymmetric respectively under interchange of the particles:

Now, applying $\mathcal{P}$ to $H(1,2)\,\psi(1,2)$ for arbitrary ψ, we obtain

$$\begin{aligned}\mathcal{P}H(1,2)\,\psi(1,2) &= H(2,1)\,\psi(2,1) = H(1,2)\,\psi(2,1)\\ &= H(1,2)\,\mathcal{P}\psi(1,2)\end{aligned} \tag{3.98}$$

where we have used the identity of particles, Eq. 3.92, in the penultimate step. Since Eq. 3.98 holds for arbitrary ψ, we have

$$\mathcal{P}H(1,2) = H(1,2)\mathcal{P} \tag{3.99}$$

Thus $\mathcal{P}$ commutes with $H(1,2)$ and hence by the results of Sec. 3.14, represents a conserved quantity. Its eigenvalue is a good quantum number, i.e. if the wave function of the system at one instant is even (or odd) under interchange of particles, in the sense of Eq. 3.97, it will remain even (or odd) for all time. This can be seen directly from the Schrödinger equation $i\hbar\,\partial\psi(1,2)/\partial t = H\psi(1,2)$, which shows that $\partial\psi/\partial t$ is even or odd according as ψ is even or odd (since H itself is even). By repeatedly differentiating the equation with respect to time we see that $\partial^2\psi/\partial t^2$ and all higher derivatives have the same property. Hence the evenness or oddness of ψ is preserved always.

These considerations can be directly extended to a system of N identical particles ($N > 2$) by defining an exchange operator $\mathcal{P}_{\alpha\beta}$ for each pair of particles (α, β). What has been proved above in regard to $\mathcal{P}$ holds also for each $\mathcal{P}_{\alpha\beta}$. In particular, $\mathcal{P}_{\alpha\beta}\,H(1, 2, \ldots, \ldots, \alpha, \ldots, \beta, \ldots, N) = H(1, 2, \ldots, \alpha, \ldots, \beta, \ldots, N)\,\mathcal{P}_{\alpha\beta}$. Since $\mathcal{P}_{\alpha\beta}$ and H thus commute, any eigenfunction ϕ of H belonging to a non-degenerate eigenvalue must, according to Sec. 3.13, be also an eigenfunction of $\mathcal{P}_{\alpha\beta}$, i.e. $\phi(1, 2, \ldots, \alpha, \ldots \beta, \ldots, N) = \lambda\phi(1, 2, \ldots \beta \ldots \alpha \ldots N)$, $\lambda = \pm 1$. We have written the eigenvalue simply as λ instead of $\lambda_{\alpha\beta}$ because it is actually independent of (α, β). This fact is easily deduced from the observation that $\mathcal{P}_{\gamma\beta}\,\mathcal{P}_{\alpha\gamma}\,\mathcal{P}_{\alpha\beta} = \mathcal{P}_{\alpha\gamma}$. Operating with both sides of this operator equation on ϕ, one gets $\lambda_{\gamma\beta}\,\lambda_{\alpha\gamma}\,\lambda_{\alpha\beta} = \lambda_{\alpha\gamma}$, so that $\lambda_{\gamma\beta}\,\lambda_{\alpha\beta} = 1$. Hence $\lambda_{\gamma\beta}$ and $\lambda_{\alpha\beta}$ must *both* be $+1$ or *both* -1 (for arbitrary α, γ) showing that $\lambda_{\alpha\beta}$ is independent of α. The reader will have no difficulty convincing himself that it is independent of β too. We conclude thus that the wave function ϕ must belong to one of two types: the *totally symmetric* (even) type which remains unchanged ($\lambda = +1$) under any interchange of particles, or the *totally antisymmetric* (odd) type which changes sign ($\lambda = -1$) with the interchange of any pair of particles.[11]

We now observe that wave functions of different symmetry types *cannot be connected* by any observable A of a system of identical particles, i.e. if ϕ is a totally symmetric/antisymmetric wavefunction, $A\phi$ is also totally symmetric/antisymmetric respectively. (This is because any observable must necessarily be a symmetric function of the variables of the particles which are indistinguishable.) Thus all the states of a given system, which are connected to one another by some observable or other, must belong to the same symmetry type. Since, as we have already seen, oddness or evenness does not change with time, it remains a permanent characteristic of the system. This property of quantum systems has no classical analogue.

It is a fact of observation that elementary particles found in nature do fall into two classes—those which always have totally antisymmetric wave functions, and those which have to be described by totally symmetric wave functions. We show below that particles which belong to the first category obey Fermi-Dirac statistics (and are hence referred to as *fermions*) while the others obey Bose-Einstein statistics (and are called *bosons*). Examples of fermions are electrons, protons and neutrons — all of which have a half-unit ($\frac{1}{2}\hbar$) of spin angular momentum — and any other particles with a half-integral value ($\frac{1}{2}, \frac{3}{2}, \ldots$) of spin.When all the dynamical variables *including the spin variable* of two electrons (or two protons...) are interchanged, the wave function of the whole system (which may include other kinds of particles also) changes sign. Particles which have no spin (e.g. pi mesons) or have an integer value of spin angular momentum (in units of $\hbar$) are found to be bosons. Photons or light quanta (insofar as they can be regarded as particles) belong to this category. Interchange of any two identical particles of this kind leaves the wave function of the whole system unchanged. The fact that half-integral-spin particles are fermions and integral spin particles are bosons is the so called *spin-statistics connection*, for which there are fundamental theoretical reasons. But these are beyond the scope of this book.

(*b*) *Relation between type of symmetry and statistics; The exclusion principle:* We have seen above that if a wave function $\psi(1, 2)$ is to describe two identical particles it should not only satisfy the general admissibility conditions of Sec. 2.8, but also the further condition $\psi(1, 2) = \pm\psi(2, 1)$ where the sign depends on the kind of particle considered. Given some wave function $\phi(1, 2)$, we can construct ϕ_A or ϕ_S having the above property by symmetrizing

[11] The arguments leading to this conclusion are clearly unaffected even if other particles (besides the identical particles on which we are focusing attention) are present in the system. The symmetry property of the wave function exists, however only with respect to interchanges of identical particles.

It is theoretically possible to conceive of identical particle systems which possess no non-degenerate energy levels and no complete set of commuting observables. [A.M.L. Messiah and O.W. Greenberg, *Phys. Rev.*, **136** *B*, 248 (1964)]. Such systems would have sets of wave functions which go over into combinations of themselves under interchange of particles, no single wave function being totally symmetric or antisymmetric. Their description cannot be done in terms of the ordinary Bose or Fermi statistics but requires the use of "parastatistics".

or antisymmetrizing ϕ. Thus $\phi_A(1, 2) \equiv \phi(1, 2) - \phi(2, 1) = -\phi_A(2, 1)$ and $\phi_s(1, 2) \equiv \phi(1, 2) + \phi(2, 1) = +\phi_s(2, 1)$.

Consider now the particular case when $\phi(1, 2)$ is the product of two single particle wave functions, say u_a and u_b, so that $\phi(1, 2) = u_a(1)\, u_b(2)$. This state can be described by saying that particle 1 has the wave function u_a and particle 2 has u_b. Such a description is not possible however if the particles are identical. If for example, we consider a system of two electrons, whose wave functions must be necessarily antisymmetric, we should use instead of ϕ,

$$\phi_A = u_a(1)\, u_b(2) - u_a(2)\, u_b(1) \tag{3.100}$$

The first term shows particle 1 as having the wave function u_a, while in the second term it is particle 2 which occupies u_a. All we can say then is that *one* of the particles occupies u_a and *the other* occupies u_b. The wave function does not permit us to specify *which* particle is where; the two particles are *indistinguishable*.

More generally, if we have a system of N electrons occupying single particle states described by wave functions $u_a, u_b, \ldots, u_g$ the state of the system as a whole must be described, not by a simple product wave function $\phi(1, 2, \ldots, N) = u_a(1)\ u_b(2) \ldots u_g(N)$ but by its antisymmetrized form which can be conveniently written as a determinant.[12]

$$\phi_A(1, 2, \ldots, N) = \begin{vmatrix} u_a(1) & u_a(2) \ldots u_a(N) \\ u_b(1) & u_b(2) \ldots u_b(N) \\ \ldots\ldots & \ldots\ldots\ldots\ldots \\ \ldots\ldots & \ldots\ldots\ldots\ldots \\ u_g(1) & u_g(2) \ldots u_g(N) \end{vmatrix} \tag{3.101}$$

Once again, it is possible to say only that one particle occupies u_a (but not *which* one), one particle occupies u_b, ...

One very important fact may now be noted. If $u_a = u_b$ in Eq. 3.100 or more generally, if any two or more of the functions $u_a, u_b, \ldots, u_g$ in Eq. 3.101 coincide, the wave function ϕ_A *of the system* vanishes — meaning that no such state exists. Thus *no two electrons can have the same wave function*. This is the famous *exclusion principle* first postulated by Pauli for explaining the periodic table of chemical elements. This principle is now seen to hold not only for electrons but for any system of particles having antisymmetric wave functions. Stating it slightly differently, *a given single particle state* (characterised by a particular wave function, say u_a) *cannot be occupied by more than one such particle*. It is a standard result in statistical mechanics that particles having this property must obey Fermi-Dirac statistics; they are fermions.

For particles which are characterized by symmetric wave functions, there is no exclusion principle. For example $u_a(1)\, u_a(2) \ldots u_a(N)$ is a perfectly good symmetric function corresponding to all N particles having the same wave function u_a. But here also it is impossible to associate a specific particle with a particular u_a or u_b etc., as may be seen from the simple example of the

[12] This wave function is unnormalized. If $u_a, u_b, \ldots$ are members of an orthonormal set, the $N!$ terms arising from the expansion of the determinant may be seen to be separately normalized and orthogonal to each other. Then we must multiply (3.101) by $(N!)^{-1/2}$ to normalize it.

two-particle wave function $\psi_s = u_a(1)\, u_b(2) + u_a(2)\, u_b(1)$. All one can say in general is that n_a particles occupy u_a, n_b occupy u_b, ... This is precisely the kind of situation which leads to Bose-Einstein statistics in statistical mechanics. Thus particles described by symmetric wave functions are bosons.

It may be noted that tightly bound aggregates of particles (such as atomic nuclei) may themselves be viewed as 'particles' for the purpose of the above discussion, provided that circumstances are such that their internal structure does not come into play. Interchange of two *identical* nuclei, each made up of A fermions (neutrons + protons) is then equivalent to A simultaneous interchanges of *corresponding* pairs of particles (a proton of one nucleus with a proton of the other, etc.). Since each fermion interchange causes a change of sign, the overall effect on the wave function is given by a factor $(-1)^A$. Thus nuclei of even atomic number (e.g. He^4, C^{12}) behave as bosons while those of odd atomic number (e.g. He^3) behave as fermions.

The requirement of evenness or oddness of wave functions under interchanges of identical particles has important consequences, which we shall point out in the relevant context (theory of atomic and molecular structure, collision theory etc.).

PROBLEMS

1. Compute $\langle x \rangle$, $\langle x^2 \rangle$, $\langle p_x \rangle$, $\langle p_x^2 \rangle$ for a particle in one of the energy eigenstates in a square well potential. Verify that $(\Delta x)(\Delta p_x) \geqslant \frac{1}{2}\hbar$, where $(\Delta x)^2 = \langle x^2 \rangle - \langle x \rangle^2$, $(\Delta p_x)^2 = \langle p_x^2 \rangle - \langle p_x \rangle^2$.
2. Verify explicitly that the different energy eigenfunctions of a particle in a square well potential are orthogonal.
3. Prove the quantum mechanical version of the virial theorem by showing that if T and V are the kinetic and potential energy operators, then $2\langle T \rangle = \langle \mathbf{x}.\nabla V \rangle$, and hence that if $V(r) = ar^n$, $2\langle T \rangle = n \langle V \rangle$.
4. If $u_1(\mathbf{x})$, $u_2(\mathbf{x})$ are degenerate eigenfunctions of the Hamiltonian $H = \mathbf{p}^2/2m + V(\mathbf{x})$, show that $\int u_1^*(\mathbf{x})\, x p_x + p_x x\, u_2(\mathbf{x})\, d\tau = 0$.
5. The wave function of a particle (in 3 dimensions) is $(x^2 + y^2)^{1/2} e^{-\alpha r}$. Normalize this wave function and calculate the expectation value of L_z.
6. Prove that for a particle in a potential V, $d\langle \mathbf{L} \rangle/dt = \langle \mathbf{T} \rangle$ where $\mathbf{T} \equiv -\mathbf{r} \times \nabla V$ is the torque. Note that this vanishes if V is a central potential.
7. Verify that for a particle in a potential, $\langle H \rangle$ can be written as $\int W d\tau$ where $W = (\hbar^2/2m)\, \nabla\psi^*.\nabla\psi + \psi^* V \psi$. Show that W satisfies the continuity equation $\partial W/\partial t + \text{div}\, \mathbf{F} = 0$, where $\mathbf{F}$ is the energy flux vector. Determine the form of $\mathbf{F}$.
8. Using the identity $[A, BB^{-1}] = 0$, show that $[A, B^{-1}] = -B^{-1}[A, B]B^{-1}$.
9. If A, B are operators such that $[A, B] = 1$, and if $Z = \alpha A + \beta B$ where α, β are numbers, verify that $[A, Z^n] = n\beta Z^{n-1}$. Hence show that $(\partial/\partial\alpha) Z^n = nAZ^{n-1} - \frac{1}{2}n(n-1)\beta Z^{n-2}$ and $(\partial/\partial\alpha)\, e^Z = (A - \frac{1}{2}\beta)\, e^Z$. By integrating this differential equation, with due attention to the order of operator factors, show that
$$e^{\alpha A + \beta B} = e^{\alpha A} e^{\beta B} e^{-\frac{1}{2}\alpha\beta}$$
10. If X and Y are any two operators, and $Z = e^{\alpha X} Y e^{-\alpha X}$, show that $dZ/d\alpha = [X, Z]$. Hence show that the Taylor expansion of Z (in powers of α) is
$$e^{\alpha X} Y e^{-\alpha X} = Y + \alpha [X, Y] + \frac{\alpha^2}{2!}[X, [X, Y]] + \frac{\alpha^3}{3!}[X, [X, [X, Y]] + \dots$$
Show how the result of Prob. 9 can be obtained from this.

4

Exactly Soluble Eigenvalue Problems

In this chapter we consider some of the most important exactly soluble eigenvalue problems of quantum mechanics.[1] They are of great intrinsic interest, and they also serve as the starting point for the approximate solution of a variety of problems for which no exact solutions can be found. Furthermore, from a detailed study of these relatively simple problems one can see clearly many properties of general validity, and this helps in the understanding of more complex systems.

Since the operators in the Schrödinger form of quantum mechanics are differential operators, the eigenvalue problems take the form of differential equations to be solved. In three dimensions, we have partial differential equations, whose exact solution depends on their separation into a set of ordinary differential equations. Once this reduction is made, the crux of the problem is to find solutions satisfying the admissibility conditions of Sec. 2.8. The condition of continuity needs to be considered only when potential functions with discontinuities (e.g. square well) are involved. The question of single-valuedness arises whenever the wave function is expressed in terms of coordinate variables which include angles (e.g. in polar coordinates). The finiteness requirement has to be enforced at infinity with respect to each variable which has an infinite range of variation, and possibly also at other points (such as the origin). Our objective in this chapter is to determine, in specific cases, the spectrum of eigenvalues permitted by these conditions, and the corresponding

[1] For further examples of exactly soluble problems, see, for instance, S. Flügge, *Practical Quantum Mechanics*, Springer Verlag, Berlin, 1971; D. ter Haar (Ed.) *Selected Problems in Quantum Mechanics*, Infosearch Ltd., London, 1964.

eigenfunctions. All the differential equations arising from the eigenvalue problems considered here turn out to be known ones, and we will make use of the known properties of their solutions in trying to understand the nature of the eigenfunctions. Derivation or discussion of these properties in the text will be kept down to the essential minimum needed for our purposes. Further results regarding the differential equations and special functions appearing in this book are listed in Appendix D.

A. THE SIMPLE HARMONIC OSCILLATOR

We have already seen in Chapter 1 the importance of the quantum properties of the oscillator in such disparate physical phenomena as black body radiation and thermal properties of solids. The wave mechanical theory of the oscillator, to be presented now, provides the basis for understanding the properties of a wide variety of systems which are analyzable in terms of harmonic oscillators. Some examples are: The vibrations of diatomic and polyatomic molecules, and oscillations of other more complicated systems expressed in terms of their normal modes (e.g. various kinds of wave motions in crystals, and electromagnetic and other fields).

4.1 The Schrödinger Equation and Energy Eigenvalues

The stationary state energies E_n and wave functions $u_n(x)$ are solutions of the eigenvalue equation, $Hu = Eu$, which is the time-independent Schrödinger equation. In the case of the simple harmonic oscillator, the Hamiltonian is of the form

$$H = \frac{p^2}{2m} + \tfrac{1}{2}Kx^2 \tag{4.1}$$

where K is the force per unit displacement of the particle from the origin. Classically, the particle oscillates with a frequency $(\omega_c/2\pi)$ where $\omega_c = (K/m)^{1/2}$. So it is convenient to write $K = m\omega_c^2$. Then the eigenvalue equation for energy is given by

$$\left(-\frac{\hbar^2}{2m}\frac{d^2}{dx^2} + \tfrac{1}{2}m\omega_c^2x^2\right)u(x) = Eu(x) \tag{4.2}$$

This equation can be rewritten in the simpler form

$$\frac{d^2u}{d\rho^2} + (\lambda - \rho^2)\,u = 0 \tag{4.3}$$

by changing over to the independent variable ρ defined by[2]

$$\rho = \alpha x, \text{ with } \alpha = (m\omega_c/\hbar)^{1/2} \tag{4.4}$$

The parameter λ in Eq. 4.3 is related to the energy E through the relation $\lambda = (2mE/\hbar^2\alpha^2) = 2E/\hbar\omega_c$, or

$$E = \tfrac{1}{2}\lambda\hbar\omega_c \tag{4.5}$$

[2] Note that α has the dimension of the inverse of length. Therefore ρ is non-dimensional.

We have to find now the solutions of Eq. 4.3 which do not diverge at infinity. Actually it can be seen by inspection that $u(\rho) = e^{-\frac{1}{2}\rho^2}$ is an exact solution[3] when $\lambda = 1$. For other values of λ also it is reasonable to expect that the dominant *asymptotic* behaviour of $u(\rho)$ would be of the same type. This is so because, as $\rho \to \pm\infty$, the constant λ becomes insignificant compared to ρ^2 in Eq. 4.3, and hence its value does not matter too much. Let us, therefore, incorporate the factor $e^{-\frac{1}{2}\rho^2}$ explicitly into u, and write

$$u(\rho) = e^{-\frac{1}{2}\rho^2}\, v(\rho) \tag{4.6}$$

Substitution of this form into Eq. 4.3 leads to

$$v'' - 2\rho v' + (\lambda - 1)\, v = 0 \tag{4.7}$$

where primes denote differentiation with respect to ρ. We shall see below that when $\rho \to \pm\infty$, the solutions of this equation diverge like e^{ρ^2} (so that u diverges like $e^{\frac{1}{2}\rho^2}$) *unless* λ has one of the values

$$\lambda = (2n + 1), \quad (n = 0, 1, 2, \ldots) \tag{4.8}$$

Therefore admissible solutions $u(\rho)$ exist only for these special values of λ. The corresponding values of E are obtained from Eq. 4.5:

$$E_n = (n + \tfrac{1}{2})\, \hbar\omega_c \tag{4.9}$$

These are the energy eigenvalues of the simple harmonic oscillator.

4.2 The Energy Eigenfunctions

(*a*) *Series Solution; Asymptotic behaviour:* Let us now verify the above statements by actually solving Eq. 4.7 by the series method. Suppose

$$v(\rho) = c_\sigma \rho^\sigma + c_{\sigma+1}\, \rho^{\sigma+1} + \ldots = \sum_{s=\sigma}^{\infty} c_s \rho^s \tag{4.10}$$

On substituting this series in Eq. 4.7 we get

$$\sum_{s=\sigma}^{\infty} c_s \left[\, s\,(s-1)\, \rho^{s-2} - 2s\rho^s + (\lambda - 1)\, \rho^s \right] = 0 \tag{4.11}$$

Since the series on the left hand side is to vanish for all ρ, the coefficient of each power of ρ must vanish separately. Taking the coefficient of ρ^s for any $s \geqslant \sigma$, we obtain

$$c_{s+2}\,(s+2)\,(s+1) - c_s\,[2s - (\lambda - 1)] = 0$$

or

$$\frac{c_{s+2}}{c_s} = \frac{2s - (\lambda - 1)}{(s+1)\,(s+2)} \tag{4.12}$$

Now, the asymptotic behaviour of a function defined by an infinite series is determined by how fast the coefficients c_s decrease as $s \to \infty$. Eq. 4.12 tells us that as $s \to \infty$,

$$\frac{c_{s+2}}{c_s} \to \frac{2}{s} \tag{4.13}$$

This behaviour of the coefficients is exactly the same as in the series for e^{ρ^2}. Hence, for large ρ, the function $v(\rho)$ behaves like e^{ρ^2}, confirming our assertion above.

[3] Though $e^{+\frac{1}{2}\rho^2}$ is also a solution, it tends to infinity as $\rho \to \pm\infty$, and is therefore unacceptable.

However, the situation is different if the series terminates after a finite number of terms. This happens in our case if $(\lambda - 1)$ *is an even integer* $(2n)$ and

$$c_{n+1} = 0, \text{ with } c_n \neq 0 \tag{4.14}$$

For, if $\lambda = 2n + 1$, the numerator in Eq. 4.12 vanishes for $s = n$, making c_{n+2} and hence $c_{n+4}, c_{n+6} \ldots$ all zero. Further, with $c_{n+1} = 0$, $c_{n+3}, c_{n+5} \ldots$ also vanish according to Eq. 4.12. In fact, by using Eq. 4.12 in reverse as $(c_{s-2}/c_s) = s(s-1)/(2s-2n-4)$, we see that $c_{n-1}, c_{n-3} \ldots$ vanish too. The same relation can be used to determine the non-vanishing coefficients c_{n-2}, c_{n-4} .. in terms of c_n. Because of the factor $s(s-1)$ in the numerator of this relation, the last of the non-zero coefficients is c_0 or c_1, according as n is even or odd. Thus the solution of Eq. 4.7 for $\lambda = 2n + 1$ is a *polynomial* of degree n. It contains only even powers of ρ, and hence is of even parity, if n is even; it is of odd parity if n is odd. It is, in fact, nothing but the *Hermite polynomial* $H_n(\rho)$, apart from an arbitrary multiplying constant; for, $H_n(\rho)$ is known to be the polynomial solution of the equation

$$H_n''(\rho) - 2\rho H_n'(\rho) + 2nH_n(\rho) = 0 \tag{4.15}$$

which has the same form as Eq. 4.7 with $\lambda = (2n + 1)$.

Thus we finally have the result that the energy eigenfunctions of the harmonic oscillator are given by

$$u_n(x) = \mathcal{N}_n e^{-\frac{1}{2}\rho^2} H_n(\rho) = \mathcal{N}_n e^{-\frac{1}{2}\alpha^2 x^2} . H_n(\alpha x), \quad n = 0, 1, 2, \ldots, \tag{4.16a}$$

where the constant $\mathcal{N}_n$ could be arbitrary, but is chosen to be

$$\mathcal{N}_n = \left(\frac{\alpha}{\sqrt{\pi} 2^n n!} \right)^{1/2} \tag{4.16b}$$

to ensure that $u_n(x)$ is normalized.

(*b*) *Orthonormality:* The functions (4.16) form an orthonormal set:

$$(u_m, u_n) \equiv \int u_m^*(x)\, u_n(x)\, dx = \delta_{mn} \tag{4.17}$$

To prove this identity, and to study other properties of the stationary state wave functions $u_n(x)$, we make use of known properties of the Hermite polynomials. In particular, we use the fact that the *generating function* $G(\rho, \xi)$ of the Hermite polynomials, defined as

$$G(\rho, \xi) \equiv \sum_{n=0}^{\infty} \frac{1}{n!} H_n(\rho) \xi^n \tag{4.18a}$$

is given by

$$G(\rho, \xi) = \exp(-\xi^2 + 2\rho\xi) \tag{4.18b}$$

Here ξ is a parameter taking arbitrary real values. Eqs. 4.18 may be considered as *defining* the Hermite polynomial: $H_n(\rho)$ is the coefficient of $(\xi^n/n!)$ in the expansion of $\exp[-\xi^2 + 2\rho\xi]$ in powers of ξ. It may be verified that the function thus defined does satisfy Eq. 4.15. The first four of the polynomials are

$$H_0(\rho) = 1,\ H_1(\rho) = 2\rho,\ H_2(\rho) = 4\rho^2 - 2,\ H_3(\rho) = 8\rho^3 - 12\rho \tag{4.19}$$

Further properties of the Hermite polynomials are given in Appendix D.

Let us now use Eqs. 4.18 to prove the orthonormality relation (4.17). Consider the integral

$$\int_{-\infty}^{\infty} G(\rho, \xi)\, G(\rho, \xi')\, e^{-\rho^2}\, d\rho \tag{4.20}$$

Equating the two alternative expressions obtained by substituting Eq. 4.18a and 4.18b respectively into Eq. 4.20, we find that

$$\sum_m \sum_n \frac{\xi^m \xi'^n}{m!n!} \int_{-\infty}^{\infty} H_m(\rho) H_n(\rho) e^{-\rho^2} d\rho = \int_{-\infty}^{\infty} e^{-\rho^2+2\rho(\xi+\xi')-(\xi^2+\xi'^2)} d\rho \tag{4.21}$$

The right hand side reduces to

$$e^{2\xi\xi'} \int_{-\infty}^{\infty} e^{-(\rho-\xi-\xi')^2} d\rho = \sqrt{\pi}\, e^{2\xi\xi'}$$

$$= \sqrt{\pi} \sum_{n=0}^{\infty} \frac{(2\xi\xi')^n}{n!} = \sqrt{\pi} \sum_{m=0}^{\infty} \sum_{n=0}^{\infty} \frac{2^n \xi^m \xi'^n}{n!} \delta_{mn} \tag{4.22}$$

If this is to be equal to the left hand side of Eq. 4.21 for all ξ and ξ', the coefficient of $\xi^m \xi'^n$ in the two expressions must be equal for each m and n. We thus get

$$\int_{-\infty}^{\infty} H_m(\rho) H_n(\rho) e^{-\rho^2} d\rho = \sqrt{\pi}\, 2^n\, n!\, \delta_{mn} \tag{4.23}$$

It is a trivial matter now to verify with the aid of this result (remembering that the $H_n(\rho)$ are real) that the functions (4.16) do have the orthonormality property (4.17). We note that from the point of view of the general theorems of Sec. 3.5, the orthogonality of u_m and u_n when $m \neq n$ is a necessary consequence of the fact that they are eigenfunctions belonging to distinct eigenvalues of the Hamiltonian operator which is self-adjoint.

Another result which we will find very useful later on is

$$(u_m, xu_n) \equiv \int_{-\infty}^{\infty} u_m^*(x)\, x\, u_n(x)\, dx = \begin{cases} \sqrt{n+1}/(\sqrt{2}\alpha), & m = n+1 \\ \sqrt{n}/(\sqrt{2}\alpha), & m = n-1 \\ 0 & \text{otherwise} \end{cases} \tag{4.24}$$

This is proved by first evaluating $\int H_m(\rho)\rho H_n(\rho) e^{-\rho^2} d\rho$, starting with $\int G^*(\rho,\xi)\, \rho\, G(\rho,\xi')\, e^{-\rho^2} d\rho$ and proceeding exactly as we did in proving Eq. 4.23. The details are left as an exercise for the reader, who may also verify in a similar fashion that

$$(u_n, x^2 u_n) = \int_{-\infty}^{\infty} u_n^*(x)\, x^2\, u_n(x)\, dx = \frac{2n+1}{2\alpha^2} \tag{4.25}$$

4.3 Properties of Stationary States

To summarize the results derived above: The stationary states of the simple harmonic oscillator are characterized by equally spaced energy levels $E_n = (n + \frac{1}{2})\hbar\omega_c$ and wave functions $u_n(x)$ given by Eq. 4.16. The constant spacing between successive levels, $\hbar\omega_c \equiv h\nu_c$, is exactly what had been postulated by Planck. However, the wave mechanical treatment gives a definite nonzero value for the lowest energy, $E_0 = \frac{1}{2}\hbar\omega_c$. This is called the *zero-point energy*. It is to be attributed to the fact that the particle is confined to a finite

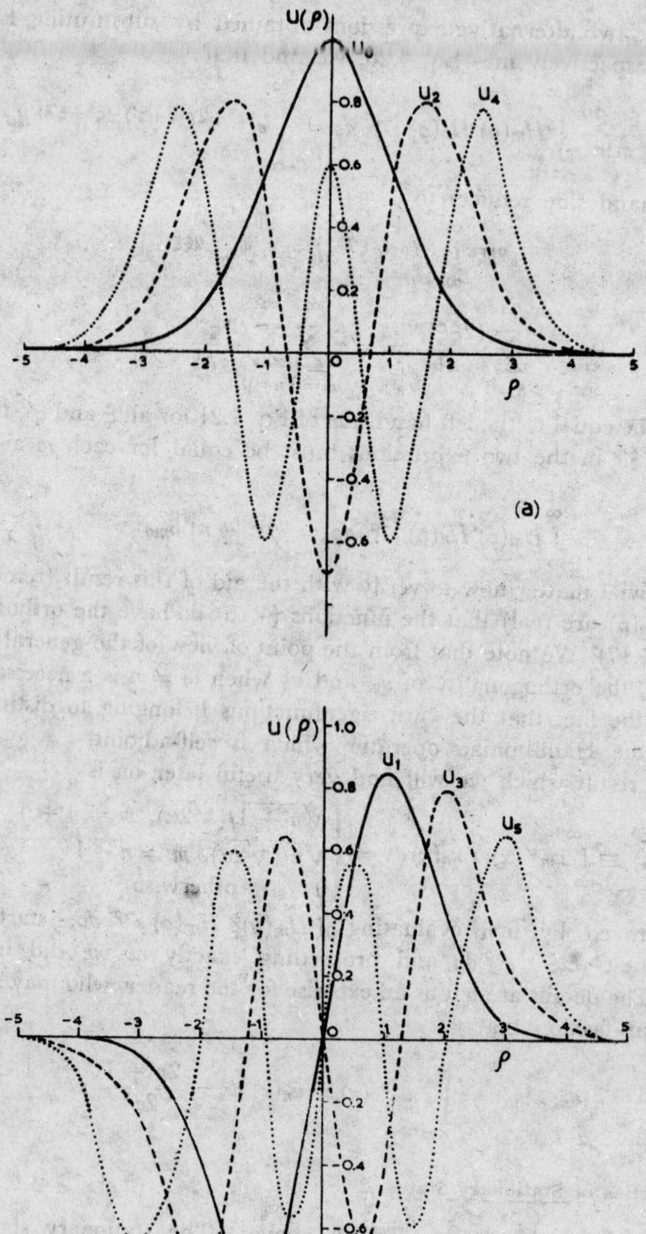

Fig. 4.1 Wave functions of stationary states of the harmonic oscillator: **(a)** Even parity wave functions; **(b)** Odd parity wave functions.

region of space by the effect of the potential, and therefore, by the uncertainty principle, its momentum cannot have a definite value. In particular the momentum cannot be zero, and hence the energy cannot vanish in any state. Actually, the wave function $u_0(x)$ of the state of lowest energy (the *ground state*) has the Gaussian form

$$u_0(x) = (\alpha^2/\pi)^{1/4} . e^{-\frac{1}{2}\alpha^2 x^2} \tag{4.26}$$

It follows then, in view of the results of Sec. 3.11, that the ground state of the harmonic oscillator is a minimum uncertainty state, wherein (Δp) (Δx) takes its minimum value $\frac{1}{2}\hbar$.

Fig. 4.1 shows a few of the eigenfunctions $u_n(x)$. It may be observed that u_n has n nodes, i.e. there are n (finite) values of x for which $u_n(x)$ vanishes. It is a useful fact to remember, as a general property of wave functions, that the higher the number of nodes, the greater the energy. This is because the kinetic energy part of the Hamiltonian contributes an amount equal to

$$\int \psi^* \frac{p^2}{2m} \psi \, d\tau = \frac{1}{2m} \int (p\psi)^* (p\psi) \, d\tau = \frac{\hbar^2}{2m} \int | \nabla\psi |^2 \, d\tau$$

to the expectation value of the energy, and large values of $|\nabla\psi|$ occur evermore extensively as the number of nodes (and hence the number of ups and downs between them) increases.

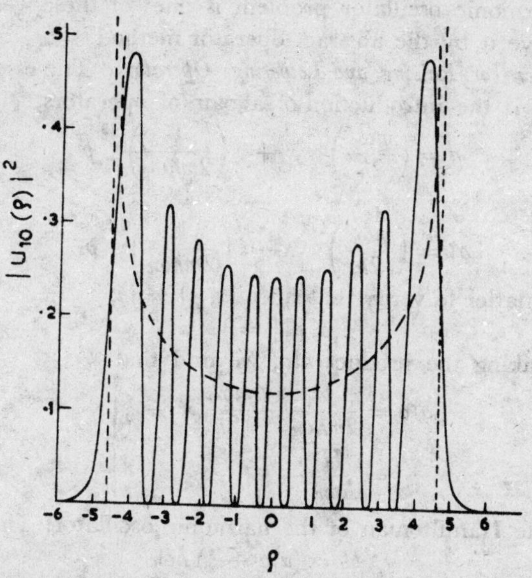

Fig. 4.2 Position probability density in the state $n = 10$

Fig. 4.2 shows the position probability density $|u_n|^2$ for $n = 10$, plotted as a function of $\rho = \alpha x$. If a classical oscillator is considered, it is easy to verify that the amount of time it spends between ρ and $\rho + d\rho$ during its motion is proportional to $(\rho_0^2 - \rho^2)^{-1/2}$, where ρ_0 is the amplitude (in units of α^{-1}) of the oscillator with the given energy. Thus the probability density for finding the oscillator in the neighbourhood of ρ is proportional to $(\rho_0^2 - \rho^2)^{-1/2}$. This is shown as a dashed curve in Fig. 4.2. This classical curve is seen to agree well with the average behaviour of the quantum mechanical curve, though the latter has rapid oscillations. This kind of correspondence does not exist for very small n, but improves rapidly with increasing n.

4.4 The Abstract Operator Method[4]

In the above treatment of the harmonic oscillator problem we have adhered strictly to the Schrödinger method of converting dynamical variables to differential operators through the replacement $p \to -i\hbar\,(d/dx)$. Actually, the fundamental feature of quantum mechanics is not this particular replacement but rather the fact that x and p have the commutator $[x, p] = i\hbar$; the quantum phenomena are consequences of this commutation rule. The Schrödinger method is the simplest way of ensuring that this rule is obeyed, and is often the only practical method for determining what the consequences are. Its chief advantage is that it enables the well developed theory of differential equations to be exploited in the solution of problems. However, there are examples where all the results can be obtained very simply and elegantly by the direct application of the abstract operator algebra (i.e., the use of the commutation rules only, without assuming any specific forms for the operators x, p). The harmonic oscillator problem is one of these, and we will now proceed to solve it by the abstract operator method.

(a) *The Ladder (or Raising and Lowering) Operators:* The essential element in the procedure is the introduction of a pair of operators,

$$a = \left(\frac{m\omega_c}{2\hbar}\right)^{1/2} x + i\left(\frac{1}{2m\hbar\omega_c}\right)^{1/2} p \tag{4.27a}$$

and its adjoint

$$a^\dagger = \left(\frac{m\omega_c}{2\hbar}\right)^{1/2} x - i\left(\frac{1}{2m\hbar\omega_c}\right)^{1/2} p \tag{4.27b}$$

It is a trivial matter to verify that since $[x, p] = i\hbar$,

$$[a, a^\dagger] = 1 \tag{4.28}$$

Further, on taking the product $a^\dagger a$, we find that

$$\left.\begin{aligned} a^\dagger a &= \frac{p^2}{2m\hbar\omega_c} + \frac{m\omega_c}{2\hbar}\,x^2 - \tfrac{1}{2} \\ &= \frac{H}{\hbar\omega_c} - \tfrac{1}{2} \end{aligned}\right\} \tag{4.29a}$$

where H is the Hamiltonian of the harmonic oscillator. Thus

$$H = (a^\dagger a + \tfrac{1}{2})\,\hbar\omega_c \tag{4.29b}$$

Since the relation between H and $a^\dagger a$ involves only ordinary numbers (no operators), their eigenvalues will also bear the same relation, and their eigenfunctions coincide. We shall now see that the eigenvalues of $a^\dagger a$ are integers n ($n = 0, 1, 2, \ldots$), so that those of the Hamiltonian are $(n + \tfrac{1}{2})\,\hbar\omega_c$, in agreement with Eq. 4.9.

[4] In this section we make essential use of the concept of the adjoint of an operator described in Sec. 3.3.

First we observe that by virtue of Eq. 4.28, $(a^\dagger a)\, a - a\, (a^\dagger a) = (a^\dagger a - aa^\dagger)a = -a$, i.e.,

$$[a^\dagger a, a] = -a \tag{4.30a}$$

Similarly

$$[a^\dagger a, a^\dagger] = a^\dagger \tag{4.30b}$$

These results imply that if u is an eigenfunction of $(a^\dagger a)$, belonging to the eigenvalue λ, i.e.,

$$(a^\dagger a)\, u = \lambda u \tag{4.31}$$

then au and $a^\dagger u$ are also eigenfunctions of $(a^\dagger a)$, belonging to the eigenvalues $(\lambda - 1)$ and $(\lambda + 1)$ respectively. For, on using Eqs. 4.28 and 4.31 we have

$$\begin{aligned}(a^\dagger a).au &= (aa^\dagger - 1)\, au = a\,(a^\dagger a - 1)\, u \\ &= a\,(\lambda - 1)\, u = (\lambda - 1)\, au\end{aligned} \tag{4.32a}$$

and similarly

$$(a^\dagger a)\, a^\dagger u = (\lambda + 1)\, a^\dagger u \tag{4.32b}$$

Starting with u, we can now construct wave functions $a^\dagger u$, $(a^\dagger)^2 u, \ldots$ belonging to eigenvalues $(\lambda + 1)$, $(\lambda + 2), \ldots$ and also au, $a^2u, \ldots$ belonging to $(\lambda - 1)$, $(\lambda - 2), \ldots$. The eigenvalues thus form a 'ladder' with unit spacing between steps, and $a^\dagger$ has the effect of raising (and a, of lowering) the eigenvalue by one step. For this reason, $a^\dagger$ and a a are called ladder operators, or more specifically, raising and lowering operators.

(*b*) *The Eigenvalue Spectrum:* As yet we know only the spacing of eigenvalues, but not the values themselves. The key to their determination is the observation that $a^\dagger a$ is a positive operator (Sec. 3.3) and, therefore, it can have no negative eigenvalues. So the sequence of eigenvalues $(\lambda - 1)$, $(\lambda - 2)\ldots$ must terminate before it reaches negative values, and correspondingly, the sequence au, a^2u, $\ldots$ must terminate. Let us call the last eigenfunction in the sequence as u_0. Then, by definition, there does not exist a state au_0. In other words,

$$au_0 = 0 \tag{4.33}$$

This implies obviously that

$$a^\dagger a\, u_0 = 0 \tag{4.34}$$

i.e. u_0 is an eigenstate of $(a^\dagger a)$, belonging to the eigenvalue 0. The other eigenstates can, now, be constructed by repeated application of $a^\dagger$ to u_0. Since the effect of $a^\dagger$ applied once is to increase the eigenvalue by unity, as seen above, we have that the eigenstates u_0, $(a^\dagger)\, u_0$, $(a^\dagger)^2\, u_0 \ldots$ belong to the eigenvalues, 0, 1, 2, $\ldots$ of $(a^\dagger a)$.

We find thus that the eigenvalue spectrum of $(a^\dagger a)$ consists of the set of non-negative integers[5] n, and hence that the eigenvalues of H are $(n + \frac{1}{2})\, \hbar\omega_c$.

(*c*) *The Energy Eigenfunctions:* The eigenstates which we have constructed above are mutually orthogonal, and the norm of $(a^\dagger)^n u_0$ is $n!$ (See Example 4.1, p. 112).

$$((a^\dagger)^m u_0, (a^\dagger)^n u_0) = n!\, \delta_{mn} \tag{4.35}$$

[5] In the eigenstate $(a^\dagger)^n u_0$ of $a^\dagger a$, belonging to the eigenvalue n, the energy of the oscillator exceeds the ground state energy by n quanta, each of magnitude $\hbar\omega_c$. Since the eigenvalue of $a^\dagger a$ gives the *number* of quanta, $a^\dagger a$ is often called the *number operator* and denoted by $\mathcal{N}$.

Therefore, the normalized eigenfunctions are given by

$$u_n = \frac{(a^\dagger)^n}{\sqrt{n!}} u_0, \; n = 0,1,2,\ldots \tag{4.36}$$

By construction, they satisfy the relation

$$(a^\dagger a)\, u_n = n u_n \tag{4.37}$$

Observe that Eq. 4.36 can be rewritten as

$$u_n = \frac{a^\dagger}{\sqrt{n}} \frac{(a^\dagger)^{n-1}}{\sqrt{(n-1)!}} u_0 = \frac{a^\dagger}{\sqrt{n}} u_{n-1}$$

Hence

$$a^\dagger u_{n-1} = \sqrt{n}\, u_n \tag{4.38a}$$

Further, by operating on this equation with a, we get $\sqrt{n}\, a u_n = a a^\dagger u_{n-1} = (a^\dagger a + 1)\, u_{n-1} = n u_{n-1}$, so that

$$a\, u_n = \sqrt{n}\, u_{n-1} \tag{4.38b}$$

From these two equations we obtain

$$(u_m, a^\dagger u_n) = \sqrt{n+1}\, \delta_{m,n+1};\; (u_m, a u_n) = \sqrt{n}\, \delta_{m,n-1} \tag{4.38c}$$

Eqs. 4.38 are of fundamental importance. They enable one to calculate the effect of any function of x and p on the energy eigenfunctions by expressing them first in terms of a and $a^\dagger$ through Eq. 4.27. The reader may verify that the result (4.24) and a corresponding expression for $(u_m, p u_n)$ follow from Eq. 4.38c

We have thus obtained the basic results for the harmonic oscillator without solving any differential equations or knowing anything about the actual form of the wave functions. We shall now see how the latter may be determined.

Consider Eq. 4.33, which, incidentally, is seen to be a special case of Eq. 4.38b. On substituting Eq. 4.27a in it and writing $p = -i\hbar d/dx$, we get the differential equation

$$\left(\frac{d}{dx} + \frac{m\omega_c}{\hbar} x\right) u_0 = 0 \tag{4.39}$$

It is easy to verify now that the solution of this equation, with proper choice of normalization constant, coincides with Eq. 4.26. All other eigenfunctions are obtained from this by using Eq. 4.36. Observing that $a^\dagger$ can be written in the form[6]

$$a^\dagger = -\frac{1}{\sqrt{2}}\left(\frac{d}{d\rho} - \rho\right) = -\frac{1}{\sqrt{2}} e^{\frac{1}{2}\rho^2} \frac{d}{d\rho} e^{-\frac{1}{2}\rho^2}$$

with ρ defined as in Eq. 4.4, and using the fact that the Hermite polynomials can be expressed as

$$H_n(\rho) = (-)^n e^{\rho^2} \frac{d^n}{d\rho^n} e^{-\rho^2} = (-)^n e^{\frac{1}{2}\rho^2}\left(e^{\frac{1}{2}\rho^2} \frac{d}{d\rho} e^{-\frac{1}{2}\rho^2}\right)^n e^{-\frac{1}{2}\rho^2} \tag{4.40}$$

we can verify that u_n defined by Eq. 4.36 coincides with Eq. 4.16.

This completes the solution of the problem.

EXAMPLE 4.1.—To prove Eq. 4.35, we first rewrite the left hand side, using the definition (3.18) of the adjoint of an operator, as $(u_0, a^m (a^\dagger)^n u_0)$. Next, observe that Eq. 4.28 implies $[a, (a^\dagger)^n] = n\, (a^\dagger)^{n-1}$, so that $a^m (a^\dagger)^n = a^{m-1} a (a^\dagger)^n$

[6] Remember that $(d/d\rho)$ is to operate not merely on $e^{-\frac{1}{2}\rho^2}$ but also on any wavefunction which appears to the right.

$= a^{m-1} \{(a^\dagger)^n a + n (a^\dagger)^{n-1}\}$. On using this expression and the fact that $au_0 = 0$, we obtain $(u_0, a^m(a^\dagger)^n u_0) = n (u_0, a^{m-1}(a^\dagger)^{n-1}u_0)$. If $m \geqslant n$, by repeating the above step n times we are led to $(u_0, a^m(a^\dagger)^n u_0) = n!\ (u_0, a^{m-n}u_0)$ $= n!\ \delta_{mn}$ which is just what Eq. 4.35 says. The validity of this result for $m < n$ can be inferred by a similar procedure starting with $((a^\dagger)^m u_0, (a^\dagger)^n u_0)$ $= (a^n(a^\dagger)^m u_0, u_0)$.

Example 4.2.— From the roles of a and $a^\dagger$ as ladder operators it is easy to see that if A is an operator which contains r factors $a^\dagger$ and s factors a (in any order), then (u_m, Au_n) vanishes unless $r - s = m - n$. This property facilitates the evaluation of matrix elements of functions of x and p. For example, to evaluate the expectation value $(u_n, x^4 u_n)$, we can use the fact that $x = (\hbar/2m\omega_c)^{1/2}$ $(a + a^\dagger)$. Retaining only those terms in the expansion of $(a + a^\dagger)^4$ which meet the above criterion, we can read off the contribution of each term with the aid of Eqs. 4.38. The result is $(u_n, x^4 u_n) = \frac{3}{2}(\hbar/m\omega_c)^2.(n^2 + n + \frac{1}{2})$. Similarly, one can see that $(u_m, x^3 u_n)$ is nonzero only for $m - n = \pm 1$ or ± 3, and that $(u_{n-1}, x^3 u_n) = (u_n, x^3 u_{n-1}) = 3\ (\hbar/2m\omega_c)^{3/2} n \sqrt{n}$, and $(u_{n-3}, x^3 u_n) =$ $(u_n, x^3 u_{n-3}) = (\hbar/2m\omega_c)^{3/2}\ [n\ (n-1)\ (n-2)]^{1/2}$.

4.5 Coherent States

We have confined our attention so far to eigenstates of the Hamiltonian, which are stationary states. We will now consider the eigenvalue problem for the lowering operator a. It has very different features from the problems considered so far, because a is not self-adjoint. In fact it turns out that *any complex number* μ *is an eigenvalue* of a. The eigenstates of a can be expressed in terms of the stationary states u_n of the harmonic oscillator, in the form

$$\phi_\mu = \sum_{n=0}^{\infty} e^{-\frac{1}{2}|\mu|^2} \cdot \frac{\mu^n}{\sqrt{n!}} u_n \tag{4.41}$$

One can easily verify, using Eq. 4.38b, that (4.41) is indeed an eigenfunction of a belonging to the eigenvalue μ:

$$a\phi_\mu = \mu\phi_\mu \tag{4.42}$$

Further, because of the orthonormality (4.17) of the functions u_n, we have

$$\begin{aligned}\int \phi_\mu^* \phi_{\mu'}\, dx &= \int \left(\sum_{m=0}^{\infty} e^{-\frac{1}{2}|\mu|^2} \frac{(\mu^*)^m}{\sqrt{m!}} u_m^*\right)\left(\sum_{n=0}^{\infty} e^{-\frac{1}{2}|\mu'|^2} \frac{(\mu')^n}{\sqrt{n!}} u_n\right) dx \\ &= \sum_{m=0}^{\infty}\sum_{n=0}^{\infty} e^{-\frac{1}{2}|\mu|^2 - \frac{1}{2}|\mu'|^2} \frac{(\mu^*)^m\ (\mu')^n}{(m!\ n!)^{1/2}} \delta_{mn} \\ &= e^{-\frac{1}{2}|\mu|^2 - \frac{1}{2}|\mu'|^2} . e^{\mu^*\mu'}\end{aligned} \tag{4.43}$$

Thus the eigenfunctions ϕ_μ, $\phi_{\mu'}$, are *not orthogonal* for $\mu \neq \mu'$. This happens because of the non-self-adjointness of a. For $\mu = \mu'$, Eq. (4.43) reduces to unity, showing that ϕ_μ is normalized.

The wave functions ϕ_μ are of great physical interest because of the way in which they change with time. We have seen in Sec. 2.9 that stationary state wave functions have a very simple time dependence: $u_E(x)$ becomes $u_E(x)e^{-iEt/\hbar}$ after the lapse of time t. In the case of the harmonic oscillator, $u_n(x) \to u_n(x) \exp[-i(n+\frac{1}{2})\omega_c t]$, since $E_n = (n+\frac{1}{2})\hbar\omega_c$. Thus, if the harmonic oscillator has the wave function (4.41) at time $t=0$, the wave function at time t will be

$$\phi_\mu(x,t) = \sum_{n=0}^{\infty} e^{-\frac{1}{2}\lambda^2} \frac{(\lambda e^{i\varkappa})^n}{\sqrt{n!}} u_n(x)\, e^{-i(n+\frac{1}{2})\omega_c t} \tag{4.44}$$

where we have written $\mu = \lambda e^{i\varkappa}$, ($\lambda, \varkappa$, real). We now substitute the explicit form (4.16) of u_n into Eq. 4.44, and also use the first of the expressions (4.40) for the Hermite polynomial. The result is that (with $\rho = \alpha x$),

$$\phi_\mu(x,t) = \left(\frac{\alpha}{\sqrt{\pi}}\right)^{1/2} . e^{\frac{1}{2}(\rho^2-\lambda^2)} . e^{-\frac{1}{2}i\omega_c t} . \sum_{n=0}^{\infty} \frac{1}{n!}\left[-\frac{\lambda}{\sqrt{2}} e^{-i(\omega_c t-\varkappa)}\right]^n \frac{d^n}{d\rho^n} e^{-\rho^2} \tag{4.45}$$

Comparing this with the Taylor series

$$f(\rho-\eta) = \sum_{n=0}^{\infty} \frac{(-\eta)^n}{n!} \frac{d^n}{d\rho^n} f(\rho) \tag{4.46}$$

we recognize that

$$\phi_\mu(x,t) = \left(\frac{\alpha}{\sqrt{\pi}}\right)^{1/2} e^{\frac{1}{2}(\rho^2-\lambda^2-i\omega_c t)}\, e^{-(\rho-\eta)^2} \tag{4.47}$$

where

$$\eta = \frac{1}{\sqrt{2}} \lambda e^{-i(\omega_c t-\varkappa)} \tag{4.48}$$

The special significance of this wave function becomes evident when we examine the position probability density,

$$|\phi_\mu(x,t)|^2 = \frac{\alpha}{\sqrt{\pi}} \exp[\rho^2-\lambda^2-(\rho-\eta)^2-(\rho-\eta^*)^2]$$

$$= \frac{\alpha}{\sqrt{\pi}} \exp[-\{\alpha x-\sqrt{2}\,\lambda\cos(\omega_c t-\varkappa)\}^2] \tag{4.49}$$

This is a Gaussian function centred at the point $x = (\sqrt{2}\lambda/\alpha).\cos(\omega_c t - \varkappa)$ which oscillates in simple harmonic fashion. The function has *no other* time dependence. Hence ϕ_μ represents a wave packet which executes simple harmonic oscillations, *moving as a whole without any change of shape*. Because of this property the state represented by ϕ_μ is called a *coherent* state. The motion of a coherent wave packet is exactly the same as that of a classical oscillator, the only difference being the nonzero size or width of the wave packet. Note that the size of the packet is the same, whatever be the value of λ; but the amplitude of oscillation is $\sqrt{2}\,\lambda/\alpha$.

EXAMPLE 4.3.—When the oscillator is in the coherent state ϕ_μ, the probability that its energy has the value $E_n = (n+\frac{1}{2})\hbar\omega_c$ is $|c_n|^2$ where c_n is the coefficient of the energy eigenfunction u_n in Eq. 4.44. We observe that $|c_n|^2 = e^{-\lambda^2}(\lambda^2)^n/n!$, which is a Poisson distribution.

B. ANGULAR MOMENTUM AND PARITY

We now turn to exactly soluble problems in three dimensions. In most of these the eigenvalue spectra and eigenfunctions associated with the angular momentum operators have a central role. So we start with the eigenvalue problem for orbital angular momentum.

4.6 The Angular Momentum Operators

Angular momentum is a vector given by $\mathbf{L} = \mathbf{x} \times \mathbf{p}$ with components

$$L_x = yp_z - zp_y,\ L_y = zp_x - xp_z,\ L_z = xp_y - yp_x \tag{4.50}$$

Their differential operator forms obtained by setting $\mathbf{p} = -i\hbar\nabla$, and the corresponding expressions in spherical polar coordinates, are given by

$$\begin{aligned}
L_x &= -i\hbar\left(y\frac{\partial}{\partial z} - z\frac{\partial}{\partial y}\right) = i\hbar\left(\sin\varphi\frac{\partial}{\partial\theta} + \cot\theta\cos\varphi\frac{\partial}{\partial\varphi}\right)\\
L_y &= -i\hbar\left(z\frac{\partial}{\partial x} - x\frac{\partial}{\partial z}\right) = i\hbar\left(-\cos\varphi\frac{\partial}{\partial\theta} + \cot\theta\sin\varphi\frac{\partial}{\partial\varphi}\right)\\
L_z &= -i\hbar\left(x\frac{\partial}{\partial y} - y\frac{\partial}{\partial x}\right) = -i\hbar\frac{\partial}{\partial\varphi}\,.
\end{aligned} \tag{4.51}$$

The square of the angular momentum vector,

$$\mathbf{L}^2 \equiv L_x^2 + L_y^2 + L_z^2 \tag{4.52}$$

can be expressed in terms of the operators of $\mathbf{x}$ and $\mathbf{p}$ using Eq. 4.50. (See Example 3.5, p. 77).

$$\mathbf{L}^2 = \mathbf{x}^2\mathbf{p}^2 - (\mathbf{x}.\mathbf{p})^2 + i\hbar\,\mathbf{x}.\mathbf{p} \tag{4.53}$$

The last term arises from the non-commutativity of $\mathbf{x}$ and $\mathbf{p}$ and is absent in classical mechanics. Setting $\mathbf{p} = -i\hbar\nabla$, we observe that

$$\mathbf{x}.\mathbf{p} = -i\hbar r\left(\frac{\mathbf{x}}{r}.\nabla\right) = -i\hbar r\frac{\partial}{\partial r} \tag{4.54}$$

Therefore, Eq. 4.53 may be rewritten as

$$\begin{aligned}
\mathbf{L}^2 &= -\hbar^2 r^2\nabla^2 + \hbar^2\left(r\frac{\partial}{\partial r}\right)^2 + \hbar^2 r\frac{\partial}{\partial r}\\
&= -\hbar^2\left[r^2\nabla^2 - \frac{\partial}{\partial r}\left(r^2\frac{\partial}{\partial r}\right)\right]
\end{aligned} \tag{4.55}$$

or

$$\nabla^2 = \frac{1}{r^2}\frac{\partial}{\partial r}\left(r^2\frac{\partial}{\partial r}\right) - \frac{1}{\hbar^2 r^2}\mathbf{L}^2 \tag{4.56}$$

Since the expression for ∇^2 in spherical polar coordinates is given by

$$\nabla^2 = \frac{1}{r^2}\frac{\partial}{\partial r}\left(r^2\frac{\partial}{\partial r}\right) + \frac{1}{r^2\sin\theta}\frac{\partial}{\partial\theta}\left(\sin\theta\frac{\partial}{\partial\theta}\right) + \frac{1}{r^2\sin^2\theta}\frac{\partial^2}{\partial\varphi^2} \tag{4.57}$$

we see from Eq. 4.55 that

$$\mathbf{L}^2 = -\hbar^2\left[\frac{1}{\sin\theta}\frac{\partial}{\partial\theta}\left(\sin\theta\frac{\partial}{\partial\theta}\right)+\frac{1}{\sin^2\theta}\frac{\partial^2}{\partial\varphi^2}\right] \tag{4.58}$$

This form will be the basis for our investigation of the eigenvalue problem for angular momentum in the next section. It could have been obtained, of course, from the polar coordinate forms in Eq. 4.51.

4.7 The Eigenvalue Equation for $\mathbf{L}^2$; Separation of Variables

The eigenvalue equation for $\mathbf{L}^2$ is, in view of Eq. 4.58, given by

$$\mathbf{L}^2 v \equiv -\hbar^2\left[\frac{1}{\sin\theta}\frac{\partial}{\partial\theta}\left(\sin\theta\frac{\partial}{\partial\theta}\right)+\frac{1}{\sin^2\theta}\frac{\partial^2}{\partial\varphi^2}\right]v = \lambda\hbar^2 v \tag{4.59}$$

where we have used the notation $\lambda\hbar^2$ for the eigenvalue parameter of $\mathbf{L}^2$. We try the method of separation of variables to solve this equation. Assuming the separable form as

$$v(\theta, \varphi) = \Theta(\theta)\,\Phi(\varphi) \tag{4.60}$$

we substitute it into Eq. 4.59, and then multiply by $(\sin^2\theta)/(\Theta\Phi)$. Noting that $(\partial/\partial\theta)$ acts on Θ only and $(\partial/\partial\varphi)$ on Φ only, we find the result to be expressible as

$$\left[\frac{\sin\theta}{\Theta}\frac{d}{d\theta}\left(\sin\theta\frac{d\Theta}{d\theta}\right)+\lambda\sin^2\theta\right] = -\frac{1}{\Phi}\frac{d^2\Phi}{d\varphi^2} \tag{4.61}$$

Now, since the left hand side is independent of φ and the right hand side is independent of θ, their equality implies that both sides are independent of both the variables which means that they have a common constant value. Denoting it by m^2, we can write Eq. 4.61 as two equations,

$$\frac{d^2\Phi}{d\varphi^2} = -m^2\Phi \tag{4.62}$$

and

$$\frac{1}{\sin\theta}\frac{d}{d\theta}\left(\sin\theta\frac{d\Theta}{d\theta}\right)+\left(\lambda-\frac{m^2}{\sin^2\theta}\right)\Theta = 0 \tag{4.63}$$

4.8 Admissibility Conditions on Solutions; Eigenvalues

As far as we know at this stage, λ and m^2 could be any (real or complex) numbers. However, restrictions on them arise as soon as the admissibility conditions for wave functions (Sec. 2.8) are imposed on Θ and Φ. Considering the latter first, we observe that Eq. 4.62 has $e^{-im\varphi}$ and $e^{im\varphi}$ as solutions. Since values of φ differing by integer multiples of 2π refer to the same physical point, these solutions will satisfy the condition of single-valuedness only if $e^{im\varphi} = e^{im(\varphi+2\pi)}$ i.e.,

$$e^{im\cdot 2\pi} = 1 \tag{4.64}$$

From the properties of the exponential function, it is well known that Eq. 4.64 will be true if and only if m is a real integer:

$$m = 0, \pm 1, \pm 2, \ldots \tag{4.65}$$

Thus the admissible solutions of Eq. 4.62 are

$$\Phi_m(\varphi) = (2\pi)^{-1/2}e^{im\varphi} \tag{4.66}$$

with m as in Eq. 4.65. Note that since both signs of m are included it is not necessary to mention $e^{-im\varphi}$ separately. The normalization constant $(2\pi)^{-1/2}$ is chosen in order that Φ be normalized to unity over the range 0 to 2π of φ. With this choice,

$$\int_0^{2\pi} \Phi^*_m(\varphi)\, \Phi_{m'}(\varphi)\, d\varphi = \delta_{mm'} \tag{4.67}$$

The quantum number m has an immediate physical significance: Φ_m, (defined by Eq. 4.66), is an eigenfunction of L_z and $m\hbar$ is the eigenvalue to which it belongs: $L_z\,\Phi_m = -\,i\hbar\,(\partial/\partial\varphi)\,\Phi_m = m\hbar\,\Phi_m$. In fact, not only Φ_m but also $\Theta\Phi_m$ is such an eigenfunction. *Thus the eigenfunctions of* $\mathbf{L}^2$ *which we get by separation of the variables* θ *and* φ *are also simultaneously eigenfunctions of* L_z. We will comment on this result in the next section.

Turning now to Eq. 4.63 for Θ, we note that the substitution

$$w = \cos\theta \tag{4.68}$$

converts it into

$$\frac{d}{dw}\left[(1-w^2)\frac{dP}{dw}\right] + \left[\lambda - \frac{m^2}{1-w^2}\right]P = 0 \tag{4.69}$$

where we have written

$$P(w) \equiv P(\cos\theta) = \Theta(\theta) \tag{4.70}$$

Assuming that P is a well-behaved function satisfying all the admissibility conditions, we infer, by multiplying Eq. 4.69 by $(1-w^2)$ and taking the limit as $w^2 \to 1$, that $m^2P(w) \to 0$ as $w \to \pm 1$. The vanishing of $P(w)$ when $w^2 = 1$, for $m \neq 0$, suggests that $P(w)$ contains a factor $(1-w^2)^a$ where $a > 0$ for $m \neq 0$. So we look for solutions of the form

$$P(w) = (1-w^2)^a\, K(w),\quad a > 0 \tag{4.71}$$

with $K(w)$ remaining finite at $w = \pm 1$. Substituting this into Eq. 4.69 we obtain

$$(1-w^2)\frac{d^2K}{dw^2} - 2\,(2a+1)w\frac{dK}{dw} + \left[\lambda - 4a^2 - 2a + \frac{4a^2 - m^2}{1-w^2}\right]K = 0 \tag{4.72}$$

The choice

$$4a^2 = m^2 \tag{4.73}$$

helps to eliminate the inconvenient last term in Eq. 4.72, and the simplified equation can be solved by assuming a series form for $K(w)$:

$$K(w) = c_\sigma w^\sigma + c_{\sigma+1}w^{\sigma+1} + \ldots = \sum_{s=\sigma}^{\infty} c_s w^s \tag{4.74}$$

Substituting this in Eq. 4.72 with Eq. 4.73 and equating the coefficient of w^r to zero, we find for $r = \sigma - 2$ and $\sigma - 1$ (the two lowest powers) that

$$\sigma(\sigma-1)\,c_\sigma = 0 \text{ and } (\sigma+1)\,\sigma c_{\sigma+1} = 0 \tag{4.75}$$

For all higher powers,

$$(r+2)\,(r+1)\,c_{r+2} - r\,(r-1)\,c_r - 2\,(2a+1)\,rc_r + (\lambda - 4a^2 - 2a)\,c_r = 0 \tag{4.76}$$

Note from Eq. 4.75 that since the coefficient of the first term, c_σ, is by definition nonzero, σ is either 0 or 1, and hence the possible values for r are positive integers. From Eq. 4.76 we have

$$\frac{c_{r+2}}{c_r} = -\,\frac{\lambda - (r+2a)\,(r+2a+1)}{(r+1)\,(r+2)} \tag{4.77}$$

For very large r,

$$\frac{c_{r+2}}{c_r} \to 1 + \frac{4a-2}{r} + \ldots \tag{4.78}$$

Since the coefficients tend to equality as $r \to \infty$, the series (4.74) evidently diverges for $w = 1$ (unless the series terminates after a finite number of terms). In fact, the asymptotic behaviour of the ratio in (4.78) is exactly the same as that of the ratio of successive coefficients in the expansion of $(1 - w^2)^{-2a}$. Therefore, in the neighbourhood of $w^2 = 1$, not only would $K(w)$ diverge like $(1 - w^2)^{-2a}$ but $P(w) = (1 - w^2)^a K(w)$ would also diverge like $(1 - w^2)^{-a}$. The only way to escape this unacceptable singularity of $P(w)$ at $w = \pm 1$ is by terminating the series (4.74) after a finite number of terms. This can be done by choosing

$$\lambda = (t + 2a)(t + 2a + 1) \tag{4.79}$$

where t may have any of the values 0, 1, 2, ... Then, with $c_t \neq 0$, we find from Eq. 4.77 that c_{t+2} and hence c_{t+4}, ... all vanish. Further, by taking $c_{t+1} = 0$ we ensure that $c_{t+3} = c_{t+5} = \ldots = 0$, and find that also $c_{t-1} = c_{t-3} = \ldots = 0$. Therefore $K(w)$ is an odd or even polynomial in w according as t in Eq. 4.79 is odd or even.

Up to this point in the solution of Eq. 4.69 we have not made any use of what we know about the admissible values of m as defined by Eq. 4.65. If we feed this information now into Eq. 4.73, we get

$$2a = |m| \tag{4.80}$$

a non-negative integer. (Remember, we have defined a to be positive for $m \neq 0$). Consequently, $t + 2a$ in Eq. 4.79 is a non-negative integer, say l:

$$t + 2a = t + |m| = l \tag{4.81}$$

Hence from Eq. 4.79, we obtain

$$\lambda = l(l + 1), \quad l = 0, 1, 2, \ldots \tag{4.82}$$

These are the admissible values of the parameter λ in (4.59), *and the eigenvalues of* $\mathbf{L}^2$ *are therefore* $\lambda\hbar^2 = l(l + 1)\hbar^2$ *with* l *as in* (4.82).

It may be observed that since t in Eq. 4.81 is non-negative, the values which m can take are limited by the value chosen for l:

$$m = l, l - 1, \ldots, -l + 1, -l \tag{4.83}$$

Since, for a given l, m can take any of these $(2l + 1)$ values, *the eigenvalue* $l(l + 1)\hbar^2$ *of* $\mathbf{L}^2$ *is* $(2l + 1)$*-fold degenerate.*

Before considering the physical significance of the quantum numbers l and m, let us complete the determination of the function $P(w)$.

4.9 The Eigenfunctions: Spherical Harmonics

The recurrence relation (4.77), with $\lambda = l(l + 1)$ and $2a = |m|$, can be trivially solved to obtain the coefficients in the polynomial expression for $K(w)$. Alternatively we can take advantage of the fact that Eq. 4.72, which we now rewrite as

$$(1-w^2)\frac{d^2K}{dw^2} - 2(|m| + 1)w\frac{dK}{dw} + \Big[l(l+1) - |m|(|m| + 1)\Big]K = 0 \tag{4.84}$$

is closely related to the well known Legendre differential equation. In fact, if $m = 0$, Eq. 4.84 reduces to the Legendre equation, whose polynomial solution is the Legendre polynomial $P_l(w)$. Explicitly the differential equation is given by

$$(1 - w^2)\frac{d^2P_l}{dw^2} - 2w\frac{dP_l}{dw} + l(l+1)P_l = 0 \tag{4.85}$$

By repeated differentiation of Eq. 4.85 it can be verified that if we set $(d^{|m|}/dw^{|m|})\,P_l(w) = K(w)$ it does satisfy Eq. 4.84. So this is the function whose power series we had determined. The function $P(w) = (1 - w^2)^{\frac{1}{2}|m|}$ $K(w)$ is now seen to be identical (apart from an arbitrary constant factor) with the associated Legendre function $P_l^m(w)$ defined by

$$P_l^m(w) = (1 - w^2)^{\frac{1}{2}m}\frac{d^m}{dw^m}P_l(w), \quad (m \geqslant 0),$$

$$P_l^m(w) = (-)^m\frac{(l+m)!}{(l-m)!}P_l^{|m|}(w), \quad (m < 0) \tag{4.86}$$

The most important formal properties of these functions are given in Appendix D. Here we will only note the fact that *for fixed m*, P_l^m and $P_{l'}^m$ are mutually orthogonal:

$$\int_{-1}^{1} P_l^m P_{l'}^m\, dw = \frac{2}{2l+1}\frac{(l+m)!}{(l-m)!}\delta_{ll'} \tag{4.87}$$

The function $\Theta(\theta)$ which we set out to determine can, now, be written down as $\Theta(\theta) \equiv P(w) = \text{const. } P_l^m(\cos\theta)$. Choosing the constant appropriately for normalization, we have for any l and m subject to the conditions (4.82) and (4.83),

$$\left.\begin{aligned} \Theta_{lm}(\theta) &= \left[\frac{2l+1}{2}\frac{(l-m)!}{(l+m)!}\right]^{1/2}(-)^m P_l^m(\cos\theta), \\ &\int_0^\pi \Theta_{lm}(\theta)\,\Theta_{l'm}(\theta)\sin\theta\, d\theta = \delta_{ll'} \end{aligned}\right\} \tag{4.88}$$

The complete eigenfunction of $\mathbf{L}^2$ corresponding to the quantum numbers l and m is thus[7]

$$\Theta_{lm}(\theta)\,\Phi_m(\varphi) = \left[\frac{2l+1}{4\pi}\frac{(l-|m|)!}{(l+|m|)!}\right]^{1/2}(-)^m P_l^m(\cos\theta)\,.\, e^{im\varphi}$$

$$\equiv Y_{lm}(\theta,\varphi) \tag{4.89}$$

The functions Y_{lm} are the well known *spherical harmonics*. The first few of these are

$$\left.\begin{aligned} Y_{00} &= \left(\frac{1}{4\pi}\right)^{1/2}, & Y_{10} &= \left(\frac{3}{4\pi}\right)^{1/2}\cos\theta \\ Y_{11} &= -\left(\frac{3}{8\pi}\right)^{1/2}\sin\theta . e^{i\varphi}, & Y_{1,-1} &= \left(\frac{3}{8\pi}\right)^{1/2} . \sin\theta . e^{-i\varphi} \\ Y_{20} &= \left(\frac{5}{16\pi}\right)^{1/2}(3\cos^2\theta - 1) \end{aligned}\right\} \tag{4.90}$$

[7] The use of the phase factor $(-)^m$ as in Eqs. 4.88 and 4.89 is a matter of convention. The spherical harmonics defined with this phase factor conform to certain sign conventions which are standard in the general theory of angular momentum (See Chapter 8).

$$\left.\begin{aligned} \Upsilon_{21} &= -\left(\frac{15}{8\pi}\right)^{1/2} \sin\theta \cos\theta . e^{i\varphi}, \quad \Upsilon_{2,-1} = \left(\frac{15}{8\pi}\right)^{1/2} \sin\theta \cos\theta . e^{-i\varphi} \\ \Upsilon_{22} &= \left(\frac{15}{32\pi}\right)^{1/2} \sin^2\theta\, e^{2i\varphi}, \quad \Upsilon_{2,-2} = \left(\frac{15}{32\pi}\right)^{1/2} \sin^2\theta\, e^{-2i\varphi} \end{aligned}\right\} \tag{4.90}$$

Their dependence on θ is shown in the polar diagrams given in Fig. 4.3. Some of their useful properties may be found in Appendix D. The functions Υ_{lm} form an orthonormal set, by virtue of Eqs. 4.67 and 4.87:

$$\int \Upsilon_{lm}^* \Upsilon_{l'm'}\, d\Omega = \int_0^{2\pi} \int_0^{\pi} \Upsilon_{lm}^* (\theta,\varphi)\, \Upsilon_{l'm'}(\theta,\varphi) \sin\theta\, d\theta\, d\varphi = \delta_{ll'}\delta_{mm'} \tag{4.91}$$

The notation $d\Omega$ emphasizes that the integration is with respect to the solid angle (i.e. directions in space) around the origin.

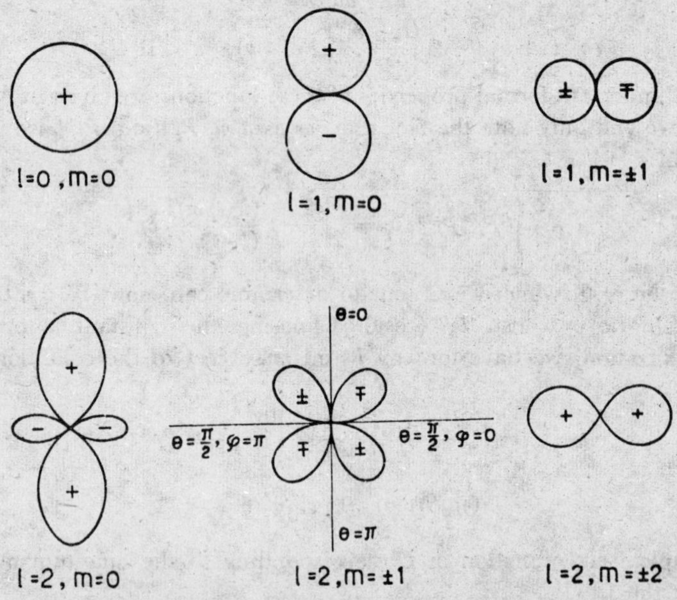

Fig. 4.3 Polar diagrams for the Υ_{lm} (θ, φ) for points in the $x-z$ plane. The right half of each diagram corresponds to φ = 0 and the left half to φ = π. The distance from the centre of the diagram to the curve, at an angle θ to the vertical, gives the magnitude of Υ_{lm} (θ, φ); the sign is indicated in the diagram. All diagrams are drawn to the same scale

4.10 Physical Interpretation

The results of our derivation may be summarized in the two equations

$$\mathbf{L}^2\, \Upsilon_{lm} (\theta,\varphi) = l\,(l + 1)\hbar^2\, \Upsilon_{lm} (\theta, \varphi), \quad l = 0, 1, 2, \ldots \tag{4.92a}$$

$$L_z\, \Upsilon_{lm} (\theta,\varphi) = m\hbar\, \Upsilon_{lm} (\theta, \varphi), \; m = l, l-1, \ldots, -l \tag{4.92b}$$

The last equation is the *quantum mechanical statement of space quantization*, namely that the z-component of the angular momentum can take only discrete values which are integer multiples of $\hbar$. More precisely, any observation designed to measure L_z can only yield the result $m\hbar$ with m as an integer. In atomic physics, the introduction of a magnetic field (whose direction is conventionally taken as the z-axis) causes the energy of the atom to change by an amount proportional to the z-component of its magnetic moment, which in turn is related to L_z. Thus space quantization manifests itself through discrete changes in atomic levels in a magnetic field (See Sec. 1.11). For this reason m is called the *magnetic quantum number.*

What we have said above regarding L_z applies equally well to any other component of **L** (along any direction, not necessarily along the coordinate axes). The reason why L_z has been in the foreground in our treatment is that we gave special prominence to the z-axis by choosing it as the direction of the polar axis in the spherical coordinate system. As far as fundamental properties like the commutation relations are concerned, L_x, L_y and L_z are all on the same footing. We recall here the commutators evaluated in Example 3.4 (p. 77):

$$[L_x, L_y] = i\hbar L_z, \quad [L_y, L_z] = i\hbar L_x, \quad [L_z, L_x] = i\hbar L_y \tag{4.93}$$

From these equations it follows that all the three components of **L** commute with $\mathbf{L}^2$:

$$[L_x, \mathbf{L}^2] = [L_y, \mathbf{L}^2] = [L_z, \mathbf{L}^2] = 0 \tag{4.94}$$

This can be seen by evaluating $[L_x, L_y^2]$, $[L_x, L_z^2]$ etc., using the reduction formula (3.15). For example, $[L_x, L_y^2] = [L_x, L_y]\, L_y + L_y\, [L_x, L_y] = i\hbar\, (L_z\, L_y + L_y\, L_z)$. It is because of the commutativity of $\mathbf{L}^2$ and L_z that they possess a *complete set* of simultaneous eigenfunctions (See Sec. 3.13). We could equally well obtain simultaneous eigenstates of $\mathbf{L}^2$ and L_x or $\mathbf{L}^2$ and L_y. Such states do not coincide with the $Y_{lm}(\theta, \varphi)$, because L_z does not commute with L_x (or L_y); as we have seen in Sec. 3.11, when two observables do not commute, a common eigenstate would be something very exceptional. However, the eigenstate of $\mathbf{L}^2$ and L_x (or $\mathbf{L}^2$ and L_y) belonging to the eigenvalues $l(l+1)\,\hbar^2$ and $\mu\hbar$ respectively can be expressed as a linear combination of the $(2l+1)$ functions $Y_{lm}(\theta, \varphi)$ characterized by the specified l. This is illustrated in Example 4.4 below, and dealt with more fully in Chapter 8. Actually, the very occurrence of the degeneracy of the eigenvalues of $\mathbf{L}^2$ ($2l+1$ — fold for any given l) is a reflection of an underlying symmetry, namely that in this problem any direction in space is as good as any other. The existence of the symmetry makes it possible to separate the operator $\mathbf{L}^2$ in many different ways into two commuting parts. The separation of variables used in our derivation was equivalent to the separation $\mathbf{L}^2 = (L_x^2 + L_y^2) + L_z^2$, but we could alternatively proceed on the basis of a separation $L^2 = L_x^2 + (L_y^2 + L_z^2)$, for example. (This would have led to simultaneous eigenfunctions of $\mathbf{L}^2$ and L_x). The link between the possibility of separating an eigenvalue equation in several ways, on the one hand, and the occurrence of degeneracy on the other, will show up again and again in our studies in the next few sections, and later in the book.

We close our discussion of angular momentum with a question. Is it possible to proceed purely by abstract operator methods, using the commutation relations (4.93) alone, as we did in the case of the harmonic oscillator? The answer is yes. But unlike in the harmonic oscillator case, Eq. 4.93 leads to something more than the states (4.92). The new element which emerges from the commutation relations (Chapter 8) is the existence of half-integral eigenvalues for angular momentum. These do not correspond to *orbital* angular momentum, however. To see what goes wrong if we assume a half-integral value for l, suppose $\lambda = \frac{3}{4}$ (corresponding to $l = \frac{1}{2}$) and $m = -\frac{1}{2}$ in Eqs. 4.69 and 4.62. It is easily verified that $v = (1 - w^2)^{1/4}\, e^{-\frac{1}{2}i\varphi}$ is the solution of these equations which is finite at $w = \pm 1$. But $L_x v$ is a singular function and $(L_x)^2 v$ is not even normalizable, as may be seen by using Eq. 4.51. So L_x, and similarly L_y, are not good operators on solutions corresponding to half-integral values of l, and such solutions have to be discarded. The single-valuedness criterion is a simple device which accomplishes this. However, half-integral values of spin angular momentum are still possible; the corresponding spin wave functions are not single valued but double-valued, as will be seen in Chapter 8.

EXAMPLE 4.4.—It is a remarkable fact that when $\mathbf{L}^2$ has the value $l(l+1)\hbar^2$, the maximum value possible for the component L_z is only l instead of $\sqrt{l(l+1)}\hbar$. If the latter value were to be achieved, the values of L_x and L_y would both have to be exactly zero; but the non-commutativity of L_x, L_y, L_z prevents them from having definite values simultaneously. The only exception to this statement is when the state is Y_{00}, as we have shown in Example 3.14, p.93.

EXAMPLE 4.5.—The three angular momentum functions corresponding to $l = 1$ in Eqs. 4.90 may be reexpressed in Cartesian coordinates as

$$Y_{1,1} = -\left(\frac{3}{8\pi}\right)^{1/2}\frac{x+iy}{r},\quad Y_{1,0} = \left(\frac{3}{4\pi}\right)^{1/2}\frac{z}{r},\quad Y_{1,-1} = \left(\frac{3}{8\pi}\right)^{1/2}\frac{x-iy}{r}.$$

From these forms one can directly write down the simultaneous eignfunctions of $\mathbf{L}^2$ and L_x (with $l = 1$) by simply making the *cyclic permutation* of variables: $x \to y,\ y \to z,\ z \to x$. Denoting them by $\phi_{1\mu}$, we have

$$\phi_{1,1} = -\left(\frac{3}{8\pi}\right)^{1/2}\frac{y+iz}{r},\quad \phi_{1,0} = \left(\frac{3}{4\pi}\right)^{1/2}\frac{x}{r},\quad \phi_{1,-1} = \left(\frac{3}{8\pi}\right)^{1/2}\frac{y-iz}{r}.$$

These functions can be written as linear combinations of the Y_{1m}. For example, $\phi_{1,1} = -\frac{1}{2}i\,(Y_{1,1}+Y_{1,-1}) - 2^{-\frac{1}{2}}i\, Y_{1,0}$. The sets Y_{1m} and $\phi_{1\mu}$ are alternative bases for the space of eigenfunctions belonging to the degenerate eigenvalue (labelled by $l = 1$) of $\mathbf{L}^2$. The student is urged to write down the eigenfunctions of L_x (or L_y) for $l = 2$ from the Cartesian forms of the Y_{2m}.

EXAMPLE 4.6.—Suppose a particle has the wave function $\phi_{1,1}$ so that it has the definite value $\hbar$ for L_x: what result would we get if we measured the z-component of its angular momentum? The answer is contained in the expansion of $\phi_{1,1}$ in terms of eigenfunctions of L_z. The coefficients of the eigenfunctions $Y_{1,1}$, $Y_{1,0}$ and $Y_{1,-1}$ in the expansion (See Example 4.5) are $(1/2i)$, $(1/\sqrt{2}i)$ and $(1/2i)$ respectively. Therefore, according to the general theory of Sec. 3.9, the absolute squares of these numbers, namely $\frac{1}{4}$, $\frac{1}{2}$ and $\frac{1}{4}$, are the respective probabilities that the result of the measurement will be

$\hbar$, 0, or $-\hbar$. The expectation value of L_z^2 in the state $\phi_{1,1}$ is $\frac{1}{4}.(\hbar)^2 + \frac{1}{2}(0)^2 + \frac{1}{4}(-\hbar)^2 = \frac{1}{2}\hbar^2$. The student may verify that this coincides with $\int \phi_{1,1}{}^* L_z{}^2 \phi_{1,1} d\Omega$.

4.11 Parity

The concept of parity was introduced in Sec. 2.11(c) as a symmetry property of wave functions. We will now define formally the *parity operator P* by specifying its effect on wave functions:

$$\psi'(\mathbf{x},t) \equiv P\psi(\mathbf{x},t) = \psi(-\mathbf{x},t) \tag{4.95}$$

i.e. the parity operator causes the function $\psi(\mathbf{x},t)$ to go over into a new function $\psi'(\mathbf{x},t)$ whose value at $\mathbf{x}$ is the same as that of ψ at $-\mathbf{x}$. Since $\mathbf{x}$ and $-\mathbf{x}$ are inverse points, equidistant from the origin but in opposite directions, this operation is also called *space inversion.* (See Fig. 4.4) It should be observed that the parity operator has no classical counterpart; there is no classical dynamical variable from which it can be obtained by the operator correspondence $\mathbf{x} \to \mathbf{x}$, $\mathbf{p} \to -i\hbar\nabla$.

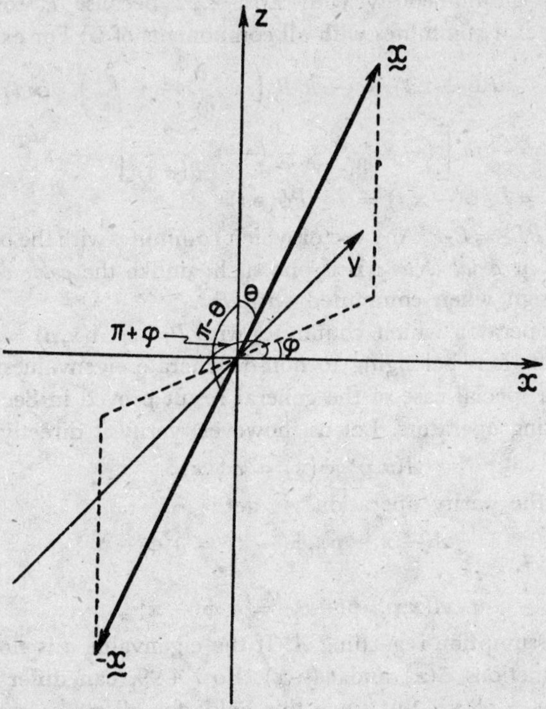

Fig. 4.4 Space Inversion — Effect on Cartesian and polar co-ordinates

The definition (4.95) implies evidently that $P^2\,\psi(\mathbf{x},t) = \psi(\mathbf{x},t)$ for any arbitrary ψ. From this, the eigenvalues of P are immediately found. For, if $\phi_\lambda\,(\mathbf{x},t)$ is an eigenstate of P, i.e. $P\,\phi_\lambda = \lambda\phi_\lambda$ then $P^2\,\phi_\lambda = \lambda^2\,\phi_\lambda$, but this must be equal to ϕ_λ since P^2 leaves any function unchanged, as noted above. Hence $\lambda^2 = 1$, and $\lambda = \pm 1$.

The parity eigenvalue $\lambda = 1$ is referred to as *even parity* since $P\phi = \phi$ means simply that $\phi(-\mathbf{x},t) = \phi(\mathbf{x},t)$, i.e. ϕ is an even function of $\mathbf{x}$. Similarly, *odd parity* functions are characterized by $\phi(-\mathbf{x},t) = -\phi(\mathbf{x},t)$ or $P\phi(\mathbf{x},t) = -\phi(\mathbf{x},t)$. We have already seen examples of functions of definite parity in one dimensional problems [Sec. 2.11(c), Sec. 4.2(a)].

The spherical harmonics are functions of definite parity. To see this we translate (4.95) into polar coordinate form:

$$P\psi(r,\theta,\varphi) = \psi(r,\pi-\theta,\ \varphi+\pi) \tag{4.96}$$

Under the change $\varphi \to \varphi + \pi$, $e^{im\varphi} \to e^{im(\varphi+\pi)} = (-)^m e^{im\varphi}$, so that Φ_m has parity $(-)^m$. Further, since $\theta \to \pi - \theta$, $\cos\theta \to \cos(\pi-\theta) = -\cos\theta$, and $\sin\theta \to \sin\theta$. Therefore, the factor $(1-w^2)^{\frac{1}{2}|m|} = \sin^{|m|}\theta$ in $P_l^m(w)$ is unaffected, while the polynomial factor acquires a sign $(-)^{l-|m|}$. Thus $Y_{lm}(\theta,\varphi) \to (-)^{l-|m|+m} Y_{lm}(\theta,\varphi) = (-)^l Y_{lm}(\theta,\varphi)$, since m is an integer. In other words, Y_{lm} has the parity $(-)^l$:

$$PY_{lm}(\theta,\varphi) = (-)^l\, Y_{lm}(\theta,\varphi) \tag{4.97}$$

Eq. 4.97 holds simultaneously with Eqs. 4.92 because P commutes with L_z and $\mathbf{L}^2$. In fact it commutes with all components of $\mathbf{L}$: For example,

$$\left.\begin{aligned} PL_z\psi(\mathbf{x},t) &= -i\hbar\, P\left(x\frac{\partial}{\partial y} - y\frac{\partial}{\partial x}\right)\psi(\mathbf{x},t) \\ &= -i\hbar\left[(-x)\frac{\partial}{\partial(-y)} - (-y)\frac{\partial}{\partial(-x)}\right]\psi(-\mathbf{x},t) \\ &= L_z\,\psi(-\mathbf{x},t) = L_z\,P\psi(\mathbf{x},t) \end{aligned}\right\} \tag{4.98}$$

for *all* ψ, so that $PL_z = L_zP$. Any vector which commutes with the parity operator is a *pseudo-vector* or *axial vector*; $\mathbf{L}$ is one such, unlike the *polar vectors* $\mathbf{x}$ and $\mathbf{p}$ which change sign when commuted with P.

If A is any operator which commutes with P, i.e. $A(\mathbf{x},\mathbf{p}) = A(-\mathbf{x},-\mathbf{p})$, then its eigenfunctions belonging to non-degenerate eigenvalues have definite parity. This is a special case of the general result proved in Sec. 3.13 for any pair of commuting operators. Let us, however, verify it directly. If

$$A(\mathbf{x},\mathbf{p})\ \phi(\mathbf{x}) = a\phi(\mathbf{x}) \tag{4.99a}$$

on performing the parity operation we get

$$A(-\mathbf{x},-\mathbf{p})\ \phi(-\mathbf{x}) = a\ \phi(-\mathbf{x}),$$

or

$$A(\mathbf{x},\mathbf{p})\ \phi(-\mathbf{x}) = a\ \phi(-\mathbf{x}) \tag{4.99b}$$

in view of our assumption regarding A. If the eigenvalue a is non-degenerate, the two eigenfunctions $\phi(\mathbf{x})$ and $\phi(-\mathbf{x})$, Eqs. 4.99, can differ at most by a constant: $\phi(-\mathbf{x}) = c\phi(\mathbf{x})$. But since this holds for all $\mathbf{x}$, we can equally well write it as $\phi(\mathbf{x}) = c\phi(-\mathbf{x})$. Putting together the two forms, we get $\phi(\mathbf{x}) = c^2\phi(\mathbf{x})$ or $c^2 = 1$. Hence $\phi(-\mathbf{x}) = \pm\,\phi(\mathbf{x})$, which is what we wanted to prove.

We see now that it is no accident that the Y_{lm} as well as the energy eigenfunctions in the harmonic oscillator and the square well potential problems have definite parity.

4.12 Angular Momentum in Stationary States of Systems with Spherical Symmetry.

The knowledge of angular momentum eigenfunctions is of immediate utility in the study of the stationary states of spherically symmetric systems. (*a*) *The Rigid Rotator:* Consider a particle which is constrained to remain at a constant distance r_0 from the origin, so that its motion is purely one of rotation about the origin. The kinetic energy of this motion is $\mathbf{L}^2/2I$, where $I = mr_0^2$ is the moment of inertia. If there are no forces affecting the motion, i.e. the rotation is completely unhindered, there is no potential energy to reckon with. The Hamiltonian is then equal to $\mathbf{L}^2/2I$. Since this differs from $\mathbf{L}^2$ only through a constant factor, we immediately obtain the stationary state wave functions to be the spherical harmonics Y_{lm}; the corresponding energy eigenvalues are

$$E_l = \frac{l(l+1)}{2I}\hbar^2, \qquad l = 0,1,2,\ldots \tag{4.100}$$

The same result holds good for a system of two particles rigidly held together (with a constant interparticle separation r_0) and rotating about the centre of mass. (This is because the motion of one particle completely determines that of the other). The only difference is that the moment of inertia I (about the centre of mass) is now μr_0^2 where $\mu = [mM/(m+M)]$, m and M being the masses of the two particles. (See Sec. 6.17). This system is called a *rigid rotator* and it serves as a good approximate model for the motion of diatomic molecules. Note the pattern of energy level spacing,

$$E_l - E_{l-1} = \frac{\hbar^2}{I}\cdot l \tag{4.101}$$

which shows an increase proportional to l in contrast with the constant spacing in the harmonic oscillator.

(*b*) *A Particle in a Central Potential; The Radial Equation:* Consider now a particle moving in a central potential, i.e. a potential $V(r)$ which is a function of the radial coordinate only. Its Hamiltonian, $H = \mathbf{p}^2/2\mu + V(r)$, commutes with $\mathbf{L}$ and hence with $\mathbf{L}^2$. (See Example 3.15, p.97; or use the expression (4.56) for ∇^2 and remember that $\mathbf{L}$ commutes with r and $\partial/\partial r$). Therefore, according to the general theory of Sec. 3.13, we can find a complete set of eigenfunctions of H which are simultaneously eigenfunctions of L_z and $\mathbf{L}^2$ also. Since the latter are the spherical harmonics $Y_{lm}(\theta,\varphi)$, it follows that we can take the energy eigenfunctions in the form

$$u(r,\theta,\varphi) = R(r)Y_{lm}(\theta,\varphi) \tag{4.102}$$

A less direct way of arriving at the same result is to write the eigenvalue equation $Hu = Eu$ explicitly as

$$\left[-\frac{\hbar^2}{2\mu}\nabla^2 + V(r)\right]u = Eu \tag{4.103}$$

and rewrite it as

$$-\frac{\hbar^2}{2\mu r^2}\left[\frac{\partial}{\partial r}r^2\frac{\partial}{\partial r} - \frac{\mathbf{L}^2}{\hbar^2}\right]u + V(r)u = Eu \tag{4.104}$$

using Eq. 4.56, and then separate it into radial and angular parts by the substitution $u = R(r)\Upsilon(\theta,\varphi)$. The separation is done exactly as in the case of Eq. 4.59. The result is that Υ is an eigenfunction of $\mathbf{L}^2$, and we thus recover (4.102). Since $\mathbf{L}^2\Upsilon_{lm} = l(l+1)\hbar^2\Upsilon_{lm}$, we obtain, on substituting, Eq. 4.102 in Eq. 4.104 the following *ordinary* differential equation for $R(r)$:

$$\frac{1}{r^2}\frac{d}{dr}\left(r^2\frac{dR}{dr}\right)+\frac{2\mu}{\hbar^2}\left[E-V-\frac{l(l+1)\hbar^2}{2\mu r^2}\right]R=0 \qquad (4.105)$$

The eigenvalue problem for a spherically symmetric potential thus reduces to determining for what values of E the *radial wave equation* (4.105) has admissible solutions, and then finding the solutions. Note that the term $l(l+1)\hbar^2/(2\mu r^2)$ in this equation appears as an addition to the potential V. It is often called the *centrifugal potential*, since its negative gradient is equal to the centrifugal force experienced by a particle moving in an orbit of radius r with angular momentum $[l(l+1)]^{1/2}\hbar$. Alternatively, it may be looked upon as the kinetic energy associated with the rotatory part of the motion, $\mathbf{L}^2/(2I)$ where $I = \mu r^2$ is the moment of inertia.

(*c*) *The Radial Wave Function:* The norm of the wave function u, when expressed in spherical polar coordinates, is evidently given by

$$\int_0^\infty\int_0^\pi\int_0^{2\pi} |u(r,\theta,\varphi)|^2\, d\varphi.\sin\theta\, d\theta.r^2dr \qquad (4.106)$$

When u is factorized into $R(r)\Upsilon_{lm}$, the norm reduces to

$$\int_0^\infty |R(r)|^2\, r^2dr \qquad (4.107)$$

since Υ_{lm} itself has unit norm. This integral has unit value if u is a normalized wave function. Then the expectation value of any operator A *which involves the radial variables only* is given by

$$\langle A\rangle = \int \{R^*(r)AR(r)\}\,.\,r^2dr \qquad (4.108)$$

It is often useful to write

$$R(r) = \frac{1}{r}\chi(r) \qquad (4.109)$$

because it enables the radial wave equation to be reduced to the simpler form

$$\frac{d^2\chi}{dr^2}+\left[\frac{2\mu}{\hbar^2}(E-V)-\frac{l(l+1)}{r^2}\right]\chi=0 \qquad (4.110)$$

The behaviour of the wave function near the origin can be readily seen from this equation. For any $l \neq 0$, the centrifugal term diverges rapidly as $r\to 0$ and dominates over the other terms (assuming that V itself is not so singular at the origin, i.e. $r^2V(r)\to 0$ as $r\to 0$). Thus in the neighbourhood of the origin we have, as a good approximation,

$$\frac{d^2\chi}{dr^2}-\frac{l(l+1)}{r^2}\chi\approx 0, \quad (r\to 0) \qquad (4.111)$$

This approximate equation admits $\chi =$ const. r^{l+1} and $\chi =$ const. r^{-l} as solutions. The latter is not acceptable since it makes $R(r)$ diverge as $r\to 0$. We are then left with the other solution, $\chi(r) \propto r^{l+1}$ which leads to

$$R(r) \approx \text{const. } r^l \quad \text{as} \quad r\to 0 \qquad (4.112)$$

Thus any acceptable solution, for angular momentum l, must behave like r^l near the origin.

It is necessary to know also the asymptotic behaviour of radial wave functions. We note that as $r \to \infty$, the centrifugal term in Eq. 4.110 becomes negligible. If V also tends to zero as $r \to \infty$, the equation reduces to

$$\frac{d^2\chi}{dr^2} + \frac{2\mu E}{\hbar^2}\chi \approx 0, \quad (r \to \infty) \tag{4.113}$$

Evidently this is satisfied by $\chi = e^{-\alpha r}$ where $\alpha^2 = -2\mu E/\hbar^2$. Actually even if we take $\chi = r^n e^{-\alpha r}$ for any positive n, $(d^2\chi/dr^2)$ and $(-2\mu E/\hbar^2)\chi$ differ only in terms involving r^{n-1} or lower powers of r. Therefore Eq. 4.113 is still satisfied by the *leading terms*, i.e. those containing the highest power (r^n) which dominates when r is very large. Thus the dominant asymptotic behaviour of R is like

$$R(r) \approx \text{const. } r^n e^{-\alpha r} \quad (r \to \infty); \quad \alpha^2 = -2\mu E/\hbar^2 \tag{4.114}$$

where n is positive but otherwise undetermined at this stage.

Finally, we observe that in the special case of s-states ($l = 0$) Eq. 4.110 is identical in form to the Schrödinger equation in one dimension. However, unlike the wave-function in one dimension which merely has to remain finite, the function $\chi = rR$ must vanish at the origin. So the s-state solutions in the three dimensional case correspond to those solutions of the one dimensional problem (with the same form for V) in which the wave function vanishes at the origin.

C. THREE-DIMENSIONAL SQUARE WELL POTENTIAL

This potential is frequently used as an approximate model in many problems, for example in nuclear physics, where the exact nature of the actual potential is not known. It is defined by $V(r) = -V_0$ for $r < a$, i.e. within a sphere of radius a, and $V(r) = 0$ for $r > a$, outside the sphere. In view of the spherical symmetry, we look for eigenfunctions in the form $R(r)Y_{lm}(\theta,\varphi)$. The radial equation (4.105) for R now takes two distinct forms in the two regions:

$$\frac{1}{r^2}\frac{d}{dr}\left(r^2\frac{dR}{dr}\right) + \frac{2\mu}{\hbar^2}\left[E + V_0 - \frac{l(l+1)}{r^2}\right]R = 0, \; r < a \tag{4.115a}$$

$$\frac{1}{r^2}\frac{d}{dr}\left(r^2\frac{dR}{dr}\right) + \frac{2\mu}{\hbar^2}\left[E - \frac{l(l+1)}{r^2}\right]R = 0, \; r > a \tag{4.115b}$$

We have to solve these equations and 'match' the solutions at the boundary between the two regions, requiring that R and dR/dr be continuous at $r = a$.

4.13 Solutions in the Interior Region

It is convenient to define

$$\beta^2 = \frac{2\mu}{\hbar^2}(E + V_0) \text{ and } \rho = \beta r \tag{4.116}$$

and rewrite the Eq. 4.115a for the interior region ($r < a$) in terms of them as

$$\frac{d^2R}{d\rho^2} + \frac{2}{\rho}\frac{dR}{d\rho} + \left[1 - \frac{l(l+1)}{\rho^2}\right]R = 0 \tag{4.117}$$

This equation has a strong resemblance to the well-known Bessel equation

(See Eq. D.19, Appendix D). In fact, it is easy to verify that Eq. 4.117 is satisfied by the *spherical Bessel function* $j_l(\rho)$ which is related to the ordinary Bessel function $J_{l+\frac{1}{2}}(\rho)$ through

$$j_l(\rho) = \left(\frac{\pi}{2\rho}\right)^{1/2} J_{l+\frac{1}{2}}(\rho) \tag{4.118}$$

As noted in Eq. D.31, this solution is proportional to ρ^l when ρ is very small and is, therefore, well-behaved or *regular* at $\rho = 0$. Eq. 4.117, being a second order equation, has another independent solution which is *singular* at the origin, being proportional to ρ^{-l-1} for very small ρ. It is denoted by

$$n_l(\rho) = (-)^{l+1} \left(\frac{\pi}{2\rho}\right)^{1/2} J_{-l-\frac{1}{2}}(\rho) \tag{4.119}$$

The nature of the j_l and n_l functions may be seen from the following examples:

$$\left.\begin{aligned} j_0(\rho) &= \frac{\sin\rho}{\rho}, \quad j_1(\rho) = \frac{\sin\rho}{\rho^2} - \frac{\cos\rho}{\rho} \\ j_2(\rho) &= \left(\frac{3}{\rho^3} - \frac{1}{\rho}\right) \sin\rho - \frac{3}{\rho^2}\cos\rho, \\ n_0(\rho) &= -\frac{\cos\rho}{\rho}, \quad n_1(\rho) = -\frac{\cos\rho}{\rho^2} - \frac{\sin\rho}{\rho} \\ n_2(\rho) &= -\left(\frac{3}{\rho^3} - \frac{1}{\rho}\right)\cos\rho - \frac{3}{\rho^2}\sin\rho \end{aligned}\right\} \tag{4.120}$$

A general solution of Eq. 4.117 would be some linear combination of $j_l(\rho)$ and $n_l(\rho)$. But since the admissibility conditions on the wave function require R to be finite everywhere, $n_l(\rho)$ must not occur in R *in the interior* region which includes $\rho = 0$ (the singular point of n_l). Thus we have, for $l = 0,1,2,\ldots,$

$$R_l(r) = A j_l(\beta r), \quad r < a \tag{4.121}$$

where A is a constant.

4.14 Solutions in the Exterior Region and Matching

Eq. 4.115b for the exterior region differs from Eq. 4.115a only in having E (which may be positive or negative) in the place of the positive quantity $(E + V_0)$. Therefore, to obtain the solutions in this case, all we have to do is to replace β^2, Eq. 4.116, by α^2 (for $E > 0$) or $-\alpha^2$ (for $E < 0$), where

$$\alpha^2 = \frac{2\mu\,|E|}{\hbar^2}, \quad \alpha > 0 \tag{4.122}$$

(a) *Nonlocalized States* ($E > 0$): In this case we have $j_l(\alpha r)$ and $n_l(\alpha r)$ as possible solutions in the exterior region. Since this region does not include the singular point of n_l, namely $r = 0$, there is no objection to the presence of n_l in the general solution. Thus for $r > a$, $R(r)$ is a linear combination of $j_l(\alpha r)$ and $n_l(\alpha r)$. One could say equally well that it is a linear combination of the *Hankel functions*, defined by

$$\begin{aligned} h_l^{(1)}(\rho) &= j_l(\rho) + i n_l(\rho) \\ h_l^{(2)}(\rho) &= j_l(\rho) - i n_l(\rho) \end{aligned} \tag{4.123}$$

The advantage of using these functions in preference to n_l and j_l will be apparent presently, when we consider the asymptotic behaviour of R_l. We note that when $\rho \to \infty$,

$$h_l^{(1)}(\rho) \approx \text{const.}\ \frac{e^{i\rho}}{\rho},\quad h_l^{(2)}(\rho) \approx \text{const.}\ \frac{e^{-i\rho}}{\rho} \tag{4.124}$$

Writing

$$R_l(r) = B_1\, h_l^{(1)}(\alpha r) + B_2\, h_l^{(2)}(\alpha r),\ r > a \tag{4.125}$$

and matching R and dR/dr between the interior and exterior regions at their common boundary $r = a$, we find that

$$(B_1/A) = [\alpha h_l^{(2)\prime}(\alpha a)\, j_l'(\beta a) - \beta j_l'(\beta a) h_l^{(2)}(\alpha a)]/D \tag{4.126a}$$

$$(B_2/A) = [\beta j_l'(\beta a)\, h_l^{(1)}(\alpha a) - \alpha h_l^{(1)\prime}(\alpha a)\, j_l(\beta a)]/D \tag{4.126b}$$

where

$$D = \alpha\, [h_l^{(1)}(\alpha a)\, h_l^{(2)\prime}(\alpha a) - h_l^{(2)}(\alpha a)\, h_l^{(1)\prime}(\alpha a)] \tag{4.126c}$$

The prime here denotes differentiation with respect to the argument of the function. Note that $h_l^{(1)}$ and $h_l^{(2)}$ are complex conjugates of each other, and so are (B_1/A) and (B_2/A). Therefore the exterior solution (4.125) is real if A is.

Thus for any energy $E > 0$, the radial wave function is uniquely determined, being given by Eqs. 4.121 and 4.125 with Eqs. 4.126. It is not normalizable, since the radial part of the integral $\int |u|^2 d\tau$ defining the norm is $\int |R_l|^2 r^2 dr$ and its integrand tends to a constant value as $r \to \infty$ in view of the asymptotic behaviour given in Eq. 4.124. Hence it is verified that in the states for $E > 0$, the particle is not localized in any finite region.

(*b*) *Bound states* ($E < 0$): We have already noted that this case differs from that for $E > 0$ only in the replacement of α^2 by $-\alpha^2$. So the general solution in the exterior region would now be given by Eqs. 4.125 and 4.126 with α replaced by $i\alpha$ i.e., $R_l(r) = B_1' h_l^{(1)}(i\alpha r) + B_2' h_l^{(2)}(i\alpha r)$ where B_1' and B_2' are obtained from Eqs. 4.126 by replacing α by $i\alpha$. However, $h^{(2)}(i\alpha r)$ diverges like $e^{\alpha r}$ for large r as can be seen by setting $\rho = i\alpha r$ in Eq. 4.124. Therefore, its coefficient B_2' must vanish in order that the solution be an admissible one:

$$\beta j_l'(\beta a)\, h_l^{(1)}(i\alpha a) - i\alpha\, h_l^{(1)\prime}(i\alpha a)\, j_l(\beta a) = 0 \tag{4.127}$$

Since α and β depend on E, this equation determines implicitly, for each l, the various energies E_{nl} for which bound states exist. Needless to say, the bound states form a discrete set. The corresponding eigenfunctions are given by

$$\left.\begin{aligned} R_{nl}(r) &= A\, j_l(\beta r) \qquad & r < a \\ &= B_1'\, h_l^{(1)}(i\alpha r) \qquad & r \geq a \end{aligned}\right\} \tag{4.128}$$

where n is a label to distinguish the different solutions of (4.127) for given l. Note that α and β are dependent on n and l through E_{nl}. So also are B_1' and A, the latter being determined by the normalization $\int |R(r)|^2 r^2 dr = 1$. The Hankel functions of imaginary argument are real apart from a factor i^l, and the reader may verify that if A is chosen real, then the radial wave function (4.128) is real everywhere.

The rather simple special case of s-states ($l = 0$) is of considerable interest. From Eqs. 4.120 we see that $j_0(\rho) = (\sin\rho)/\rho$ and $h_0^{(1)}(\rho) = e^{i\rho}/(i\rho)$. On using these forms in Eq. 4.127 and simplifying, it reduces to $\beta \cot \beta a = -\alpha$. This is identical with Eq. 2.66a which determines the odd parity levels of the square well potential in one dimension. Thus the energy levels of s-states of a three dimensional square well of depth V_0 and radius a coincide with the odd parity levels of a one-dimensional square well having the same depth and half-width a. (This result provides an illustration of the general remarks made at the end of Sec. 4.12c, and could be obtained directly and simply from Eq. 4.110. From Fig. 2.3 and the discussion following Eq. 2.71, it is clear that if at least one such bound state is to exist, we must have

$$V_0 > \pi^2 \Delta \quad \text{or} \quad V_0 a^2 > (\pi^2 \hbar^2 / 2\mu) \tag{4.129}$$

Unlike the one-dimensional case where a bound state exists however small V_0 may be, in three dimensions the well must have a minimum size characterized by (4.129) if there is to be any bound state. (This condition was derived for s-states; for bound states with $l > 0$, the well must be still larger.)

Fig. 4.5 shows the energy levels in a three-dimensional square well, as a function of (V_0/Δ). The number of levels for a given depth is of course finite. (Only the two lowest energy levels for a given l are given in the figure).

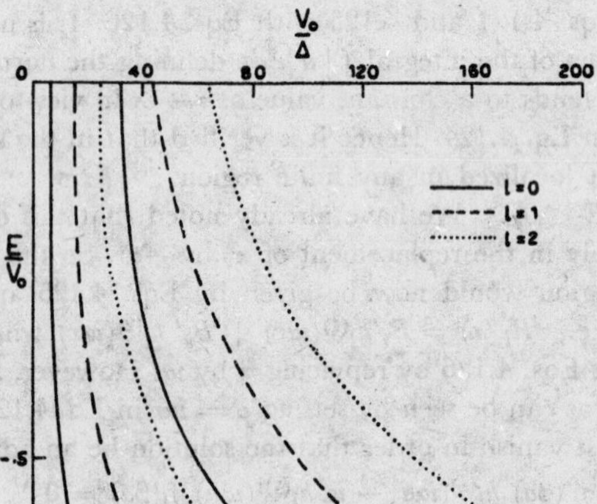

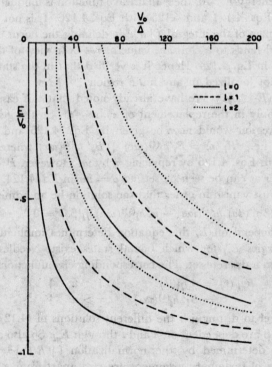

Fig. 4.5 Energies of the lowest two s, p and d states in a three-dimensional square well, as a function of well size.

D. THE HYDROGEN ATOM

We now take up the wave equation for a one-electron atom (hydrogen, singly ionized helium, doubly ionized lithium etc.). This is a two-particle system, consisting of the atomic nucleus of charge Ze, and the electron of charge $-e$. But if the nucleus is supposed to remain static, the Schrödinger equation is just that for a single particle, the electron, moving in a potential $V(r) = -Ze^2/r$. (This is just the electrostatic potential energy of interaction of the two charges). Actually when the atom as a whole is at rest, it is not the nucleus, but the centre of mass of the two-particle system which remains static. However, this fact can be rigorously taken into account by using, in the kinetic energy term in the Hamiltonian, the 'reduced mass' μ instead of the electron mass m_e. We shall prove this in Sec. 6.17. The reduced mass is given by the relation $\mu = m_e m_n/(m_e + m_n)$ where m_n is the mass of the nucleus.

4.15 Solution of the Radial Equation; Energy Levels

Since the potential is spherically symmetric, our task is just to solve the radial wave equation (4.105), which in the present case becomes

$$\frac{1}{r^2}\frac{d}{dr}\left(r^2\frac{dR}{dr}\right) + \frac{2\mu}{\hbar^2}\left[E + \frac{Ze^2}{r} - \frac{l(l+1)\hbar^2}{2\mu r^2}\right] R = 0 \tag{4.130}$$

Let us consider bound states, for which $E < 0$, and define the positive real parameters α and λ through

$$\alpha^2 = -\frac{8\mu E}{\hbar^2} \quad \text{and} \quad \lambda = \frac{2\mu Ze^2}{\alpha\hbar^2} = \frac{Ze^2}{\hbar}\left(\frac{\mu}{-2E}\right)^{1/2} \tag{4.131}$$

Dividing Eq. 4.130 throughout by α^2, we rewrite it in dimensionless form as

$$\frac{1}{\rho^2}\frac{d}{d\rho}\left(\rho^2\frac{d\mathcal{R}}{d\rho}\right) + \left[\frac{\lambda}{\rho} - \frac{1}{4} - \frac{l(l+1)}{\rho^2}\right]\mathcal{R} = 0 \tag{4.132}$$

where we have written

$$\mathcal{R}(\rho) = R(r) \text{ with } \rho = \alpha r \tag{4.132a}$$

As in the harmonic oscillator problem, we examine first the behaviour of $\mathcal{R}$ in the asymptotic region. It is clear that as $\rho \to \infty$, the term $-\frac{1}{4}$ is dominant within the square brackets. This suggests that the dominant behaviour of $\mathcal{R}$ for large ρ is like $e^{-\frac{1}{2}\rho}$. So we could write $\mathcal{R}(\rho) = K(\rho)\, e^{-\frac{1}{2}\rho}$ and try to determine $K(\rho)$. However, it simplifies matters somewhat if we remember that for small ρ, $\mathcal{R}(\rho)$ behaves like ρ^l [as shown in Sec. 4.12(b)] and incorporate such a factor also explicitly into $\mathcal{R}(\rho)$. So we write

$$\mathcal{R}(\rho) = \rho^l e^{-\frac{1}{2}\rho} L(\rho) \tag{4.133}$$

Substituting this into Eq. 4.132, we obtain the following differential equation for $L(\rho)$:

$$\rho\frac{d^2L}{d\rho^2} + [2(l+1) - \rho]\frac{dL}{d\rho} + [\lambda - (l+1)]\, L = 0 \tag{4.134}$$

To solve this equation, let us assume a solution for $L(\rho)$ in the series form:

$$L(\rho) = c_0 + c_1\rho + c_2\rho^2 + \ldots = \sum_{s=0}^{\infty} c_s\rho^s \tag{4.135}$$

Since we know that $\mathcal{R}(\rho)$ behaves like ρ^l for small ρ, $L(\rho)$ must tend to a constant as $\rho \to 0$, according to Eq. (4.133). This is why the series has been taken to start with a constant term. On substituting Eq. 4.135 into Eq. 4.134, we obtain the recurrence relation

$$c_{s+1} = \frac{s + l + 1 - \lambda}{(s+1)\,(s+2l+2)}\, c_s \tag{4.136}$$

We must ensure that the series (4.135) determined by this relation behaves asymptotically in such a way as not to spoil the exponential decrease of $\mathcal{R}(\rho)$ which is built into Eq. 4.133. The dominant asymptotic properties are determined by the limiting behaviour of (c_{s+1}/c_s) as $s \to \infty$. From Eq. 4.136 we find that for large s,

$$(c_{s+1}/c_s) \to (1/s) \tag{4.137}$$

which is just the ratio of successive coefficients in the expansion of e^ρ. Thus, if the series for $L(\rho)$ does not terminate, then

$$L(\rho) \xrightarrow[\rho\to\infty]{} \text{const. } e^\rho,\ \mathcal{R}(\rho) \to \text{const. } \rho^l\, e^{\frac{1}{2}\rho} \tag{4.138}$$

i.e., it diverges. To avoid this, we have to make the series terminate. This can be done by choosing

$$\lambda = n = \text{integer} \tag{4.139}$$

so that when s reaches the value n' given by

$$n' = n - l - 1 \tag{4.140}$$

the factor $(s + l + 1 - \lambda)$ in Eq. 4.136 vanishes. Consequently $c_{n'+1}$ and all higher coefficients become zero, and $L(\rho)$ becomes a polynomial. The degree of the polynomial, n', must of course be a non-negative integer. Hence it follows from Eq. 4.140 that the possible values of $n = n' + l + 1$ are the positive integers, and that l cannot exceed $n - 1$:

$$n = 1,2,3\ldots;\ l = 0,1,\ldots, n-1 \tag{4.141}$$

The condition (4.139) which determines the admissible solutions also gives us the energy eigenvalues E_n: on substituting $\lambda = n$ in Eq. 4.131 we get

$$E_n = -\ |E_n| = -\frac{\mu Z^2 e^4}{2\hbar^2 n^2},\quad n = 1,2,3,\ldots \tag{4.142}$$

4.16 Stationary State Wave Functions

It is customary to express the polynomial solutions of (4.134) (for $\lambda = n = 1, 2\ldots$) in terms of the *associated Laguerre polynomials*[8] $L_q^p(\rho)$. These are defined through the generating function given in Appendix D (Eq. D.14), and are known to satisfy the equation:

$$\rho \frac{d^2}{d\rho^2} L_q^p + (p + 1 - \rho) \frac{d}{d\rho} L_q^p + (q - p)\, L_q^p = 0 \tag{4.143}$$

Our equation (4.134) (with $\lambda = n$) has precisely this form, with $p = 2l + 1$ and $q = n + l$. Therefore, its polynomial solution is the associated Laguerre polynomial, $L_{n+l}^{2l+1}(\rho)$, apart from an arbitrary constant factor, say N_{nl}. Correspondingly, $\mathcal{R}(\rho)$ for a given n and l is obtained from Eq. 4.133 as

$$\mathcal{R}_{ln}(\rho) = N_{nl}\, \rho^l e^{-\frac{1}{2}\rho} L_{n+l}^{2l+1}(\rho) \tag{4.144}$$

where

$$\rho = \alpha r = \frac{2Zr}{na} \quad \text{with } a = \frac{\hbar^2}{\mu e^2} \tag{4.145}$$

[8] They are also a special case of the confluent hypergeometric functions. See the next section.

This expression for ρ is obtained by introducing E_n [Eq. 4.142], for E in the definition of α. It is important to note that ρ is not the same in all the wave functions, since it is dependent on n.

The choice of the constant $\mathcal{N}_{nl}$ in Eq. 4.144 for normalization of the wave function is dictated by the fact that

$$\int_0^\infty e^{-\rho}\,\rho^{p+1}\left[L_q{}^p(\rho)\right]^2 d\rho = \frac{(2q-p+1)(q!)^3}{(q-p)!} \tag{4.146}$$

Eq. (4.146) may be verified with the aid of the generating function (D.14). The evaluation is similar to the proof of orthonormality of the harmonic oscillator wave functions [Sec. 4.2(b)] and will not be detailed here. Using Eqs. 4.144 and 4.146 we obtain the radial wave functions $R_{nl}(r)$ with the normalization $\int |R_{nl}(r)|^2 r^2 dr = 1$ as

$$R_{nl}(r) = -\left\{\left(\frac{2Z}{na}\right)^3 \frac{(n-l-1)!}{2n\,[(n+l)!]^3}\right\}^{1/2} \times e^{-Zr/na}\left(\frac{2Zr}{na}\right)^l L_{n+l}^{2l+1}\left(\frac{2Zr}{na}\right) \tag{4.147}$$

The complete hydrogen atom wave functions are then given by

$$u_{nlm}(r,\theta,\varphi) = R_{nl}(r)\,Y_{lm}(\theta,\varphi) \tag{4.148}$$

In particular

$$u_{100}(r,\theta,\varphi) = \frac{1}{\sqrt{\pi}}\left(\frac{Z}{a}\right)^{3/2} e^{-(Zr/a)} \tag{4.148a}$$

The first few radial wave functions and some of their useful properties are given in Tables 4.1 and 4.2, p. 136. Their behaviour is graphically illustrated in Fig. 4.6. The probability that the electron is within the spherical shell between radii r and $r + dr$ from the nucleus is given by $|R_{nl}(r)|^2 r^2 dr \equiv \mathcal{P}_{nl}(r)dr$, say. Some of the functions $\mathcal{P}_{nl}(r)$ are shown in Fig. 4.7.

4.17 Discussion of Bound States

Returning now to the energy eigenvalues, we observe that E_n can be written as

$$E_n = -\frac{1}{2}\frac{Ze^2}{(a/Z)}\cdot\frac{1}{n^2} \tag{4.149}$$

where $a \equiv (\hbar^2/\mu e^2)$ differs from the Bohr radius a_0 of Sec. 1.9 (where the hydrogen nucleus was assumed to remain fixed) in having the reduced mass μ in the place of the electron mass m. Note that (a/Z) is the first Bohr radius in an atom with charge Ze. The ground state energy E_1 is just half the potential energy of an electron at this distance from the nucleus. As n increases, $|E_n|$ decreases as $(1/n^2)$ and correspondingly $E_n = -|E_n|$ increases converging to the limit $E = 0$ as $n \to \infty$. The occurrence of an infinite number of levels (in constrast to the case of the square well potential) is noteworthy. The reason is that the magnitude of the Coulomb potential (proportional to $1/r$) decreases very slowly[9] at large r.

[9] Another way of saying the same thing is that the 'width' of the potential region (within

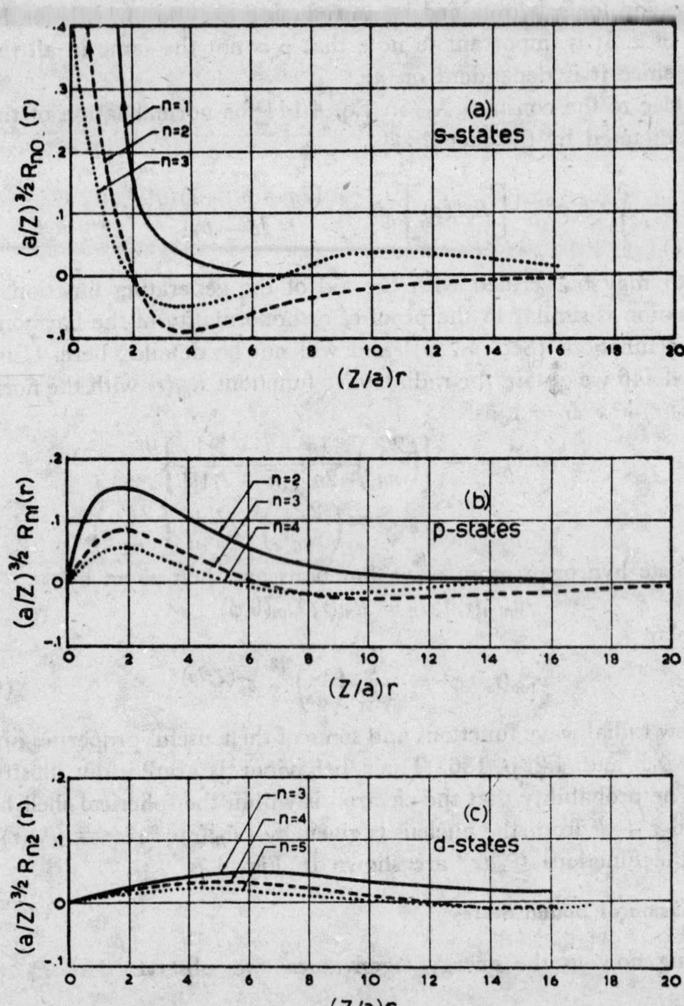

Fig. 4.6 Radial wave functions of hydrogen-like atoms. If the curves for $n = 1$ and $n = 2$ in (**a**) are extended down to $r = 0$, they would rise to 2 and $1/\sqrt{2}$ respectively.

Each of the energy levels E_n (except the lowest one) is *degenerate* since the energy depends only on the quantum number n and not on l and m. All the eigenfunctions u_{nlm} for given n have the same energy. As noted in Eq. 4.141, l can take all integer values from 0 to $(n - 1)$, and for each l there are $(2l + 1)$ states labelled by different values of the magnetic quantum number m. Thus the total number of states belonging to E_n (i.e. its degree of degeneracy) is given by

which a classical particle of energy E can move) increases rapidly to infinity as E approaches zero from below. We have seen from the example of the square well that the wider the well (for given depth), the smaller the level spacing and the larger the number of levels.

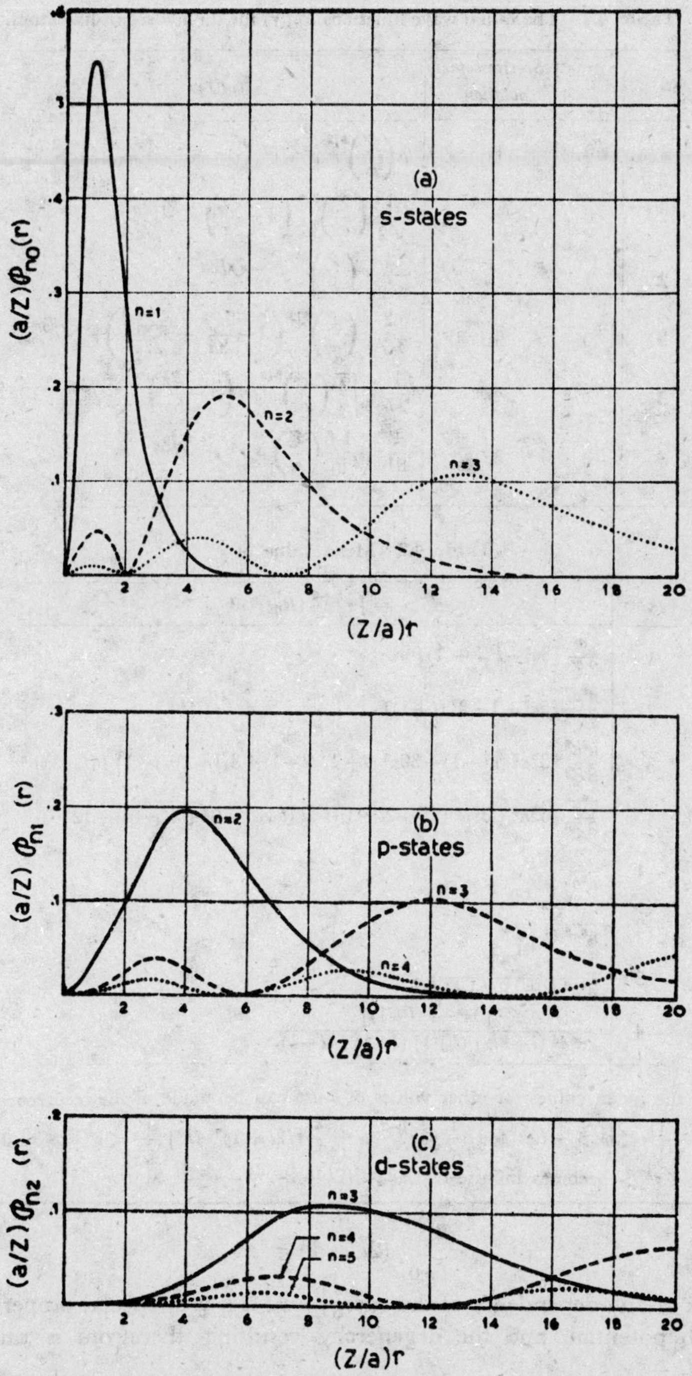

Fig. 4.7 Electron density distribution in the radial variable: $\mathscr{P}_{nl}(r) = r^2 \mid R_{nl}(r) \mid^2$.

Table 4.1 The radial wave functions $R_{nl}(r)$ for the hydrogen-like atom.

n	l	*Spectroscopic notation*	$R_{nl}(r)$
1	0	1s	$2\left(\frac{Z}{a}\right)^{3/2} e^{-Zr/a}$
2	0	2s	$\frac{1}{\sqrt{2}}\left(\frac{Z}{a}\right)^{3/2}\left(1-\frac{Zr}{2a}\right) e^{-Zr/2a}$
2	1	2p	$\frac{1}{\sqrt{24}}\left(\frac{Z}{a}\right)^{5/2} r\, e^{-Zr/2a}$
3	0	3s	$\frac{2}{3\sqrt{3}}\left(\frac{Z}{a}\right)^{3/2}\left(1-\frac{2Zr}{3a}+\frac{2Z^2r^2}{27a^2}\right) e^{-Zr/3a}$
3	1	3p	$\frac{4}{27}\sqrt{\frac{2}{3}}\left(\frac{Z}{a}\right)^{5/2} r\left(1-\frac{Zr}{6a}\right) e^{-Zr/3a}$
3	2	3d	$\frac{4}{81}\cdot\frac{1}{\sqrt{30}}\left(\frac{Z}{a}\right)^{7/2} r^2\, e^{-Zr/3a}$

Table 4.2 Mean value of r^k

k	$\langle r^k\rangle = \int r^{k+2}(R_{nl})^2\,dr$
1	$\frac{a}{2Z}[3n^2-l(l+1)]$
2	$\frac{n^2a^2}{2Z^2}[5n^2+1-3l(l+1)]$
3	$\frac{n^2a^3}{8Z^3}[35n^2(n^2-1)-30n^2(l+2)(l-1)+3(l+2)(l+1)l(l-1)]$
4	$\frac{n^4a^4}{8Z^4}[63n^4-35n^2(2l^2+2l-3)+5l(l+1)(3l^2+3l-10)+12]$
−1	$\frac{Z}{n^2a}$
−2	$\frac{Z^2}{n^3a^2(l+\frac{1}{2})}$
−3	$\frac{Z^3}{n^3a^3(l+1)(l+\frac{1}{2})l}$
−4	$\frac{Z^4[3n^2-l(l+1)]}{2n^5a^4(l+3/2)(l+1)(l+\frac{1}{2})l(l-\frac{1}{2})}$

To obtain the mean values for other values of k use can be made of the recurrence relation:

$$\frac{(k+1)}{n^2}\langle r^k\rangle-(2k+1)\frac{a}{Z}\langle r^{k-1}\rangle+\frac{k}{4}[(2l+1)^2-k^2]\frac{a^2}{Z^2}\langle r^{k-2}\rangle=0$$

Note that $\langle r^k\rangle$ becomes infinite for $k \leqslant -(2l+3)$.

$$\sum_{l=0}^{n-1}(2l+1)=n^2 \tag{4.150}$$

The lack of any dependence of the energy on l is a very special property of the Coulomb potential, and the degeneracy resulting therefrom is said to be *accidental*.

We have already noted that the radial wave functions R_{nl} behave like r^l near the origin and hence vanish at $r = 0$ (the position of the nucleus) for all

$l \neq 0$. In s-states ($l = 0$) both R and (dR/dr) are nonzero at the origin; in fact as one approaches the origin, the rate of increase of the wave function tends to a nonzero limit and on passing through the origin to the opposite side, the wave function abruptly starts decreasing at the same rate. Thus the gradient of the wave function is *discontinuous* at the origin (for $l = 0$). This is a consequence of the singularity of the Coulomb potential at $r = 0$.

It is instructive to note that the probability distribution of the radial position of the electron, $\mathcal{P}_{nl}(r) \equiv R_{nl}^2.r^2$ rises to a maximum at $r = a$ in the ground state ($n = 1$, $l = 0$). In this sense the Bohr radius a is significant even in the wave mechanical context though orbits of the type envisaged in the Old Quantum Theory (Sec. 1.9) do not exist.

4.18 Solution in Terms of Confluent Hypergeometric Functions; Nonlocalized States

(*a*) *The Confluent Hypergeometric Functions:* We now return to Eq. 4.134 and observe that much of the labour of the last two sections could be short-circuited by recognizing that this equation is really the *confluent hypergeometric equation* whose standard form is

$$\rho \frac{d^2W}{d\rho^2} + (c - \rho) \frac{dW}{d\rho} - aW = 0 \tag{4.151}$$

Two independent solutions of this equation are given by

$$F(a;c;\rho) = 1 + \frac{a}{1.c}\rho + \frac{a(a+1)}{1.2.c(c+1)}\rho^2 + \dots \tag{4.152a}$$

$$G(a;c;\rho) = \rho^{1-c} F(a - c + 1;\ 2 - c;\rho) \tag{4.152b}$$

While the former is regular at $\rho = 0$, the latter becomes singular if $c > 1$. As regards asymptotic behaviour, it is best displayed in terms of another pair of solutions W_1, W_2 defined through

$$F = W_1 + W_2, \quad G = W_1 - W_2 \tag{4.153}$$

Their asymptotic behaviour is known to be as follows:

$$W_1(a;c;\rho) = \frac{\Gamma(c)}{\Gamma(c-a)} (-\rho)^{-a}\, g(a;\ a - c + 1; -\rho)$$

$$W_2(a;c;\rho) = \frac{\Gamma(c)}{\Gamma(a)} e^{\rho} \rho^{a-c}\, g(1 - a;\ c - a;\rho) \tag{4.154}$$

$$g(\alpha;\beta;\rho) \underset{\rho\to\infty}{\longrightarrow} 1 + \frac{\alpha\beta}{1!\rho} + \frac{\alpha(\alpha+1)\beta(\beta+1)}{2!\ \rho^2} + \dots$$

Since the asymptotic form of W_2 contains a factor e^{ρ}, $F(a;c;\rho)$ diverges exponentially at infinity *unless* W_2 vanishes identically. Inspection of Eq. 4.154 shows that W_2 does vanish if a has any of the values 0,—1,—2,... (since $\Gamma(a)$ then becomes infinite). For these special values of a, $F(a;c;\rho)$ reduces to a polynomial of degree $(-a)$ as can be seen directly from Eq. 4.152.

(*b*) *Bound States:* Now, comparing Eqs. 4.134 and 4.151, and recalling that $L(\rho)$ has to be regular at the origin, we obtain

$$L(\rho) = \text{const.}\ F(l + 1 - \lambda;\ 2l + 2;\rho) \tag{4.155}$$

(The second solution G diverges at $\rho = 0$ in our case, since $c = 2l + 2 > 1$). We also require that $\mathcal{R}(\rho) = e^{-\frac{1}{2}\rho} \rho^l L(\rho)$ should go to zero as $\rho \to \infty$. This will be satisfied if and only if the W_a part of F vanishes identically. According to the discussion just given, this happens if $l+1-\lambda = -n'$ $(n' = 0,1,2\ldots)$, or $\lambda = n' + l + 1 =$ integer n. This is precisely our Eq. 4.139 which directly gives the energy levels. The corresponding radial functions are seen to be

$$\mathcal{R}_{nl}(\rho) = \text{const. } e^{-\frac{1}{2}\rho}\, \rho^l F(l+1-n;\, 2l+2;\, \rho) \qquad (4.156)$$

wherein the F function is a polynomial. In fact it is the associated Laguerre polynomial found in the last section.

(*c*) *Non-localized States* $(E > 0)$: The above considerations need only slight modifications in order to go over from the case of bound states to non-localized states of a particle in a Coulomb potential. Referring to Eq. 4.131 we observe that when E is positive, both λ and α are imaginary. But the reasoning which led us to assume the form (4.133) for $\mathcal{R}(\rho)$ still holds good, and we are again led to the confluent hypergeometric function (4.155) as the solution for $L(\rho)$. Writing the value of α from Eq. 4.131, for $E > 0$, as

$$\alpha = -2ik, \quad k = (2mE/\hbar^2)^{1/2} \qquad (4.157)$$

and correspondingly,

$$\lambda = i\nu, \quad \nu = (\mu Ze^2/\hbar^2 k) \qquad (4.158)$$

we now have from Eqs. 4.133 and (4.155), the following explicit expression for the radial wave function:

$$R_l(r;k) = C_l\, r^l\, e^{ikr} F(l+1-i\nu;\, 2l+2;\, -2ikr) \qquad (4.159)$$

Here C_l is an arbitrary constant [into which we have absorbed a factor $(-2ik)^l$ coming from ρ^l]. As we have observed repeatedly before, the part of the energy spectrum to which non-localized states belong is a continuum (here, all $E > 0$). The counterpart here of the quantum number n of the bound state case is ν. But it is preferable to use k which is uniquely determined by ν through Eq. 4.158 as a label, since it has an immediate physical interpretation: $\hbar k$ is the momentum of the particle when it is very far from the origin. For under such circumstances, the potential energy is practically zero so that the energy E is wholly kinetic, and by Eq. 4.157, $E = (\hbar k)^2/2m$. We have shown k as an argument of R_l since it is a parameter on which the wave function depends; but it is usually suppressed, and we shall often do so later.

Before finally accepting the solution (4.159), we have to verify that it does not diverge when $r \to \infty$ despite the presence of the factor r^l. On substituting for F its asymptotic form from Eqs. 4.153 and 4.154, it is readily seen that at large distances,

$$R_l(r;k) \to \frac{C_l \exp\,[i\eta_l - \frac{1}{2}\nu\pi]}{(2k)^l\, \Gamma(l+1-i\nu)}\, \frac{\Gamma(2l+2)}{kr} \cdot \sin\left(kr - \frac{l\pi}{2} + \nu\log 2kr + \eta_l\right), \qquad (4.160)$$

$$\eta_l = \arg \Gamma(l+1-i\nu)$$

This asymptotic behaviour is clearly acceptable, and so Eq. 4.159 does give the stationary wave function for non-localized states.

The results of this section will prove to be useful in the study of scattering by the Coulomb potential (Sec. 6.16).

4.19 Solution in Parabolic Coordinates

The Schrödinger equation for hydrogen-like atoms is separable in more than one kind of coordinate system. This fact is related to the occurrence of the accidental degeneracy noted in Sec. 4.17. Specifically, it turns out to be possible to form independent linear combinations of the wave functions u_{nlm} belonging to a given degenerate energy level (i.e., fixed n) in such a way that the new combinations are factorizable in terms of new coordinate variables, e.g. parabolic. The parabolic coordinates are related to the spherical polar coordinates through the relations

$$\begin{aligned} \xi &= r - z = r\,(1 - \cos\theta) \\ \eta &= r + z = r(1 + \cos\theta) \\ \varphi &= \varphi \end{aligned} \tag{4.161}$$

The surfaces of constant ξ are all paraboloids with common focus (at the origin). They are paraboloids of revolution about the z- axis, open in the direction of positive z, while the surfaces of constant η are similar paraboloids open in the negative z-direction. The surfaces for given ξ and η intersect in a circle, and the point where this circle penetrates the surface for given φ (a half plane with the z-axis as its edge) is the point (ξ,η,φ).

In terms of parabolic coordinates, the wave equation for the hydrogen-like atom is

$$-\frac{\hbar^2}{2\mu}\left\{\frac{4}{\xi+\eta}\left[\frac{\partial}{\partial\xi}\left(\xi\frac{\partial u}{\partial\xi}\right)+\frac{\partial}{\partial\eta}\left(\eta\frac{\partial u}{\partial\eta}\right)\right]+\frac{1}{\xi\eta}\frac{\partial^2 u}{\partial\varphi^2}\right\}-\frac{2Ze^2}{\xi+\eta}u = Eu \tag{4.162}$$

Substituting

$$u(\xi,\eta,\varphi) = f(\xi)g(\eta)\Phi(\varphi) \tag{4.163}$$

in Eq. 4.162 and dividing throughout by u, we obtain

$$\frac{4\xi\eta}{\xi+\eta}\left[\frac{1}{f}\frac{d}{d\xi}\left(\xi\frac{df}{d\xi}\right)+\frac{1}{g}\frac{d}{d\eta}\left(\eta\frac{dg}{d\eta}\right)\right] + \frac{4\mu Ze^2\xi\eta}{\hbar^2\,(\xi+\eta)}+\frac{2\mu E\,\xi\eta}{\hbar^2} = -\frac{1}{\Phi}\frac{d^2\Phi}{d\varphi^2} \tag{4.164}$$

Since the two sides depend on different variables, each must be equal to a constant, m^2. The now-familiar criterion of single-valuedness of Φ restricts the admissible solutions to

$$\Phi_m(\rho) = (2\pi)^{-1/2}\,e^{im\varphi},\ (m = 0, \pm 1,\ldots) \tag{4.165}$$

Now, on equating the left hand side of Eq. 4.164 to the same constant m^2, and multiplying throughout by $(\xi+\eta)/(\xi\eta)$, we obtain

$$\left[\frac{1}{f}\frac{d}{d\xi}\left(\xi\frac{df}{d\xi}\right)-\frac{m^2}{4\xi}+\frac{\mu E}{2\hbar^2}\xi\right]+\left[\frac{1}{g}\frac{d}{d\eta}\left(\eta\frac{dg}{d\eta}\right)-\frac{m^2}{4\eta}+\frac{\mu E}{2\hbar^2}\eta\right] = -\frac{\mu Ze^2}{\hbar^2} \tag{4.166}$$

Since each of the square-bracketed expressions depends on one of the variables only, their sum can be a constant only if each is equal to a constant. Thus we can separate this equation into two parts:

$$\frac{1}{f}\frac{d}{d\xi}\left(\xi\frac{df}{d\xi}\right)-\frac{m^2}{4\xi}+\frac{\mu F}{2\hbar^2}\xi=-\sigma_1$$
$$\frac{1}{g}\frac{d}{d\eta}\left(\eta\frac{dg}{d\eta}\right)-\frac{m^2}{4\eta}+\frac{\mu E}{2\hbar^2}\eta=-\sigma_2 \tag{4.167}$$

where σ_1, σ_2 are constants such that

$$\sigma_1+\sigma_2=\frac{\mu Ze^2}{\hbar^2} \tag{4.168}$$

By defining

$$\alpha=(-2\mu E/\hbar^2)^{1/2} \text{ and } \lambda_1=(\sigma_1/\alpha) \tag{4.169}$$

and introducing $\zeta=\alpha\xi$, we can rewrite the first of Eqs. 4.167 in the dimensionless form

$$\frac{1}{\zeta}\frac{d}{d\zeta}\left(\zeta\frac{df}{d\zeta}\right)+\left[\frac{\lambda_1}{\zeta}-\frac{1}{4}-\frac{m^2}{4\zeta^2}\right]f=0 \tag{4.170}$$

An exactly similar form obtains for the second equation (for g) too, except that instead of λ_1 we have $\lambda_2=(\sigma_2/\alpha)$. By Eq. (4.168),

$$\lambda_1+\lambda_2=\frac{\sigma_1+\sigma_2}{\alpha}=\frac{Ze^2}{\hbar}\left(\frac{\mu}{-2E}\right)^{1/2} \tag{4.171}$$

(*a*) *Bound States:* In this case $(E<0)$, α defined by Eq. 4.169 is real. Eq. 4.170 can now be solved in the same way as the radial wave equation (4.132). As before, from the dominance of the term $-\frac{1}{4}$ in square brackets when $\zeta\to\infty$, we conclude that the dominant asymptotic behaviour of f is like $e^{\pm\frac{1}{2}\zeta}$, and we reject $e^{\frac{1}{2}\zeta}$ as unacceptable since it diverges as $\zeta\to\infty$. Further, one can easily check that for small ζ, when the term $-m^2/4\zeta^2$ is the dominant one (except for $m=0$), f behaves like $\zeta^{\frac{1}{2}|m|}$ or $\zeta^{-\frac{1}{2}|m|}$; the latter is uninteresting since it is singular at $\zeta=0$. Incorporating the behaviour for $\zeta\to\infty$ and for $\zeta\to 0$ explicitly into the form of f, we write

$$f=e^{-\frac{1}{2}\zeta}\,\zeta^{\frac{1}{2}|m|}L(\zeta) \tag{4.172}$$

On feeding this into Eq. 4.170 we find that $L(\zeta)$ satisfies the equation

$$\zeta L''+(|m|+1-\zeta)\,L'+[\lambda_1-\tfrac{1}{2}(|m|+1)]\,L=0 \tag{4.173}$$

This equation has just the same form as Eq. 4.134 and exactly as in that case, we find that the power series for L would behave like e^{ζ} and, therefore, make f diverge as $\zeta\to\infty$ unless the series terminates. This happens when

$$n_1=\lambda_1-\tfrac{1}{2}(|m|+1) \tag{4.174a}$$

is a non-negative integer, and then L turns out to be the associated Laguerre polynomial $L^{|m|}_{n_1+|m|}(\alpha\xi)$.

The solution of the equation for g on the same lines leads to the condition that

$$n_2=\lambda_2-\tfrac{1}{2}(|m|+1) \tag{4.174b}$$

be a positive integer or zero. From Eqs. 4.174,

$$\lambda_1+\lambda_2=n_1+n_2+|m|+1=n,\quad n\geqslant|m|+1=1,2,3\ldots \tag{4.175}$$

This result combined with Eq. 4.171 determines the energy eigenvalues to be

$$E_n=-\frac{\mu Z^2e^4}{2\hbar^2n^2} \tag{4.176}$$

in agreement with Eq. 4.142. The degeneracy of the energy level is evident, since there are several states with different sets of values of n_1, n_2 and m, but the same value of n and hence the same energy. In fact, for a given n, m can take values $n-1, n-2, \ldots -(n-1)$, according to Eq. 4.175; and given any such value of m, there are evidently $n-|m|$ ways in which two non-negative integers n_1, n_2 can add up to $n-|m|-1$. Thus the total number of states labelled by the various values of n_1, n_2 and m for given n is

$$\sum_{m=-(n-1)}^{(n-1)} (n-|m|) = n(2n-1) - 2\sum_{m=1}^{n-1} m = n^2 \tag{4.177}$$

This is just what we had found from the spherical polar coordinate approach earlier.

We note finally that the energy eigenfunctions are given by Eq. 4.163 with

$$\left.\begin{aligned} f(\xi) &= \mathcal{N}_1 e^{-\frac{1}{2}\alpha\xi} \xi^{\frac{1}{2}|m|} L^{|m|}_{n_1+|m|}(\alpha\xi), \\ g(\eta) &= \mathcal{N}_2 e^{-\frac{1}{2}\alpha\eta} \eta^{\frac{1}{2}|m|} L^{|m|}_{n_2+|m|}(\alpha\eta), \\ \alpha &= (\mu Z e^2/n\hbar^2) \end{aligned}\right\} \tag{4.178}$$

By evaluating the Jacobian of the transformation from (r, θ, φ) to (ξ, η, φ), it can be verified that the norm is

$$\int_0^\infty d\xi \int_0^\infty d\eta \int_0^{2\pi} d\varphi \; \tfrac{1}{4}(\xi+\eta)\, |u_{n_1n_2m}(\xi\eta\varphi)|^2 \tag{4.179}$$

This integral can be evaluated with the help of Eqs. 4.149. The constants $\mathcal{N}_1$, $\mathcal{N}_2$ can then be chosen so as to make the norm unity.

(*b*) *Non-localized States:* The wave functions of the non-localized eigenstates (corresponding to $E > 0$) can be inferred from the above derivation. Considering solutions of the same form $u = f(\xi)\, g(\eta)\, \Phi(\varphi)$ as before, we obtain f once again from Eqs. 4.172 and 4.173 wherein $\zeta = \alpha\xi$. The essential difference now is that α is purely imaginary. Because of this it is no longer necessary to require that the series solution for $L(\zeta)$ should terminate after a finite number of terms. The asymptotic behaviour like $e^\zeta = e^{\alpha\xi}$ is not dangerous when α is purely imaginary, since the function merely oscillates without diverging as $\xi \to \infty$. Therefore λ_1 may be any complex number (and so can λ_2, for similar reasons). But the sum $(\lambda_1 + \lambda_2)$ is fixed by Eq. 4.168 since, by definition, $(\lambda_1 + \lambda_2) = (\sigma_1 + \sigma_2)/\alpha$. Thus, writing

$$\alpha = -ik, \quad k = (2\mu E/\hbar^2)^{1/2} \tag{4.180}$$

we have

$$\lambda_1 + \lambda_2 = \frac{\mu Z e^2}{\hbar^2 \alpha} = \frac{i\mu Z e^2}{\hbar^2 k} \tag{4.181}$$

Comparing Eq. 4.173 for arbitrary λ_1 with the confluent hypergeometric equation (4.151), we find that for $E > 0$, $L(\zeta)$ is given by $F(\frac{1}{2}|m| + \frac{1}{2} - \lambda_1, |m| + 1; \zeta)$ and hence

$$f(\xi) = \mathcal{N}_1 e^{\frac{1}{2}ik\xi} \xi^{\frac{1}{2}|m|} F(\tfrac{1}{2}|m| + \tfrac{1}{2} - \lambda_1, |m| + 1; -ik\xi) \tag{4.182}$$

Similarly

$$g(\eta) = \mathcal{N}_2 e^{\frac{1}{2}ik\eta} \eta^{\frac{1}{2}|m|} F(\tfrac{1}{2}|m| + \tfrac{1}{2} - \lambda_2, |m| + 1; -ik\eta) \tag{4.183}$$

with $\lambda_1 + \lambda_2$ determined by the energy E through Eqs. 4.181 and 4.180. These expressions, together with $\Phi_m = (2\pi)^{-1/2} e^{im\varphi}$, give the complete wave function

$$u_{\lambda_1\lambda_2 m}(\zeta\eta\varphi) = f_{\lambda_1 m}(\zeta)\, g_{\lambda_2 m}(\eta)\, \Phi_m(\varphi) \qquad (1.101)$$

apart from the constants N_1, N_2 which are arbitrary.

E. OTHER PROBLEMS IN THREE DIMENSIONS

4.20 The Anisotropic Oscillator

As a further example of exactly soluble problems, we consider the harmonic oscillator in three dimensions, with the Hamiltonian

$$H = \frac{p^2}{2m} + \frac{1}{2} m\, (\omega_1^2\, x^2 + \omega^2_2\, y^2 + \omega_3^2\, z^2) \qquad (4.185)$$

This Hamiltonian can be broken up into a sum, $H^{(1)} + H^{(2)} + H^{(3)}$ of Hamiltonians of three independent simple harmonic oscillators:

$$H = \left(\frac{p_x^2}{2m} + \tfrac{1}{2}m\omega_1^2\, x^2\right) + \left(\frac{p_y^2}{2m} + \tfrac{1}{2}m\omega_2^2\, y^2\right) + \left(\frac{p_z^2}{2m} + \tfrac{1}{2}m\omega_3^2\, z^2\right) \qquad (4.186)$$

Therefore, according to the general theory of Sec. 3.15, a complete set of eigenfunctions of H may be found in the form

$$u_{n_1n_2n_3}(x,y,z) = u^{(1)}_{n_1}(x)\; u^{(2)}_{n_2}(y)\; u^{(3)}_{n_3}(z) \qquad (4.187)$$

where $u^{(1)}_{n_1}$, $u^{(2)}_{n_2}$, $u^{(3)}_{n_3}$ are eigenfunctions of the three different oscillators, given by expressions similar to Eqs. 4.16. The eigenvalue to which $u_{n_1n_2n_3}$ belongs is given by

$$\begin{aligned} E_{n_1n_2n_3} &= E^{(1)}_{n_1} + E^{(2)}_{n_2} + E^{(3)}_{n_3} \\ &= [(n_1 + \tfrac{1}{2})\, \hbar\omega_1 + (n_2 + \tfrac{1}{2})\, \hbar\omega_2 + (n_3 + \tfrac{1}{2})\, \hbar\omega_3], \\ &\qquad (n_1, n_2, n_3 = 0,1,2,\ldots) \end{aligned} \qquad (4.188)$$

These results could of course be obtained by a more elaborate derivation starting with the Schrödinger equation and its separation with respect to the x, y and z variables.

4.21 The Isotropic Oscillator

When the oscillator is *isotropic*, $(\omega_1 = \omega_2 = \omega_3 = \omega$, say) the energy eigenvalues reduce to

$$E_{n_1n_2n_3} = (n + \tfrac{3}{2})\hbar\omega, \quad n = n_1 + n_2 + n_3 \qquad (4.189)$$

Since the energy depends only on the sum $n = n_1 + n_2 + n_3$ in this case, the levels are degenerate. The degree of degeneracy of E_n for given n may be easily seen to be $\tfrac{1}{2}(n + 1)\,(n + 2)$. As we have noted in connection with the hydrogen atom, the existence of degeneracy is an indication that the wave equation can be separated in more than one way. For the isotropic oscillator, an obvious possibility is separation in spherical polar coordinates, since the potential energy is given by

$$V = \tfrac{1}{2}m\omega^2\, (x^2 + y^2 + z^2) = \tfrac{1}{2}m\omega^2 r^2 \qquad (4.190)$$

Thus, instead of the complete set of eigenfunctions (4.187) one could have the alternative set

$$u_{nlm}(r,\theta,\varphi) = R_{nl}(r)\ \Upsilon_{lm}(\theta,\varphi) \tag{4.191}$$

where R satisfies the radial wave equation (4.105) with $V = \frac{1}{2}m\omega^2r^2$. In terms of the dimensionless variable

$$\rho = \alpha r \text{ with } \alpha = (m\omega/\hbar)^{1/2} \tag{4.192}$$

it takes the form

$$\frac{1}{\rho^2}\frac{d}{d\rho}\left(\rho^2\frac{d\mathcal{R}}{d\rho}\right) + \left[\lambda - \rho^2 - \frac{l(l+1)}{\rho^2}\right]\mathcal{R} = 0 \tag{4.193}$$

where

$$\mathcal{R}(\rho) = R(r) \text{ and } \lambda = \frac{2mE}{\hbar^2\alpha^2} = \frac{2E}{\hbar\omega} \tag{4.194}$$

The solution of Eqs. (4.193) follows the now-familiar pattern. Observing that $\mathcal{R}(\rho)$ behaves like ρ^l for small ρ and $e^{-\frac{1}{2}\rho^2}$ for large ρ, one makes the substitution:

$$\mathcal{R} = \rho^l e^{-\frac{1}{2}\rho^2} K \tag{4.195}$$

The resulting equation for K can be reexpressed as

$$\xi\frac{d^2K}{d\xi^2} + (l + \tfrac{3}{2} - \xi)\frac{dK}{d\xi} + \tfrac{1}{4}(\lambda - 3 - 2l)K = 0 \tag{4.196}$$

where $\xi = \rho^2$. Since this equation is identical in form to Eq. 4.134, we can see straightaway by comparison with our treatment of that equation, that only *finite* series (polynomial) solutions are admissible, and that such solutions exist only when

$$\tfrac{1}{4}(\lambda - 3 - 2l) = n', \quad n' = 0,1,2,\ldots \tag{4.197a}$$

or

$$\lambda = 2n + 3, \quad n = l + 2n' \tag{4.197b}$$

The energy eigenvalues are now obtained from Eq. 4.194 as $E_n = (n + \frac{3}{2})\hbar\omega$, $n = 0,1,2,\ldots$, in argreement with Eq. 4.189. It should be observed that for given n, the possible values of l are $n, n-2, n-4\ldots.$

As for the polynomial solution of Eq. 4.196 with $\lambda = 2n + 3$, we find by comparison with Eq. 4.143 that it is the associated Laguerre polynomial $L^p_q(\xi)$ with $p = l + \frac{1}{2}$ and $q = \frac{1}{2}(n + l + 1) = n' + l + \frac{1}{2}$. Remembering that $\xi = \rho^2 = \alpha^2r^2$ we obtain finally the normalized radial wave functions for the isotropic harmonic oscillator as

$$R_{nl}(r) = \mathcal{N}_{nl}\, r^l e^{-\frac{1}{2}\alpha^2r^2} L^{l+\frac{1}{2}}_{\frac{1}{2}(n+l+1)}(\alpha^2r^2), \tag{4.198}$$

$$\mathcal{N}_{nl} = \left\{\frac{[\frac{1}{2}(n+l)]!\ (n-2l)!\ 2^{n+l+2}\ \alpha^{2l+3}}{(n+l+1)!\ \pi^{\frac{1}{2}}}\right\}^{1/2}$$

As a final remark, we note that the harmonic oscillator has no non-localized states, since the potential function does not 'level off' at large distances but keeps on increasing indefinitely.

4.22 Normal Modes of Coupled Systems of Particles

It is a well-known fact that when a system of mutually interacting particles has an equilibrium configuration, small oscillations of the particles about their equilibrium positions can be analyzed into normal modes. If q_1, q_2, $q_3\ldots$ are the changes (due to small displacements from equilibrium) in the coordinates of the particles, the corresponding increase in potential energy can be closely approximated by a quadratic form, $\frac{1}{2}\Sigma K_{ij}\, q_i q_j$. Writing the total energy then as

$$\tfrac{1}{2}\sum_i \dot{q}_i'^2 + \tfrac{1}{2}\sum_{ij} K'_{ij} q_i' q'_j,$$

$$q_i' = m_i^{1/2} q_i, \; K'_{ij} = (m_i m_j)^{1/2} K_{ij} \tag{4.199}$$

one observes that any *orthogonal* transformation

$$q_i' = \sum_k a_{ik}\, Q_k,\; Q_k = \sum_j (a^{-1})_{kj}\, q_j' = \sum_j a_{jk}\, q_j' \tag{4.200}$$

takes $\Sigma \dot{q}'^2_i$ into $\Sigma \dot{Q}_i^2$, and further that a particular orthogonal transformation can always be found which diagonalizes the (quadratic) potential term in Eq. 4.199.

$$\sum_{ij} K'_{ij} q_i'\, q_j' = \sum_k \omega_k^2\, Q_k^2 \tag{4.201}$$

The ω_k are the eigenvalues of the matrix having elements K'_{ij}.

In terms of the normal *coordinates* Q_k, the energy (4.199) reduces to a sum of Hamiltonians of harmonic oscillators of unit mass:

$$H = \tfrac{1}{2}\sum_k (\dot{Q}_k^2 + \omega_k^2 Q_k^2) = \tfrac{1}{2}\sum_k (P_k^2 + \omega_k^2 Q_k^2), \tag{4.202}$$

$$P_k = \dot{Q}_k = \Sigma\, a_{jk}\dot{q}_j' = \Sigma\, a_{jk}. m_j^{-1/2} p_j \tag{4.203}$$

where $p_j = m_j\, \dot{q}_j = m_j^{1/2} \dot{q}_j'$.

From the above classical analysis one proceeds to the quantum counterpart by imposing the familiar commutation rule $[q_i, p_j] = i\hbar\delta_{ij}$, from which we have

$$[Q_k, P_l] = [\sum_j a_{jk}\, m_j^{\frac{1}{2}}\, q_j,\; \sum_{j'} a_{j'l}\, m_{j'}^{-\frac{1}{2}}\, p_{j'}]$$

$$= \sum_{jj'} a_{jk}\, a_{j'l}\, i\hbar\, \delta_{jj'} = i\hbar\, \delta_{kl} \tag{4.204}$$

(In the last step we have used the fact that the a_{jk} are the elements of an orthogonal matrix). Thus Q_k, P_k are canonically conjugate quantum operators obeying $[Q_k, P_k] = i\hbar$, and different pairs (Q_k, P_k) and (Q_l, P_l), $k \neq l$, commute. With this property, $H = \frac{1}{2}\Sigma(P_k^2 + \omega_k^2\, Q_k^2)$ is the Hamiltonian of an assembly of independent quantum harmonic oscillators. The kth oscillator is characterized by the classical frequency $(\omega_k/2\pi)$ and has energy levels $E_{n_k}^{(k)} = (n_k + \frac{1}{2})\, \hbar\omega_k$, $n_k = 0, 1, 2, \ldots\ldots$ Given the quantum numbers $n_1, n_2, \ldots$ of the various normal mode oscillators, one gets the eigenvalue of the total Hamiltonian as $E_{n_1 n_2} = \Sigma\, (n_k + \frac{1}{2})\, \hbar\omega_k$, and the corresponding eigenfunction as a product $u_{n_1}^{(1)}(Q_1)\, u_{n_2}^{(2)}(Q_2)\ldots$ of the eigenfunctions of the individual oscillators.

These results are of wide applicability. For example, the system of particles may be the atoms constituting a molecule. Then the $E_{nk}^{(k)}$ for given k are the energy levels associated with the kth normal mode of vibration of the molecule. Another important example is that of a regular lattice of atoms forming a crystalline solid. The displacements of the atoms in any normal mode vary simple harmonically with position along any direction in the lattice. (At low frequencies ω_k, these harmonic waves are just sound waves). The quantum excitations of these modes are called *phonons*. If the quantum number denoting the degree of excitation of the kth mode has the value n_k, one says that n_k phonons are present in that mode. Photons also are quanta of 'oscillators' associated with normal modes of the electromagnetic field (Sec. 9.19).

4.23 A Charged Particle in a Uniform Magnetic Field

(*a*) *Reduction to Harmonic Oscillator Problem; The Energy-Spectrum:* The Hamiltonian of a particle of charge e in the presence of a magnetic field $\mathcal{H} \equiv \text{curl } \mathbf{A}$ is

$$H = \frac{\pi^2}{2m} \quad \text{where } \pi = \mathbf{p} - \frac{e}{c}\mathbf{A}(\mathbf{x}) \tag{4.205}$$

The eigenvalue problem for this Hamiltonian,

$$\frac{\pi^2}{2m} u = Eu \tag{4.206}$$

can be solved very simply by an algebraic method analogous to that of Sec. 4.4 if $\mathcal{H}$ is independent of $\mathbf{x}$. We observe first of all that since $[p_x, A_y] = -i\hbar\, \partial A_y/\partial x$ etc., the commutators of the components of π reduce to

$$[\pi_x,\pi_y] = \frac{ie\hbar}{c}\mathcal{H}_z,\quad [\pi_y,\pi_z] = \frac{ie\hbar}{c}\mathcal{H}_x,\quad [\pi_z,\pi_x] = \frac{ie\hbar}{c}\mathcal{H}_y \tag{4.207}$$

In the case of a homogeneous magnetic field in the z-direction ($\mathcal{H}_x = \mathcal{H}_y = 0$, $\mathcal{H}_z = \mathcal{H} = \text{const.}$), these commutators simplify to[10]

$$[(c/e\mathcal{H})^{1/2}\,\pi_x,\ (c/e\mathcal{H})^{1/2}\,\pi_y] = i\hbar,\quad [\pi_z,\pi_x] = [\pi_z,\pi_y] = 0 \tag{4.208}$$

Here we have transferred certain constant factors to the left hand side in the first commutator, to bring out its similarity to the commutator of x and p. Exploiting this similarity, we introduce operators

$$b = \left(\frac{c}{2\hbar e\mathcal{H}}\right)^{1/2}(\pi_x + i\pi_y),\quad b^\dagger = \left(\frac{c}{2\hbar e\mathcal{H}}\right)^{1/2}(\pi_x - i\pi_y) \tag{4.209}$$

which are analogues of the ladder operators a, $a^\dagger$ of the harmonic oscillator:

$$[b, b^\dagger] = 1 \tag{4.210}$$

The Hamiltonian $H = \pi^2/2m$ now becomes

$$H = (b^\dagger b + \tfrac{1}{2})\frac{e\hbar\mathcal{H}}{mc} + \frac{\pi_z^2}{2m} \equiv H_\perp + H_\parallel \tag{4.211}$$

where the symbols $\perp$ and $\parallel$ refer to the fact that the first part of H arises from π_x and π_y (perpendicular to $\mathcal{H}$) while the second part pertains to the component of motion parallel to $\mathcal{H}$. We observe that $\mathcal{H}_\perp$ has the same structure as the Hamiltonian (4.29b) of the harmonic oscillator. So its eigenvalues are given by

$$(n + \tfrac{1}{2})\frac{e\hbar\mathcal{H}}{mc},\quad n = 0,1,2,\ldots \tag{4.212}$$

and the eigenfunctions are determined by

$$u_{n\perp} = (n!)^{-1/2}\,(b^\dagger)^n\,u_{0\perp} \text{ with } bu_{0\perp} = 0 \tag{4.213}$$

These equations are the counterparts of Eqs. 4.36 and 4.33. As for $H_\parallel$, it simplifies to $(p_z^2/2m)$ on putting $A_z = 0$ corresponding to $\mathcal{H}$ being in the

[10] If e is negative, the first of Eqs. 4.208 would be replaced by $[(c/|e|\,\mathcal{H})^{1/2}\,\pi_y,\ (c/|e|\,\mathcal{H})^{1/2}\,\pi_x] = i\hbar$ and corresponding changes made in the definitions of b and $b^\dagger$.

z-direction. Its eigenvalue spectrum is continuous, being given by $(P^2/2m)$, where P, the eigenvalue of p_z, takes any value from $-\infty$ to $+\infty$. The corresponding eigenfunction of course, is $(2\pi)^{-1/2}\, e^{iPz/\hbar}$. Since H is the sum of the two commuting parts $H_{\parallel}$ and $H_{\perp}$ describing independent degrees of freedom, we immediately obtain its eigenfunctions and eigenvalues, as

$$u_{n\perp}\,(x,y)\,(2\pi)^{-1/2}\, e^{iPz/\hbar} \text{ and } (n+\tfrac{1}{2})\,\frac{e\hbar\mathcal{H}}{mc}+\frac{P^2}{2m} \tag{4.214}$$

respectively (Sec. 3.15).

(*b*) *The Eigenfunctions:* So far we have not had to specify the forms of A_x and A_y. We can get a homogeneous $\mathcal{H}$ in the z-direction by taking

$$A_x = -\tfrac{1}{2}y\mathcal{H},\; A_y = \tfrac{1}{2}x\mathcal{H},\; A_z = 0 \tag{4.215}$$

Using this in the definition of π and introducing in Eq. (4.209), we obtain b explicitly as

$$\begin{aligned} b &= \left(\frac{c}{2e\hbar\mathcal{H}}\right)^{1/2}\left[-i\hbar\frac{\partial}{\partial x}+\frac{e\mathcal{H}}{2c}y+\hbar\frac{\partial}{\partial y}-\frac{ie\mathcal{H}}{2c}x\right] \\ &= -\tfrac{1}{2}i\left(\frac{\partial}{\partial\xi}+i\eta+i\frac{\partial}{\partial\eta}+\xi\right) \end{aligned} \tag{4.216}$$

where we have introduced the dimensionless variables

$$\xi = \varkappa x,\; \eta = \varkappa y,\; \varkappa = (e\mathcal{H}/2\hbar c)^{1/2} \tag{4.217}$$

In view of the above form of b, it is clear that the equation $bu_{0\perp}=0$ can be readily solved by separation of variables, assuming $u_{0\perp}=f(\xi)\,g(\eta)$. The equations for $f(\xi)$ and $g(\eta)$ are

$$\left(\frac{\partial}{\partial\xi}+\xi-\alpha\right)f(\xi)=0,\;\left(\frac{\partial}{\partial\eta}+\eta-i\alpha\right)g(\eta)=0 \tag{4.218}$$

where α is the (arbitrary) separation constant. The solution is immediate, and we obtain

$$u_{0\perp}=f(\xi)\,g(\eta) = \text{const.}\; e^{-\frac{1}{2}\xi^2+\alpha\xi}\, e^{-\frac{1}{2}\eta^2+i\alpha\eta} \tag{4.219}$$

This function is obviously well behaved everywhere including at infinity, whether α be real or complex. Writing $\alpha=\xi_0+i\eta_0$, we can reexpress Eq. 4.219 in the suggestive form

$$u_{0\perp} = \text{const.}\,\exp\{-\tfrac{1}{2}[(\xi-\xi_0)^2+(\eta-\eta_0)^2]\}\, e^{i(\eta_0\xi+\xi_0\eta)} \tag{4.220}$$

From this it is seen that the probability density in the x-y plane, which is $\propto |u_{0\perp}|^2$, is a function only of the square of the distance of (x,y) from the point $(x_0, y_0)\equiv(\xi_0/\varkappa,\ \eta_0/\varkappa)$. The probability distribution has circular symmetry around this point. This fact reminds one of the classical situation where the helical trajectory of a charged particle in a magnetic field, when projected on to the plane perpendicular to the field, is a circle. The position of the centre of the circle is arbitrary. In the quantum mechanical case one does not have a sharply defined trajectory, but the point (x_0, y_0) has an obvious correspondence to the centre of the classical orbit.

The solutions (4.220) for all ξ_0 and η_0 are *not* linearly independent. In fact if we go back to Eq. (4.219) we see that the part of $u_{0\perp}$ which depends on ξ_0 and η_0 is $e^{\alpha(\xi+i\eta)}$, which can be written as a linear combination of $(\xi+i\eta)^s$, $s=0, 1,2,\ldots$. So we may take as independent solutions the functions

$$u_{0\perp,s} = \text{const. } e^{-\frac{1}{2}(\xi^2+\eta^2)}.\ (\xi + i\eta)^s,\quad s = 0,1,2,\ldots \tag{4.221}$$

We have thus found an infinite number of eigenfunctions belonging to the lowest eigenvalue ($n = 0$). We can get similar (infinite) sets of eigenfunctions belonging to any higher level, using the first of Eqs. 4.213. The simplest way of evaluating these is by expressing everything in terms of the new (complex) variables

$$\zeta = \xi + i\eta,\quad \zeta^* = \xi - i\eta \tag{4.222}$$

The student may verify that the operator $b^\dagger$, which differs from b only in the signs of the last two terms in Eq. 4.216, reduces now to the simple form[11]

$$b^\dagger = -i\left(\frac{\partial}{\partial\zeta} - \tfrac{1}{2}\zeta^*\right) = -i\, e^{\frac{1}{2}\zeta\zeta^*}\frac{\partial}{\partial\zeta}\, e^{-\frac{1}{2}\zeta\zeta^*} \tag{4.223}$$

Also, $u_{0\perp,s} = \text{const. } e^{-\frac{1}{2}\zeta\zeta^*}.\zeta^s$. Hence the first of Eqs. 4.213 becomes

$$u_{n\perp,s} = \text{const.}\left(e^{\frac{1}{2}\zeta\zeta^*}\frac{\partial}{\partial\zeta}\, e^{-\frac{1}{2}\zeta\zeta^*}\right)^n.\, e^{-\frac{1}{2}\zeta\zeta^*}.\,\zeta^s$$

$$= \text{const.}\, e^{\frac{1}{2}\zeta\zeta^*}\frac{\partial^n}{\partial\zeta^n}\,(e^{-\zeta\zeta^*}.\,\zeta^s) \tag{4.224}$$

This form reminds one of the associated Laguerre functions. In fact, using Eqs. D.13a and D.13b of Appendix D, we obtain

$$u_{n\perp,s} = \text{const. } \zeta^{s-n}\, e^{-\frac{1}{2}\zeta\zeta^*}\, L_s^{s-n}\,(\zeta\zeta^*),\quad s > n \tag{4.225a}$$

$$= \text{const. } (\zeta^*)^{n-s}\, e^{-\frac{1}{2}\zeta\zeta^*}\, L_n^{n-s}\,(\zeta\zeta^*),\quad s < n \tag{4.225b}$$

Finally, putting $s - n = m$, and noting that $\zeta = x + iy = \rho e^{i\varphi}$ and $\zeta^* = \rho e^{-i\varphi}$ where (ρ,φ) are polar coordinates in the $x-y$ plane, we can combine the two forms (4.225) into

$$u_{n\perp,m}\ (\rho,\varphi) = \mathcal{N}_{n,m}\ (\varkappa\rho)^{|m|}\, e^{-\frac{1}{2}\varkappa^2\rho^2}\, L^{|m|}_{n+\frac{1}{2}(m+|m|)}\,(\varkappa^2\rho^2)\ e^{im\varphi} \tag{4.226}$$

For normalization, $\iint |u_{n\perp,m}|^2\ \rho d\rho\ d\varphi = 1$, it can be shown that $\mathcal{N}_{n,m}$ must be equal to

$$[(n + \tfrac{1}{2}m - \tfrac{1}{2}|m|)!/\pi\,\{(n + \tfrac{1}{2}m + \tfrac{1}{2}|m|)!\}^3]^{\frac{1}{2}}$$

The expressions obtained for the energy eigenvalues and eigenfunctions hold good for negatively charged particles, with the simple replacement of e by $|e|$ in (4.209) and in the definition (4.217) of $\varkappa$.

(*c*) *Gauge Transformations:* It is well known that the vector and scalar potentials $\mathbf{A}$ and ϕ of electromagnetic fields are not uniquely determined by the knowledge of $\mathcal{E}(\mathbf{x}, t)$ and $\mathcal{H}(\mathbf{x}, t)$. The defining equations of the potentials are

$$\mathcal{H} = \text{curl }\mathbf{A} \text{ and } \mathcal{E} = -\text{ grad }\phi - \frac{1}{c}\frac{\partial \mathbf{A}}{\partial t} \tag{4.227}$$

If, in these equations, we replace $\mathbf{A}$ and ϕ by

$$\mathbf{A}' = \mathbf{A} + \text{grad }\chi \text{ and } \phi' = \phi - \frac{1}{c}\frac{\partial\chi}{\partial t} \tag{4.228}$$

[11] Use $\dfrac{\partial}{\partial\xi} = \dfrac{\partial\zeta}{\partial\xi}\cdot\dfrac{\partial}{\partial\zeta} + \dfrac{\partial\zeta^*}{\partial\xi}\cdot\dfrac{\partial}{\partial\zeta^*}$ etc., treating ζ and ζ^* as independent variables. The differential operator $(\partial/\partial\zeta)$ does not act on ζ^*.

where χ $(\mathbf{x}, t)$ is an arbitrary function, the fields $\mathcal{E}$ and $\mathcal{H}$ remain unchanged. So this replacement should have no physical consequences. It is called a *gauge transformation.*

Consider now the Schrödinger equation for a particle of charge e in electromagnetic fields:

$$\left(i\hbar\frac{\partial}{\partial t} - e\phi\right)\psi = \frac{1}{2m}\left(\mathbf{p} - \frac{e}{c}\mathbf{A}\right)^2\psi + V\psi \tag{4.229}$$

(Here V $(\mathbf{x}, t)$ represents any potential energy which the particle may have, other than that due to the electiomagnetic fields we are considering). If we replace $\mathbf{A}$, ϕ in this equation by $\mathbf{A}$,', ϕ' of Eq. 4.228, then corresponding to any solution ψ of Eq. 4.229 the new equation has the solution

$$\psi' = \psi e^{ie\chi/\hbar c} \tag{4.230}$$

The wavefunction ψ' in the presence of potentials $\mathbf{A}', \phi'$ must describe exactly the same physical situation as ψ in the presence of $\mathbf{A}$ and ϕ, because both describe the same particle in the same electric and magnetic fields $\mathcal{E}$ and $\mathcal{H}$.

In the special case of the homogeneous magnetic field which we considered earlier in this section, one could very well take $A_x = -y\mathcal{H}$, $A_y = 0$, $A_z = 0$. This choice differs from (4.215) by a gauge transformation with $\chi = -\frac{1}{2}xy\mathcal{H}$. With the new choice, the solutions for $u_{n\perp}$ appear to have a very different form from those obtained above; but they are necessarily linear combinations of the old solutions belonging to the particular n, multiplied by $e^{ie\chi/\hbar c} = \exp\ [-\frac{1}{2}ie\ \mathcal{H}\ xy/\hbar c]$.

PROBLEMS

1. Discuss the manner in which the harmonic oscillator eigenfunctions approach those of the free particle in the limit $\omega \to 0$ [Use the asymptotic form (D. 11)].
2. Prove (4.24) and (4.25). Determine (u_m, x^2u_n) for arbitrary m, n.
   If $\omega_1 = \omega_2$ in (4.185), one obtains an axially symmetric oscillator. Solve its eigenvalue problem by separation of variables in cylindrical polar coordinates.
3. From (4.25) and the value of (u_n, Hu_n) for the harmonic oscillator, deduce the value of (u_n, p^2u_n). Show that $(\Delta x)(\Delta p) = (n + \frac{1}{2})\hbar$ in the state u_n.
4. A system of two equally massive particles (in one dimension) is given to have the potential energy function $\frac{1}{2}\omega^2(x_1^2 + x_2^2 + 2\lambda x_1x_2)$. Show that the system is equivalent to two harmonic oscillators, and determine its energy levels.
5. Evaluate $\langle xp + px \rangle$ explicitly for any stationary state of the harmonic oscillator. (*cf.* Prob. 4, Chapter 3).
6. If an isotropic harmonic oscillator of charge e is subjected to a uniform electric field of strength $\mathcal{E}$, show that its energy levels are displaced by $(e^2\mathcal{E}^2/2m\omega^2)$.
7. Write down the Schrödinger equation for the harmonic oscillator in the momentum representation, and obtain its solutions by comparison with Eq. 4.2.
   If a harmonic oscillator is prepared to have the unnormalized wave function $(\alpha/\sqrt{\pi})^{\frac{1}{2}}(2\alpha^2x^2 - 2\alpha x + 1)\,e^{-\frac{1}{2}\alpha^2x^2}$, determine its normalized wave function in the energy representation and hence obtain the expectation value of energy in this state.
8. Show that the wave function $R_l(r)\ Y_{lm}(\theta, \varphi)$ goes over into $f_l(k)\ Y_{lm}(\theta', \varphi')$ in the

momentum representation, where (k, θ', φ') are the polar coordinates of $\mathbf{k}$. (Use Eqs. D.49 and D.48). Evaluate $f_l(k)$ if $R_l(r) = j_l(\alpha r)$.

9. Solve the eigenvalue problem for a particle in a spherical cavity with impenetrable walls, $V(r) = 0$ for $r < a$, $V(r) = \infty$ for $r > a$.
10. Show that $E_n = \frac{1}{2}\langle V \rangle$ in the stationary states of the hydrogen atom.
11. Separate the Schrödinger equation for the hydrogen atom in an external electric field in parabolic coordinates. What is the effect of the field on the ground state?
12. Show that the coherent state wave function ϕ_μ, Eq. 4.41, is equal to $e^{-\frac{1}{2}|\mu|^2} \exp(\mu a\dagger)\, u_0$. Using the result of Prob. 9 Chapter 3, prove also that $\phi_\mu = \exp(\mu a\dagger - \mu^* a)\, u_0$.
13. If a, $a\dagger$ were required to satisfy the anticommutation rule $\{a, a\dagger\} = 1$ instead of (4.28), and also $\{a, a\} = \{a\dagger, a\dagger\} = 0$, Eqs. 4.30 would still hold. However, the operator $N = a\dagger a$ would have only 0 and 1 as eigenvalues. Verify these statements. (Note: $\{A, B\}$ means $AB + BA$).
14. For the isotropic oscillator we can construct three sets of ladder operators $\mathbf{a}$, $\mathbf{a}\dagger$ where $\mathbf{a} = 2^{-1/2}(\mathbf{X} + i\mathbf{P})$, $\mathbf{a}\dagger = 2^{-1/2}(\mathbf{X} - i\mathbf{P})$ with $\mathbf{X} = (m\omega_c/\hbar)^{1/2}\mathbf{x}$, $\mathbf{P} = (m\omega_c\hbar)^{-1/2}\mathbf{p}$. Show that the combinations $(a_x + ia_y)$, $(a_x\dagger + ia_y\dagger)$ are ladder operators changing the orbital angular momentum by one unit, and that $\mathbf{a}^2$ changes the energy while leaving the angular momentum unchanged.
15. Let a_1, $a_1\dagger$ and a_2, $a_2\dagger$ be the ladder operators for two harmonic oscillators, i.e. $[a_r, a_s\dagger] = \delta_{rs}$, $[a_r, a_s] = [a_r\dagger, a_s\dagger] = 0$, $(r, s = 1, 2)$. If $J_1 = \frac{1}{2}(a_2\dagger a_1 + a_1\dagger a_2)$, $J_2 = \frac{i}{2}(a_2\dagger a_1 - a_1\dagger a_2)$, $J_3 = \frac{1}{2}(a_1\dagger a_1 - a_2\dagger a_2)$, $S = \frac{1}{2}(a_1\dagger a_1 + a_2\dagger a_2)$, show that J_1, J_2, J_3 obey commutation relations of the same form as the components of angular momentum, and further that $\mathbf{J}^2 = S(S+1)$ and $[S, J_i] = 0$.
16. Show that the Runge-Lenz vector $\mathbf{A} = (2\mu)^{-1}[\mathbf{L} \times \mathbf{p} - \mathbf{p} \times \mathbf{L}] - e^2\mathbf{x}/r$ is a constant of the motion for the hydrogen atom.

5 Approximation Methods for Stationary States

In all but a limited set of problems such as the ones considered in the last chapter, it is impossible to obtain exact solutions, and one has to resort to approximations. In this chapter we present three of the most important types of approximation methods which are useful under different conditions.

A. PERTURBATION THEORY FOR DISCRETE LEVELS

Suppose all the eigenvalues E_n and the eigenfunctions u_n (forming a complete orthonormal set) of the Hamiltonian H_0 of a system are known[1]

$$H_0 u_n = E_n u_n; \quad (u_m, u_n) = \delta_{mn} \tag{5.1}$$

If the system is *perturbed* (by the action of an external field, for example), so that the Hamiltonian changes to H, what will the energy levels and stationary wave functions of the perturbed system be? They are, of course, solutions of the eigenvalue equation

$$Hv = Wv \tag{5.2}$$

of the *perturbed* Hamiltonian. This equation may not be (and usually is not) exactly soluble. The Rayleigh-Schrödinger perturbation theory provides a systematic method of successive approximations to any *perturbed* eigenfunction v and eigenvalue W *belonging to the discrete part of the spectrum*[2] in terms of the

[1] We use here the scalar product notation introduced in Chapter 3: $(\psi, \chi) = \int \psi^* \chi \, d\tau$, $(\psi, A\chi) = \int \psi^* A\chi \, d\tau$ etc.

[2] In the case of states belonging to energies in the continuum, the question of calculating the change in an energy level does not arise. One simply asks: given an eigenfunction $u_0(x)$ of H_0

unperturbed eigenfunctions u_n and eigenvalues E_n. The basic idea is to split up H into the *unperturbed Hamiltonian* H_0 plus a *perturbation part* $\lambda H'$,

$$H = H_0 + \lambda H' \tag{5.3}$$

and to seek expressions for v and W in the form of power series in the perturbation. It is to be expected that the successive terms in the power series will decrease rapidly if the perturbation is small. (For example, when the unperturbed system consists of charged particles, e.g. an atom, and the perturbation is due to electromagnetic fields acting on it, the parameter λ which characterizes the strength of the perturbation is the fine structure constant $e^2/c\hbar = 1/137$). Then it would be adequate to calculate the first few terms in the series to get a good approximation to W and v, and this is where the method is really useful.

5.1 Equations in Various Orders of Perturbation Theory[8]

We now develop the perturbation theory, assuming that W and v can be expressed as

$$W = \sum_{n=0}^{\infty} \lambda^n W^{(n)}, \quad v = \sum_{n=0}^{\infty} \lambda^n v^{(n)} \tag{5.4}$$

The terms proportional to a particular power λ^n in the above series are called the nth *order* terms; in any such term, the perturbation $\lambda H'$ is involved n times. Substituting Eqs. 5.4 and 5.3 in Eq. 5.2 and grouping terms of the same order together, we have

$$(H_0 - W^{(0)})\, v^{(0)} + [(H_0 - W^{(0)})\, v^{(1)} + (H' - W^{(1)})\, v^{(0)}]\, \lambda + [(H_0 - W^{(0)})\, v^{(2)} + (H' - W^{(1)})\, v^{(1)} - W^{(2)}\, v^{(0)}]\, \lambda^2 + \ldots = 0 \tag{5.5}$$

We want this equation to be valid for any arbitrary (but small) strength of the perturbation, i.e., for any λ. This can happen only if the coefficient of each power of λ vanishes. Thus we should have

$$(H_0 - W^{(0)})\, v^{(0)} = 0 \tag{5.6}$$

$$(H_0 - W^{(0)})\, v^{(1)} + (H' - W^{(1)})\, v^{(0)} = 0, \tag{5.7}$$

$$(H_0 - W^{(0)})\, v^{(2)} + (H' - W^{(1)})\, v^{(1)} - W^{(2)}\, v^{(0)} = 0 \tag{5.8}$$

etc. These are the zeroth order, first order, second order,... equations of perturbation theory. The zeroth order equation (5.6) evidently implies that $W^{(0)}$ is one of the unperturbed eigenvalues, say

$$W^{(0)} = E_m \tag{5.9}$$

Then $v^{(0)}$ is an eigenfunction of H_0 belonging to this eigenvalue. Now there are two cases to be considered:

(i) If the unperturbed level E_m is *non-degenerate*, then $v^{(0)}$ is uniquely determined:

$$v^{(0)} = u_m \tag{5.10}$$

for a particular energy, what is the eigenfunction of $H = H_0 + \lambda H'$ which would reduce to $u_0(x)$ if the limit $\lambda \to 0$ were taken? The perturbation theory appropriate to this situation is considered in the context of the scattering problem in Sec. 6.6.

[8] What is presented here is the Rayleigh-Schrödinger perturbation theory (E. Schrödinger, *Ann. d. Physik*, **80**, 437, 1926). For other perturbation methods, See C. H. Wilcox (Ed.) *Perturbation Theory and its Applications in Quantum Mechanics*, John Wiley and Sons, Inc., New York, 1966.

(ii) If E_m is *degenerate*, and the eigenspace belonging to this level is spanned by $u_{m1}, u_{m2}, \ldots, u_{mr}$, (Cf. Sec. 3.4) then all we can say about $v^{(0)}$ is that

$$v^{(0)} = \sum_{\alpha=1}^{r} a_\alpha u_{m\alpha} \tag{5.11}$$

where $a_1, a_2, \ldots, a_r$ are constants which are quite arbitrary at this stage. The two cases will now be taken up separately.

5.2 The Non-Degenerate Case

To solve the first and higher order equations, we expand $v^{(1)}, v^{(2)}, \ldots$ in terms of the complete set of unperturbed wave functions u_n,

$$v^{(1)} = \Sigma\, c_n^{(1)} u_n, \quad v^{(2)} = \Sigma\, c_n^{(2)} u_n, \ldots \tag{5.12}$$

and then seek to determine the coefficients $c_n^{(1)}, c_n^{(2)}, \ldots$. The use of these expansions enables us to replace the operator H_0 by its eigenvalues in the Eqs. 5.7, 5.8 etc.

$$H_0 v^{(s)} = \sum_n c_n^{(s)} H_0 u_n = \sum_n c_n^{(s)} E_n u_n \tag{5.13}$$

With this reduction, Eq. 5.7, taken with Eqs. 5.9 and 5.10, becomes

$$\sum_n (E_n - E_m)\, c_n^{(1)} u_n + (H' - W^{(1)})\, u_m = 0 \tag{5.14}$$

If we now take the scalar product of u_k with Eq. 5.14 and use the fact that $(u_k, u_n) = \delta_{kn}$, the first term reduces to $\Sigma\, (E_n - E_m)\, c_n^{(1)}\, \delta_{kn} = (E_k - E_m)\, c_k^{(1)}$. So the equation becomes

$$(E_k - E_m)\, c_k^{(1)} + H'_{km} - W^{(1)}\delta_{km} = 0 \tag{5.15}$$

Here

$$H'_{km} = (u_k, H' u_m) \tag{5.16}$$

It is called the *matrix element of* H' between the states u_k and u_m.

Again, by substituting Eq. 5.12 in the second order Eq. 5.8 and taking the scalar product of u_k with the resulting equation, we obtain

$$(E_k - E_m)\, c_k^{(2)} + \sum_n (H'_{kn} - W^{(1)}\delta_{kn})\, c_n^{(1)} - W^{(2)}\, \delta_{km} = 0 \tag{5.17}$$

We can now solve Eq. 5.15 for $W^{(1)}$ and the $c_k^{(1)}$, and then use these in Eq. 5.17 to find $W^{(2)}$ and the $c_k^{(2)}$, and so on.

(*a*) *The First Order:* We obtain $W^{(1)}$ immediately from Eq. 5.15 by taking $k = m$, so that $E_k - E_m = 0$ and $\delta_{km} = 1$. We have

$$W^{(1)} = H'_{mm} \tag{5.18}$$

Thus the *first order change in the energy* is just the *expectation value of the perturbation*, taken with respect to the *unperturbed* wave function u_m. For any $k \neq m$, Eq. 5.15 tells us that

$$c_k^{(1)} = -\frac{H'_{km}}{E_k - E_m} \qquad (k \neq m) \tag{5.19a}$$

The only thing that remains undetermined is $c_m^{(1)}$. To find it we use the condition that the perturbed wave function v be normalized. Taking v to the first order,

$$v \approx u_m + \lambda\, \Sigma\, c_n^{(1)} u_n = (1 + \lambda\, c_m^{(1)})\, u_m + \Sigma'\, \lambda\, c_n^{(1)} u_n$$

where the prime on the summation sign indicates that $n = m$ is to be excluded. The normalization condition is clearly

$$|1 + \lambda c_m^{(1)}|^2 + \Sigma' |\lambda c_n^{(1)}|^2 = 1.$$

We are justified in keeping only terms of the first order in λ since the second order terms are not complete without the inclusion of contributions from $\lambda^2 v^{(2)}$. On ignoring the second order terms, we get

$$\lambda(c_m^{(1)} + c_m^{(1)*}) = 0$$

which means that $c_m^{(1)}$ must be purely imaginary, say $c_m^{(1)} = i\alpha$. The coefficient of u_m in v is then $(1 + i\alpha\lambda) \approx e^{i\alpha\lambda}$ to the first order in λ. But this is simply a phase factor, and since the wave function for any state is arbitrary to the extent of a phase factor, we can redefine our u_m to include this factor. With this redefinition, the coefficient of u_m in v becomes unity, which amounts to saying that

$$c_m^{(1)} = 0 \tag{5.19b}$$

From Eqs. 5.12 and 5.19 we finally have

$$v^{(1)} = - \sum_n{}' u_n \frac{H'_{nm}}{E_n - E_m} \tag{5.20}$$

The usefulness of perturbation theory depends on the smallness of the change $\lambda v^{(1)}$ in the wave function, which clearly requires that

$$|\lambda H'_{nm}| \ll |E_n - E_m| \quad \text{for all } n \neq m \tag{5.21}$$

This gives a quantitative *criterion for the smallness of the perturbation*. Another useful (and self evident) criterion is that the energy change $\lambda W^{(1)} \equiv \lambda H'_{mm}$ should be small compared to the spacing between E_m and the levels nearest to it.

(*b*) *The Second Order:* The second order equation (5.17) can be solved in the same way. First, by setting $k = m$ we obtain

$$W^{(2)} = \sum_n (H'_{mn} - W^{(1)} \delta_{mn})\, c_n^{(1)}$$
$$= \sum_n{}' H'_{mn} c_n^{(1)} = - \sum_n{}' \frac{H'_{mn} H'_{nm}}{E_n - E_m} \tag{5.22}$$

An immediate observation is that since H' is a self adjoint (Hermitian) operator, $H'_{mn} = (H'_{nm})^*$, and hence the numerator in the above expression is the *nonnegative* quantity $|H'_{mn}|^2$. Hence, *if the unperturbed state is the ground state*, so that $(E_n - E_m) \equiv (E_n - E_o) > 0$ for all n, it follows that $W^{(2)}$ *is necessarily negative.*

Returning now to Eq. 5.17 and setting $k \neq m$, we obtain $c_k^{(2)}$ to be

$$c_k^{(2)} = - \frac{1}{E_k - E_m} \left[\sum_n H'_{kn} c_n^{(1)} - W^{(1)} c_k^{(1)} \right], \ (k \neq m) \tag{5.23a}$$

To determine $c_m^{(2)}$ we once again appeal to the normalization of v:

$$v \approx u_m + \lambda v^{(1)} + \lambda^2 v^{(2)} = (1 + \lambda^2 c_m^{(2)})\, u_m + \sum_n{}' (\lambda c_n^{(1)} + \lambda^2 c_n^{(2)})\, u_n$$

The normalization condition is

$$|1 + \lambda^2 c_m^{(2)}|^2 + \sum_n{}' |\lambda c_n^{(1)} + \lambda^2 c_n^{(2)}|^2 = 1$$

which, on retaining only terms to the second order, becomes

$$\lambda^2 [c_m^{(2)} + c_m^{(2)*} + \sum_n{}' \mid c_n^{(1)} \mid^2] = 0$$

This condition determines the real part of $c_m^{(2)}$. The imaginary part, as before, can be absorbed into u_m by a further redefinition of its phase factor, and can therefore be taken to be zero. We then have, from the above equation,

$$c_m^{(2)} = c_m^{(2)*} = -\tfrac{1}{2} \Sigma' \mid c_n^{(1)} \mid^2 \tag{5.23b}$$

On substituting Eqs. 5.23 in $v^{(2)} = \Sigma\, c_k^{(2)}\, u_k$ and using the already calculated values for $c_n^{(1)}$ and $W^{(1)}$, we get

$$v^{(2)} = \sum_k{}' \left\{ \sum_n{}' \frac{u_k H'_{kn} H'_{nm}}{(E_k - E_m)(E_n - E_m)} - \frac{u_k H'_{km} H'_{mm}}{(E_k - E_m)^2} - \tfrac{1}{2} u_m \frac{\mid H'_{km} \mid^2}{(E_k - E_m)^2} \right\} \tag{5.24}$$

The increase in complexity of the expressions as one goes to higher orders is already evident.

EXAMPLE 5.1.—*The anharmonic oscillator.* The Hamiltonian $H = (p^2/2m) + \frac{1}{2} m\omega^2 x^2 + \lambda x^4$ differs from that of the harmonic oscillator $(H_0 = p^2/2m + \frac{1}{2} m\omega^2 x^2)$ by the anharmonic term λx^4. Taking λ to be very small and using first order perturbation theory with $\lambda H' = \lambda x^4$, we obtain the change $\triangle E^{(1)} = \lambda W^{(1)}$ in the nth harmonic oscillator level $E_n = (n + \frac{1}{2})\hbar\omega$:

$$\triangle E_n^{(1)} = \lambda H'_{nn} = \lambda(u_n, x^4 u_n) = \tfrac{3}{2}\lambda \left(\frac{\hbar}{m\omega}\right)^2 (n^2 + n + \tfrac{1}{2})$$

The value of $(u_n, x^4 u_n)$ is taken from Example 4.2, p. 113. If instead of λx^4 we had a cubic term $\varkappa x^3$, there would be no energy change in the first order, since $(u_n, x^3 u_n) = 0$ for parity reasons. It is then necessary to make a second order calculation using Eq. 5.22. For this we need the matrix elements $(x^3)_{mn}$, which are given in Example 4.2, p. 113. Only those with $(m - n) = \pm 1$, ± 3 are nonzero, and we obtain (for cubic anharmonicity)

$$\begin{aligned}\triangle E^{(2)} = \varkappa^2 W^{(2)} &= \varkappa^2 \left(\frac{\hbar}{2m\omega}\right)^3 \left\{ \frac{1}{3\hbar\omega} [n(n-1)(n-2) \right. \\ &\quad \left. - (n+3)(n+2)(n+1)] + \frac{9}{\hbar\omega}[n^3 - (n+1)^3] \right\} \\ &= -\frac{15}{4} \frac{\varkappa^2 \hbar^2}{m^3 \omega^4} \left(n^2 + n + \frac{11}{30} \right).\end{aligned}$$

These calculations are made on the tacit assumption that perturbed levels do exist. Actually, with a purely cubic anharmonicity the potential falls to $-\infty$ as $x \to -\infty$ (or as $x \to +\infty$ if $\varkappa$ is negative); no bound states can exist at all in such a potential. The same is true of a quartic potential with $\lambda < 0$. Perturbation theory (at least in the lowest order) does not reveal this fact. Secondly, even if H has bound states (quartic anharmonicity, $\lambda > 0$) the calculated $\triangle E^{(1)}$ makes sense *only for low lying levels:* as n increases, $\triangle E^{(1)}$ would increase rapidly and exceed the unperturbed level spacing $\hbar\omega$. Going to higher orders in the perturbation only makes matters worse, since they bring in higher powers of n^2.

5.3 The Degenerate Case — Removal of Degeneracy

Let us return to Eq. 5.6 and suppose that $W^{(0)} = E_m$ is a degenerate level of the unperturbed system. For clarity we shall first consider the simple case of a doubly degenerate level. Let u_{m1}, u_{m2} be two orthonormal eigenfunctions belonging to the level E_m:

$$H_0 u_{m\alpha} = E_m u_{m\alpha} \quad (\alpha = 1,2) \tag{5.25a}$$

$$\int u_{m\alpha}^* u_{m\beta}\, d\tau \equiv (u_{m\alpha}, u_{m\beta}) = \delta_{\alpha\beta} \tag{5.25b}$$

Then the zeroth order equation (5.6) tells us that $v^{(0)}$ must be some linear combination of u_{m1} and u_{m2}

$$v^{(0)} = a_1^{(0)} u_{m1} + a_2^{(0)} u_{m2} \tag{5.26}$$

where the constants $a_1^{(0)}$, $a_2^{(0)}$ are arbitrary at this stage. However, the first order equation (5.7) places constraints on them, and also determines $W^{(1)}$. To see this, let us expand $v^{(1)}$ as

$$v^{(1)} = (a_1^{(1)} u_{m1} + a_2^{(1)} u_{m2}) + \sum_n{}' c_n^{(1)} u_n \tag{5.27}$$

The eigenfunctions u_n belonging to all the levels $E_n \neq E_m$ of H_0, together with u_{m1} and u_{m2}, form a complete set and so an expansion of this form is always possible. On substituting Eqs. 5.26 and 5.27 in Eq. 5.7, we obtain

$$\sum_n{}' (E_n - E_m)\, c_n^{(1)} u_n + (H' - W^{(1)})\,(a_1^{(0)} u_{m1} + a_2^{(0)} u_{m2}) = 0. \tag{5.28}$$

Now, we recall that u_{m1} and u_{m2} are mutually orthogonal and also orthogonal to all the u_n (which belong to $E_n \neq E_m$). Therefore, if we take the scalar product of u_{m1} with (5.28), it would reduce to

$$(h_{11} - W^{(1)})\, a_1^{(0)} + h_{12}\, a_2^{(0)} = 0 \tag{5.29}$$

where

$$h_{\alpha\beta} = (u_{m\alpha}, H' u_{m\beta}) \tag{5.30}$$

Similarly, taking the scalar product of u_{m2} with (5.28), we get

$$h_{21}\, a_1^{(0)} + (h_{22} - W^{(1)})\, a_2^{(0)} = 0 \tag{5.31}$$

Eqs. 5.29 and 5.31 are two homogeneous equations for $a_1^{(0)}$ and $a_2^{(0)}$, and they can have solutions (other than the trivial one $a_1^{(0)} = a_2^{(0)} = 0$) only if the determinant of coefficients vanishes, i.e.

$$\begin{vmatrix} (h_{11} - W^{(1)}) & h_{12} \\ h_{21} & (h_{22} - W^{(1)}) \end{vmatrix} = 0 \tag{5.32}$$

This is called the *secular equation* for the problem. It is a second degree algebraic equation for $W^{(1)}$. Explicitly,

$$(W^{(1)})^2 - (h_{11} + h_{22})\, W^{(1)} + (h_{11} h_{22} - h_{12} h_{21}) = 0;$$

$$W^{(1)} = \tfrac{1}{2}(h_{11} + h_{22}) \pm \tfrac{1}{2}\left[(h_{11} - h_{22})^2 + 4\, h_{12} h_{21}\right]^{1/2} \tag{5.33}$$

Thus there are two possible values of $W^{(1)}$, say $W_1^{(1)}$ and $W_2^{(1)}$ and consequently the original energy level splits into two, namely $E_m + W_1^{(1)}$ and $E_m + W_2^{(1)}$, as a result of the perturbation. We say that the perturbation *removes the degeneracy*. This is a general consequence of perturbations which is of the utmost importance. (It may happen however that for some perturbations $W_1^{(1)} = W_2^{(1)}$, in which case the degeneracy would still remain.)

Now, let the solution of Eqs. 5.29 and 5.31 with $W^{(1)}$ set equal to $W_\mu^{(1)}$ ($\mu = 1$ or 2) be $a^{(0)}_{1(\mu)}$, $a^{(0)}_{2(\mu)}$.

Actually only the ratio of these two is obtained but a common multiplying constant can be chosen so as to make $|a^{(0)}_{1(\mu)}|^2 + |a^{(0)}_{2(\mu)}|^2 = 1$. When $a^{(0)}_{1(\mu)}$ and $a^{(0)}_{2(\mu)}$, so determined, are introduced in Eq. 5.26, we get the normalized *unperturbed* wave function $v_\mu^{(0)}$ associated with the level $(E_m + W_\mu^{(1)})$.

EXAMPLE 5.2—The two-dimensional harmonic oscillator, with the Hamiltonian $H_n = (p_x^2 + p_y^2)/2m + \frac{1}{2}\, m\omega^2\, (x^2 + y^2)$, has a level with energy $E = 2\hbar\omega$, which is doubly degenerate. The eigenfunctions

$$u_{\rm I} = u_0\,(x)\, u_1\,(y) \text{ and } u_{\rm II} = u_1(x)\, u_0(y)$$

both belong to this level. Here $u_n(x)$ and $u_n(y)$ are eigenfunctions of simple harmonic oscillators in the x and y directions respectively. Suppose now that this oscillator is subjected to a perturbation $\lambda H' = \lambda m\omega^2 xy$. Then we have

$$h_{\rm I,I} = m\omega^2 \iint u_0(x)\, u_1(y)\; xy\; u_0(x)\, u_1(y)\; dx\, dy = 0$$

and similarly $h_{\rm II,II} = 0$, for parity reasons. It is easily verified, however, that

$$h_{\rm I,II} = m\omega^2\; (u_0(x),\, x\, u_1\,(x))\;\; (u_1(y),\, y\, u_0(y)) = \tfrac{1}{2}\,\hbar\omega = h_{\rm II,I}$$

Introducing these in Eq. 5.33 — note that we are using I, II instead of 1,2 as subscripts now — we find that the two values of $W^{(1)}$ are $+\, h_{\rm I,II}$ or $-\, h_{\rm I,II}$. Thus the degenerate energy level splits into the doublet:

$$2\hbar\omega + \tfrac{1}{2}\lambda\hbar\omega \text{ and } 2\hbar\omega - \tfrac{1}{2}\lambda\hbar\omega$$

The zeroth order eigenfunctions associated with the first of these may be found by setting $W^{(1)} = +\, h_{\rm I,\,II}$ in Eq. 5.29, whence $a_{\rm I}^{(0)}/a_{\rm II}^{(0)} = +\,1$. Thus the normalized eigenfunction is $(u_{\rm I} + u_{\rm II})/\sqrt{2}$. For the second (lower) member of the doublet we find from Eq. 5.29 with $W^{(1)} = -\, h_{\rm I,II}$, that $a_{\rm I}^{(0)}/a_{\rm II}^{(0)} = -\,1$ so that the zeroth order wave function is $(u_{\rm I} - u_{\rm II})/\sqrt{2}$.

* * *

The generalization of the above treatment to the case of r-fold degeneracy is straightforward. Now one has

$$v^{(0)} = \sum_{\alpha=1}^{r} a_\alpha\, u_{m\alpha} \tag{5.34}$$

and Eqs. 5.29 and 5.31 generalize to r equations for $a_1, a_2, \ldots, a_r$. These can be written in matrix form as

$$\begin{pmatrix} h_{11} & h_{12} & \cdot\;\cdot\;\cdot & h_{1r} \\ h_{21} & h_{22} & \cdot\;\cdot\;\cdot & h_{2r} \\ \cdot & \cdot & & \cdot \\ \cdot & \cdot & & \cdot \\ \cdot & \cdot & & \cdot \\ h_{r1} & h_{r2} & \cdot\;\cdot\;\cdot & h_{rr} \end{pmatrix} \begin{pmatrix} a_1 \\ a_2 \\ \cdot \\ \cdot \\ \cdot \\ a_r \end{pmatrix} = W^{(1)} \begin{pmatrix} a_1 \\ a_2 \\ \cdot \\ \cdot \\ \cdot \\ a_r \end{pmatrix} \tag{5.35}$$

This is obviously an eigenvalue equation for the $r \times r$ matrix appearing on the left, which is Hermitian $(h_{ij} = h_{ji}^*)$. We recognize that $W^{(1)}$ can be any of the r real eigenvalues, say $W_\mu^{(1)} (\mu = 1, 2, \ldots, r)$, of this matrix. Hence the original energy level E_m splits (in general) into an r-plet, $(E_m + \lambda W_1^{(1)})$, $(E_m + \lambda W_2^{(1)}), \ldots, (E_m + \lambda W_r^{(1)})$. If the column eigenvector belonging to the eigenvalue $W_\mu^{(1)}$ in Eq. 5.35 has the elements $a_{1(\mu)}, a_{2(\mu)}, \ldots, a_{r(\mu)}$ (and it is normalized so that $\sum_\alpha |a_{\alpha(\mu)}|^2 = 1$), then by substituting these values for the a_α in Eq. 5.34 we get the unperturbed eigenfunction $v_\mu^{(0)}$ associated with the μth member $(E_m + \lambda W_\mu^{(1)})$ of the multiplet:

$$v_\mu^{(0)} = \sum_\alpha a_{\alpha(\mu)}\, u_{m\alpha} \tag{5.36}$$

The process of solving the matrix eigenvalue problem (5.35) would be greatly simplified if the choice of the basis functions $u_{m\alpha}$ could be made in such a way that many of the $h_{\alpha\beta}$ vanish. In practice one tries to accomplish this by finding an observable F which commutes with *both* H_0 and H'. If such an F is available, one takes the $u_{m\alpha}$ to be simultaneous eigenfunctions of H_0 and F:

$$[F, H'] = 0, \quad Fu_{m\alpha} = f_\alpha\, u_{m\alpha} \tag{5.37}$$

Then all $h_{\alpha\beta}$ vanish except those for which $f_\alpha = f_\beta$. This follows from the fact that

$$\begin{aligned} 0 &= (u_{m\alpha}, (FH' - H'F)\, u_{m\beta}) = (Fu_{m\alpha}, H'u_{m\beta}) - (u_{m\alpha}, H'Fu_{m\beta}) \\ &= (f_\alpha - f_\beta)\, h_{\alpha\beta} \end{aligned} \tag{5.38}$$

The set of normalized column eigenvectors of the (Hermitian) matrix in (5.35) forms an orthonormal set in the sense that

$$\sum_\alpha a^*_{\alpha(\mu)}\, a_{\alpha(\nu)} = \delta_{\mu\nu} \tag{5.39}$$

As a consequence, the r functions $v_\mu^{(0)}$ also form an orthonormal function set:

$$\begin{aligned} (v_\mu^{(0)}, v_\nu^{(0)}) &= \sum_{\alpha\beta} a^*_{\alpha(\mu)}\, a_{\beta(\nu)}\, (u_{m\alpha}, u_{m\beta}) \\ &= \sum_{\alpha\beta} a^*_{\alpha(\mu)}\, a_{\beta(\nu)}\, \delta_{\alpha\beta} = \delta_{\mu\nu} \end{aligned} \tag{5.40}$$

The $v_\mu^{(0)}$, being linear combinations of the $u_{m\alpha}$, obviously belong to the eigenvalue E_m of H_0, and can be used as a basis for the eigenspace belonging to E_m. Unlike the basis which we started with (consisting of the $u_{m\alpha}$) *the new basis is one which is adapted to the nature of the perturbation H'*. In fact it has the property that

$$(v_\mu^{(0)}, H'v_\nu^{(0)}) = W_\mu^{(1)}\, \delta_{\mu\nu} \tag{5.41}$$

For, on using the definition (5.36), the left hand side becomes

$$\begin{aligned} &\Big(\sum_\alpha a_{\alpha(\mu)}\, u_{m\alpha}, H' \sum_\beta a_{\beta(\nu)}\, u_{m\beta}\Big) \\ &= \sum_{\alpha\beta} a^*_{\alpha(\mu)}\, a_{\beta(\nu)}\, (u_{m\alpha}, H'u_{m\beta}) = \sum_{\alpha\beta} a^*_{\alpha(\mu)}\, h_{\alpha\beta}\, a_{\beta(\nu)} \\ &= W_\nu^{(1)} \sum_\alpha a^*_{\alpha(\mu)}\, a_{\beta(\nu)} = W_\nu^{(1)}\, \delta_{\mu\nu}. \end{aligned}$$

In the last three steps we have used Eqs. 5.30, 5.35 and 5.39 respectively.

Eq. 5.41 states that the matrix whose (μ, ν) element is $(v_\mu^{(0)}, H'v_\nu^{(0)})$ is diagonal, and that the diagonal elements are the first order changes $W_\mu^{(1)}$ which we are seeking. Thus the first order perturbation theory of a degenerate level is completely equivalent to finding a *particular basis* $v_\mu^{(0)}$ $(\mu = 1, 2, \ldots, r)$ for the eigenspace of this level, *with respect to which the matrix of the perturbation is diagonal.*

5.4 The Effect of an Electric Field on the Energy Levels of an Atom (Stark Effect).

The calculation of this physically important effect provides a good illustration of perturbation techniques.

When a homogeneous electric field $\mathcal{E}$ is applied to an atom (or other system of charged particles), the system acquires an electrostatic potential energy

$$H' = -\sum_{\alpha} e_\alpha \mathbf{x}_\alpha . \mathcal{E} = -\sum_{\alpha} d_\alpha . \mathcal{E} = -\mathbf{d}.\mathcal{E} \tag{5.42}$$

Here e_α is the charge on the αth particle which is at $\mathbf{x}_\alpha$, and $\mathbf{d}_\alpha = e_\alpha \mathbf{x}_\alpha$ is the electric dipole moment associated with it. H' may be treated as a perturbation on the free atom. Since H' is of odd parity, its matrix elements between states of equal parity vanish; in particular, since atomic states may be chosen to be of definite parity, the expectation values H'_{mm} in such states vanish. As a result, there is no first order Stark effect in the ground state of an atom; but splitting of *degenerate* levels does often take place in the first order.

(*a*) *The Ground State of the Hydrogen Atom:* Since there is no first order effect in this case, as already mentioned, we have to make a second order calculation using Eq. 5.22. The evaluation of the infinite sum would seem quite cumbersome; but in the present case this calculation can be short-circuited by a method due to Dalgarno and Lewis,[4] which we first explain.

The method depends on finding an operator F such that

$$H'u_m = (H_0F - FH_0)\, u_m \tag{5.43}$$

for the particular state u_m whose perturbation is being considered. By multiplying Eq. 5.43 by u_n^* and integrating, one finds that[5] $H'_{nm} = (E_n - E_m) F_{nm}$. On substituting this in Eq. 5.22, we get

$$W^{(2)} = -\sum_n{}' \frac{H'_{mn} H'_{nm}}{E_n - E_m} = -\sum_n H'_{mn} F_{nm} = -(H'F)_{mm} \tag{5.44}$$

There is no longer any sum over intermediate states.

In the Stark effect of the hydrogen atom,

$$H_0 = -(\hbar^2/2m)\nabla^2 - e^2/r \text{ and } H' = -e\mathcal{E}z \tag{5.45}$$

it being assumed that the electric field is in the z-direction. On using these and taking u_m as the ground state u_{100} (which we abbreviate here to u_0), Eq. 5.43 becomes

$$-\frac{\hbar^2}{2m}\nabla^2(Fu_0) - \left(\frac{e^2}{r} + E_0\right) Fu_0 = -e\mathcal{E}zu_0 \tag{5.46}$$

where E_0 is the ground state energy. Since the angular part of the right hand side comes from z only and is proportional to the spherical harmonic Y_{10}, it is clear that F also must be proportional to $Y_{10} \propto z$. (Remember Y_{10} is an eigenfunction of ∇^2 and there are no angle-dependent factors multiplying F on the left hand side). Hence, assuming $F = f(r)\, z$ and substituting in Eq. 5.46, and making use of the explicit form [Eq. 4.148a] of u_0, we get $F = -(mae\mathcal{E}/\hbar^2)(\frac{1}{2}r + a)\, z$. Noting that $\hbar^2/me^2 = a$, we then have

$$W^{(2)} = -(H'F)_{00} = -\mathcal{E}^2 (u_0, (\tfrac{1}{2}r + a)\, z^2 u_0) \tag{5.47}$$

[4] A. Dalgarno and J. T. Lewis, *Proc. Roy. Soc., Lond.*, **A233**; 70, 1955.

[5] This relation requires that $H'_{mm} = 0$. If it is not zero, say $H'_{mm} = \varepsilon$, we can take $H' - \varepsilon$ as a new H', and use this for further calculations. It is evident that the matrix elements H'_{nm}, $n \neq m$, which appear in $W^{(2)}$ are unaffected by this change.

Now in the integral which defines this expectation value, all factors except z^2 are functions of r only, independent of direction; and so it would make no difference to its value if z^2 were replaced by x^2 or y^2, or better, $\frac{1}{3}(x^2+y^2+z^2) = \frac{1}{3}r^2$. Taking advantage of this, we obtain, on using Eq. 4.148a for u_0,

$$W^{(2)} = -\frac{4\mathcal{E}^2}{a^3}\int_0^\infty (\tfrac{1}{2}r + a)\,\tfrac{1}{3}r^2 e^{-2r/a}\,4r^2\,dr = -\tfrac{9}{4}a^3\mathcal{E}^2 \tag{5.48}$$

This energy change can be pictured as being due to the atom becoming electrically 'polarized' (i.e., acquiring a dipole moment $\alpha\mathcal{E}$, where α is the atomic polarizability). The magnitude of the energy of polarization is $\frac{1}{2}\alpha\mathcal{E}^2$. Comparison with Eq. 5.48 shows that the polarizability (α) of the hydrogen atom in its ground state is equal to $(9/2)\ a^3$.

(*b*) *The First Excited Level of the Hydrogen Atom:* There are four states u_{2lm} ($l = 0, m = 0; l = 1, m = 1, 0, -1$) belonging to this level. So we have to use the perturbation theory of a degenerate level. To construct the secular determinant for this problem, we need the matrix elements $(u_{2l'm'}, H'u_{2lm})$. Now, since H', (Eq. 5.45), is of odd parity, it has nonvanishing matrix elements only between states of opposite parities. In our case this means that all matrix elements in which $l = l' = 0$ or $l = l' = 1$ are zero. The remaining matrix elements are $(u_{200}, H'u_{21m})$ and $(u_{21m'}, H'u_{200})$. When the integrals which define these matrix elements are written in terms of polar coordinates (r, θ, φ), the only φ-dependent factors in the integrand are $e^{im\varphi}$ and $e^{-im'\varphi}$ respectively in the two cases (since $H' \propto z \propto \cos\theta$ is independent of φ). So these integrals (from 0 to 2π) vanish except when $m = m' = 0$. Thus, out of 16 matrix elements we are finally left with just two nonvanishing ones, $(u_{210}, H'u_{200})$ and $(u_{200}, H'u_{210})$. On substituting the wave functions from Eq. 4.148 we get

$$\begin{aligned}(u_{210}, H'u_{200}) &= (u_{200}, H'u_{210})^* \\ &= -\frac{e\mathcal{E}}{16a^4}\int_0^\infty\int_0^\pi r^2\left(2-\frac{r}{a}\right).e^{-r/a}\cos^2\theta.r^2\sin\theta\,d\theta.dr \\ &= -3e\mathcal{E}a.\end{aligned} \tag{5.49}$$

Thus, if the zeroth order wave function is written as

$$v^{(0)} = (a_{00}u_{200} + a_{10}u_{210} + a_{11}u_{211} + a_{1,-1}u_{21,-1}) \tag{5.50}$$

Eq. 5.35 for the present case becomes

$$\begin{bmatrix} 0 & -3e\mathcal{E}a & 0 & 0 \\ -3e\mathcal{E}a & 0 & 0 & 0 \\ 0 & 0 & 0 & 0 \\ 0 & 0 & 0 & 0 \end{bmatrix}\begin{bmatrix} a_{00} \\ a_{10} \\ a_{11} \\ a_{1,-1}\end{bmatrix} = W^{(1)}\begin{bmatrix} a_{00} \\ a_{10} \\ a_{11} \\ a_{1,-1}\end{bmatrix} \tag{5.51}$$

The secular determinant is the determinant of the above matrix with the zeroes along the diagonal replaced by $-\lambda$. Equating the value of the determinant to zero, we get the secular equation $\lambda^2[\lambda^2 - (3e\mathcal{E}a)^2] = 0$. The possible values of $W^{(1)}$ (which are the roots of this equation) and the corresponding sets of values of the coefficients a_{lm} are easily found to be

$$\begin{array}{ccccc} W^{(1)} & = \quad 0 \quad , & 0 \quad , & -3e\mathcal{E}a \quad , & 3e\mathcal{E}a \end{array}$$

$$\begin{pmatrix} a_{00} \\ a_{10} \\ a_{11} \\ a_{1,-1} \end{pmatrix} = \left\{ \begin{pmatrix} 0 \\ 0 \\ 1 \\ 0 \end{pmatrix}, \begin{bmatrix} 0 \\ 0 \\ 0 \\ 1 \end{bmatrix}, \begin{bmatrix} 1/\sqrt{2} \\ 1/\sqrt{2} \\ 0 \\ 0 \end{bmatrix}, \begin{bmatrix} 1/\sqrt{2} \\ -1/\sqrt{2} \\ 0 \\ 0 \end{bmatrix} \right\} \tag{5.52}$$

Note that the $n = 2$ level is split into *three*, of which the middle one (unshifted, $W^{(1)} = 0$) is still doubly degenerate. The other two levels of this triplet have $W^{(1)} = + 3e\mathcal{E}a$, $v^{(0)} = 2^{-\frac{1}{2}} (u_{200} + u_{210})$, and $W^{(1)} = - 3e\mathcal{E}a$, $v^{(1)} = 2^{-\frac{1}{2}} (u_{200} - u_{210})$ respectively. These energy changes coincide with the energies of interaction of permanent electric dipoles of magnitude $(3ea)$ oriented antiparallel or parallel to the electric field. Examine the charge distributions in the four zeroth order wave functions and try to understand why two of these behave as if they have permanent electric dipole moments.

5.5 Two-Electron Atoms[6]

The Hamiltonian of two electrons moving in the field of a fixed nucleus of charge Ze is

$$H = -\frac{\hbar^2}{2m}(\nabla_1^2 + \nabla_2^2) - \frac{Ze^2}{r_1} - \frac{Ze^2}{r_2} + \frac{e^2}{r_{12}} \tag{5.53}$$

$$= H(1) + H(2) + \frac{e^2}{r_{12}} \tag{5.54}$$

where $H(1) \equiv - (\hbar^2/2m) \nabla_1^2 - Ze^2/r_1$ involves the coordinates of electron 1 only, and similarly for $H(2)$ (The quantities r_1, r_2, r_{12} are indicated in Fig. 5.1). Note that $H(1)$ and $H(2)$ are the Hamiltonians of two *independent* hydrogen-like (one-electron) atoms with nuclear charge Ze. We denote these 'atoms' by A and B respectively.[7] Therefore, the eigenvalue problem for $H_0 \equiv H(1) + H(2)$ is exactly soluble. If $H_0\, u(1, 2) = E\, u\, (1, 2)$, then by Sec. 3.15,

$$u\,(1, 2) = u_n\,(1)\, u_{n'}\,(2), \quad E = E_n + E_{n'} \tag{5.55}$$

where $u_n\,(1)$ is an abbreviation for the eigenfunction $u_{nlm}\,(\mathbf{x}_1)$ of 'atom' A involving electron 1 (described by $H(1)$), and $u_{n'}(2)$ means $u_{n'l'm'}\,(\mathbf{x}_2)$. The actual wave functions and corresponding energies $(E_n = - Z^2e^2/2an^2)$ are known from Sec. 4.16 We can, therefore, treat the term (e^2/r_{12}) in H as a perturbation H' which couples the two hitherto-independent 'atoms'.

Consider now the unperturbed level having the lowest energy $E = 2E_0$,

[6] This is our first example of a system of identical particles. We have observed in Sec. 3.16 that any wave function of a two electron system (or of any larger system of which the electrons form a part) has to be such that it changes sign if the position *and* spin variables of one electron were interchanged with those of the other. However, we shall ignore this requirement here and elsewhere in this Chapter, since we have not yet effected the introduction of spin into the formalism. Also we shall continue for the present to ignore contributions to the Hamiltonian arising from spin. After the inclusion of spin is accomplished, we shall comment in Sec. 8.8 on the consequences of the identity of particles in various problems (including that of the two-electron atom).

[7] In reality, both the electrons move in the field of the same nucleus. When the interaction between the two electrons is disregarded, each electron behaves as if it has exclusive possession of the nucleus. In this approximation the system behaves as if it consists of two independent atoms, with both nuclei at the same position.

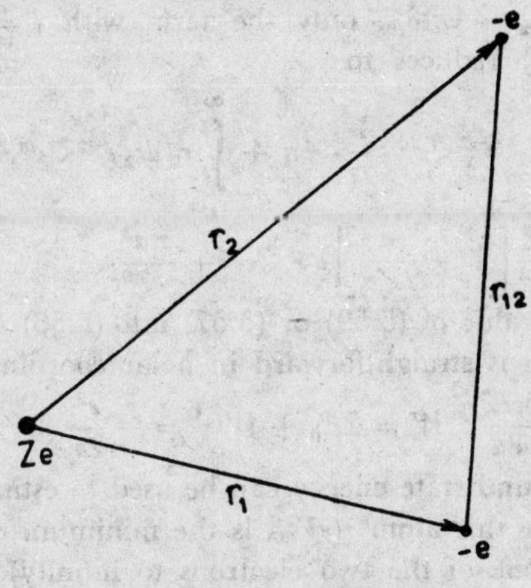

Fig. 5.1 Two-electron atom.

the corresponding wave function being $u_0(1)\, u_0(2)$. The change in energy of this level due to the perturbation H' is, in the first order,

$$W^{(1)} = \int\int u_0^*(1)\, u_0^*(2)\, \frac{e^2}{r_{12}}\, u_0(1)\, u_0(2)\, d\tau_1\, d\tau_2$$
$$= \left(\frac{Z}{a}\right)^6 \frac{e^2}{\pi^2} \iint \frac{1}{r_{12}} \exp\left[-\frac{2Z}{a}(r_1 + r_2)\right] d^3x_1 d^3x_2 \qquad (5.56)$$

where we have used Eq. 4.148a for the wave function $u_0(\mathbf{x}) = u_{100}(r\theta\varphi)$. Consider first the integration over $\mathbf{x}_2$, keeping $\mathbf{x}_1$ fixed. The relevant part of the integral is

$$\int \frac{1}{r_{12}} \exp\left[-\frac{2Z}{a} r_2\right] d^3x_2 = \int_0^\infty r_2^2\, dr_2 \int d\Omega_2\, e^{-2Zr_2/a} \cdot \frac{1}{r_{12}} \qquad (5.57)$$

where we have introduced polar coordinates $(r_2, \theta_2, \varphi_2)$ for $\mathbf{x}_2$, and $d\Omega_2$ is the element of solid angle, $\sin\theta_2\, d\theta_2\, d\varphi_2$. To evaluate (5.57), use the expansion

$$\left.\begin{aligned} \frac{1}{r_{12}} &= \frac{1}{r_1} \sum_{l=0}^{\infty} \left(\frac{r_2}{r_1}\right)^l P_l(\cos\theta_{12}), \quad r_1 > r_2 \\ &= \frac{1}{r_2} \sum_{l=0}^{\infty} \left(\frac{r_1}{r_2}\right)^l P_l(\cos\theta_{12}), \quad r_1 < r_2 \end{aligned}\right\} \qquad (5.58)$$

and also choose the polar axis to be along the direction of $\mathbf{x}_1$ so that the polar angle θ_2 is just the angle θ_{12} between $\mathbf{x}_1$ and $\mathbf{x}_2$. Then (5.57) becomes

$$\int_0^{r_1} r_2^2\, dr_2 \int d\Omega_2\, e^{-2Zr_2/a} \left[\frac{1}{r_1} \sum_{l=0}^{\infty} \left(\frac{r_2}{r_1}\right)^l P_l(\cos\theta_2)\right]$$
$$+ \int_{r_1}^{\infty} r_2^2\, dr_2 \int d\Omega_2\, e^{-2Zr_2/a} \left[\frac{1}{r_2} \sum_{l=0}^{\infty} \left(\frac{r_1}{r_2}\right)^l P_l(\cos\theta_2)\right] \qquad (5.59)$$

Since $\int P_l (\cos\theta_2)\, d\Omega_2 = 4\pi\, \delta_{l0}$, only the term with $l = 0$ survives in each integral. Thus (5.59) reduces to

$$\int_0^{r_1} dr_2\, r_2^2\, e^{-2Zr_2/a} \cdot \frac{1}{r_1} \cdot 4\pi + \int_{r_1}^{\infty} r_2\, dr_2\, e^{-2Zr_2/a}.4\pi$$

$$= -\frac{\pi a^2}{Z^2}\left[1 + \frac{a}{2Zr_1}\right] e^{-2Zr_1/a} + \frac{\pi a^3}{Z^3 r_1} \cdot \tag{5.60}$$

On substituting this value of (5.59) or (5.57) into (5.56) and carrying out the $\mathbf{x}_1$ integration (which is straightforward in polar coordinates), we finally get

$$W^{(1)} = \frac{5e^2 Z}{8a};\; W \approx 2E_0 + W^{(1)} = -\frac{e^2}{2a}\left(2Z^2 - \tfrac{5}{4} Z\right) \tag{5.61}$$

This value of the ground state energy can be used to estimate the first ionization potential V_{ion} for this atom[8] (eV_{ion} is the minimum energy that must be supplied to remove one of the two electrons to infinity). The energy of the remaining one-electron atom (namely E_o) is $W + eV_{ion}$. Hence

$$V_{ion} = (E_o - W)/e = \frac{e}{2a}\left(Z^2 - \tfrac{5}{4} Z\right) \tag{5.62}$$

For the helium atom ($Z = 2$) this works out to be 20.33 volts, compared to the experimental value 24.43 volts.

B. THE VARIATION METHOD

5.6 Upper Bound on Ground State Energy

The variation method[9] is used primarily for the estimation of the ground state energy. It is based on the fact that the lowest value which the expectation value of the Hamiltonian can take is the ground state energy E_0:

$$W \equiv \int \psi^* H\psi\, d\tau \geqslant E_0 \tag{5.63}$$

for any arbitrary state ψ of the system. This inequality is easy to prove. If ψ is imagined to be expanded in terms of the orthonormal set of eigenfunctions u_k of H as

$$\psi = \Sigma\, c_k u_k,\quad Hu_k = E_k u_k,\quad \int u_k^*\, u_l\, d\tau = \delta_{kl} \tag{5.64}$$

it follows that

$$\begin{aligned} W &= \int \sum_k c_k^*\, u_k^*\, H \sum_l c_l\, u_l\, d\tau = \sum_{kl} c_k^*\, c_l\, E_l\, \delta_{kl} \\ &= \sum_k |c_k|^2\, E_k \geqslant \sum_k |c_k|^2\, E_0 = E_0 \end{aligned} \tag{5.65}$$

We have made use of the fact that $E_k \geqslant E_0$ for all k, and that $|\sum_k c_k|^2 = 1$. (ψ is assumed to be normalized).

Now Eq. 5.63 shows that the lower the value of $W \equiv \langle H \rangle$, the closer it is to E_0. Therefore, we can try to minimize W by varying ψ, and take the

[8] The two-electron atom we are considering may be already an ionized one: We have neutral helium, singly ionized lithium, doubly ionized beryllium,... when $Z = 2,3,4...$ respectively.

[9] This method is originally due to Lord Rayleigh who applied it in 1873 to classical vibration problems (*Theory of Sound*, Macmillan, London, 1937, Vol I, Sec. 88) and J. Ritz, *Reine u. angew. Math.*, **135**, 1, 1908.

minimum value of W as an estimate for E_0. The variation of ψ is performed by first choosing for ψ a 'suitable' functional form depending on a number of parameters $a_1, a_2, \ldots, a_r$. When this is done,

$$\int \psi^* (\mathbf{x}; a_1, a_2, \ldots, a_r) H\psi(\mathbf{x}; a_1, a_2, \ldots, a_r)\, d\tau \equiv W(a_1, a_2, \ldots, a_r) \quad (5.66)$$

As the parameters are varied, the value of W also varies. Minimization of W is then accomplished by finding those values of $a_1, a_2, \ldots, a_r$ for which

$$\frac{\partial W}{\partial a_1} = \frac{\partial W}{\partial a_2} = \ldots = \frac{\partial W}{\partial a_r} = 0 \quad (5.67)$$

These equations may have several solutions, leading to corresponding values of W: The lowest of the values thus got for W gives the best estimate that can be obtained for E_0 starting from the particular form tried for ψ (which is called the *trial wave function*). With a different form for the trial wave function (e.g. one which has more parameters) it may be possible to get a still better estimate. In principle, there is no limit to the accuracy one can obtain by this method, provided one is willing to undertake the labour involved in minimizing with respect to a large number of parameters. However, the efficiency of the method depends very much on one's ability to employ physical intuition to guess the appropriate form for the trial wave function to be used in a particular problem.

As already noted, W is an upper bound to the value of E_0. It is possible to place limits on the error in the estimate, i.e. on $(W - E_0)$, by considering $\int \psi^* (H - W)^2 \psi\, d\tau$ and making use of Eq. 5.64.

$$\left.\begin{aligned} \int \psi^* (H - W)^2 \psi\, d\tau &= \sum_{kl} c_k^* c_l \int u_k^* (H - W)^2 u_l\, d\tau \\ &= \Sigma\, c_k^* c_k (E_k - W)^2 \\ &= (E_0 - W)^2 + \sum_k |c_k|^2 \{(E_k - W)^2 - (E_0 - W)^2\} \end{aligned}\right\} \quad (5.68)$$

The last term is non-negative, *provided the estimate W is closer to E_0 than to any excited level E_k*. This condition must certainly be met if the estimate is to be considered any good at all. Once this is ensured, we have, from Eq. 5.68, that $\int \psi^* (H - W)^2 \psi d\tau \geqslant (E_0 - W)^2$, or

$$(E_0 - W)^2 \leqslant \langle H^2 \rangle_\psi - W^2 \quad (5.69)$$

Since $(W - E_0)$ is known to be positive, we finally have

$$(W - E_0) \leqslant [\langle H^2 \rangle_\psi - W^2]^{1/2} \quad (5.70)$$

5.7 Application to Excited States

To apply the variation method to obtain an upper bound for the mth excited level E_m, all one has to do is to choose a trial wave function ψ which is orthogonal to the wave functions $u_0, u_1, \ldots, u_{m-1}$ belonging to all the lower levels. In the expansion (5.64) of such a ψ, the coefficients $c_1, c_2, \ldots, c_{m-1}$ would obviously vanish. So the lowest energy which appears in the expression $\Sigma\, |c_k|^2 E_k$ for W (Eq. 5.65), will be E_m; we can conclude then that $W \geqslant E_m$.

It is often the case that there is some observable F (e.g. angular momentum) which commutes with H and whose eigenfunctions are well known. If u_m is known to belong to a different eigenvalue (say f) of F than the ones to which

$u_0, u_1, \ldots, u_{m-1}$ belong, then by constructing a trial wave function ψ out of eigenfunctions of F belonging to this particular eigenvalue f, one automatically ensures that ψ is orthogonal to the *exact* eigenfunctions $u_0, u_1, \ldots, u_{m-1}$. In the absence of such simplifying circumstances, one would have to know the actual functions $u_0, u_1, \ldots, u_{m-1}$ (either exactly or to a sufficiently good approximation) in order to construct a ψ orthogonal to them. If ϕ is any arbitrary trial function

$$\psi \equiv \phi - \sum_{i=1}^{m-1} u_i \int u_i^* \phi d\tau$$

will have the desired orthogonality property, as may be directly verified. (This is the Schmidt orthogonalization process. See Example 3.11, p. 83.)

EXAMPLE 5.3—THE GROUND STATE OF A TWO-ELECTRON ATOM:[10] To estimate E_0 for a two-electron atom whose nucleus has charge Ze, by the variation method, let us use a trial wave function

$$\psi(1,2) = u_{100}(\mathbf{x}_1, Z')\, u_{100}(\mathbf{x}_2, Z')$$

Here $u_{100}(\mathbf{x}, Z')$ is the ground state wave function of an electron in the field of a nucleus of charge $Z'e$, and Z' will be used as the variational parameter. (The atom would have the wave function ψ if the presence of each electron affected the other one only through a partial 'screening' of the charge of the nucleus which reduces the effective nuclear charge as seen by each electron to $Z'e < Ze$. Clearly, this trial wave function satisfies the relation

$$\left(-\frac{\hbar^2}{2m}\nabla_1^2 - \frac{Z'e^2}{r_1} - \frac{\hbar^2}{2m}\nabla_2^2 - \frac{Z'e^2}{r_2}\right)\psi = 2\varepsilon_0(Z')\,\psi$$

where $\varepsilon_0(Z') = -Z'e^2/2a$ is the ground state energy of a hydrogen-like atom. Therefore, $H\psi$ (with H given by Eq. 5.53, the actual Hamiltonian of the two-electron atom), reduces to

$$H\psi = 2\varepsilon_0(Z')\psi + \left[(Z' - Z)\left(\frac{e^2}{r_1} + \frac{e^2}{r_2}\right) + \frac{e^2}{r_{12}}\right]\psi$$

Hence W (Eq. 5.63) becomes

$$W = 2\varepsilon_0(Z') + \iint \psi^* \left[(Z' - Z)\left(\frac{e^2}{r_1} + \frac{e^2}{r_2}\right) + \frac{e^2}{r_{12}}\right]\psi\, d\tau_1\, d\tau_2.$$

The last term in this integral is identical with (5.56) except that Z has to be replaced by Z'; from (5.61) its value is $(5e^2Z'/8a)$. The integrals involving $(1/r_1)$ and $(1/r_2)$ have equal values.

$$\iint \psi^* \frac{1}{r_1}\psi\, d\tau_1\, d\tau_2 = \int u^*_{100}(\mathbf{x}_1)\frac{1}{r_1}u_{100}(\mathbf{x}_1)\, d\tau_1 = \frac{Z'}{a}.$$

Putting together these results, we obtain

$$W = 2\varepsilon_0(Z') + 2(Z' - Z)\, e^2 \cdot \frac{Z'}{a} + \frac{5e^2Z'}{8a}$$
$$= \left(Z'^2 - 2ZZ' + \frac{5Z'}{8}\right)\frac{e^2}{a}.$$

[10] See footnote 8, p. 162.

This is a minimum when the free parameter Z' is chosen so that $(\partial W/\partial Z')=0$, which yields

$$Z' = Z - \frac{5}{16} \cdot$$

Hence the minimum value of W, which is our estimate for E_0, is

$$W_{min} = -\left(Z - \frac{5}{16}\right)^2 \frac{e^2}{a} \cdot$$

This is lower (and therefore a better estimate) than the value (5.61) of W obtained from perturbation theory.

5.8 Trial Function Linear in Variational Parameters

In many problems, especially in molecular physics, trial wave functions are chosen as linear combinations of known functions, say

$$\psi = a_1\chi_1 + a_2\chi_2 + \ldots + a_r\chi_r \tag{5.71}$$

with the coefficients $a_1, a_2, \ldots, a_r$ treated as variational parameters. The χ_i may not be normalized or mutually orthogonal. With such a choice of ψ,

$$W = \sum_{i,j=1}^{r} a_i^* a_j H_{ij}, \quad H_{ij} = \int \chi_i^* H \chi_j \, d\tau \tag{5.72}$$

We have to minimize W with respect to the a_i subject to the constraint that ψ be normalized, i.e.

$$\int \psi^*\psi d\tau \equiv \Sigma\, a_i^* a_j \Delta_{ij} = 1, \quad \Delta_{ij} = \int \chi_i^* \chi_j \, d\tau \tag{5.73}$$

The method of Lagrangian multipliers may be used to take account of this constraint. Thus we require that the quantity

$$I = \Sigma\, a_i^* a_j H_{ij} - \varepsilon\, (\Sigma\, a_i^* a_j \Delta_{ij} - 1) \tag{5.74}$$

be a minimum, where ε is the (undetermined) Lagrangian multiplier. For this it is necessary that the derivatives of I with respect to the real and imaginary parts of the a_i (or equivalently, with respect to the a_i and a_i^* treated as independent complex parameters) be zero. By considering $\partial I/\partial a_i^* = 0$, we get

$$\sum_{j=1}^{r} (H_{ij} - \varepsilon\Delta_{ij})\; a_j = 0, \quad (i = 1,2,\ldots,r) \tag{5.75}$$

The equations $\partial I/\partial a_i = 0$ are simply complex conjugates of (5.75) and give nothing new. Eqs. 5.75 form a set of r homogeneous linear equations for the r unknown a_i. If any nontrivial solution of this set is to exist, the determinant of the matrix of coefficients must vanish, i.e.

$$\begin{vmatrix} H_{11} - \varepsilon\Delta_{11} & H_{12} - \varepsilon\Delta_{12} \cdots & H_{1r} - \varepsilon\Delta_{1r} \\ H_{21} - \varepsilon\Delta_{21} & H_{22} - \varepsilon\Delta_{22} \cdots & H_{2r} - \varepsilon\Delta_{2r} \\ \cdots\cdots & \cdots\cdots & \cdots\cdots \\ H_{r1} - \varepsilon\Delta_{r1} & H_{r2} - \varepsilon\Delta_{r2} \cdots & H_{rr} - \varepsilon\Delta_{rr} \end{vmatrix} = 0 \tag{5.76}$$

This is evidently an algebraic equation of degree r in ε with r roots $\varepsilon^{(1)}, \varepsilon^{(2)}, \ldots, \varepsilon^{(r)}$. *These are the possible values* of W, for, by Eqs. 5.72, 5.75 and 5.73,

$$W = \sum_{ij} H_{ij}\; a_i^* a_j = \varepsilon \sum_{ij} \Delta_{ij}\; a_i^* a_j = \varepsilon \tag{5.77}$$

Of the values $\varepsilon^{(1)}, \varepsilon^{(2)}, \ldots, \varepsilon^{(r)}$, the lowest one then gives the minimum value of W, and hence provides our best estimate for the ground state energy E_0.

5.9 The Hydrogen Molecule[11]

We shall now make an estimation of the ground state energy of the hydrogen molecule as $W = \int \psi^* H \psi d\tau$, where ψ is a trial wave function suggested by the supposition that the ground state wave function does not differ much from that of a pair of independent (non-interacting) hydrogen atoms. Assuming that the nuclei A,B of the two hydrogen atoms remain at fixed positions, the Hamiltonian H may be written as

$$H = -\frac{\hbar^2}{2m}\nabla_1^2 - \frac{\hbar^2}{2m}\nabla_2^2 - \frac{e^2}{r_{1A}} - \frac{e^2}{r_{1B}} - \frac{e^2}{r_{2A}} - \frac{e^2}{r_{2B}} + \frac{e^2}{r_{12}} + \frac{e^2}{R} \quad (5.78)$$

where the labels 1 and 2 identify the two electrons, and the distances between pairs of particles are as shown in Fig. 5.2. The two atoms could be thought of

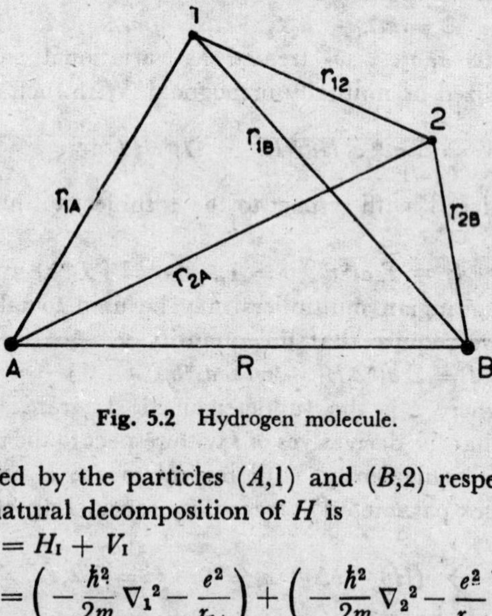

Fig. 5.2 Hydrogen molecule.

as being constituted by the particles $(A,1)$ and $(B,2)$ respectively. From this point of view, a natural decomposition of H is

$$\left.\begin{aligned} H &= H_{\text{I}} + V_{\text{I}} \\ H_{\text{I}} &= \left(-\frac{\hbar^2}{2m}\nabla_1^2 - \frac{e^2}{r_{1A}}\right) + \left(-\frac{\hbar^2}{2m}\nabla_2^2 - \frac{e^2}{r_{2A}}\right) \\ V_{\text{I}} &= -\frac{e^2}{r_{1B}} - \frac{e^2}{r_{2A}} + \frac{e^2}{r_{12}} + \frac{e^2}{R} \end{aligned}\right\} \quad (5.79)$$

H_{I} describes the two separate atoms while V_{I} constitutes the interaction between them. On the other hand, one could equally well think of the two atoms as consisting of (A,2) and (B,1) respectively. In this case the appropriate decomposition of H would be

$$\left.\begin{aligned} H &= H_{\text{II}} + V_{\text{II}} \\ H_{\text{II}} &= \left(-\frac{\hbar^2}{2m}\nabla_1^2 - \frac{e^2}{r_{1B}}\right) + \left(-\frac{\hbar^2}{2m}\nabla_2^2 - \frac{e^2}{r_{2A}}\right) \\ V_{\text{II}} &= -\frac{e^2}{r_{1A}} - \frac{e^2}{r_{2B}} + \frac{e^2}{r_{12}} + \frac{e^2}{R} \end{aligned}\right\} \quad (5.80)$$

If the interaction between the two atoms were ignored, the ground state wave

[11] See footnote 6, p. 160.

function, according to the two alternate points of view would be either $u_{\rm I}$ or $u_{\rm II}$, where

$$u_{\rm I} = u_{\rm A}(1)u_{\rm B}(2), \quad H_{\rm I}\, u_{\rm I} = 2\varepsilon u_{\rm I} \tag{5.81a}$$

$$u_{\rm II} = u_{\rm B}(1)u_{\rm A}(2), \quad H_{\rm II}\, u_{\rm II} = 2\varepsilon u_{\rm II} \tag{5.81b}$$

Here $u_{\rm A}(1)$ is the normalized ground state wave function of the atom formed by electron 1 with nucleus A, and so on; and ε is the ground state energy of a hydrogen atom:

$$\varepsilon = -\frac{e^2}{2a}, \quad u_{\rm A}(1) = \left(\frac{1}{\pi a^3}\right)^{1/2} e^{-r_{1\rm A}/a}, \text{ etc.} \tag{5.82}$$

Since both the points of view I and II are equally possible, it would be appropriate to use a linear combination of $u_{\rm I}$ and $u_{\rm II}$, say

$$\psi = a_{\rm I} u_{\rm I} + a_{\rm II} u_{\rm II} \tag{5.83}$$

as a trial wave function in a variational calculation of the lowest eigenvalue E_0 of H. According to the method of the last section, the best estimate for E_0 (based on the above trial wave function) would be the lower of the two solutions of the determinantal equation,

$$\begin{vmatrix} H_{\rm I,I} - E\Delta_{\rm I,I} & H_{\rm I,II} - E\Delta_{\rm I,II} \\ H_{\rm II,I} - E\Delta_{\rm II,I} & H_{\rm II,II} - E\Delta_{\rm II,II} \end{vmatrix} = 0 \tag{5.84}$$

Now, we shall see below that

$$H_{\rm I,I} = H_{\rm II,II}, \; H_{\rm I,II} = H_{\rm II,I}; \tag{5.85a}$$

$$\Delta_{\rm I,I} = \Delta_{\rm II,II} = 1, \; \Delta_{\rm I,II} = \Delta_{\rm II,I} \tag{5.85b}$$

If we assume this for the moment, Eq. 5.84 is seen to reduce to $(H_{\rm I,I} - E)^2 - (H_{\rm I,II} - E\Delta_{\rm I,II})^2 = 0$, with the solutions $E = E_{\rm S}$, $E_{\rm A}$ where

$$E_{\rm S} = \frac{H_{\rm I,I} + H_{\rm I,II}}{1 + \Delta_{\rm I,II}}, \quad E_{\rm A} = \frac{H_{\rm I,I} - H_{\rm I,II}}{1 - \Delta_{\rm I,II}} \tag{5.86}$$

On choosing one of these values, the ratio of coefficients $(a_{\rm I}/a_{\rm II})$ in the trial wave function also gets determined in accordance with Eq. 5.75. In the present case, $a_{\rm I}(H_{\rm I,I} - E) + a_{\rm II}(H_{\rm I,II} - E\,\Delta_{\rm I,II}) = 0$, whence we have, on using Eq. 5.86:

For $E = E_{\rm S}$

$$a_{\rm I} = a_{\rm II} \equiv a_{\rm S}; \quad \psi = \psi_{\rm S} = a_{\rm S}(u_{\rm I} + u_{\rm II}) \tag{5.87a}$$

For $E = E_{\rm A}$

$$a_{\rm I} = -a_{\rm II} \equiv a_{\rm A}; \quad \psi = \psi_{\rm A} = a_{\rm A}(u_{\rm I} - u_{\rm II}) \tag{5.87b}$$

Now, it is clear from the definitions (5.81) that if the electrons 1 and 2 are interchanged, $u_{\rm I} \to u_{\rm II}$ and $u_{\rm II} \to u_{\rm I}$. Therefore, under this interchange,

$$\psi_{\rm S} \to \psi_{\rm S} \text{ and } \psi_{\rm A} \to -\psi_{\rm A} \tag{5.88}$$

This symmetry/antisymmetry property of these functions is what is indicated by the subscripts S and A. We can now see also why Eqs. 5.85 hold. For example, if in $H_{\rm I,I} \equiv \iint u_{\rm I} H u_{\rm I}\, d\tau_1 d\tau_2$ we relabel $\mathbf{x}_2$ as $\mathbf{x}_1$ and vice versa, the value of the integral is not altered since these are merely dummy variables of integration. However under this relabelling $u_{\rm I} \to u_{\rm II}$ and $u_{\rm II} \to u_{\rm I}$, and H (being symmetric with respect to 1 and 2) remains unchanged. Thus we find $H_{\rm I,I} = \iint u_{\rm II} H u_{\rm II}\, d\tau_2\, d\tau_1 = H_{\rm II,II}$. The other relations in (5.85) follow in a similar fashion. Note that

$$\Delta_{I,I} = \iint (u_I)^2 \, d\tau_1 d\tau_2 - \int (u_A(1))^2 \, d\tau_1 \int (u_B(2))^2 d\tau_2 -$$

and

$$\Delta_{I,II} = \iint u_I \; u_{II} \; d\tau_1 d\tau_2$$
$$= \iint u_A(1) \; u_B(2) \; u_A(2) \; u_B(1) d\tau_1 d\tau_2 = S^2 \qquad (5.89)$$

where

$$S = \int u_A(\mathbf{x}) u_B(\mathbf{x}) d\tau \qquad (5.90)$$

What remains to be done is to evaluate $H_{I,I}$, $H_{II,I}$ and S, and hence determine E_S and E_A associated with the symmetric/antisymmetric wave functions ψ_S, ψ_A. The lower of these would be our estimate for the ground state energy of the molecule. Now, in view of Eqs. 5.79 and 5.81a, we can write

$$H_{I,I} = \iint u_I \; (H_I + V_I) \; u_I \; d\tau_1 d\tau_2$$
$$= \iint u_I \; (2\varepsilon + V_I) \; u_I \; d\tau_1 d\tau_2 = 2\varepsilon + K, \qquad (5.91)$$

$$K = \iint [e u_A(1)]^2 \; [e u_B(2)]^2 \left[-\frac{1}{r_{A2}} - \frac{1}{r_{B1}} + \frac{1}{R} + \frac{1}{r_{12}} \right] d\tau_1 d\tau_2 \qquad (5.92)$$

Similarly,

$$H_{II,I} = \iint u_{II} \; (H_I + V_I) \; u_I \; d\tau_1 d\tau_2$$
$$= \iint u_{II} \; (2\varepsilon + V_I) \; u_I \; d\tau_1 d\tau_2 = 2\varepsilon S^2 + J \qquad (5.93)$$

$$J = \iint [e^2 u_A(1) u_B(1)] \; [e^2 u_A(2) u_B(2)] \left[-\frac{1}{r_{A2}} - \frac{1}{r_{B1}} + \frac{1}{R} + \frac{1}{r_{12}} \right] d\tau_1 d\tau_2 \qquad (5.94)$$

On introducing Eqs. 5.89, 5.91 and 5.93 into Eqs. 5.86, they become

$$E_S = 2\varepsilon + \frac{K + J}{1 + S^2}, \quad E_A = 2\varepsilon + \frac{K - J}{1 - S^2} \qquad (5.95)$$

The integrals K, J and S can be evaluated exactly,[12] but the expressions are cumbersome, involving various transcendental functions, and we do not present them here. The final results for E_A and E_S, as functions of the internuclear distance R, are shown in Fig. 5.3. It is seen that E_A exceeds the energy 2ε of two independent hydrogen atoms for all R, which means that the two atoms cannot occur bound (as a molecule) in the antisymmetric state. But $(E_S - 2\varepsilon)$ does become negative, with a minimum at a certain $R = R_0$ which should be the equilibrium internuclear distance in the hydrogen molecule. The theoretical value $R_o = 0.77$Å compares with 0.74Å obtained from spectroscopic data. Further, the maximum value of $(2\varepsilon - E_S)$ gives the binding energy of the molecule. Its calculated value, 30·75 eV, leads to a heat of dissociation[13] 86·24 kcal/mole as compared with the experimental value, 102·3 kcal/mole. Considering the simple-minded nature of the theoretical calculation, the agreement must be considered quite good.

5.10 Exchange Interaction

It is instructive to note from Eqs. 5.95 that the difference between E_A and E_S arises from the non-vanishing of J and S. In the absence of these E_A and E_S

[12] See, e.g. S. Flügge, Practical Quantum Mechanics, Springer Verlag, Berlin, 1971, Vol. II, p. 87.

[13] This is the amount of heat required to break up 1 mole ($N = 6{\cdot}022 \times 10^{23}$ molecules) of hydrogen into separate atoms.

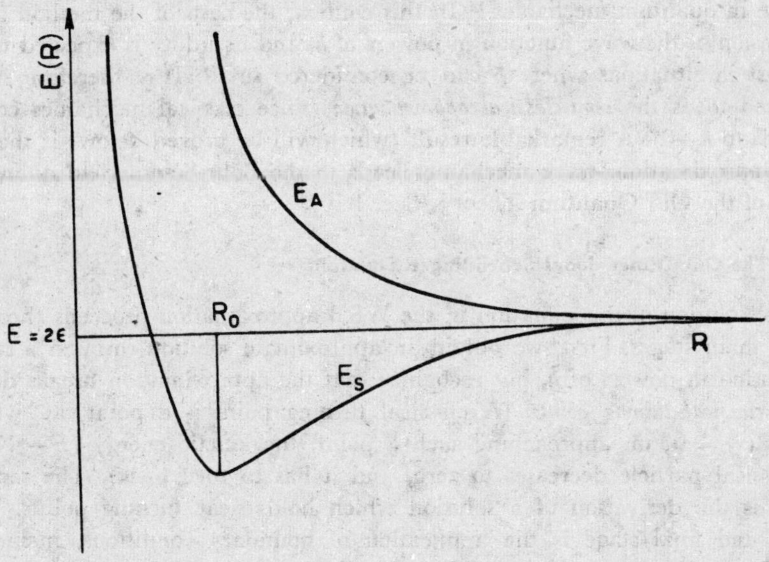

Fig. 5.3 Energy of the hydrogen molecule as a function of interatomic distance.

would reduce to $2\varepsilon + K$. Inspection of Eq. 5.92 shows that K is just the energy of Coulomb (electrostatic) interaction between the electronic charge distribution (with charge density $[eu_A(1)]^2$) and nuclear charge of atom A, with those of atom B. For this reason K is called the *Coulomb integral.* On the other hand, both J and S involve products $u_A(1)u_B(1)$ and/or $u_A(2)u_B(2)$. These would not have arisen but for the fact that the wave function (whether ψ_A or ψ_S) has *two* terms u_I and u_{II} wherein the positions of the electrons are interchanged. For this reason J is called the *exchange integral*; and S is called the *overlap integral,* since it directly measures the degree of overlap between the wave functions $u_A(1)$ and $u_B(1)$, i.e. between the ground state atomic wave functions centred at A and B respectively. A high degree of overlap, and the rather large value of the 'exchange interaction' resulting therefrom, are clearly necessary for the observed degree of binding of the molecule. It should be borne in mind that the exchange interaction is not a new type of interaction but only a quantum mechanical manifestation (through the possibility of particle-exchange in the wave function) of the basic Coulomb interaction.

C. THE WKB APPROXIMATION

This is an approximate method of solution of ordinary differential equations of the type $d^2u/dx^2 + Q(x)u = 0$. Though it has been known for a long time as a general mathematical method and has in particular, been studied extensively by Jeffreys, it is named after Wentzel, Kramers and Brillouin who pioneered its use in quantum mechanics.[14] In this context, the basis of the method is an expansion of the wave function in powers of $\hbar$, and its utility is expected to be highest in situations where $\hbar$ can be considered small. It is, therefore, often referred to as the *semi-classical approximation* (since classical mechanics corresponds to $\hbar \to 0$). A remarkable result (which will be proved below) is that in this approximation, wave mechanics leads to the Bohr-Sommerfeld quantum rules of the Old Quantum Theory (Sec. 1.10).

5.11 The One-Dimensional Schrödinger Equation

The solution of this equation in the WKB approximation proceeds through three main stages. First, we obtain an approximate solution through a series expansion in powers of $\hbar$, but recognize that the approximation breaks down near *classical turning points.* (A classical turning point is a point at which $E - V(x) = 0$; on approaching such a point, the kinetic energy $(E - V)$ of a classical particle decreases to zero, and it has to turn back). The second stage is the derivation of a solution which holds near turning points. The third and final stage is the application of boundary conditions, including matching of the expressions valid in the different regions. It is this last step which brings out quantum conditions and other typical quantum effects, as we already know from the simple problem of square well potentials and barriers.

(*a*) *The Asymptotic Solution:* The Schrödinger equation

$$\frac{d^2u}{dx^2} + \frac{2m}{\hbar^2}(E - V)\, u = 0 \tag{5.96}$$

has the approximate solution $A \exp\, [i\, \{2m(E - V)\}^{1/2}\, x/\hbar]$ in any region over which V is sensibly constant. This suggests that in the general situation it would be useful to write u in the form

$$u(x) = A(x)\, e^{iS(x)/\hbar} \tag{5.97}$$

where A and S are real functions. On substituting this into Eq. 5.96 and separating real and imaginary parts, we obtain

$$\hbar^2 A'' + [2m(E - V) - S'^2]\, A = 0 \tag{5.98}$$

$$2A'S' + S''A = 0 \tag{5.99}$$

Eq. 5.99 has the immediate solution

$$A = \text{const.}\ (S')^{-1/2} \tag{5.100}$$

For Eq. 5.98, which is not exactly soluble, let us try a series solution

$$A = A_0 + \hbar^2 A_2 + \ldots,\quad S = S_0 + \hbar^2 S_2 + \ldots \tag{5.101}$$

On introducing these in 5.98 and equating first the terms independent of $\hbar$, we get

$$A_0\, [2m(E - V) - (S_0')^2] = 0 \tag{5.102a}$$

so that

$$S_0' = \pm\, [2m\, (E - V)]^{1/2}, \text{ and } A_0 = \text{const.}\ (S_0')^{-1/2} \tag{5.102b}$$

Now, equating coefficients of $\hbar^2$ and using Eq. 5.102a, we obtain

$$A_0'' - 2S_0'S_2'A_0 = 0$$

[14] G. Wentzel, *Z. Physik*, **38**, 518, 1926; H. A. Kramers, *Z. Physik*, **39**, 828, 1926; L. Brillouin *Comptes Rendus*, **183**, 24, 1926.

and hence, with the aid of Eq. 5.102b we get

$$S_2' = \frac{[(S_0')^{-1/2}]''}{2(S_0')^{1/2}} \tag{5.103}$$

Let us consider the lowest order approximation ($A \approx A_0$, $S \approx S_0$). Then

$$u(x) = A_0 e^{iS_0/\hbar} = a_\pm k^{-\frac{1}{2}} \exp \left[\pm i \int kdx \right] \tag{5.104}$$

where a_+, a_- are constants, and A_0, S_0, (Eq. 5.102b), have been expressed in terms of $k(x)$ defined by

$$k(x) = + [2m(E - V)/\hbar^2]^{1/2} \tag{5.105}$$

Here and in the following, it is to be understood that when $(E - V)$ is negative,

$$k(x) \to i\varkappa(x), \ \varkappa = + [2m(V - E)/\hbar^2]^{1/2} \tag{5.106}$$

The two independent solutions distinguished by the + and − signs in the exponent in Eq. 5.104 may have arbitrary constant coefficients, a_+ and a_-. It may be observed that $\hbar k(x)$, for $(E - V) > 0$, can be thought of as the momentum of the particle at x; provided it does not vary rapidly with x, we can associate a local de Broglie wave length $\lambda = 2\pi/k(x)$ with the particle.

If the above approximation is to be good in some region of x, it is necessary that $(\hbar^2 S_2'/S_0') \ll 1$ in that region.[15] We can express this in physical terms by using Eq. 5.103 and substituting $S_0' = \hbar k = h/\lambda$, whereupon the condition reduces to $|\lambda'^2 - 2\lambda\lambda''| \ll 32\pi^2$. It is satisfied if λ', $\lambda\lambda''$ are not too large, which requires that the potential function be slowly and smoothly varying over a distance of several wave lengths. [Note that λ' is the fractional change in λ over one wave length: $\lambda' = \{(d\lambda/dx)\}/\lambda$]. This requirement is *not* met when x is near a classical turning point. The reason is that as x approaches a point x_0 where $E - V(x_0) = 0$, the wavelength λ itself, being proportional to $[E - V(x)]^{-1/2}$, tends to infinity. Clearly, λ' and λ'' then become unbounded too. In fact if $E - V(x)$ behaves like $(x - x_0)$ near x_0, it is easy to verify that $|\lambda'^2 - 2\lambda\lambda''| \propto (x - x_0)^{-3}$.

The solution (5.104) is, therefore, an *asymptotic* solution in the sense that it is good far from classical turning points (provided that the variation of the potential there is slow, as is usually the case). Knowledge of such solutions, valid in different regions separated by turning points, is not by itself adequate in the quantum mechanical context. As we already know, the imposition of boundary conditions, including the 'matching' of solutions across boundaries between different regions, is essential to get a complete wave function which is acceptable. In order to perform such matching, we have to obtain solutions of our equation which are valid in the neighbourhood of the classical turning points. We consider this aspect of the problem now.

(*b*) *Solution Near a Turning Point:* Consider a particular turning point occurring for a given energy E. We can take the origin of the x coordinate at this point, without loss of generality, so that $(E - V) = 0$ at $x = 0$. Making

[15] It is not meaningful to compare S_0 and S_2 because they are determined by Eqs. 5.102b and 5.103 only to within arbitrary constants of integration. These constants lead merely to a constant phase factor in $u(x)$, and therefore have no physical significance.

a Taylor expansion of $(E - V)$, or equivalently of $k^2(x)$, about this point, we write

$$k^2(x) = \rho x^n \; (1 + \alpha x + \beta x^2 + \ldots), \; \rho > 0 \qquad (5.107)$$

where n is a positive integer ($n > 0$) and $\rho, \alpha, \beta \ldots$ are constants. For definiteness we have assumed that $(E - V)$ is positive to the right of the origin, and this fixes the sign of ρ as in Eq. 5.107.

If we ignore all but the leading term in the expansion of k^2, the wave equation for $x > 0$ becomes $(d^2u/dx^2) + \rho x^n u = 0$. This equation is known to be exactly soluble; the solutions are

$$u(x) = c_{\pm} \xi^{1/2} k^{-1/2} J_{\pm m} (\xi), \; m = \frac{1}{n+2} \qquad (5.108)$$

where

$$\xi = \int_0^x k \, dx \qquad (5.109)$$

and $J_m(x)$ is the Bessel function of order m. When $k^2(x)$ does not have the exact form ρx^n, Eq. 5.108 is no longer a solution of the equation $d^2u/dx^2 + k^2 u = 0$. By actual differentiation of u given by Eq. 5.108, one finds that it satisfies the modified equation

$$\frac{d^2u}{dx^2} + (k^2 - \chi) \; u = 0 \qquad (5.110)$$

where

$$\chi = \frac{3k'^2}{4k^2} - \frac{k''}{2k} + (m^2 - \tfrac{1}{4}) \frac{k^2}{\xi^2} \qquad (5.111)$$

Nevertheless, we can still consider Eq. 5.108 to give solutions of the Schrödinger equation near $x = 0$ to a good approximation, provided χ is very small in this region. By substituting Eq. 5.107 in Eq. 5.111 one finds that

$$\chi(0) = \frac{3(n+5)\alpha^2}{2(n+4)\,(n+6)} - \frac{3\beta}{n+6} \qquad (5.112)$$

Thus χ is of the second order in the deviation of k^2 from ρx^n, and will be small if V is slowly varying near the turning point. Eq. 5.108 gives two independent solutions which are good to this degree of approximation. It is a remarkable fact that the asymptotic form of either of these solutions goes over directly into a linear combination of the asymptotic solutions (5.104). This fact simplifies the 'matching' problem immensely.

(*c*) *Matching at a Linear Turning Point:* Let us now consider the most commonly occurring situation, of a linear turning point, i.e. one near which the variation of $(E - V)$ or k^2 with x is linear (Fig. 5.4). This corresponds to taking $n = 1$ in Eq. 5.107. In considering the solutions (5.108) it is convenient from here on to take explicit account of the following fact: on going across the turning point from positive to negative x, $(E - V)$ changes from positive to negative and hence k and ξ change from real to pure imaginary. So we make the following definitions:

For $x > 0$ $(E > V)$: $$\xi_{>} = \xi = \int_0^x k dx \qquad (5.113a)$$

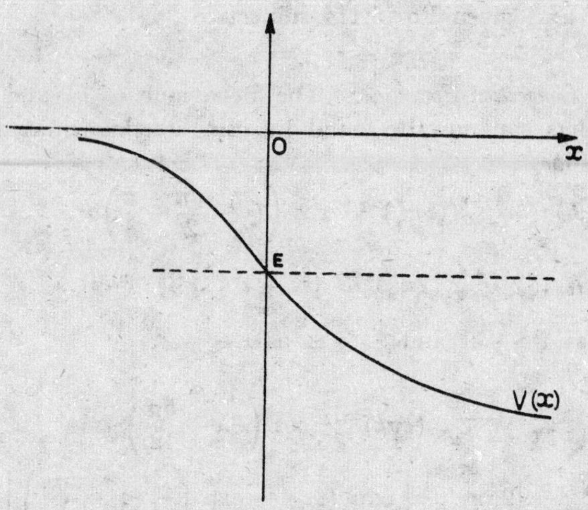

Fig. 5.4 A linear turning point.

For $x < 0 \ (E < V)$: $$k = i\varkappa, \quad \xi_< = i\xi = \int_x^0 \varkappa dx \tag{5.113b}$$

Both $\xi_<$ and $\xi_>$ are real and positive, and increase with distance from the turning point. In terms of these, Eq. 5.108 with $n = 1$ becomes

$$u_{>,\pm}(x) = c_{>,\pm}\ \xi_>^{1/2}\ k^{-1/2}\ J_{\pm 1/3}\ (\xi_>), \quad (x > 0) \tag{5.114a}$$
$$u_{<,\pm}(x) = c_{<,\pm}\ \xi_<^{1/2}\ \varkappa^{-1/2}\ I_{\pm 1/3}\ (\xi_<), \quad (x < 0) \tag{5.114b}$$

We have introduced here the standard notation I_m for Bessel functions of imaginary argument.

To match $u_>$ and $u_<$ at $x = 0$, we need to know their behaviour near this point. Since for small x, the dominant term in k^2 is ρx, we have

$$k \approx \rho^{1/2} x^{1/2}, \ \xi_> \approx \tfrac{2}{3}\rho^{1/2} x^{3/2}, \ \varkappa \approx \rho^{1/2} | x |^{1/2}, \ \xi_< \approx \tfrac{2}{3}\rho^{1/2} | x |^{3/2} \tag{5.115}$$

Further, for small ζ,

$$J_m(\zeta) \approx I_m(\zeta) \approx \frac{(\frac{1}{2}\zeta)^m}{\Gamma(1+m)} \tag{5.116}$$

On introducing these in Eqs. 5.114 we find that near $x = 0$,

$$u_{>,+} \approx \frac{c_{>,+}(\frac{2}{3})^{1/2}\ (\frac{1}{3}\rho^{\frac{1}{2}})^{1/3}}{\Gamma(\frac{4}{3})}\ x, \ u_{>,-} \approx \frac{c_{>,-}(\frac{2}{3})^{1/2}\ (\frac{1}{3}\rho^{\frac{1}{2}})^{-1/3}}{\Gamma(\frac{2}{3})}$$
$$u_{<,+} \approx \frac{c_{<,+}(\frac{2}{3})^{1/2}\ (\frac{1}{3}\rho^{\frac{1}{2}})^{1/3}}{\Gamma(\frac{4}{3})}\ | x |, \ u_{<,-} \approx \frac{c_{<,-}(\frac{2}{3})^{1/2}\ (\frac{1}{3}\rho^{\frac{1}{2}})^{-1/3}}{\Gamma(\frac{2}{3})} \tag{5.117}$$

It is clear from this that $u_{>,+}$ and $u_{<,+}$ join together smoothly (i.e. their values as well as first derivatives match), provided

$$c_{>,+} = -\,c_{<,+} \equiv c_+ \tag{5.118}$$

Thus one properly matched solution of the equation is given by u_+:

$$u_+ = u_{>,+}, \ (x > 0); \quad u_+ = u_{<,+}, \ (x < 0) \tag{5.119}$$

with $u_{>,+}$, $u_{<,+}$ given by Eqs. 5.114 subject to (5.118). Similarly, another independent solution is given by u_-:

$$u_- = u_{>,-}, \quad (x > 0); \quad u_- = u_{<,-}, \quad (x < 0) \tag{5.120}$$

with $u_{>,-}$ and $u_{<,-}$ given Eq. 5.114 subject to

$$c_{>,-} = c_{<,-} \equiv c_- \tag{5.121}$$

(*d*) *Asymptotic Connection Formulae:* The behaviour of u_+ and u_- far from the turning points can now be found by using the known asymptotic forms of Bessel functions:

$$J_m(\zeta) \xrightarrow[\zeta\to\infty]{} (\tfrac{1}{2}\pi\zeta)^{-1/2} \cos\left(\zeta - \frac{m\pi}{2} - \frac{\pi}{4}\right) \tag{5.122a}$$

$$I_m(\zeta) \xrightarrow[\zeta\to\infty]{} (2\pi\zeta)^{-1/2} \, [e^{\zeta} + e^{-\zeta} . e^{-(\frac{1}{2}+m)\pi i}] \tag{5.122b}$$

Since $\xi_> \to \infty$ as $x \to +\infty$ and $\xi_< \to \infty$ as $x \to -\infty$, we find that

$$u_+ \xrightarrow[x\to+\infty]{} (\tfrac{1}{2}\pi k)^{-1/2} \cos\left(\xi_> - \frac{5\pi}{12}\right) \tag{5.123a}$$

$$u_+ \xrightarrow[x\to-\infty]{} -(2\pi\varkappa)^{-1/2} \left[\exp(\xi_<) + \exp\left(-\xi_< - \frac{5\pi i}{6}\right)\right] \tag{5.123b}$$

$$u_- \xrightarrow[x\to+\infty]{} (\tfrac{1}{2}\pi k)^{-1/2} \cos\left(\xi_> - \frac{\pi}{12}\right) \tag{5.124a}$$

$$u_- \xrightarrow[x\to-\infty]{} (2\pi\varkappa)^{-1/2} \left[\exp(\xi_<) + \exp\left(-\xi_< - \frac{\pi i}{6}\right)\right] \tag{5.124b}$$

By forming suitable linear combinations of the independent solutions u_+ and u_- we can form particular solutions with specified asymptotic properties in either of the regions ($x > 0$ or $x < 0$) and find the behaviour of the solution in the other region. For example, from Eqs. 5.123b and 5.124b it is evident that it is the combination $(u_+ + u_-)$ which *decreases* exponentially on going into the interior of the classically forbidden region ($x \to -\infty$); on noting its asymptotic form as $x \to +\infty$ from Eqs. 5.123a and 5.124a, we see that it behaves like $k^{-1/2} \cos(\xi_> - \frac{1}{4}\pi)$. Thus we have the following connection formula from the asymptotic region $x \to -\infty$ to the region $x \to +\infty$:

$$\tfrac{1}{2}\varkappa^{-1/2} \exp(-\xi_<) \to k^{-1/2} \cos(\xi_> - \tfrac{1}{4}\pi) \tag{5.125}$$

The formula is not valid in the reverse direction, since a slight error in the phase of the cosine function is enough to introduce a small admixture of the exponentially increasing function, which will overwhelm $\exp(-\xi_<)$ for large negative x (i.e., large $\xi_<$).

Similarly, by considering the combination $\sin\frac{1}{6}\pi . \cos\eta . (u_+ - u_-) - \cos\frac{1}{6}\pi . \sin\eta . (u_+ - u_-)$ we obtain, on neglecting $\exp(-\xi_<)$ in comparison with $\exp(\xi_<)$, the connection formula

$$\sin\eta . \varkappa^{-1/2} \exp(\xi_<) \leftarrow k^{-1/2} \cos(\xi_> - \tfrac{1}{4}\pi + \eta) \tag{5.126}$$

This is valid provided $\sin\eta$ is not so small as to reduce the left hand side to the same order of magnitude as the neglected $\exp(-\xi_<)$ term. Once again the formula holds only in the direction indicated by the arrow, since a small admixture of the $\exp(-\xi_<)$ term on the left hand side is enough to change the phase on the right hand side drastically.

5.12 The Bohr-Sommerfeld Quantum Condition

As an illustration of the use of the WKB method, we consider the stationary states of a particle in a simple potential well, and show that the application of the asymptotic connection formulae in this case leads to one of the Bohr-Sommerfeld quantum rules. Let x_1, x_2 be the classical turning points corresponding to some energy E (Fig. 5.5). The regions $x < x_1$ and $x > x_2$ are classically

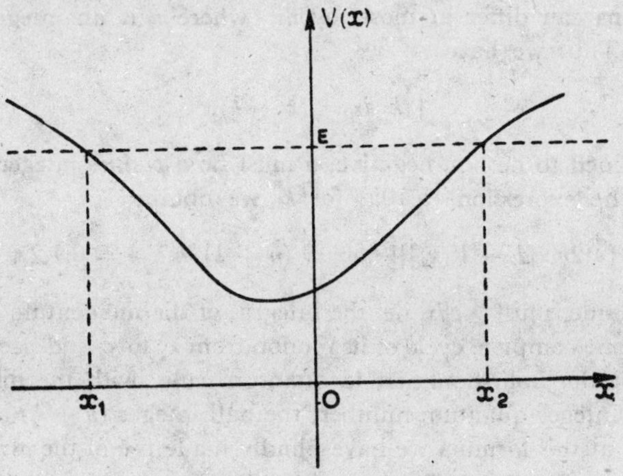

Fig. 5.5 The turning points for a particle in a potential well

forbidden, and we know that only the decreasing exponential WKB solutions $\propto \exp(-\xi_<)$ are admissible in these regions if the boundary conditions at infinity are not to be violated. Therefore, the wave function in the interior of the classically allowed region ($x_1 < x < x_2$) can be obtained by using the connection formula (5.125), starting from either of the forbidden regions. However, it must be recalled at this point that both $\xi_>$ and $\xi_<$ were defined to be positive and to be increasing from zero as one moves away from a specified turning point. Accordingly, it is to be understood that with reference to the turning point x_1,

$$\xi_< = \int_x^{x_1} \varkappa\, dx \text{ and } \xi_> = \int_{x_1}^{x} k\, dx \tag{5.127}$$

Therefore, on using the asymptotic connection formula (5.125) across x_1, we obtain the wave function in the allowed region to the right of x_1 as

$$u = c\, k^{-1/2} \cos\left(\int_{x_1}^{x} k\, dx - \frac{\pi}{4}\right) \tag{5.128}$$

where c is an arbitrary constant. Again, with reference to the turning point x_2,

$$\xi_< = \int_{x_2}^{x} \varkappa\, dx \text{ and } \xi_> = \int_{x}^{x_2} k\, dx \tag{5.129}$$

With this understanding, (5.125) applied across x_2 gives the wave function in the allowed region to the left of x_2 as

$$u = c'k^{-1/2} \cos\left(\int_x^{x_2} k\,dx - \frac{\pi}{4}\right) = c'k^{-1/2} \cos\left(\frac{\pi}{4} - \int_x^{x_2} k\,dx\right) \tag{5.130}$$

where c' is a constant. The last step is written so that the argument of the cosine function increases with x, as in (5.128). Note that Eqs. 5.128 and 5.130 are both obtained for the *same* region $x_1 < x < x_2$, so that they must coincide. Evidently this requires that the arguments of the cosine functions in these equations can differ at most by $n\pi$ where n is an integer, and that $c' = (-)^n c$: Thus we have

$$\int_{x_1}^{x_2} k\,dx = (n + \tfrac{1}{2})\pi \tag{5.131}$$

Since k is defined to be non-negative, n must be a *positive* integer. Finally, on substituting the expression (5.105) for k, we obtain

$$2\int_{x_1}^{x_2} \{2m\,[E - V(x)]\}^{1/2}\,dx = (n + \tfrac{1}{2})\,h, \quad n = 0,1,2,\ldots \tag{5.132}$$

The left hand side is just $\oint p dx$, i.e. the integral of the momentum of a classical particle over one complete cycle of its motion from x_1 to x_2 and back to x_1. Thus (5.132) is just the Bohr-Sommerfeld quantum rule, with the difference that instead of an integer quantum number, the half integer $(n + \frac{1}{2})$ appears here.

In arriving at this formula we have blindly made use of the *asymptotic* WKB solutions which are expected to be good only several wavelengths away from the turning points. Now it is easy to see from the nature of the wave function (5.128) that n defined by Eq. 5.131 is the number of nodes of $u(x)$ between x_1 and x_2. Therefore if this region is to include several wavelengths, n must be large, and it is under this condition that the quantum rule (5.132) is expected to be really good. In practice, the results obtained from the WKB approximation are found to be quite good even for low quantum numbers in the case of many systems. In particular, it may be verified that Eq. 5.132 reproduces the energy levels of the simple harmonic oscillator exactly.

5.13 WKB Solution of the Radial Wave Equation

Let us now consider Eq. 4.110 when it is satisfied by $\chi(r) = r\,R(r)$ where $R(r)$ is the radial wave function in a potential with spherical symmetry. This equation viz.,

$$\frac{d^2\chi}{dr^2} + \frac{2m}{\hbar^2}\left[E - V(r) - \frac{\hbar^2\, l(l+1)}{2m\,r^2}\right]\chi = 0 \tag{5.133}$$

has the same general form as Eq. 5.96 but the boundary conditions here are specified at the values 0 and $+\infty$ of the independent variable r instead of at $-\infty$ and $+\infty$ as in the one-dimensional case. To bring the conditions in line with those of the one dimensional problem. we make the transformation of variables

$$r = e^s, \quad \chi(r) = e^{\frac{1}{2}s}\,\phi(s) \tag{5.134}$$

which maps the points $r = 0, +\infty$ to $s = -\infty, +\infty$ respectively. It is a straightforward matter to verify then that $(d^2\chi/dr^2)$ reduces to $e^{-3s/2}$ $[(d^2\phi/ds^2) - \frac{1}{4}\phi]$; hence Eq. 5.133 becomes

$$\frac{d^2\phi}{ds^2} + \frac{2m}{\hbar^2}\left[\{E - V(e^s)\}\, e^{2s} - \frac{\hbar^2}{2m}(l + \tfrac{1}{2})^2 \right] \phi = 0 \qquad (5.135)$$

The boundary conditions on $\phi(s)$ — which is nothing but $r^{1/2}R(r)$ — are now

$$\phi(s) \to 0 \text{ as } s \to \pm\infty \qquad (5.136)$$

Eqs. 5.135 and 5.136 correspond exactly to the one-dimensional case, and therefore the solution obtained in the previous sections can be directly taken over. In particular, for the case of a simple potential well we have as the counterpart of Eq. 5.132

$$2\int_{s_1}^{s_2} [2m\{E - V(e^s)\}\, e^{2s} - \hbar^2 (l + \tfrac{1}{2})^2]^{1/2}\, ds = (n + \tfrac{1}{2})\, h \qquad (5.137)$$

where s_1, s_2 are the limits of the region within which the square-bracketed quantity is positive. Since $s = \log r$, we can re-express the above condition in terms of r as

$$2\int_{r_1}^{r_2} [2m\{E - V(r)\} - \frac{\hbar^2}{r^2}(l + \tfrac{1}{2})^2]^{1/2}\, dr = (n + \tfrac{1}{2})\, h \qquad (5.138)$$

It is noteworthy that except for the appearance of $(l + \frac{1}{2})^2$ instead of $l(l + 1)$, we would have got the same result if (5.132) had been applied directly to Eq. 5.133.

PROBLEMS

1. The potential energy of an electron near a finite nucleus (with uniform charge distribution) differs from that due to a point nucleus by $\Delta V(r) = 0$ $(r > R)$, $\Delta V(r) = (Ze^2/R)[(r^2/2R^2) - 3/2 + R/r]$, $r < R$. The nuclear radius R is $r_0 A^{1/3}$, $r_0 = 1{\cdot}2 \times 10^{-13}$ cm.
   (a) Treating ΔV as a perturbation, calculate the isotope shift (i.e. dependence on A) of the energy of the K electron ($1s$ state) for $Z = 81$.
   (b) Explain why this would be a poor approximation if the electron were replaced by a mu meson (Muon mass = 207 × electron mass).
2. The kinetic energy of a relativistic particle is $(c^2\mathbf{p}^2 + m^2c^4)^{1/2} - mc^2 \approx (\mathbf{p}^2/2m) - (\mathbf{p}^2)^2/8m^3c^2$. Adding the last term as a relativistic correction H' to the Hamiltonian of the hydrogen atom, show from first order perturbation theory that the resulting change in the energy of the state (nlm) is $\Delta E^{(1)}{}_{nlm} = \left[\frac{3}{8n^4} - \frac{1}{(2l+1)\,n^3}\right]\frac{me^4}{\hbar^2}\left(\frac{e^2}{\hbar c}\right)^2$.
3. A slightly anisotropic three-dimensional harmonic oscillator $(\omega_1 \approx \omega_2 = \omega_3)$ has charge e and is subjected to a uniform magnetic field in the z-direction. Assuming that the effects of the anisotropy and the magnetic field are of the same order but small compared to $\hbar\omega$, calculate to first order the energies of the components into which the first excited state is split.
4. Estimate the ground state energy of the anharmonic oscillator $(V = \frac{1}{2}m\omega^2x^2 + \lambda x^4)$ by the variation method, taking the trial wave function to be a linear combination of the $n = 0$ and $n = 2$ eigenfunctions of a harmonic oscillator of angular frequency ω'. Compare the result with that obtained from the perturbation theory.

5. Estimate the ground state energy for a particle in the potential $V(r) = -V_0 e^{-\mu r}$, using the variation method with a trial wave function proportional to (i) e^{-ar}, (ii) $e^{-b^2r^2}$; which is the better estimate? Obtain an upper bound on the error in this case.

6. For the wave function of an electron in the potential due to all the atoms of a molecule (the 'molecular orbital'), the LCAO (Linear Combination of Atomic Orbitals) approximation is often made. In the case of hydrogen, the molecular orbital is the wave function of the ion H_2^+. Discuss the ground state of H_2^+, using the variational method with the LCAO function $c_A u_A(\mathbf{x}) + c_B u_B(\mathbf{x})$ as trial wave function (u_A, u_B are the hydrogen atom ground state functions centred at nuclei A, B)

7. Show that the WKB approximation gives the eigenvalues of the harmonic oscillator and of the hydrogen atom correctly, on using Eqs. 5.132 and 5.138 respectively in the two cases.

8. Calculate the probability for transmission of a particle through a potential barrier of arbitrary shape, using the WKB method.

6

Scattering Theory

This chapter is devoted to the quantum mechanical treatment of the scattering of a particle by a potential, and of the equivalent problem of the mutual scattering of two particles.[1] We shall be concerned for the most part with approximate methods, suitable under various circumstances, for the evaluation of scattering cross-sections. As may be anticipated, scattering by the Coulomb potential is one of the very few excactly soluble cases. This is dealt with towards the end of the chapter since it exhibits peculiarities not shared by other (shorter range) potentials. The last part of the chapter, on two-particle collisions, includes a discussion of the specifically quantum mechanical interference effects which arise when the colliding particles are identical.

A. THE SCATTERING CROSS-SECTION: GENERAL CONSIDERATIONS

6.1 Kinematics of the Scattering Process: Differential and Total Cross-Sections

Experiments on the scattering of particles usually employ the following kind of arrangements: A *parallel beam* of particles of given momentum is directed towards a target which deflects or scatters the particles in various directions. The scattered particles diverge. Eventually, at large distances from the target

[1] Only a limited selection of scattering problems will be dealt with here. For a more complete treatment, see for instance R. G. Newton, *Scattering Theory of Waves and Particles*, McGraw-Hill Book Company, New York, 1966.

(where they are detected by suitable instruments), their motion is directed *radially outwards*. The theory of scattering is developed against the background of this general picture of the scattering process. It is usual and convenient to choose a coordinate system with the origin at the position of the target or *scattering centre*, and with the z-axis in the direction of the incident beam (Fig. 6.1). The direction of any scattered particle is indicated by polar

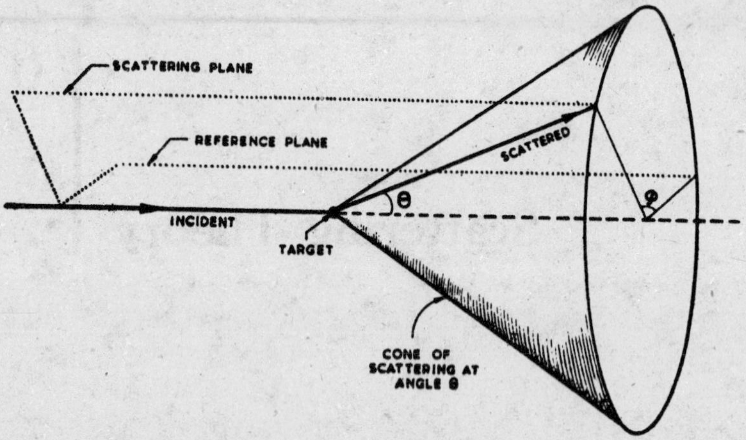

Fig. 6.1 Schematic diagram of a scattering event.

angles (θ,φ), with the z-axis taken as the polar axis. Then θ is the *angle of scattering*, i.e. the angle between the scattered and incident directions. These two directions together define the *plane of scattering*. The azimuthal angle φ specifies the orientation of this plane with respect to some reference plane containing the z-axis.

Let us consider a scattering experiment in which a steady incident beam is maintained for an indefinitely long time, i.e. the incident flux F (the number of particles crossing unit area taken normal to the beam direction per unit time) is independent of time. There will then be a steady stream of scattered particles too. Let $\Delta\mathcal{N}$ be the number of particles scattered into a small solid angle $\Delta\Omega$ about the direction (θ,φ) in time Δt. Evidently $\Delta\mathcal{N}$ must be proportional to $\Delta\Omega\,\Delta t$ and to the incident flux[2] F. The proportionality factor, which depends in general on θ and φ, is called the *differential scattering cross-section*, and is denoted by $(d\sigma/d\Omega)$:

$$\Delta\mathcal{N} = \frac{d\sigma(\theta,\varphi)}{d\Omega}\cdot \Delta t\,\Delta\Omega.F \tag{6.1}$$

It is easily verified that $(d\sigma/d\Omega)$ has the dimensions of an area, which justifies the term 'cross-section'. It depends only on the parameters of the incident particle and nature of the target. Given this information, the aim of scattering

[2] It is assumed that the flux is low enough that any interaction of the incident particles with one another is completely negligible, i.e. the scattering of each particle is independent of others.

theory is to determine $(d\sigma/d\Omega)$ as a function of θ and φ. The *total scattering cross-section* σ may be obtained from it by integration over all directions:

$$\sigma = \int \left(\frac{d\sigma}{d\Omega}\right) d\Omega = \int_0^{2\pi}\int_0^{\pi} \frac{d\sigma}{d\Omega} \cdot \sin\theta\, d\theta\, d\varphi \tag{6.2}$$

In most of the cases we consider $(d\sigma/d\Omega)$ is independent of φ; then Eq. 6.2 simplifies to

$$\sigma = \int \frac{d\sigma}{d\Omega} \cdot 2\pi \sin\theta\, d\theta \tag{6.3}$$

6.2 Wave Mechanical Picture of Scattering: The Scattering Amplitude

When the particles involved in the scattering process are quantum mechanical objects, we must describe them by a wave function. The wave function will be a *stationary* one if we assume as in the last section that a steady (time-independent) incident beam is maintained for all time. The phenomenon of scattering is manifested as a distortion in the stationary wave pattern, caused by the presence of the scattering centre (just as the diffraction of light is manifested through particular stationary patterns of light waves). *At large distances* from the scattering centre $(r \to \infty)$, the form of the wave function $u(\mathbf{x})$ must be such as to conform to the general picture outlined at the beginning of the last section: it must consist of a part u_{inc} corresponding to the parallel beam of incident particles, and another part u_{sc} representing the scattered particles moving radially outwards from the centre:

$$u(\mathbf{x}) \underset{r\to\infty}{\longrightarrow} u_{inc} + u_{sc} \tag{6.4}$$

We shall now analyze this requirement in more detail, and thereby arrive at an expression for the scattering cross-section in terms of the asymptotic $(r \to \infty)$ form of the wave function.

The beam of incident particles with momentum $p = \hbar k$ along the z-axis must evidently be described by the plane wave (momentum eigenfunction) $u_{inc} = e^{ikz}$. According to the interpretation of unnormalizable wave functions given in Sec. 2.5, $|u_{inc}|^2$ is to be understood as the number of incident particles per unit volume. The incident flux is obtained by multiplying this quantity by the particle velocity v.

$$F = |u_{inc}|^2 v = \hbar k/m \tag{6.5}$$

As far as the scattered particles are concerned, we shall assume that they have the same momentum as the incident ones, i.e. that the scattering is *elastic*. Then the wave u_{sc} representing them must have the same propagation constant k; but it must be a *spherical wave* since these particles move radially. The only such waves are e^{ikr} and e^{-ikr}; the latter is an incoming wave[3]

[3] A simple way of seeing this is by looking at the wave function including the time-dependent factor $e^{-i\omega t}$, $(\omega = E/\hbar)$. In $e^{ikr-i\omega t}$, the surfaces of constant phase, $kr - \omega t =$ const, move outwards from the origin since $r = (\text{const} + \omega t)/k$ increases with time. One sees directly in a similar fashion that $e^{-ikr-i\omega t}$ represents incoming waves.

(contracting towards the origin). Since the scattered particles move outwards, we must choose the *outgoing* wave, $u_{sc} \propto e^{ikr}$. Further, the flux of scattered particles, $|u_{sc}|^2 (\hbar k/m)$ must evidently decrease as $(1/r^2)$ as r increases. Hence $u_{sc} \propto (1/r)$ also, and we write

$$u_{sc} = f(\theta,\varphi)\, \frac{e^{ikr}}{r} \tag{6.6}$$

The dependence of the proportionality factor f on θ, φ allows for the fact that the scattered flux is, in general, direction-dependent. We observe now that $\Delta\mathcal{N}$ in Eq. 6.1 is nothing but the radial flux times $\Delta S\, \Delta t$, where $\Delta S \equiv r^2 d\Omega$ is the element of area (normal to the radial direction) covered by the solid angle $\Delta\Omega$. Thus

$$\begin{aligned} \Delta\mathcal{N} &= |u_{sc}|^2 (\hbar k/m).r^2\, \Delta\Omega\, \Delta t \\ &= |f(\theta,\varphi)|^2 (\hbar k/m)\, \Delta\Omega\, \Delta t \end{aligned} \tag{6.7}$$

On substituting this together with (6.5) in (6.1), we obtain

$$\frac{d\sigma(\theta,\varphi)}{d\Omega} = |f(\theta,\varphi)|^2 \tag{6.8}$$

In view of this relation, $f(\theta,\varphi)$ is called the *scattering amplitude.*

It is clear now that in order to calculate the scattering cross-section what we have to do is to determine the particular stationary wave function $u(\mathbf{x})$ which has the following asymptotic form demanded by Eqs. 6.4 and 6.6:

$$u(\mathbf{x}) \xrightarrow[r\to\infty]{} e^{ikz} + f(\theta,\varphi)\, \frac{e^{ikr}}{r} \tag{6.9}$$

Since any stationary wave function must satisfy the time independent Schrödinger equation, we must have

$$-\frac{\hbar^2}{2m} \nabla^2 u(\mathbf{x}) + V(\mathbf{x})\, u(\mathbf{x}) = E\, u(\mathbf{x}), \quad E = \frac{\hbar^2 k^2}{2m} \tag{6.10}$$

Here $V(\mathbf{x})$ is the potential energy function for the projectile particle in the force field of the scattering centre. Note that the eigenvalue E has to be taken equal to the kinetic energy of the particle in the asymptotic region (where $V = 0$). If the solution of Eq. 6.10 subject to the asymptotic form (6.9) can be obtained, then the coefficient of (e^{ikr}/r) in the latter gives the scattering amplitude: However, it is only in very exceptional cases (such as the Coulomb potential) that this can be done exactly.

6.3 Green's Functions; Formal Expression for Scattering Amplitude

It is possible to get a formal solution for the scattering amplitude by transforming Eq. 6.10 into an integral equation. To accomplish this we make use of the Green's function method.

Let $\Omega(\mathbf{x},\nabla)$ be any *linear* differential operator, i.e.

$$\Omega(c_1u_1 + c_2u_2) = c_1\Omega u_1 + c_2\Omega u_2 \tag{6.11}$$

for arbitrary functions u_1, u_2 and constants c_1, c_2. Then any function $G(\mathbf{x},\mathbf{x}')$ such that

$$\Omega(\mathbf{x},\nabla)\, G(\mathbf{x},\mathbf{x}') = \delta(\mathbf{x} - \mathbf{x}') \tag{6.12}$$

is said to be a *Green's Function* for the operator Ω. It is obvious that if we add to a particular Green's function G any solution S of the *homogeneous* equation $\Omega S = 0$, the result $G + S$ is still a Green's function: $\Omega(G + S) = \delta(\mathbf{x} - \mathbf{x}')$. Conversely, the difference between any two Green's functions must be a solution of the homogeneous equation.

The general solution of any inhomogeneous equation

$$\Omega(\mathbf{x},\nabla)\, u(\mathbf{x}) = F(\mathbf{x}) \tag{6.13}$$

can be written down immediately in terms of Green's functions by first expanding $F(\mathbf{x})$ as a linear combination of delta functions:

$$F(\mathbf{x}) = \int \delta(\mathbf{x} - \mathbf{x}')\, F(\mathbf{x}')\, d\tau' \tag{6.14}$$

In fact, if $u_0(\mathbf{x})$ is any function such that $\Omega u_0(\mathbf{x}) = 0$, then

$$u(\mathbf{x}) = u_0(\mathbf{x}) + \int G(\mathbf{x},\mathbf{x}')\, F(\mathbf{x}')\, d\tau' \tag{6.15}$$

is a solution of (6.13). This may be easily verified by substituting this expression in Eq. 6.13 and noting that

$$\Omega \int G(\mathbf{x},\mathbf{x}')\, F(\mathbf{x}')\, d\tau' = \int \Omega G(\mathbf{x},\mathbf{x}')\, F(\mathbf{x}')\, d\tau' = \int \delta(\mathbf{x} - \mathbf{x}')\, F(\mathbf{x}')\, d\tau' = F(\mathbf{x}).$$

To obtain a particular solution, one has to make a specific choice of the Green's function G and of the function $u_0(\mathbf{x})$ in Eq. 6.15.

Let us now specialize to the equation of interest to us, namely the Schrödinger equation (6.10). We rewrite it as

$$(\nabla^2 + k^2)\, u(\mathbf{x}) = U(\mathbf{x})\, u(\mathbf{x}), \quad U(\mathbf{x}) = 2m\, V(\mathbf{x})/\hbar^2. \tag{6.16}$$

This has the form (6.13) with

$$\Omega = \nabla^2 + k^2, \quad F(\mathbf{x}) = U(\mathbf{x})\, u(\mathbf{x}) \tag{6.17}$$

The Green's functions for $(\nabla^2 + k^2)$ can be deduced from the well-known fact that

$$\nabla^2 \frac{1}{r} = -4\pi\, \delta(\mathbf{x}) \tag{6.18}$$

On using this we find immediately that $(\nabla^2 + k^2)\,(e^{\pm ikr}/r) = -4\pi\delta(\mathbf{x})$. By simply replacing $\mathbf{x}$ by $\mathbf{x} - \mathbf{x}'$ and r by $|\mathbf{x} - \mathbf{x}'|$ in this equation, we get two Green's functions G^+, G^- for $(\nabla^2 + k^2)$, namely

$$G^{\pm}(\mathbf{x}, \mathbf{x}') = \frac{\exp\,[\pm ik\,|\mathbf{x} - \mathbf{x}'|]}{-4\pi\,|\mathbf{x} - \mathbf{x}'|} \tag{6.19}$$

In view of Eqs. 6.15, 6.17, and 6.19 we may write down a 'solution' of the Schrödinger equation as

$$u(\mathbf{x}) = e^{ikz} - \frac{1}{4\pi} \int \frac{e^{ik|\mathbf{x}-\mathbf{x}'|}}{|\mathbf{x} - \mathbf{x}'|} \cdot U(\mathbf{x}')\, u(\mathbf{x}')\, d\tau' \tag{6.20}$$

The quotation marks on 'solution' are in recognition of the fact that the right hand side also contains (within the integral) the function u which we are seeking to determine. Thus Eq. 6.20 is really an *integral equation* for u. It will be observed that we have made a particular choice e^{ikz} for the function $u_0(\mathbf{x})$, and also chosen the Green's function as G^+. These choices have been made because we are interested in the scattering problem and in this context $u(\mathbf{x})$ is required to have the asymptotic form (6.9). We shall now see that (6.20) does have this asymptotic behaviour.

We observe first of all that the potential function is a factor in the integrand in Eq. 6.20, so that regions of $\mathbf{x}'$ where $V(\mathbf{x}')$ is very small do not contribute appreciably to the integral. Therefore, though the integration is in principle carried over all $\mathbf{x}'$, it extends effectively only over the region of $\mathbf{x}'$ near the origin (where $V(\mathbf{x}')$ is large). This means that in considering the asymptotic form ($|\mathbf{x}| = r \to \infty$) of Eq. 6.20 we can take $|\mathbf{x}| \gg |\mathbf{x}'|$ and hence (Fig. 6.2),

$$|\mathbf{x} - \mathbf{x}'| \approx r - \mathbf{x}'.\mathbf{x}/r = r - \mathbf{x}'.\hat{\mathbf{k}} \tag{6.21}$$

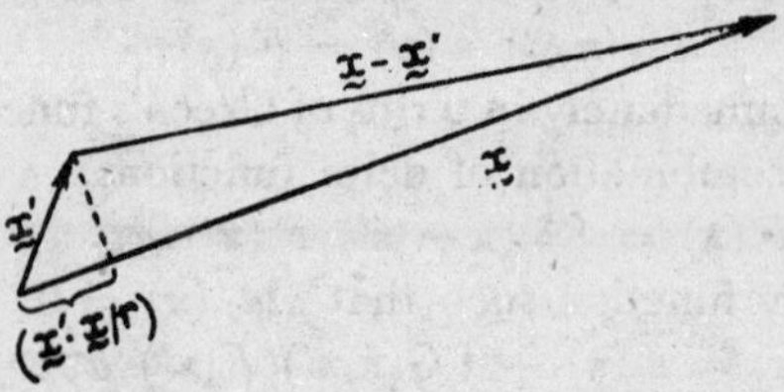

Fig. 6.2 The approximation of $|\mathbf{x}-\mathbf{x}'|$ for large r.

In the last step we have used the fact that scattered particles observed at $\mathbf{x}$ have momentum $\mathbf{k}$ in the direction of the vector $\mathbf{x}$, so that the unit vector $\hat{\mathbf{k}} = (\mathbf{x}/r)$. The term $\mathbf{x}'.\hat{\mathbf{k}}$ in Eq. 6.21 is very small in comparison with r and may be neglected in substituting for $|\mathbf{x} - \mathbf{x}'|$ in the denominator in Eq. 6.20. But in the case of $\exp[ik|\mathbf{x} - \mathbf{x}'|] \approx e^{ikr} \exp[-i\mathbf{x}'.\mathbf{k}]$, the absolute value of $\mathbf{x}'.\mathbf{k}$ is clearly important though it is small *relative to kr*. With due regard to these observations we substitute Eq. 6.21 into Eq. 6.20 to obtain the asymptotic form[4] of $u(\mathbf{x})$:

$$u(\mathbf{x}) \xrightarrow[r\to\infty]{} e^{ikz} - \frac{e^{ikr}}{4\pi r} \int e^{-i\mathbf{k}.\mathbf{x}'} U(\mathbf{x}')\, u(\mathbf{x}')\, d\tau' \tag{6.22}$$

Comparison of this form with Eq. 6.9 (which defines the scattering amplitude) shows that

$$f(\theta,\varphi) = -\frac{1}{4\pi} \int e^{-i\mathbf{k}.\mathbf{x}}\, U(\mathbf{x})\, u(\mathbf{x})\, d\tau,$$

$$= -\frac{m}{2\pi\hbar^2} \int e^{-i\mathbf{k}.\mathbf{x}}\, V(\mathbf{x})\, u(\mathbf{x})\, d\tau \tag{6.23}$$

where we have dropped the primes on the dummy variable of integration.

Eq. 6.23 is *not* a formula from which $f(\theta,\varphi)$ can be immediately calculated, since $u(\mathbf{x})$ which appears in the integral is still not known everywhere. However, it suggests a very useful approximation which we now consider.

B. THE BORN AND EIKONAL APPROXIMATIONS

6.4 The Born Approximation[5]

Suppose that the deviation $g(\mathbf{x})$ of $u(\mathbf{x})$ from the free-particle wave function e^{ikz} is relatively very small, i.e.

[4] The reader may verify that if we had chosen the Green's function G^- instead of G^+ in (6.20), we would have been led to the *incoming* spherical wave e^{-ikr} (which is unacceptable) instead of e^{ikr} in (6.22).

[5] M. Born, *Z. Physik*, **38**, 803, 1926.

$$|g(\mathbf{x})| \equiv |u(\mathbf{x}) - e^{ikz}| \ll |e^{ikz}| = 1 \qquad (6.24)$$

Then we can replace the unknown $u(\mathbf{x})$ in Eq. 6.23, to a good approximation, by $e^{ikz} \equiv e^{i\mathbf{k}_0.\mathbf{x}}$ where k_0 is the vector of magnitude k in the incident direction. The resulting expression is called the *first Born approximation* to $f(\theta,\varphi)$ and is given by

$$f_B(\theta,\varphi) = -(4\pi)^{-1} \int e^{-i\mathbf{K}.\mathbf{x}}\, U(\mathbf{x})\, d\tau \qquad (6.25)$$

The scattering cross-section in this approximation is

$$\left(\frac{d\sigma}{d\Omega}\right)_B = (4\pi)^{-2} \left| \int e^{-i\mathbf{K}.\mathbf{x}}\, U(\mathbf{x})\, d\tau \right|^2. \qquad (6.25a)$$

In Eqs. 6.25, $\mathbf{K} = \mathbf{k} - \mathbf{k}_0$. It may be noted that $\hbar\mathbf{K}$ is the momentum transferred to the particle in its encounter with the potential. Since $|k_0| = |\mathbf{k}| = k$, it follows (See Fig. 6.3) that

$$K \equiv |\mathbf{K}| = 2k \sin \tfrac{1}{2}\theta \qquad (6.26)$$

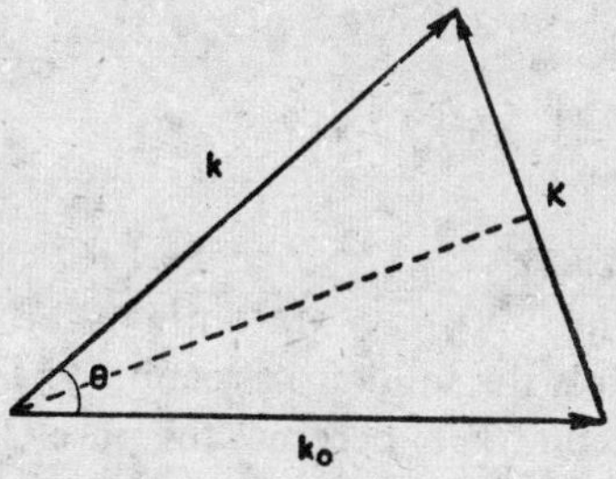

Fig. 6.3 The momentum transfer in an elastic collision.

The scattering amplitude (6.25) in the Born approximation, considered as a function of $\mathbf{K}$, is the *Fourier transform* of the potential (apart from constant factors).

In the most important special case when V is spherically symmetric, $V(\mathbf{x}) = V(r)$, we can reduce (6.25) to an integral over r alone, by going over to spherical polar coordinates (r,α,β) with the direction of $\mathbf{K}$ chosen as the polar axis. Then $\mathbf{K}.\mathbf{x} = Kr \cos\alpha$ and on carrying out the angular integrations one gets

$$f_B(\theta) = -K^{-1} \int_0^\infty r \sin Kr\; U(r)\, dr \qquad (6.27)$$

As is to be expected, the scattering amplitude in this case is independent of φ.

It is an important feature of the Born approximation formula (6.25) that f depends on the momentum *transfer* only, and not on the initial momentum and the angle of scattering separately. Another noteworthy feature is that the amplitude for a momentum transfer $\hbar\mathbf{K}$ is (apart from numerical factors) the value of the Fourier transform of the potential function $V(\mathbf{x})$ at this particular $\mathbf{K}$. Unlike the exact scattering amplitude, $f_B(\theta,\varphi)$ is *real* whenever the potential has inversion symmetry, $V(\mathbf{x}) = V(-\mathbf{x})$, as the reader may verify.

Example 6.1—The Screened Coulomb Potential. An electron incident on a neutral atom sees the nuclear charge as shielded or screened by the surrounding atomic electrons. The potential energy varies from $(-Ze^2/r)$ very near the nucleus — where there is no screening — to practically zero just outside the atom. It is convenient to represent such a *screened Coulomb potential* by the function[6]

$$V(r) = -\frac{Ze^2}{r}\, e^{-\varkappa r}$$

where $(1/\varkappa)$ has the dimensions of length and is a measure of the 'radius' of the atom.

The Born approximation amplitude $f_B(\theta)$ in this case is obtained from Eqs. 6.25 or 6.27 as

$$f_B(\theta) = \frac{2mZe^2}{4\pi\hbar^2}\int_0^\infty e^{-i\mathbf{K}.\mathbf{x}}\,\frac{e^{-\varkappa r}}{r}\,d\tau$$

$$= \frac{2mZe^2}{\hbar^2 K}\int_0^\infty \sin Kr\, e^{-\varkappa r}\,dr$$

$$= \frac{2mZe^2}{\hbar^2(K^2+\varkappa^2)} = \frac{2mZe^2}{4p^2\sin^2\frac{1}{2}\theta + \hbar^2\varkappa^2}$$

It is finite for all angles and so is the total cross-section $\int |f_B(\theta)|^2 d\Omega$.

However, if we let $\varkappa \to 0$ (so that $V(r)$ becomes the pure Coulomb potential), $f_B(\theta)$ diverges violently as $\theta \to 0$. This behaviour is attributed to the 'long-range' nature of the Coulomb potential. On account of its slow variation over an infinite range, the Fourier components with $\mathbf{K} \to 0$ have overwhelming dominance in its Fourier transform (which is proportional to the Born approximation amplitude). It may be noted that the cross-section for $\varkappa = 0$ is

$$\left(\frac{d\sigma}{d\Omega}\right)_{\text{Coul}} = \left|f_B(\theta)\right|^2_{\varkappa=0} = \frac{m^2Z^2e^4}{4p^4\sin^4\frac{1}{2}\theta}.$$

It coincides with the *exact* classical (Rutherford) formula for Coulomb scattering. It turns out that the rigorous quantum mechanical calculation also leads to precisely the same expression (See Sec. 6.16). The Coulomb potential is something very special, and it is the only potential for which this kind of coincidence occurs.

6.5 Validity of the Born Approximation

We now return to the assumption (6.24) on which the Born approximation is based, and examine its implications in one or two simple cases. Since the distortion $g(\mathbf{x}) \equiv u(\mathbf{x}) - e^{ikz}$ of the wave by the potential would be expected to be largest at or near the origin, let us apply the above condition at $\mathbf{x} = 0$. Noting that $g(\mathbf{x})$ is given by the integral term in Eq. 6.20, we can write Eq. 6.24 at $\mathbf{x} = 0$ as

$$\frac{1}{4\pi}\left|\int \frac{e^{ikr'}}{r'}\cdot\frac{2m}{\hbar^2}\,V(r')\,e^{ikr'\cos\theta'}\,d\tau'\right| \ll 1 \qquad (6.28)$$

[6] In the context of nuclear physics, a potential of the form *const.* $r^{-1}e^{-\varkappa r}$ is referred to as the Yukawa potential.

Here we have replaced $u(\mathbf{x}')$ by $e^{ikz'} = e^{ikr'\cos\theta'}$ in the spirit of the approximation, and taken V to be spherically symmetric. On integration over the polar coordinates (θ', φ') of $\mathbf{x}'$ the above inequality reduces to

$$\frac{m}{\hbar^2 k}\left|\int_0^\infty (e^{2ikr}-1)\, V(r)\, dr\right| \ll 1 \tag{6.29}$$

Let us see what this condition means in the particular case of a square well potential of depth V_0 and range a. In this case Eq. 6.29 becomes

$$\frac{mV_0}{\hbar^2 k}\left|\int_0^a (e^{2ikr}-1)\, dr\right| = \frac{mV_0}{2\hbar^2 k^2}\left|e^{2ika}-2ika-1\right|$$

$$= \frac{mV_0}{2\hbar^2 k^2}(\rho^2 - 2\rho\sin\rho - 2\cos\rho + 2)^{1/2} \ll 1, \quad \rho = 2ka \tag{6.30}$$

The quantity under square root is $\approx \frac{1}{4}\rho^4$ for $\rho \ll 1$, i.e. for $(4\pi a/\lambda) \ll 1$. In this limit when the de Broglie wave length is much greater than the range of the potential, the condition (6.30) becomes $(mV_0 a^2/\hbar^2) \ll 1$. This means, in view of the results of Sec. 4.14, that the potential should be much too weak to have any bound state. If the potential is not so weak, the Born approximation may still be valid provided the particle momentum is sufficiently large. Under these circumstances $(\rho \gg 1)$, the condition (6.30) becomes

$$(mV_0\,\rho/2\hbar^2 k^2) = (V_0\, a/\hbar v) \ll 1 \tag{6.31}$$

Thus, for any given depth and range of the potential well, the approximation becomes better as the incident particle velocity is increased.

This is just the kind of result one would expect on physical grounds. The approximation consists in assuming that the deviation of $u(\mathbf{x})$ from the free particle wave function is small, i.e. that the effect of the potential on the incident particle is small. One does not expect this to be the case for slow particles; but as the incident kinetic energy is increased, the potential becomes more and more in the nature of a small perturbation and the Born approximation should improve correspondingly. Actually the matter is not quite so simple, as the validity criterion (6.31) for the square well potential involves the velocity (rather than the kinetic energy) of the incident particle. (The same thing happens in the case of the screened Coulomb potential, to be considered now). It remains true however that in general, the Born approximation is a high energy approximation.

As a second example, we consider the screened Coulomb potential of Example 6.1, p. 185. With this potential, the left hand side of (6.29) becomes $(Ze^2m/\hbar^2 k)\ |I|$, where

$$I = \int_0^\infty (e^{2ikr}-1)\frac{e^{-\varkappa r}}{r}\, dr = \int_0^\infty (e^{2ikr}-1)\int_\varkappa^\infty e^{-\varkappa' r}\, d\varkappa'\, dr \tag{6.32}$$

$$= \int_\varkappa^\infty \left[\frac{1}{\varkappa' - 2ik} - \frac{1}{\varkappa'}\right] d\varkappa' = -\ln(1 - 2ik/\varkappa)$$

When $2(k/\varkappa) \gg 1$, $|I|$ tends to $\ln(2k/\varkappa)$, and the condition (6.29) for the validity of the Born approximation then becomes

$$\frac{Ze^2m}{\hbar^2k} \cdot \ln\left(\frac{2k}{\varkappa}\right) \ll 1. \tag{6.33}$$

This criterion can also be written as $(Z\alpha/\beta)\ln(2k/\varkappa) \ll 1$ where $\beta = v/c = (\hbar k/mc)$, and $\alpha = e^2/c\hbar = 1/137$. It turns out that essentially the same condition holds even if the particle is relativistic ($\beta \to 1$). Even with this maximum value of β the above condition is only poorly satisfied for heavy nuclei. At the other extreme of low velocities, $|I| = |\ln(1 - 2ik/\varkappa)| \approx (2k/\varkappa)$ and hence we need $(2Z e^2m/\hbar^2\varkappa) \ll 1$. An estimate of the atomic radius $(1/\varkappa)$ in the case of heavy atoms, according to the Thomas-Fermi statistical theory (Appendix E) is $(\hbar^2/me^2Z^{1/3})$. Thus the condition for the validity of the Born approximation becomes $2Z^{2/3} \ll 1$ which cannot be satisfied at all.

6.6 The Born Series

We have so far considered the first Born approximation to the scattering amplitude, obtained by taking $u(\mathbf{x})$ in Eq. 6.23 to be e^{ikz}. Actually one can solve the integral equation (6.20) for $u(\mathbf{x})$ as an infinite series. Rewriting it more generally as

$$u(\mathbf{x}) = u_0(\mathbf{x}) + \int G^+(\mathbf{x}, \mathbf{x}')\, U(\mathbf{x}')\, u(\mathbf{x}')\, d\tau' \tag{6.34}$$

we substitute for $u(\mathbf{x}')$ within the integral the entire expression on the right hand side (with $\mathbf{x}$ replaced by $\mathbf{x}'$ and the integration variable $\mathbf{x}'$ replaced by $\mathbf{x}''$). We then get

$$u(\mathbf{x}) = u_0(\mathbf{x}) + \int G^+(\mathbf{x}, \mathbf{x}')\, U(\mathbf{x}')\, u_0(\mathbf{x}')\, d\tau' + \iint G^+(\mathbf{x}, \mathbf{x}')\, U(\mathbf{x}')\, G^+(\mathbf{x}', \mathbf{x}'')\, u(\mathbf{x}'')\, d\tau' d\tau''.$$

Once again we can substitute from Eq. 6.34 for $u(\mathbf{x}'')$, which appears in the last integral, and keep on repeating this process. What we get then is the infinite series

$$\begin{aligned} u(\mathbf{x}) = u_0(\mathbf{x}) &+ \int G^+(\mathbf{x}, \mathbf{x}')\, U(\mathbf{x}')\, u_0(\mathbf{x}')\, d\tau' \\ &+ \iint G^+(\mathbf{x}, \mathbf{x}')\, U(\mathbf{x}')\, G^+(\mathbf{x}', \mathbf{x}'')\, U(\mathbf{x}'')\, u_0(\mathbf{x}'')\, d\tau'' d\tau' \\ &+ \iiint G^+(\mathbf{x}, \mathbf{x}')\, U(\mathbf{x}')\, G^+(\mathbf{x}', \mathbf{x}'')\, U(\mathbf{x}'')\, G^+(\mathbf{x}'', \mathbf{x}''') \times \\ &\qquad \times U(\mathbf{x}''')\, u_0(\mathbf{x}''')\, d\tau''' d\tau'' d\tau' + \dots \end{aligned} \tag{6.35}$$

This expression is evidently a perturbation expansion of $u(\mathbf{x})$ with the potential V treated as a perturbation. The $(n+1)^{\text{th}}$ term in the series is of the nth order in the perturbation, since n factors of $U \equiv 2mV/\hbar^2$ appear in it. Thus the first Born approximation, wherein the first term $u_0(\mathbf{x}) = e^{ikz}$ alone is used for $u(\mathbf{x})$ in Eq. 6.23, is really a lowest order perturbation calculation. If the entire series (6.35) were used in Eq. 6.23 one would get an expression for $f(\theta, \varphi)$ which is, in principle, exact (provided of course that the series converges). However, the difficulty of evaluation of higher order terms is such that they are rarely considered in practice.

6.7 The Eikonal Approximation

Besides the systematic but difficult process of evaluating the Born series, there are other procedures for improving upon the first Born approximation. One of these is to use for $u(\mathbf{x})$ in Eq. 6.23 a form which takes into account the fact that when the incident particles enter the potential region, their de Broglie wave length gets modified from $(2\pi/k)$ to $2\pi/[k^2 - U(\mathbf{x})]^{1/2}$. One of the consequences of the position-dependence of this wavelength is that wave fronts get distorted, and the rays cease to be strictly parallel to the z-axis. When $k^2 \gg U(\mathbf{x})$ this effect is small. Let us ignore it and focus our attention on the effect of the wavelength variation on the *phase* of the wave as it moves in the z-direction. The difference in phase of the wave between two points (x, y, z) and (x, y, z_0) is no longer $k(z - z_0)$, but $\int_{z_0}^{z} [k^2 - U(x, y, z')]^{1/2} dz'$. Of course this effect begins to manifest itself only when the wave enters the region where the potential is nonzero. *Before* the potential region (which is around the origin) is reached, at some point (x, y, z_0) with $z_0 < 0$, let the phase of the wave be kz_0. As one advances in the positive z-direction to another point (x, y, z) within the potential region or beyond, the phase becomes

$$kz_0 + \int_{z_0}^{z} [k^2 - U(x, y, z')]^{1/2} dz'$$

$$= kz + \int_{z_0}^{z} \{[k^2 - U(x, y, z')]^{1/2} - k\} dz' \tag{6.36}$$

instead of just kz. On assuming $|U| \ll k^2$ everywhere, so that $[k^2 - U]^{1/2} \approx k[1 - U/2k^2]$, the above expression reduces to

$$kz - \frac{1}{2k}\int_{z_0}^{z} U(x, y, z')\, dz' = kz - \frac{1}{\hbar v}\int_{z_0}^{z} V(x, y, z')\, dz' \tag{6.37}$$

Since all that is demanded of z_0 is that it should refer to a point outside the potential region and before this region is reached $(z_0 < 0)$, we can conveniently take it to be $-\infty$. With this choice, the new approximate wave function to be used for $u(\mathbf{x})$ in Eq. 6.23 (improving on the Born approximation e^{ikz}) is

$$\exp\left\{i\left[kz - \frac{1}{\hbar v}\int_{-\infty}^{z} V(x, y, z')\, dz'\right]\right\}.$$

With this, (6.23) yields what is called the *eikonal approximation* for the scattering amplitude:

$$f(\theta, \varphi) \approx -\frac{m}{2\pi\hbar^2}\int V(\mathbf{x}) \exp.\left\{i\left[-\mathbf{K}.\mathbf{x} - \frac{1}{\hbar v}\int_{-\infty}^{z} V(x, y, z')dz'\right]\right\} d\tau \tag{6.38}$$

where $\mathbf{K} = \mathbf{k} - \mathbf{k}_0$, $\mathbf{k}_0$ being the initial propagation vector $(\mathbf{k}_0.\mathbf{x} = kz)$. The Born approximation becomes comparably good only under such circumstances that

$$\left|\frac{1}{\hbar v}\int_{-\infty}^{z} V(x, y, z')\, dz'\right| \ll 1 \text{ for all } \mathbf{x} \tag{6.39}$$

This is a criterion for the validity of the Born approximation, which is more general than what is obtained in Sec. 6.5.

We have already noted that the eikonal approximation is a high energy approximation. At high energies, scattering takes place predominantly in near-forward directions (i.e. at small θ). When θ is small, $\mathbf{K}$ is nearly perpendicular to the incident direction (see Fig. 6.3); its longitudinal component (K_z) is smaller than its transverse components by approximately a factor θ. If we neglect it and take $\mathbf{K}.\mathbf{x}$ to be $K_x x + K_y y$, the integrand in (6.38) can be written as $\hbar v \ (\partial Q/\partial z) \exp(iQ)$ where $Q = -(K_x x + K_y y) - (\hbar v)^{-1} \int_{-\infty}^{z} V(x, y, z')\, dz'$. Integration of this quantity over z directly yields $(\hbar v/i)$ $[(e^{iQ})_{z=+\infty} - (e^{iQ})_{z=-\infty}]$ and hence we obtain (with $mv/\hbar = k$).

$$f(\theta,\varphi) \approx \frac{ik}{2\pi} \int_{-\infty}^{+\infty} dx \int_{-\infty}^{+\infty} dy \, e^{-i(K_x x + K_y y)} . \left\{1 - \exp\left[-\frac{i}{\hbar v} \int_{-\infty}^{\infty} V(x, y, z)\, dz\right]\right\} \tag{6.40}$$

This simplified form has a very interesting physical interpretation. Observing that only the transverse coordinates x, y appear in the integration, we can think of each pair (x, y) as defining the classical trajectory of a particle moving parallel to the z-axis (while approaching the potential). The scattering amplitude appears then as a sum of contributions from all possible classical trajectories. If the potential is symmetric about the z-axis, the distance $b = (x^2 + y^2)^{1/2}$ of a trajectory from the axis of symmetry is known as the *impact parameter*. The amplitude can, in this case, be written as a simple integral over the impact parameter. Introducing polar coordinates (b, θ) in the x-y plane, and writing $K_x x + K_y y = Kb \cos\theta$ (remembering $K_z \approx 0$), one obtains with the aid of a well known integral representation (Eq. D.24) for Bessel functions,

$$f(\theta) \approx ik \int_0^{\infty} J_0(bK) \left\{1 - \exp\left[-\frac{i}{\hbar v} \int_{-\infty}^{\infty} V(b, z)\, dz\right]\right\} bdb \tag{6.41}$$

Note that for small θ, for which this expression is valid, $K = 2k \sin \frac{1}{2}\theta \approx k\theta$.

Finally, it may be observed that the eikonal approximation to $u(\mathbf{x})$ is very closely related to the WKB approximation.

C. PARTIAL WAVE ANALYSIS

While the Born approximation is basically a truncation of a perturbation expansion of $u(\mathbf{x})$, the method of partial waves is based upon an expansion of $u(\mathbf{x})$ in terms of angular momentum eigenfunctions. It is applicable if the potential is spherically symmetric. The new expansion of $u(\mathbf{x})$ must of course be such as to be consistent with the asymptotic form (6.9). When this requirement is imposed, one gets a formal expression for the scattering amplitude as an infinite series; each term of the series is the contribution to $f(\theta,\varphi)$ from a *partial wave* characterized by a particular angular momentum. We shall see that the first few terms in the series approximate $f(\theta,\varphi)$ if the incident particle is of low energy. Thus the partial wave method leads to a low energy approximation which complements the Born approximation (good at high energies).

6.8 Asymptotic Behaviour of Partial Waves: Phase Shifts

(*a*) *Partial Waves:* We have seen in Chapter 4 that when the potential has spherical symmetry we can separate the Schrödinger equation into radial and angular parts, and obtain solutions in the form $R_l(r)\ Y_{lm}(\theta,\varphi)$, where the radial wave functions $R_l(r)$ satisfy the equation

$$\frac{1}{r^2}\frac{d}{dr}\left(r^2\frac{dR_l}{dr}\right)+\frac{2\mu}{\hbar^2}\left[E-V-\frac{l(l+1)\hbar^2}{2\mu r^2}\right]R_l=0 \qquad (6.42)$$

The general solution of the Schrödinger equation can be written as a linear combination of such solutions. In the scattering problem which we are considering, only the function $R_l(r)\ Y_{l0}(\theta,\varphi) \propto R_l(r)\ P_l(\cos\theta)$ appears in the linear combination. This is because we have assumed the particle to be travelling in the z-direction initially, so that the z-component of its angular momentum is zero; it continues to be zero because a spherically symmetric potential does not disturb the angular momentum. Thus we can write

$$u(\mathbf{x})=\sum_{l=0}^{\infty} R_l(r)\ P_l(\cos\theta) \qquad (6.43)$$

The term corresponding to a particular l in this series is called the lth *partial wave.*

(*b*) *Asymptotic form of radial function:* We must now require that $u(\mathbf{x})$ should have the asymptotic form (6.9). This condition places constraints on the radial wave functions $R_l(r)$. To see what they are, we need to have the asymptotic ($r\to\infty$) form of $R_l(r)$. We can infer it from the radial wave equation (6.42) which we rewrite as

$$\left.\begin{aligned}&\frac{d^2\chi_l}{dr^2}+\left[k^2-U(r)-\frac{l(l+1)}{r^2}\right]\chi_l=0,\\&\chi_l=r\,R_l(r),\ k^2=(2mE/\hbar^2),\ U(r)=2mV/\hbar^2\end{aligned}\right\} \qquad (6.44)$$

In the asymptotic region, both $U(r)$ and the centrifugal potential ($\propto 1/r^2$) are very small. On neglecting these, in comparison with k^2, we obtain the approximate asymptotic solution $\chi_l(r)\propto e^{\pm ikr}$. To improve this approximation, let us suppose that

$$\chi_l(r)=v_l(r)\,e^{\pm ikr} \qquad (6.45)$$

where $v_l(r)$ is expected to be very slowly varying in the asymptotic region. On introducing this into Eq. 6.44 we get

$$d^2v_l/dr^2 \pm 2ik\,dv_l/dr-[U+l(l+1)/r^2]\,v_l=0$$

Since v_l is nearly constant for large r, it would be expected that v_l'' is much less than the first derivative term. On solving this equation neglecting v_l'', we get

$$\ln v_l \approx \mp \frac{i}{2k} \int^r \left[U + \frac{l(l+1)}{r^2} \right] dr \tag{6.46}$$

In the case of the Coulomb potential $U \propto r^{-1}$, the right hand side is proportional to[7] $\ln r$ (disregarding other terms which are relatively small as $r \to \infty$). For any potential decreasing faster than $(1/r)$ at infinity (i.e: $rU(r) \to 0$ as $r \to \infty$) the integral in Eq. 6.46 is convergent, which means that *if r is large enough* the value of the integral is effectively independent of r and hence $v_l(r)$ is a constant in the asymptotic region. In the following, we will confine our attention to such potentials, leaving the Coulomb case to be dealt with later (Sec. 6.16). Then the asymptotic form of $\chi_l(r)$ is, in general, some linear combination of the two solutions in Eq. 6.45 with v_l constant, i.e. of e^{ikr} and e^{-ikr}. Without loss of generality, we can write any such combination in the form

$$\chi_l \to C_l \sin(kr + \Delta_l) \tag{6.47}$$

where C_l, Δ_l are constants. Since all quantities appearing in the differential equation satisfied by χ_l are real, and since χ_l is required to vanish at the origin (Sec. 4.12c), it is easy to convince oneself that χ_l must be real everywhere, apart from an overall factor which is arbitrary and may be complex. Thus Δ_l in (6.47) must be *real*, though C_l may be complex.

(*c*) *Phase Shifts:* Let us compare the above asymptotic behaviour with that of the partial waves for a free particle ($V(r) = 0$). We have already seen (Sec. 4.13) that in any range of r over which $V(r)$ is constant, the solutions of the radial wave equation are the spherical Bessel functions j_l or n_l. In particular, if $V = 0$ everywhere, the only admissible solution is $j_l(kr)$; n_l is ruled out because it becomes infinite at $r = 0$. Thus $R_l(r) =$ const. $j_l(kr)$ in the free-particle case. Since the asymptotic forms of the spherical Bessel functions are known to be given by

$$j_l(kr) \to (kr)^{-1} \sin(kr - \tfrac{1}{2}l\pi) \tag{6.48a}$$

$$n_l(kr) \to -(kr)^{-1} \cos(kr - \tfrac{1}{2}l\pi) \tag{6.48b}$$

(as $r \to \infty$), it follows that $\chi_l(r) = rR_l(r) \to$ const. $\sin(kr - \frac{1}{2}l\pi)$. Hence χ_l in this case has the asymptotic form (6.47) with the particular value $-\frac{1}{2}l\pi$ for Δ_l.

The effect of the potential on the partial waves in the asymptotic region is, therefore, simply to change the phase from $-\frac{1}{2}\pi l$ to some other value Δ_l. This change,

$$\delta_l = \Delta_l + \tfrac{1}{2}l\pi \tag{6.49}$$

is called the *phase shift* in the lth partial wave.

This result may be expressed in a slightly different fashion which is instructive from a physical point of view. Observe that the asymptotic form (6.48a) in the absence of any potential is $\propto (e^{-ikr} - e^{-il\pi} e^{ikr})$, while in the presence of a potential we have (6.47) which is $\propto (e^{-ikr} - e^{-2i\Delta_l} e^{ikr})$. In the former case, the *incoming spherical wave* e^{-ikr} (converging on to the origin) is accompanied by an *outgoing* spherical wave $e^{-il\pi} e^{ikr}$; when there is a potential, the

[7] The radial wave function in a Coulomb potential is determined exactly in Sec. 4.18. Its asymptotic form, Eq. 4.160, does exhibit the logarithmic term in the phase.

same incoming wave gives rise to the outgoing wave $e^{2i\Delta_l}e^{ikr}$ which differs from the free case by the factor

$$S_l = e^{2i\delta_l} \tag{6.50}$$

The effect of the potential on *spherical waves* is contained entirely in this factor. However, scattering experiments do not use converging beams. Since in practice one employs parallel beams, represented by *plane* waves, we need to know the effect of the potential on such waves. We shall now see that this effect, which is described by the scattering amplitude $f(\theta)$, can also be expressed in terms of the phase shifts δ_l.

6.9 The Scattering Amplitude in Terms of Phase Shifts

From (6.47) and (6.49) the asymptotic form of $R_l(r)$ is

$$R_l(r) \to A_l\,(kr)^{-1} \sin\,(kr - \tfrac{1}{2}l\pi + \delta_l) \tag{6.51}$$

where $A_l = kC_l$. Hence the asymptotic form of $u(\mathbf{x})$ is

$$u(\mathbf{x}) \to \sum_l A_l(kr)^{-1} \sin\,(kr - \tfrac{1}{2}l\pi + \delta_l).\; P_l\,(\cos\theta) \tag{6.52}$$

We have to choose the constants A_l in such a way that this form agrees with (6.9). To do this we need the following expansion:

$$e^{ikz} = e^{ikr\cos\theta} = \sum_{l=0}^{\infty} (2l+1) i^l j_l\,(kr)\, P_l\,(\cos\theta)$$

$$\xrightarrow[r\to\infty]{} \sum_{l=0}^{\infty} (2l+1)\, i^l\,(kr)^{-1} \sin\,(kr - \tfrac{1}{2}l\pi)\, P_l\,(\cos\theta) \tag{6.53}$$

From (6.52) and (6.53) we have that

$$\{u\,(\mathbf{x}) - e^{ikz}\} \longrightarrow$$

$$(2ikr)^{-1}\, e^{ikr} \sum_{l=0}^{\infty} \left\{ A_l\, e^{i(\delta_l - \frac{1}{2}l\pi)} - (2l+1)\, i^l\, e^{-i\frac{1}{2}l\pi} \right\} P_l\,(\cos\theta)$$

$$+\,(2ikr)^{-1}\, e^{-ikr} \sum_{l=0}^{\infty} \left\{ -A_l\, e^{-i(\delta_l - \frac{1}{2}l\pi)} + (2l+1)\, i^l\, e^{i\frac{1}{2}l\pi} \right\} P_l\,(\cos\theta) \tag{6.54}$$

According to (6.9) the right hand side should contain only outgoing spherical waves ($\propto e^{ikr}$). The terms proportional to e^{-ikr} should vanish. This can happen for all θ only if each term in the second sum vanishes, since the $P_l\,(\cos\theta)$ are orthogonal functions. Hence we must have

$$A_l = (2l+1)\, i^l\, e^{i\delta_l} \tag{6.55}$$

On substituting this in (6.54) and equating the resulting expression (which now contains only e^{+ikr}) to $f(\theta)\, e^{ikr}/r$, we get

$$f(\theta) = (2ik)^{-1} \sum_{l=0}^{\infty} (2l+1)\,(e^{2i\delta_l} - 1)\, P_l\,(\cos\theta) \tag{6.56a}$$

$$= k^{-1} \sum_{l=0}^{\infty} (2l+1)\, e^{i\delta_l} \sin\delta_l\, P_l(\cos\theta) \tag{6.56b}$$

This gives $f(\theta)$ as a sum of contributions from partial waves with $l = 0$ to ∞.

6.10 The Differential and Total Cross-Sections; Optical Theorem

The differential cross-section $d\sigma/d\Omega = |f(\theta)|^2$ is given now by

$$\frac{d\sigma(\theta)}{d\Omega} = \frac{1}{k^2}\sum_{l=0}^{\infty}\sum_{l'=0}^{\infty}(2l+1)(2l'+1)\,e^{i(\delta_l-\delta_{l'})}\sin\delta_l\sin\delta_{l'}\times$$
$$\times P_l(\cos\theta)\,P_{l'}(\cos\theta) \quad (6.57)$$

Note that this expression contains terms (those with $l \neq l'$) representing interference between different partial waves. These terms do not contribute to the total scattering cross-section, in view of the fact that $\int d\Omega\, P_l(\cos\theta)\, P_{l'}(\cos\theta) = 2\pi\int P_l(\cos\theta)\, P_{l'}(\cos\theta).\sin\theta\, d\theta = 4\pi\,\delta_{ll'}/(2l+1)$. Thus

$$\sigma = 2\pi\int_0^{\pi}\frac{d\sigma}{d\Omega}\,.\,\sin\theta\, d\theta = \frac{4\pi}{k^2}\sum_{l=0}^{\infty}(2l+1)\sin^2\delta_l \quad (6.58)$$

The contribution to σ from the lth partial wave, namely $\sigma_l \equiv (4\pi/k^2)(2l+1)\sin^2\delta_l$, may be called the *partial cross-section* of this wave.

The interference terms in Eq. 6.57 do influence the angular distribution—in fact quite strongly. Consider for instance a situation where all phase shifts except δ_0 and δ_1 are negligibly small. Then Eq. 6.57 reduces to

$$\frac{d\sigma(\theta)}{d\Omega} = \frac{1}{k^2}\left[\sin^2\delta_0 + 9\sin^2\delta_1\cos^2\theta + 6\sin\delta_0\sin\delta_1\cos(\delta_0-\delta_1)\cos\theta\right] \quad (6.59)$$

The first term gives the contribution of the $l = 0$ partial wave (*s-wave*), which is independent of θ and hence spherically symmetrical. The second term due to the *p-wave* ($l = 1$) alone is proportional to $\cos^2\theta$ and is, therefore, symmetric between forward and backward directions (i.e. has the same value at θ and $\pi - \theta$). The third (interference) term, $\propto \cos\theta$, gives rise to a forward-backward asymmetry which may be very large even if $\sin\delta_1 \ll \sin\delta_0$. Suppose for example, that $\delta_0, \delta_1 \ll 1$ so that $\cos(\delta_0 - \delta_1) \approx 1$, and that $\sin\delta_1 = (1/10)\sin\delta_0$. Then the ratio of the contributions to the *total* cross-section from the *p*- and *s*-waves is only $\sigma_1/\sigma_0 = 3\sin^2\delta_1/\sin^2\delta_0 = 0{\cdot}03$, i.e. 3%. Nevertheless the difference in $d\sigma/d\Omega$ between the forward ($\theta = 0°$) and backward ($\theta = 180°$) directions, which according to Eq. 6.59 is $\approx 12\sin\delta_0\sin\delta_1$, is *greater* than (about 1.2 times) the forward cross-section itself. The important role of the interference terms in determining the angular distributions is clearly brought out by this example.

The *optical theorem*

$$\sigma = \frac{4\pi}{k}\,\text{Im}\, f(0) \quad (6.60)$$

relating the total cross-section to the imaginary part of the *forward* ($\theta = 0$) scattering amplitude, follows directly from Eqs. 6.56 and 6.58. For, from (6.56b), $\text{Im}\, f(0) = k^{-1}\,\Sigma\,(2l+1)\sin^2\delta_l\, P_l(1)$, which reduces to $(k/4\pi)\sigma$ since $P_l(1) = 1$ for all l.

Incidentally, since the amplitude $f_B(\theta)$ is real in the Born approximation, this approximation violates the optical theorem.

6.11 Phase Shifts: Relation to the Potential

Though the phase shift δ_l depends only on the asymptotic form of R_l, to determine it exactly one has to solve the radial equation completely. Of course, this is possible only for very special potentials. However it is possible, from general considerations, to deduce certain important features of δ_l and to obtain a formal expression for it which gives very useful information.

(*a*) *Sign of* δ_l: If the potential V is an attractive one ($V < 0$ for all r), the 'classical turning point' r_0, where $k^2 - U(r)$ in Eq. 6.44 overcomes the centrifugal repulsion, occurs closer to the origin than in the absence of the potential. Further, the local wave number $[k^2 - U(r) - l(l+1)/r^2]^{1/2}$ is greater than it is when $V = 0$. Both these effects cause the number of oscillations of the wave function between the origin and any point r to be more in the case of an attractive potential than in the absence of any potential. Consequently, the phase of the wave is more advanced (greater) than when $V = 0$. In other words, the *phase shift is positive.* Exactly the reverse happens in the case of a repulsive potential ($V > 0$): the wave function gets 'pushed' outwards by the potential, and develops slower. So the phase shift is *negative for* $V > 0$.

(*b*) *Expression for the Phase Shift:* To express the phase shift in terms of the potential, we compare χ_l with the corresponding function $\chi_l^{(0)} \propto r j_l(kr)$ of the potential-free case, starting from their differential equations:

$$\chi_l'' + [k^2 - U - l(l+1)/r^2]\,\chi_l = 0 \tag{6.61}$$

$$\chi_l^{(0)\prime\prime} + [k^2 - l(l+1)/r^2]\,\chi_l^{(0)} = 0, \quad \chi_l^{(0)} \propto r j_l(kr) \tag{6.62}$$

Multiplying these equations by $\chi_l^{(0)}$ and χ_l respectively and taking the difference, we obtain

$$[\chi_l^{(0)}\chi_l' - \chi_l\chi_l^{(0)\prime}]' - U\chi_l^{(0)}\chi_l = 0 \tag{6.63}$$

We integrate this equation from 0 to r, and recall that χ_l and χ_l^0 have to vanish at the origin (Sec. 4.12c). Then

$$\chi_l^{(0)}(r)\,\chi_l'(r) - \chi_l(r)\chi_l^{(0)\prime}(r) = \int_0^r U(r')\,\chi_l^{(0)}(r')\,\chi_l(r')\,dr' \tag{6.64}$$

This relation is clearly independent of the normalization of χ_l and $\chi_l^{(0)}$. Let us normalize them such that the constant C_l in the asymptotic form (6.47) is unity, i.e.

$$\chi_l \to \sin(kr + \Delta_l), \quad \chi_l^{(0)} \to \sin(kr - \tfrac{1}{2}l\pi) \tag{6.65}$$

as $r \to \infty$. This means that $\chi_l^{(0)} = kr\,j_l(kr)$, in view of (6.48a). With this normalization, the left hand side of Eq. 6.64 becomes, for $r \to \infty$,

$k[\sin(kr - \tfrac{1}{2}l\pi)\cos(kr + \Delta_l) - \sin(kr + \Delta_l)\cos(kr - \tfrac{1}{2}l\pi)] = -k\sin(\tfrac{1}{2}l\pi + \Delta_l) = -k\sin\delta_l$.

Thus as $r \to \infty$, Eq. 6.64 reduces to

$$k\sin\delta_l = -\int_0^\infty U(r')\,\chi_l^{(0)}(r')\,\chi_l(r')\,dr'$$

or

$$\sin\delta_l = -\int_0^\infty U(r)\,r j_l(kr)\,\chi_l(r)\,dr \tag{6.66}$$

χ_l being normalized according to (6.65).

The expression (6.66) for the phase shift is exact, but is purely formal inasmuch as we do not know $\chi_l(r)$ for all r. Nevertheless, it is useful as a starting point for approximate evaluation. Suppose, for instance, that χ_l differs very little from $\chi_l^{(0)} \equiv kr\, j_l\,(kr)$. Then

$$\sin \delta_l \approx -k \int_0^\infty U(r)\, r^2 j_l^2\,(kr)\, dr \tag{6.67}$$

This is the *Born approximation* for *phase shifts*. For this approximation to be valid it is clearly necessary that the effect of the potential term $U(r)$ in Eq. 6.61 be very small. This can happen if either the kinetic energy term k^2 or the centrifugal potential dominates over $U(r)$. In the former case, δ_l is expected to be small for *all* l. Then we can take $(e^{2i\delta_l} - 1)$ in Eq. 6.56a to be $\approx 2i\delta_l \approx 2i \sin \delta_l$ and substitute for this quantity from (6.67). The result is

$$f(\theta) \approx -\sum_{l=0}^{\infty} (2l+1) \int_0^\infty U(r)\, r^2 j_l^2(kr)\, dr\, P_l\,(\cos\theta)$$

It reduces to the Born approximation formula (6.27) since $\Sigma\ (2l+1)\ j_l^2(kr)\ P_l\,(\cos\theta) = (\sin Kr)/Kr$, with $K = 2k \sin\frac{1}{2}\theta$. It is to be noted, however, that in writing $(e^{2i\delta_l} - 1) \approx 2i\delta_l$ and thereby making $f(\theta)$ real, we violate the optical theorem. Therefore, the use of (6.67) directly in Eqs. 6.56, avoiding the above step, is in principle an improvement over the Born approximation (6.27) for the complete amplitude $f(\theta)$. This procedure is also found to give better results in practice, especially in cases where the phase shifts are not all very small.

6.12 Potentials of Finite Range

(*a*) *Phase Shifts for Large l.* We shall now use (6.67) which is good for large l, to show that in the case of any potential which vanishes beyond some finite range a, the phase shifts decrease rapidly as l increases beyond ka. This can be seen qualitatively from the fact that $j_l(kr)$ has its first and largest maximum roughly at $r = l/k$, so that if $l \gg ka$, the maximum occurs well outside the range of the potential. This would mean that $j_l(kr)$ is still small within the region of integration in (6.67) which now extends only from 0 to a. Hence $\sin\delta_l$ is small, and gets smaller as l keeps increasing. To get a quantitative estimate of $\sin\delta_l$ for large l, we consider $l \gg (ka)^2$. Then the approximation (D.31) for $j_l\,(kr)$, namely $j_l\,(kr) \approx (kr)^l/1{\cdot}3{\cdot}5\ldots(2l+1)$, becomes good, so that

$$\sin\delta_l \approx \frac{-\,k^{2l+1}}{[1.3.5\ldots(2l+1)]^2} \int_0^a U(r)\, r^{2l+2}\, dr \tag{6.68}$$

The magnitude of the integral in (6.68) cannot exceed $a^{2l+3}\,|U|_{max}/(2l+3)$, where $|U|_{max}$ is the maximum value[8] of $|U|$. Thus

[8] This assumes that U is a finite potential. If U diverges as cr^{-s} $(s < 2)$ at the origin, the contribution from this singularity to the integral would have $|c|\ a^{2l+3-s}/(2l+3-s)$ instead of $a^{2l+3}|U|_{max}/(2l+3)$ in (6.69). The l-dependence therefore remains unaltered, and the conclusions drawn therefrom are unaffected. Attractive potentials more singular than r^{-2} raise pathological problems which are beyond our scope. Repulsive potentials diverging faster than r^{-2} suppress the wave function near the origin so strongly that $|\sin\delta_l|$ in this case is even smaller than is indicated by (6.69).

$$|\sin\delta_l| \leqslant \frac{(ka)^{2l+1}a^2|U|_{max}}{[1.3.5.\;.(2l+1)]^2(2l+3)} < \frac{a^2|U|_{max}}{4}\left(\frac{2ka}{l}\right)^{2l+1} \quad (6.69)$$

where we have used the fact that $1.3.5\ldots(2l+1) = (2l+1)!/2^l l! = (l+1)(l+2)\ldots(2l+1)/2^l > l^l/2^l$. It is evident that the phase shift decreases extremely fast as l increases.

Classically, a particle of momentum $\hbar k$ and angular momentum $l\hbar$ is one whose line of approach would pass the origin at a distance (l/k). If this distance exceeds the range a of the potential, i.e. $l > ka$ the particle would not be scattered at all. In quantum mechanics particle trajectories are not sharp, and one does not expect an abrupt disappearance of the scattering when l exceeds ka, but a fast decrease of δ_l should take place, and this is just what we have found.

The computation of phase shifts is a laborious process and the method of partial waves is, therefore, not very useful if a large number of partial waves contribute to $f(\theta)$. If ka is small, the result derived above shows that only a small number of partial waves (with $l \lesssim ka$) contribute significantly. Therefore, this method is most useful in the case of low energy scattering. Before we go on to consider particular examples, we derive another formal expression for the phase shifts, valid for potentials of finite range. It is more suitable than Eq. 6.66 for the discussion of scattering at low energies.

(*b*) *A Formal Expression for δ_l:* Since we assume that $U(r) = 0$ for $r > a$, in this *exterior region* the wave function must be a linear combination of $j_l(kr)$ and $n_l(kr)$. In fact,

$$R_l \propto [j_l(kr)\cos\delta_l - n_l(kr)\sin\delta_l], \quad r > a \quad (6.70)$$

For, as can be seen from the asymptotic forms (6.48), it is this combination which leads to (6.47) with $\Delta_l = \delta_l - \frac{1}{2}\pi l$. We do not know R_l in the *interior region* $(r < a)$, but we know that it has to match with the exterior wave function at $r = a$. In particular, if

$$\frac{1}{R_l}\frac{dR_l}{dr} \to \frac{1}{a\gamma_l(k)} \quad (6.71)$$

as $r \to a$ from the interior region, the ratio $(1/R_l)(dR_l/dr)$ for the exterior region, obtained from Eq. 6.70, must also tend to $(1/a\gamma_l)$ as $r \to a$.

$$\frac{k[j_l'(ka)\cos\delta_l - n_l'(ka)\sin\delta_l]}{j_l(ka)\cos\delta_l - n_l(ka)\sin\delta_l} = \frac{1}{a\gamma_l(k)} \quad (6.72)$$

Hence $\tan\delta_l$ can be formally expressed in terms of γ_l as

$$\tan\delta_l = \frac{ka\,\gamma_l(k)\,j_l'(ka) - j_l(ka)}{ka\,\gamma_l(k)\,n_l'(ka) - n_l(ka)} \quad (6.73)$$

The parameter γ_l can be determined exactly only by solving the wave equation in the interior region.

6.13 Low Energy Scattering

If the bombarding energy is low enough that $ka \ll 1$, the approximations (D.31) and (D 32) for j_l and n_l become good. On using them in Eq. 6.73, it reduces to

$$\tan \delta_l \approx \frac{(ka)^{2l+1}\,(2l+1)}{[1.3.5\ldots(2l+1)]^2}\left\{\frac{l\,\gamma_l(k)-1}{(l+1)\,\gamma_l(k)+1}\right\} \tag{6.74}$$

for all l. This is really a special case of (6.68), appropriate in the low energy limit. In the following, we shall also use the fact that

$$\gamma_l(k) = \gamma_l(0) + \beta_l a^2 k^2 \tag{6.75}$$

where β_l is positive and effectively constant at low energies. This can be verified by a comparison of Eq. 6.61 for $\chi_l(r;\,k)$ with the corresponding equation for $\chi_l(r;\,0)$ obtained by setting $k = 0$. Starting from these two equations and proceeding as in the derivation of Eq. 6.64, we find that

$$\chi_l(r;0)\,\chi_l'(r;k) - \chi_l(r;k)\,\chi_l'(r;0) = -k^2\int_0^r \chi_l(r';0)\,\chi_l(r';k)\,dr',$$

or since $\chi_l = rR_l$

$$R_l(r;0)\,R_l'(r;k) - R_l(r;k)\,R_l'(r;0) = -\frac{k^2}{r^2}\int_0^r r'^2\,R_l(r';0)\,R_l(r';k)\,dr' \tag{6.76}$$

Dividing throughout by $R_l'(r;0)\,R_l'(r;k)$, then setting $r = a$ and recalling that $R_l(a;k)/R_l'(a;k) = a\gamma_l(k)$, we obtain

$$\gamma_l(k) - \gamma_l(0) = \frac{k^2}{a^3 R_l'(a;0)\,R_l'(a;k)}\int_0^a r'^2 R_l(r';0)\,R_l(r';k)\,dr' \tag{6.77}$$

This result is exact, and is just what Eq. 6.75 says. If, on the right hand side, we make the low energy approximation $R(r;k) \approx R(r;0)$ for $0 \leqslant r \leqslant a$, the coefficient of k^2 which is defined as $\beta_l a^2$ is clearly seen to be positive.

(*a*) *Resonant and Nonresonant Scattering:* Inspection of Eq. 6.74 reveals that $\tan \delta_l$ may become arbitrarily large (even for $ka \ll 1$, despite the factor $(ka)^{2l+1}$) if it so happens that, for this particular l, $[(l+1)\,\gamma_l(k) + 1]$ vanishes near $k = 0$. Such a phenomenon is described as a *resonance* in the lth partial wave.

In the absence of any resonance, $\tan\delta_l \propto (ka)^{2l+1}$ is very small at low energies for all l. The *partial cross-section* $\sigma_l = (4\pi/k^2)\,\sin^2\delta_l$ (which is the contribution of the lth partial wave to the total cross-section) is then $\approx (4\pi/k^2)\tan^2\delta_l \propto k^{4l}$. Hence *the s-wave partial cross-section σ_0 is independent of energy*, while σ_l for $l > 0$ falls off to zero very fast (like k^{4l}) as $k \to 0$. Clearly then low energy scattering is dominated by the *s*-wave, and is isotropic, i.e. $f(\theta) \approx e^{i\delta_0}\sin\delta_0$ is independent of θ.

If a resonance occurs in the lth partial wave, it changes the energy dependence of σ_l drastically. In particular, σ_l (considered as a function of the energy) acquires a sharp maximum at some energy, called the resonance energy. The enhancement of the scattering amplitude in this partial wave at energies near resonance may even be sufficient (if $l \neq 0$) to make it comparable to the *s*-wave amplitude. In such a case the isotropy of the scattering would also be destroyed. To see in more quantitative terms what happens, let us consider the partial wave cross-section

$$\sigma_l = \frac{4\pi}{k^2}\sin^2\delta_l = \frac{4\pi}{(k\cot\delta_l)^2 + k^2} \tag{6.78}$$

in the resonant situation where $k \cot\delta_l$ vanishes at some very low energy, or more generally, when $(k\cot\delta_l)^2$ ceases to be the dominant term in the denominator of Eq. 6.78 in some part of the low energy region. The essential features can be seen from a study of s-wave and p-wave scattering.

(*b*) *s-Wave Resonance: Scattering Length and Effective Range:* For $l = 0$, it is customary to write

$$k\cot\delta_0 = -\left(L^{-1} + \tfrac{1}{2}\, r_{\text{eff}} \cdot k^2\right) \tag{6.79}$$

where L, the zero energy limit of $(-\tan\delta_0/k)$, is called the *scattering length*[9] and r_{eff} is called the *effective range.* We can express L and r_{eff} in terms of γ_0 using (6.74) and (6.75), which give $k\cot\delta_0 = -a^{-1}[1 + \gamma_0(k)] = -a^{-1}[1 + \gamma_0(0) + \beta_0 a^2k^2]$. Thus

$$L^{-1} = a^{-1}[1 + \gamma_0(0)], \text{ and } r_{\text{eff}} = (2\beta_0 a) \tag{6.80}$$

Note that r_{eff} is positive, while L may be positive or negative. The partial cross-section σ_0 may now be written as $\sigma_0 = 4\pi\sin^2\delta_0/k^2 = 4\pi/[k^2(\cot^2\delta_0 + 1)]$ and hence, using Eq. 6.79,

$$\sigma_0 = \frac{4\pi}{(L^{-1} + \frac{1}{2}\, r_{\text{eff}} \cdot k^2)^2 + k^2} \tag{6.81}$$

When L^{-1} is not very small, it dominates the denominator, and then $\sigma_0 \approx 4\pi L^2$. This corresponds to nonresonant scattering. Resonance occurs when L^{-1} is nearly zero, i.e. $1 + \gamma_0(0) = 0$. The maximum of σ_0 should then occur where the denominator is a minimum, which is easily seen to be at $k^2 = -2(1 + L^{-1} r_{\text{eff}})/r_{\text{eff}}^2 \approx -2/r_{\text{eff}}^2$ since $L^{-1} \approx 0$. This *negative* value of k^2 is not in the physical scattering region. It follows that σ_0 is largest at $k = 0$ (with the value $4\pi L^2$ which is now very large since $|L|$ itself is), and decreases monotonically as k increases.

(*c*) *p-Wave Resonance:* When $l = 1$, we have from (6.74) that $k\cot\delta_1 = (3/k^2a^3)(2\gamma_1 + 1)/(\gamma_1 - 1) \approx -(2/k^2a^3)(\varepsilon_1 + 2\beta_1 a^2k^2)$. We have used Eq. 6.75 and written ε_1 for the quantity $2\gamma_1(0) + 1$ which has to be nearly zero if resonance is to occur. Hence

$$\sigma_1 \approx \frac{4\pi a^2}{4(\varepsilon_1/k^2a^2 + 2\beta_1)^2 + k^2a^2} \tag{6.82}$$

This expression has a sharp maximum at $ka \approx (|\varepsilon_1|/2\beta_1)^{1/2}$ if ε_1 is negative, and a much lower maximum at $ka \approx (\varepsilon_1/\beta_1)^{1/4}$ if ε_1 is positive. The occurrence of a sharp maximum at a nonzero energy is a feature of resonance scattering in any partial wave with $l > 0$, in contrast with the s-wave case where the cross-section is highest at zero energy.

(*d*) *Physical Explanation:* In the example of the square well potential, to be considered below, we shall see explicitly that the vanishing of $1 + \gamma_0(0)$ happens when the parameters of the potential are such that a bound state with $l = 0$ occurs exactly at the 'brim' of the potential well, i.e. with zero energy. Under such circumstances, the wave function of an incident particle tends to get closer and closer to the (localized) bound state wave function as the energy E approaches zero, and the consequent distortion (as compared

[9] The symbol a is usually employed for the scattering length, but we have reserved it here for the range of the potential.

to the free particle wave function) manifests itself in a large amplitude for scattering, increasing monotonically as $E \to 0$. Precisely the same explanation of *s*-wave resonance holds for arbitrary attractive potentials.

In the case of higher partial waves, the particle moves in an effective potential, including the centrifugal potential, which has the form shown in Fig. 6.4 for the square well case. Suppose the potential is such that it would

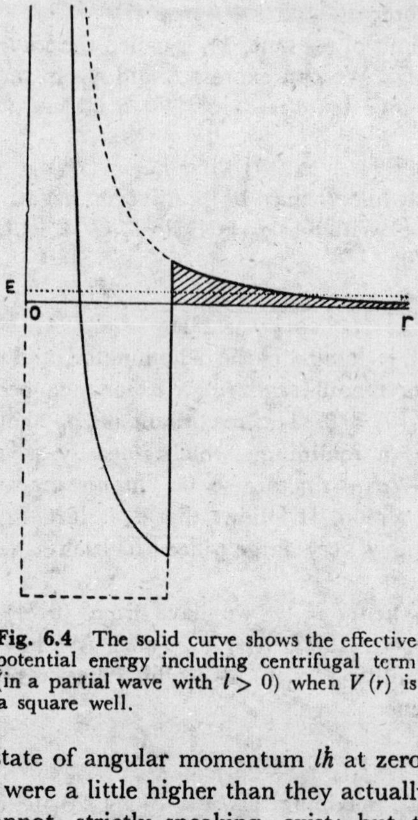

Fig. 6.4 The solid curve shows the effective potential energy including centrifugal term (in a partial wave with $l > 0$) when $V(r)$ is a square well.

have had a bound state of angular momentum $l\hbar$ at zero energy if the depth or width of the well were a little higher than they actually are. In such a case this bound state cannot, strictly speaking, exist; but the existence of the (centrifugal) potential barrier extending above $E = 0$ makes it possible for this state to appear with energy slightly above zero. This *virtual level* can accommodate a particle for a relatively long time though the particle will eventually 'tunnel' through the barrier and escape. Conversely, a particle incident on the potential with an enegy near that of the virtual level tends to be captured in this level. The consequent concentration of the wave function near the potential well is responsible for the resonance in the scattering at this energy.

(*e*) *The Ramsauer-Townsend Effect:* A rather strange phenomenon occurs if the depth of the potential is large enough that $\delta_0 = 180°$ at some small value of k for which $ka \ll 1$. The latter condition means that all phase shifts for $l > 0$ are practically zero; and with $\sin\delta_0$ also vanishing, *no scattering takes place.*

This is the *Ramsauer-Townsend effect*, actually observed in the scattering of electrons by certain rare gas atoms. Such atoms have a closed shell electronic configuration, so that the orbital electrons are compactly packed within a well-defined spherical region. The combined electrostatic forces due to the nucleus and orbital electrons are practically zero beyond this radius, and hence the incident electron 'sees' a strong attractive potential well of finite range. The observed drop-off of the scattering cross-section to extremely small values (despite the strong potential), at about 0.7 eV incident energy, has been explained in terms of the above effect. It should be observed that this kind of an effect cannot be produced by a repulsive potential, which suppresses the wave fuction in the interior region. This means that in order to make $\delta_0 = 180°$, the potential would have to be very strong; for, the phase shift is determined by the product of the potential and the interior wave function, as may be seen from Eq. 6.66. Such a strong potential would produce non-negligible phase shifts in the higher partial waves too, so that the cross-section does not vanish even with $\delta_0 = 180°$.

D. EXACTLY SOLUBLE PROBLEMS

6.14 Scattering by a Square Well Potential

This is an exactly soluble problem, and we shall use it to illustrate some of the general results derived so far. Since the potential in the interior region $r < a$ is a constant, $V = -V_0$, the radial wave function in this region is immediately obtained as

$$R_l(r) = \text{const. } j_l(\alpha r), \quad \alpha^2 = 2m\,(E + V_0)/\hbar^2 \tag{6.83}$$

(No admixture of n_l would be admissible since it diverges at $r = 0$ which lies within this region). Hence we have

$$\gamma_l(k) = \frac{j_l(\alpha a)}{\alpha a\, j_l'(\alpha a)} \tag{6.84}$$

The exact phase shifts are then obtained by using this result in Eq. 6.73.

In the particular case of the s-wave we have $j_0(\rho) = \rho^{-1} \sin\rho$, $n_0(\rho) = -\rho^{-1} \cos\rho$ and hence from Eq. 6.84, $\gamma_0(k) = (\alpha a \cot\alpha a - 1)^{-1}$. The scattering length is then found from Eq. 6.80 to be $L = a(1 - \tan\alpha^{(0)}a/\alpha^{(0)}a)$ where $\alpha^{(0)}$ stands for $\alpha(k)$ at $k = 0$, i.e. $\alpha^{(0)} = (2mV_0/\hbar^2)^{1/2}$. We have seen in the last section that an s-wave resonance occurs when $L \to \infty$. In the present case this happens obviously when $\alpha^{(0)}a = (n + \frac{1}{2})\pi$, $n = 0,1,2\ldots$ This is also precisely the condition for the occurrence of a bound state with $l = 0$ at zero energy. This observation confirms the explanation of resonance given in the last section. If we imagine $V_0\, a^2$ to be gradually increased, a new $l = 0$ level appears at $E = 0$ every time $[(\alpha^{(0)}a/\pi) - \frac{1}{2}]$ passes through an integer value and this is accompanied by a resonance in the low energy s-wave scattering. The discussion of the $l = 1$ partial wave and consideration of the exact expression for σ_l following from Eqs. 6.78, 6.73, and 6.84 for arbitrary energies, are left as exercise for the reader.

6.15 Scattering by a Hard Sphere

The case of the hard sphere ($V = +\infty$ for $r < a$, $V = 0$ for $r > a$) is simpler, but merits separate mention because of the special nature of the boundary condition at $r = a$. As may be seen from Sec. 2.8, $R_l'(r)$ is *not* continuous at $r = a$ where V has an infinite discontinuity. Further, $R_l(r)$ vanishes in the interior region where $V = +\infty$, and its continuity with the exterior wave function (6.70) leads to

$$\tan\delta_l = \frac{j_l(ka)}{n_l(ka)} \tag{6.85}$$

This is equivalent to letting $\gamma_l(k) \to 0$ in Eq. 6.73; then the factor in curly brackets in the low energy approximation (6.74) becomes -1. There is no possibility of resonance, and the low energy scattering cross-section σ becomes $\approx \sigma_0 = 4\pi a^2$. This is *four times* the classical cross-section presented by a sphere of radius a.

6.16 Scattering by a Coulomb Potential

This is a problem which is exactly soluble in closed form. We recall that what one has to do to solve the scattering problem is to determine those solutions of the Schrödinger equation which obey the right boundary conditions (incident plane wave + outgoing spherical wave). In the case of the Coulomb potential we already have complete sets of exact solutions of the Schrödinger equation, both in spherical polar coordinates (Sec. 4.18c)and in parabolic coordinates (Sec. 4.19). While the superposition of an infinite number of partial waves (spherical polar coordinates) is necessary to get the right asymptotic behaviour, it turns out that the desired behaviour is possessed by individual solutions (4.184) of the parabolic coordinate case (with suitable choice of the quantum numbers). We shall now verify that this is so, and obtain the scattering amplitude from the asymptotic form of (4.184).

(*a*) *Solution in Parabolic Coordinates.* Observe first of all that since the potential is spherically symmetric, we need consider only solutions independent of φ, i.e. those with $m = 0$. We reproduce them here for ease of reference, (dropping the index $m = 0$).

$$\left.\begin{aligned} u_{\lambda_1\lambda_2}(\xi,\eta) &= f_{\lambda_1}(\xi) f_{\lambda_2}(\eta), \\ f_{\lambda_1}(\xi) &= N_1\, e^{i\frac{1}{2}k\xi} F(\tfrac{1}{2} - \lambda_1,\ 1; -ik\xi) \\ f_{\lambda_2}(\eta) &= N_2\, e^{\frac{1}{2}ik\eta} F(\tfrac{1}{2} - \lambda_2,\ 1; -ik\eta) \end{aligned}\right\} \tag{6.86}$$

where[10]

$$\xi = r - z,\ \eta = r + z,\ i(\lambda_1 + \lambda_2) = \frac{\mu Z Z' e^2}{\hbar^2 k} \equiv \nu \tag{6.87}$$

The leading terms in the asymptotic form of the confluent hyper-geometric function $F(a,b;\rho)$, as seen from Eq. 4.154 are

$$F(a,b;\rho) \sim \frac{\Gamma(b)(-\rho)^{-a}}{\Gamma(b-a)} + \frac{\Gamma(b)\, \rho^{a-b} e^{\rho}}{\Gamma(a)} \tag{6.88}$$

[10] For the present purposes we take the Coulomb potential energy to be $ZZ'e^2/r$. In the case of the hydrogen-like atom, we had $Z' = -1$.

Therefore the asymptotic form of $u_{\lambda_1\lambda_2}$ as $|\xi|$, $|\eta| \to \infty$, is proportional to

$$e^{i\frac{1}{2}k(\eta+\xi)} \cdot \left\{\frac{(ik\xi)^{\lambda_1-\frac{1}{2}}}{\Gamma(\lambda_1+\frac{1}{2})} + \frac{(-ik\xi)^{-\frac{1}{2}-\lambda_1}}{\Gamma(\frac{1}{2}-\lambda_1)}e^{-ik\xi}\right\} \times$$
$$\times \left\{\frac{(ik\eta)^{\lambda_2-\frac{1}{2}}}{\Gamma(\lambda_2+\frac{1}{2})} + \frac{(-ik\eta)^{-\frac{1}{2}-\lambda_2}}{\Gamma(\frac{1}{2}-\lambda_2)}e^{-ik\eta}\right\} \tag{6.89}$$

We recall now that the only exponential functions appearing in the required asymptotic form (6.9) are $e^{ikz} \equiv e^{\frac{1}{2}ik(\eta-\xi)}$ and $e^{ikr} \equiv e^{\frac{1}{2}ik(\eta+\xi)}$. If (6.89) is to have this property, the coefficient of $e^{-ik\eta}$ in the last term must evidently vanish. This can happen only if $\Gamma(\frac{1}{2}-\lambda_2) = \infty$, and hence $\frac{1}{2}-\lambda_2 = -n$, $(n = 0, 1, 2, \ldots)$. With this, we have, from (6.87), $\lambda_1 = -i\nu - \lambda_2 = -i\nu - n - \frac{1}{2}$, so that $\xi^{-\frac{1}{2}-\lambda_1}$ in the first factor in (6.89) becomes $\xi^{i\nu+n}$. But with a positive real part n in the exponent, $\xi^{i\nu+n}$ would diverge at infinity; so the only value admissible for n is $n = 0$. Thus we finally conclude that $u_{\lambda_1\lambda_2}$ of (6.86) has the desired asymptotic behaviour if and only if

$$\lambda_1 = -\tfrac{1}{2} - i\nu, \quad \lambda_2 = \tfrac{1}{2} \tag{6.90}$$

When these are substituted in (6.86), the wave function becomes [in view of the fact that $F(0,b;\rho) = 1$],

$$u = \mathcal{N}e^{ikr}\, F(1+i\nu;\, 1;\, -ik\xi), \quad \xi = (r-z) = 2r\sin^2\tfrac{1}{2}\theta \tag{6.91}$$

Using (6.88) we find the leading terms in the asymptotic form of u to be

$$\frac{\mathcal{N}e^{\frac{1}{2}\nu\pi}}{\Gamma(1+i\nu)}\left\{e^{i[+\nu\ln k(r-z)]} + \frac{\Gamma(1+i\nu)}{i\Gamma(-i\nu)}\frac{e^{i[kr-\nu\ln k(r-z)]}}{k(r-z)}\right\} \tag{6.92}$$

where we have used the fact that $(-i)^i = (i)^{-i} = e^{\frac{1}{2}\pi}$. The logarithmic terms in the exponents come from $(k\xi)^{\pm i\nu}$. They are a consequence of the long 'reach' of the Coulomb potential, which prevents the phase of plane or spherical waves from becoming $(kz + \text{const.})$ or $(kr + \text{const.})$ even at infinite distances. For pure spherical waves, the r-dependent part of the phase in the asymptotic region is known from the solution in spherical polar coordinates. From Eq. 4.160 we find this to be $(kr - \nu\ln 2kr)$. So the Coulomb scattering amplitude is to be defined by writing the outgoing wave term in curly brackets as $r^{-1}f(\theta)\exp[ikr - i\nu\ln 2kr]$. On doing this we obtain

$$f(\theta) = \frac{\Gamma(1+i\nu)}{i\Gamma(-i\nu)}\frac{e^{-i\nu\ln(\sin^2\frac{1}{2}\theta)}}{2k\sin^2\frac{1}{2}\theta}$$
$$= \frac{\nu}{2k\sin^2\frac{1}{2}\theta}e^{-i\nu\ln(\sin^2\frac{1}{2}\theta)+i\pi+2i\eta_0} \tag{6.93}$$

where $\eta_0 = \arg\Gamma(1+i\nu)$. Hence the differential cross-section is found to be

$$\frac{d\sigma(\theta)}{d\Omega} = |f(\theta)|^2 = \frac{\nu^2}{4k^2\sin^4\frac{1}{2}\theta} = \left(\frac{ZZ'e^2}{2\mu v^2}\right)^2\operatorname{cosec}^4\tfrac{1}{2}\theta \tag{6.94}$$

This is just the Rutherford formula, derived by him from classical mechanics. It may be recalled that we obtained the same expression from the Born approximation (Example 6.1, p. 185). However, in the scattering of identical particles, the θ-dependent phase factor in Eq. 6.93 enters the cross-section in an essential fashion (Sec. 6.19).

(*b*) ***Partial Wave Expansion:*** Finally, the expansion of the Coulomb wave function (6.91) in terms of partial waves may be noted for completeness. Since we already know that the radial wave functions are given by Eq. 4.159 we have

$$u(r,\theta) = \Sigma R_l(r) P_l(\cos\theta) = \Sigma C_l\, r^l\, e^{ikr} F(l + 1 + i\nu,\, 2l + 2;\, -2ikr) P_l(\cos\theta) \quad (6.95)$$

The coefficients C_l are to be so chosen that Eq. 6.95 coincides with Eq. 6.91. To do this it is enough to compare typical terms in the two expressions. Ignoring the factor e^{ikr} which is common to both, we observe that in Eq. 6.95, the coefficient of $r^l P_l(\cos\theta)$ is C_l. (The F function contributes only to higher powers of r). In Eq. 6.91, after excluding e^{ikr}, the term involving r^l is obtained using the series expansion (4.152a) of the F function, as

$$\frac{\mathcal{N}\,\Gamma(1 + i\nu + l)\,(-ikr)^l}{\Gamma(1 + i\nu)\,(l!)^2}\left[\frac{(-)^l 2^l (l!)^2}{(2l)!} P_l(\cos\theta) + \ldots\right]$$

The square bracketed quantity stands for $(1 - \cos\theta)^l$. The factor multiplying $P_l(\cos\theta)$ therein is determined — knowing the expansion (D. 37) of $P_l(\cos\theta)$ — so as to get the correct coefficient $(-)^l$ for $\cos^l\theta$; the dots represent Legendre polynomials of lower degree which are of no interest here. The coefficient of $r^l P_l(\cos\theta)$ can be read off from the above expression; it must be equal to C_l. Substituting this value of C_l in Eq. 6.95, we finally have

$$u(r,\theta) = \sum_l \frac{\mathcal{N}\Gamma(1+i\nu+l)}{\Gamma(1+i\nu)(2l)!} (2ikr)^l\, e^{ikr}\, F(l+1+i\nu;\, 2l+2;\, -2ikr) P_l(\cos\theta) \quad (6.96)$$

Example 6.2—An interesting application of the Coulumb wave function (6.91) is in estimating the rate of nuclear reactions induced by bombarding heavy nuclei with charged particles (e.g. some other nuclei, like alpha particles). The reaction rate is proportional to the probability that the two nuclei find themselves close enough to interact. When the wavelength $\lambda = (2\pi/k)$ is much larger than the nuclear radii (low bombarding energy) the value of u within the interaction region does not differ appreciably from $u(0)$, and so the probability is proportional to $|u(0)|^2$. The reaction rate *per unit incident flux* is, therefore, proportional to $|u(0)|^2 = \mathcal{N}^2$ where $\mathcal{N}$ is presumed to be chosen so that the incident flux is indeed unity. Now, the flux is v times the absolute square of the coefficient of the incident part of the asymptotic wave function (6.92). So we must have

$$v\left|\frac{\mathcal{N}e^{\frac{1}{2}\nu\pi}}{\Gamma(1 + i\nu)}\right|^2 = 1 \text{ or } \mathcal{N}^2 = \frac{|\Gamma(1 + i\nu)|^2}{v\, e^{\nu\pi}} = \frac{2\nu\pi}{v\,(e^{2\nu\pi} - 1)} \quad (6.97)$$

In the last step we have used the standard result that $|\Gamma(1 + ix)|^2 = \pi x/\sinh \pi x$. The above value of $\mathcal{N}^2$ determines the reaction rate. When ν is positive (i.e. the colliding particles have the same sign of charge) and $\nu\pi \gg 1$, we have $\mathcal{N}^2 \approx (2\nu\pi/v)\, e^{-2\nu\pi}$. The exponential $e^{-2\nu\pi} = \exp\,[-2\pi ZZ'e^2/\hbar v]$ depends very sensitively on v. This factor, called the *Gamow factor*, has the dominant role in determining the rate of many nuclear reactions at low energies of bombardment.

E. MUTUAL SCATTERING OF TWO PARTICLES

6.17 Reduction of the Two-Body Problem: The Centre of Mass Frame

It is a well-known fact in classical mechanics that if a system of particles is not subject to any external forces (i.e. forces other than the mutual interactions of the particles themselves) then the centre of mass of the system moves with constant velocity $\mathbf{v}_c$. In a reference frame which moves along with the same velocity $\mathbf{v}_c$, the centre of mass of the particle-system appears to remain at rest. Any such reference frame is called a *centre of mass frame*. In the particular case of a two-body system, the motion as seen from the centre of mass frame is especially simple since, with the centre of mass remaining fixed, the instantaneous position of either particle completely determines that of the other. The two-body problem then becomes effectively a one-body problem. The same simplification is possible in quantum mechanics too, as we shall now see.

Consider a two-particle system with the Hamiltonian

$$H = -\frac{\hbar^2}{2m_1}\nabla_1^2 - \frac{\hbar^2}{2m_2}\nabla_2^2 + V(\mathbf{x}_1 - \mathbf{x}_2) \tag{6.98}$$

In analogy with the classical case, let us define the *centre of mass* and *relative* coordinate variables as

$$\mathbf{X} = \frac{m_1\mathbf{x}_1 + m_2\mathbf{x}_2}{m_1 + m_2} \text{ and } \mathbf{x} = \mathbf{x}_1 - \mathbf{x}_2 \tag{6.99}$$

respectively. We denote the corresponding gradient operators by

$$\nabla_X \equiv \left(\frac{\partial}{\partial X}, \frac{\partial}{\partial Y}, \frac{\partial}{\partial Z}\right) \text{ and } \nabla = \left(\frac{\partial}{\partial x}, \frac{\partial}{\partial y}, \frac{\partial}{\partial z}\right) \tag{6.100}$$

respectively. Both $\nabla_1 \equiv (\partial/\partial x_1, \partial/\partial y_1, \partial/\partial z_1)$ and ∇_2 can be expressed in terms of these.

$$\frac{\partial}{\partial x_1} = \frac{\partial X}{\partial x_1}\frac{\partial}{\partial X} + \frac{\partial x}{\partial x_1}\frac{\partial}{\partial x} = \frac{m_1}{m_1 + m_2}\frac{\partial}{\partial X} + \frac{\partial}{\partial x}$$

$$\frac{\partial^2}{\partial x_1^2} = \left(\frac{m_1}{m_1 + m_2}\right)^2 \frac{\partial^2}{\partial X^2} + \frac{2m_1}{m_1 + m_2}\frac{\partial^2}{\partial X \partial x} + \frac{\partial^2}{\partial x^2}, \text{ etc.} \tag{6.101}$$

so that

$$\nabla_1^2 = \left(\frac{m_1}{m_1 + m_2}\right)^2 \nabla_X^2 + \frac{2m_1}{m_1 + m_2}\nabla_X \cdot \nabla + \nabla^2 \tag{6.102}$$

and similarly

$$\nabla_2^2 = \left(\frac{m_2}{m_1 + m_2}\right)^2 \nabla_X^2 - \frac{2m_2}{m_1 + m_2}\nabla_X \cdot \nabla + \nabla^2 \tag{6.103}$$

On substituting Eqs. 6.102 and 6.103 into Eq. 6.98 it becomes

$$H = -\frac{\hbar^2}{2M}\nabla_X^2 - \frac{\hbar^2}{2\mu}\nabla^2 + V(\mathbf{x}) \tag{6.104}$$

with

$$M = m_1 + m_2, \quad \mu = \frac{m_1 m_2}{m_1 + m_2} \tag{6.105}$$

In this form, H is the sum of two commuting parts depending on independent sets of variables. The first part, $H_{cm} = -(\hbar^2/2M)\nabla_X^2$, describes the free motion of the centre of mass; it represents the kinetic energy of the system (with total mass M) as a whole. The second part, $H_{rel} = -\frac{\hbar^2}{2\mu}\nabla^2 + V(\mathbf{x})$ depends on the relative variable $\mathbf{x}$ only, and is identical in form with the Hamiltonian of a single particle of mass μ moving in a potential $V(\mathbf{x})$. We call $\mu \equiv m_1m_2/(m_1 + m_2)$ the *reduced mass* of the system.

A complete set of eigenfunctions of H is now given by

$$(2\pi)^{-3/2} \exp(i\mathbf{K}.\mathbf{X})\, u_n(\mathbf{x}) \tag{6.106}$$

(for all $\mathbf{K},n$). The factor $(2\pi)^{-3/2} \exp[i\mathbf{K}.\mathbf{X}]$ describes the motion of the centre of mass with momentum $\hbar\mathbf{K}$, and the $u_n(\mathbf{x})$ are eigenfunctions of H_{rel}[11] given by

$$-\frac{\hbar^2}{2M}\nabla_X{}^2[(2\pi)^{-3/2}\exp(i\mathbf{K}.\mathbf{X})] = \frac{\hbar^2K^2}{2M}[(2\pi)^{-3/2}\exp(i\mathbf{K}.\mathbf{X})] \tag{6.107}$$

$$\left[-\frac{\hbar^2}{2\mu}\nabla^2 + V(\mathbf{x})\right]u_n(\mathbf{x}) = E_nu_n(\mathbf{x}) \tag{6.108}$$

Thus the problem reduces essentially to that of solving the single-particle Schrödinger equation (6.108). We have already referred to this fact in Chapter 4 in connection with the treatment of the hydrogen atom problem. Our concern now is to see how, once the solution of the reduced problem is obtained, the results for the two-particle problem may be recovered from it. More specifically, we show how the cross-sections for the mutual scattering of two particles interacting through a potential $V(\mathbf{x}_1 - \mathbf{x}_2)$ may be found, when the results for the scattering of one particle by a fixed potential $V(\mathbf{x})$ are known.

6.18 Transformation from Centre of Mass to Laboratory Frame of Reference

Let us consider the usual situation when one of the two colliding particles, say of mass m_2, is initially at rest. This 'target' is hit by the 'projectile' (of mass m_1) whose initial speed will be denoted by v_0 (Fig. 6.5). The centre of mass moves in the same direction as m_1 with speed

$$v_c \equiv \frac{(m_1v_0 + m_2.0)}{m_1 + m_2} = \frac{\mu v_0}{m_2} \tag{6.109}$$

Its motion remains undisturbed by the collision. When viewed from a reference frame moving along with the centre of mass, the initial speed of m_1 is $v_0 - v_c = m_2v_0/(m_1 + m_2) = \mu v_0/m_1$, so that its momentum is μv_0. The other particle has exactly the same momentum, but in the opposite direction (Fig. 6.5). After collision the particles move directly away from each other, at an angle θ to the respective initial directions, while the magnitudes of their momenta remain unchanged. The total energy in the centre of mass frame is

$$\tfrac{1}{2}(\mu v_0)^2/m_1 + \tfrac{1}{2}(\mu v_0)^2/m_2 = \tfrac{1}{2}\mu v_0{}^2$$

Quantum mechanically, the motion in the centre of mass frame is described by the Schrödinger equation (6.108) in the relative coordinate $\mathbf{x}$, as already noted. Since this equation is *identical in form* to the equation (6.10) for a particle in a fixed potential $V(\mathbf{x})$, all the formulae for $f(\theta)$, $d\sigma/d\Omega$ etc. obtained so far in

[11] In the centre of mass frame, $\mathbf{K} = 0$ by definition, and then Eq. 6.106 reduces to $u_n(\mathbf{x})$. Thus the $u_n(\mathbf{x})$ are eigenfunctions of the two-particle system as seen in the centre of mass frame.

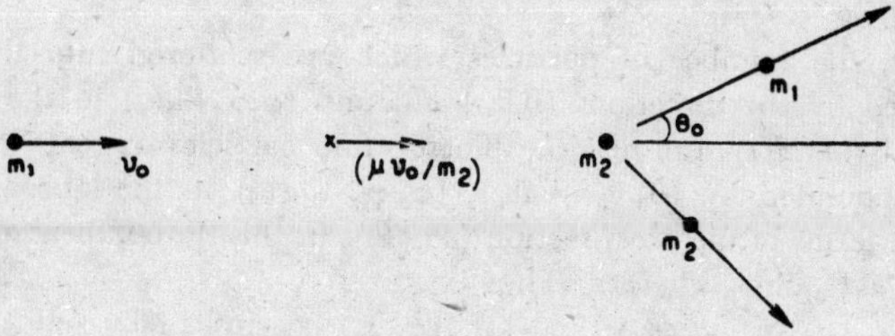

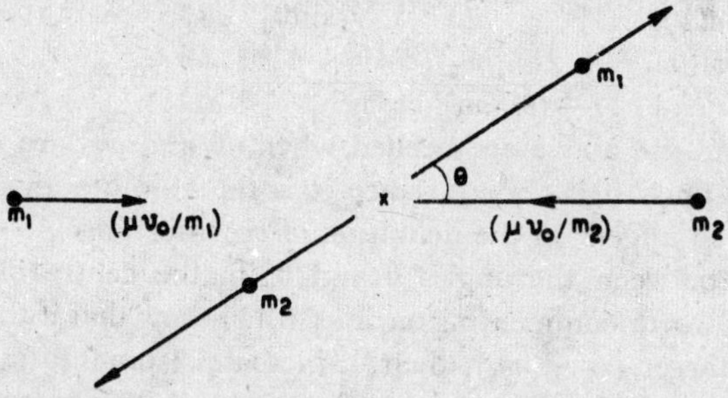

Fig. 6.5 The parameters of a scattering event in the laboratory frame (above) and the centre of mass frame (below).

this chapter are seen to hold equally well for the *mutual scattering* of two particles, as seen in the *centre of mass frame*, provided that m in Eq. 6.10 is taken to be the *reduced mass* μ of the two particles. The values to be used for the wave number k and the energy E, in these formulae, are determined by the momentum and energy in the centre of mass frame, found in the last pargaraph. We have

$$k = (\mu v_0/\hbar) \text{ and } E = \tfrac{1}{2}\mu v_0^2 = \hbar^2k^2/2\mu \tag{6.110}$$

Now that the cross-sections in the centre of mass frame are known, the next question is how to obtain these in the laboratory frame in which the measurements are actually made. For this we need the relation between the scattering angles θ_0 and θ in the two frames. The key to obtaining this is the observation that in going from the centre of mass frame to the laboratory, velocity components *transverse* (perpendicular) to the initial direction of m_1 remain *unchanged*, while the *longitudinal* component (in the initial direction of m_1) is increased by the speed $(\mu v_0/m_2)$ of the centre of mass. After scattering through an angle θ in the centre of mass frame, particle 1 has transverse and longitudinal velocity components equal to $(\mu v_0/m_1)\sin\theta$ and $(\mu v_0/m_1)\cos\theta$ respectively. The corresponding quantities in the laboratory frame are, therefore, $(\mu v_0/m_1)\sin\theta$ and $(\mu v_0/m_2) + (\mu v_0/m_1)\cos\theta$. The ratio of these should of course be $\tan\theta_0$, where θ_0 is the scattering angle in the laboratory. Thus

$$\tan\theta_0 = \frac{\sin\theta}{(m_1/m_2) + \cos\theta} \tag{6.111a}$$

and

$$\frac{d\theta}{d\theta_0} = \frac{(m_1/m_2)^2 + 2(m_1/m_2)\cos\theta + 1}{(m_1/m_2)\cos\theta + 1} \tag{6.111b}$$

The azimuthal angles in the two frames are however equal: $\varphi_0 = \varphi$.

Now, the number of particles which are scattered into the solid angle bounded by the directions $(\theta, \theta + d\theta)$ and $(\varphi, \varphi + d\varphi)$ in the centre of mass frame is $(d\sigma/d\Omega) \sin\theta\, d\theta\, d\varphi$. These same particles appear within the solid angle bounded by $(\theta_0, \theta_0 + d\theta_0)$, $(\varphi_0, \varphi_0 + d\varphi_0)$ in the laboratory frame, so that in terms of the cross-section $(d\sigma/d\Omega)_0$ in this frame, their number is equal to $(d\sigma/d\Omega)_0 \sin\theta_0\, d\theta_0 d\varphi_0$. Thus

$$\left(\frac{d\sigma}{d\Omega}\right)_0 (\theta_0, \varphi_0) = \frac{d\sigma}{d\Omega}(\theta, \varphi) \cdot \frac{\sin\theta}{\sin\theta_0} \left| \frac{d\theta}{d\theta_0} \right|$$

$$= \frac{[(m_1/m_2)^2 + 2(m_1/m_2) \cos\theta + 1]^{3/2}}{|1 + (m_1/m_2) \cos\theta|} \frac{d\sigma}{d\Omega}(\theta, \varphi) \tag{6.112}$$

The modulus signs in the above are needed when $d\theta$ and $d\theta_0$ are of opposite signs (which can happen if $m_1 > m_2$), since it is the absolute magnitudes of the solid angles which appear in the definition of cross-sections.

The relationship between the angles θ and θ_0 in the centre of mass and laboratory frames is worth commenting on. Eqs. 6.111 show that if the projectile is lighter than the target $(m_1 < m_2)$, then θ_0 increases from 0 to π monotonically with θ passing the value $\frac{1}{2}\pi$ when $\cos\theta = -(m_1/m_2)$. The case $m_1 = m_2$ is of particular interest in connection with the scattering of identical particles. Eq. 6.111 then yields $\theta_0 = \frac{1}{2}\theta$, so that it never exceeds $\frac{1}{2}\pi$. Further Eq. 6.112 simplifies to

$$\left(\frac{d\sigma}{d\Omega}\right)_0 (\theta_0, \varphi_0) = 4 \cos\tfrac{1}{2}\theta \cdot \frac{d\sigma}{d\Omega}(\theta, \varphi), \quad \theta_0 = \tfrac{1}{2}\theta, \ \varphi_0 = \varphi \tag{6.112a}$$

When $(m_1/m_2) > 1$, θ_0 increases to a maximum and then decreases to 0 as θ goes from 0 to π. The maximum value, $(\tan\theta_0)_{\max} = [(m_1/m_2)^2 - 1]^{1/2}$, occurs when $\cos\theta = -(m_2/m_1)$. Though the relations between the angles and between the differential cross-sections themselves are rather complex, it should be noted that the *total* cross-section is the *same* in the two reference frames, since the total number of collisions is independent of the manner in which the process is described.

6.19 Collisions Between Identical Particles

(*a*) *Classical:* When the colliding particles are identical, one cannot decide simply from the observation of a particle reaching a recording device whether it is the incident particle (after scattering) or the recoiling target particle. Classical mechanics permits one, nevertheless, to follow the trajectories of the two particles throughout the scattering process and thus decide which of them has reached the recorder. Ultimately if one is not interested in keeping the distinction between the two, one can simply add up the numbers of projectile and target particles received by the device. Evidently, the cross-section for the observation of *either* the projectile *or* the recoil particle in some direction (θ, φ) is then just the sum of the two separate cross-sections, and is given by

$$\frac{d\sigma}{d\Omega}(\theta, \varphi) + \frac{d\sigma}{d\Omega}(\pi - \theta, \varphi + \pi) \tag{6.113}$$

in the centre of mass frame (since the two particles go off in opposite directions).

(*b*) *Quantum Mechanical:* The situation is quite different in quantum mechanics. There is no possibility here of distinguishing between the two particles by means of their trajectories, since any attempt to make the trajectories sufficiently well-defined would seriously disturb the scattering process. (See Sec. 1.18). Therefore there is no reason why the cross-section should be separable into a sum of contributions from incident and struck particles as in (6.113). In fact, it is not. This is a consequence of the requirement that the wave function of a system of identical particles should be either symmetric or antisymmetric under interchange of the particles (Sec. 3.16). Now, the interchange $\mathbf{x}_1 \longleftrightarrow \mathbf{x}_2$ results in the transformation $\mathbf{x} \rightarrow -\mathbf{x}$ of the relative coordinate. Thus the wave function $u(\mathbf{x})$ in the centre of mass system must be such that $u(-\mathbf{x}) = \pm u(\mathbf{x})$. If it does not already have this property, it must be symmetrized or antisymmetrized, i.e. in the case of identical particles we can only accept the combinations $u(\mathbf{x}) \pm u(-\mathbf{x})$. Correspondingly, the asymptotic form goes over from (6.9) to its symmetrized or antisymmetrized form:

$$(e^{ikz} \pm e^{-ikz}) + [f(\theta,\varphi) \pm f(\pi - \theta, \ \varphi + \pi)]\, r^{-1} e^{ikr} \tag{6.114}$$

It is the quantity

$$f_\pm(\theta,\varphi) \equiv f(\theta,\varphi) \pm f(\pi - \theta, \ \varphi + \pi) \tag{6.115}$$

which now gives the amplitude for the scattering of either the incident or the struck particle in the direction (θ,φ). The differential scattering cross-section is the absolute square of this quantity:

$$\frac{d\sigma}{d\Omega}(\theta,\varphi) = |f_\pm(\theta)|^2 = |f(\theta,\varphi)|^2 + |f(\pi - \theta, \ \varphi + \pi)|^2$$
$$\pm 2Re\,[f(\theta,\varphi) f^*(\pi - \theta, \ \varphi + \pi)] \tag{6.115a}$$

The last term represents a truly quantum mechanical effect, which may be described as an interference between the amplitudes for scattering of the projectile and target in the direction (θ,φ). Under conditions where this term is negligible, the cross-section consisting of the remaining two terms coincides with the classical expression (6.113), as it should.

Finally, it should be observed that to calculate the total cross-section from (6.115a), integration should be carried out only over directions in the forward hemisphere, wherein θ varies from 0 to $\frac{1}{2}\pi$, (*or* over the backward hemisphere). Since the counting of both the projectile and target particles is covered by (6.115a) — or (6.113) in the classical case — integration over *all* directions would obviously amount to double counting. Incidentally, since (6.115a) is symmetric with respect to $\theta = \frac{1}{2}\pi$, it may be verified with the aid of (6.112a) that the cross-section per unit θ_0 in the laboratory frame, namely $(d\sigma/d\theta_0) \equiv (d\sigma/d\Omega)_0 \sin\theta_0$, is symmetric with respect to $\theta_0 = \frac{1}{4}\pi$.

(*c*) *Spinless Particles and Spin-½ Particles:* In the case of spinless identical particles we recall from Sec. 3.16 that the wave function must be symmetric. The choice of the positive sign in (6.115a) is then mandatory.

To see the essential features of the case of identical particles with spin, consider the example of proton-proton scattering. At low energies the electrostatic repulsion prevents the protons from approaching close enough to feel the specifically nuclear part of the interaction. The scattering is then due to the Coulomb interaction, which is independent of spin. Nevertheless, the existence of spin affects the cross-section through the requirement of anti-symmetrization of the wave function. Anticipating the results of Sec. 8.8b, we observe that the spin part of the wave function of a two-proton system can be either symmetric or anti-symmetric under interchange of the protons: there are *three* symmetric spin functions and *one* antisymmetric.[12] With any of the former one has to associate an antisymmetric orbital wave function, corresponding to which the scattering cross-section (in the centre of mass frame) is $|f_-(\theta)|^2$. With the latter, one has to associate a symmetric orbital wave function, and cross-section $|f_+(\theta)|^2$. Now, except in experiments with polarized proton beams and targets, there is no reason why any of the four spin configurations should be preferred to any of the others: they should occur equally often. So the scattering cross-section to be expected when nothing is specified about the spin orientations is

$$\frac{d\sigma}{d\Omega}(\theta) = \tfrac{3}{4}\,|f_-(\theta)|^2 + \tfrac{1}{4}\,|f_+(\theta)|^2 \tag{6.116}$$

On substituting for $f(\theta)$ the Coulomb scattering amplitude (6.93) with μ = half the proton mass and $\nu = e^2/\hbar v$, we obtain

$$\frac{d\sigma}{d\Omega}(\theta) = \left(\frac{e^2}{2\mu v^2}\right)^2 \{\operatorname{cosec}^4 \tfrac{1}{2}\theta + \sec^4 \tfrac{1}{2}\theta - \operatorname{cosec}^2 \tfrac{1}{2}\theta \sec^2 \tfrac{1}{2}\theta \cos[(e^2/\hbar v) \ln (\tan^2 \tfrac{1}{2}\theta)]\} \tag{6.117}$$

This is the Mott scattering formula.

PROBLEMS

1. Give a qualitative discussion of scattering by a lattice of identical atoms, using the Born approximation.
2. Find a general expression for the phase shifts produced by a potential $V(r) = a/r^2$ when $a > 0$. Determine whether the total cross-section is finite. Comment on the case when $a < 0$.
3. Assuming the interaction of an electron with a rare gas atom to be given by a square-well potential of radius 10^{-8} cm, calculate the effective depth V_0 of the well in order that the Ramsauer-Townsend effect should take place (in the zero energy limit). What is the phase shift δ_1 for this potential if the electron energy is 0·7 eV?
4. The deuteron is the only bound state of the neutron-proton system, and its spin wave function is triplet. Assuming the neutron-proton interaction in triplet spin states to be given by a square well potential of radius $a = 2{\cdot}1 \times 10^{-13}$ cm, show that the depth of the well should be 34 MeV to account for the binding energy (2·23 MeV) of the deuteron.
5. The observed zero-energy neutron-proton scattering cross-section (which includes contributions from both singlet and triplet spin states) is $20{\cdot}4 \times 10^{-24}$ cm². Evaluate the triplet contribution using the potential of Prob. 4, and hence determine the singlet cross-section σ_s.

[12] These are referred to as 'triplet' and 'singlet' respectively, and are identified by subscripts t, s, as in σ_t, σ_s.

6. If the potential $V(r)$ is complex with negative imaginary part, there is absorption of particles (Prob. 2, Chapter 2). The phase shifts δ_l in this case are complex, and the elastic scattering cross-section is $\sigma_{el} = (\pi/k^2)\,\Sigma\,(2l+1)\,|\,1 - e^{2i\delta_l}\,|^2$. Verify this. Defining the absorption cross-section as the net inward flux,

$$\sigma_{abs} = -v^{-1} \lim_{r \to \infty} \int S_r r^2 d\Omega$$

where $\qquad S_r = (\hbar/2im)\,[u^*\,\partial u/\partial r - (\partial u^*/\partial r)\,u]$,

show that $\qquad \sigma_{abs} = (\pi/k^2)\,\Sigma\,(2l+1)\,(1 - |\,e^{2i\delta_l}\,|^2)$.

Also show that the differential cross-section for elastic cross-section in the forward direction cannot be less than $(k\sigma_{tot}/4\pi)^2$ where $\sigma_{tot} = \sigma_{el} + \sigma_{abs}$.

7. The asymptotic form (6.51) of $R_l(r)$ holds also for negative energies (say $k = i\varkappa$, $\varkappa > 0$) but with δ_l complex. Show that the bound state energies (for which $R_l \to$ const. $e^{-\varkappa r}$) are determined by $e^{-i\delta_l} = 0$. If there is an s-wave bound state of nearly zero energy, show by applying (6.79) for $k = i\varkappa$ corresponding to this state that the relation $(\varkappa L)^{-1} + \frac{1}{2}\varkappa r_{eff} = 1$ holds between the bound state and scattering parameters.

8. Show that for a central potential $V(r)$, the phase shifts in the WKB approximation are given by

$$\delta_l = \lim_{r \to \infty} \left\{ \int_{r_1}^{r} \left[k^2 - \frac{2mV(r')}{\hbar^2} - \frac{(l+\frac{1}{2})^2}{r'^2} \right]^{1/2} dr' - kr \right\} + (l + \tfrac{1}{2})\,\frac{\pi}{2}$$

where r_1 is the classical turning point at energy $E = (\hbar^2 k^2/2m)$ in the effective potential including the centrifugal term.

7 Representations, Transformations and Symmetries

We return now to further development of the formalism of quantum mechanics. The particular aspects which will concern us here are, firstly the geometrical concept of quantum states as vectors in an infinite-dimensional vector space, and secondly their representations with reference to bases in the space. Changes of representation, especially those induced by transformations of coordinates in the ordinary space, are of considerable importance and are dealt with in the latter half of the chapter.

7.1 Quantum States; State Vectors and Wave Functions

It was observed in Sec. 3.2 that the values of the Schrödinger wave function $\psi(\mathbf{x})$ at various points $\mathbf{x}$ are akin to the components of a vector. While the components are *numbers* which depend on the choice of coordinate system or basis in the vector space, the vector itself is an abstract geometrical object with an existence and significance independent of coordinate systems. In the quantum mechanical context, the abstract *mathematical* object which is associated with the *physical* entity, the quantum state, is the *state vector*. The different kinds of wave functions for a given state (in the configuration space, in momentum space, in the 'A-representation' mentioned at the end of Sec. 3.9, ...) are simply the sets of components of the state vector as referred to different 'coordinate sytems' or bases in the state vector space. The notation devised by Dirac (for abstract state vectors, their scalar products, etc.) greatly facilitates the explicit demonstration of this fact as well as the exploration of the interrelation of different representations. The notation is very general, and is particularly suitable for the handling of systems which have no classical counterpart, e.g. spin angular momentum. The next two sections present, in this abstract language, some aspects of the formalism of quantum mechanics which had been introduced earlier (Chapter 3) in terms of the Schrödinger representation.

7.2 The Hilbert Space of State Vectors; Dirac Notation

(*a*) *State Vectors and their Conjugates:* Dirac introduced the symbol $|\psi\rangle$ to denote an abstract state vector as distinct from its representation, the wave function. (This is the counterpart of the notation **V** for a vector in three dimensions). The state vectors constitute a complex vector space, i.e. if $|\psi\rangle, |\chi\rangle$ are state vectors, so is $c\,|\psi\rangle + c'\,|\chi\rangle$ for any complex numbers c and c'. All vectors $c\,|\psi\rangle$, $c \neq 0$, which have the same 'direction' as $|\psi\rangle$, correspond to the *same* physical state. The set of all such vectors constitute what is called a *ray* in the space. Therefore, we say that any physical state is represented by a ray (rather than a single vector) in the vector space. Nevertheless, it is customary to use the word vector (rather imprecisely) for ray, and we shall follow this custom.

Corresponding to every vector $|\psi\rangle$ is defined a *conjugate vector* $\overline{|\psi\rangle}$, for which Dirac used the notation $\langle\psi|$.

$$\langle\psi| \equiv \overline{|\psi\rangle}, \quad \overline{\langle\psi|} \equiv |\psi\rangle \tag{7.1}$$

The relation between any vector and its conjugate may be visualized readily if we consider examples of their representations, anticipating later developments. In the Schrödinger representation $|\psi\rangle \to \psi(\mathbf{x})$, $\langle\psi| \to \psi^*(\mathbf{x})$; and in an 'A-representation' where $|\psi\rangle$ is represented by an infinite-component *column* with elements c_a, $\langle\psi|$ goes over into a *row* with elements c_a^*. Clearly, the correspondence between $|\psi\rangle$ and $\langle\psi|$ is not a linear one, i.e. if c is a complex number, the conjugate of $c\,|\psi\rangle$ is *not* $c\,\langle\psi|$. Instead,

$$\overline{c\,|\psi\rangle} = c^*\,\overline{|\psi\rangle} = c^*\,\langle\psi| \tag{7.2}$$

(*b*) *Norm and Scalar Product:* The squared length or norm of the vector $|\psi\rangle$ is denoted by $\langle\psi|\psi\rangle$. By definition, $\langle\psi|\psi\rangle$ is a real non-negative number:

$$\langle\psi|\psi\rangle \geqslant 0, \quad \langle\psi|\psi\rangle = 0 \text{ if and only if } |\psi\rangle = 0. \tag{7.3a}$$

More generally, when any vector $|\psi\rangle$ and the conjugate $\langle\phi|$ of some vector $|\phi\rangle$ are juxtaposed in such a way as to form a *closed* bracket, $\langle\phi|\psi\rangle$, this symbol is understood to mean a single (real or complex) number which is defined to be the *scalar product* of $|\phi\rangle$ and $|\psi\rangle$, taken in that order.

Vectors of the type $\langle\psi|$ which go to make up the *first half* of such *closed brackets* are called *bra vectors* and those of the type $|\psi\rangle$ constituting the second half are called *ket vectors*. The conjugate of a bra vector is a ket vector, and vice versa.

In the Schrödinger representation, $\langle\phi|\psi\rangle \to \int \phi^*(\mathbf{x})\,\psi(\mathbf{x})d\tau$. Note that $\langle\phi|\psi\rangle \neq \langle\psi|\phi\rangle$; instead

$$\langle\phi|\psi\rangle = \langle\psi|\phi\rangle^* \tag{7.3b}$$

The scalar product of two vectors is linear in the second vector:

$$\langle \phi | (c|\psi\rangle + c'|\chi\rangle) = c\langle \phi|\psi\rangle + c'\langle \phi|\chi\rangle \tag{7.3c}$$

Any vector space (in general infinite dimensional) such that between any two vectors of the space a scalar product with the properties (7.3) is defined, is called a *Hilbert Space*. The space of ket vectors (or bra vectors) representing quantum mechanical states is thus a Hilbert space.[1]

(*c*) *Basis in Hilbert Space:* Let $|\phi_1\rangle, |\phi_2\rangle, \ldots$ be a linearly independent set of states in the Hilbert space, i.e. no linear combination of these vectors vanishes except when all the coefficients vanish. Let the set be also a *complete* one, in the sense that any arbitrary vector $|\psi\rangle$ of the space can be expressed in the form

$$|\psi\rangle = c_1|\phi_1\rangle + c_2|\phi_2\rangle + \ldots = \sum_i c_i|\phi_i\rangle \tag{7.4}$$

Then the set $|\phi_1\rangle, |\phi_2\rangle, \ldots$ can be chosen as a *basis* or 'coordinate system' for the Hilbert space (just as one can use as a basis for the ordinary three-dimensional space any set of three linearly independent, i.e. noncoplanar, vectors). The set of numbers $c_1, c_2, \ldots$ in Eq. 7.4 are the 'components' of $|\psi\rangle$ with respect to this basis, and it provides a *representation of* $|\psi\rangle$.

7.3 Dynamical Variables and Linear Operators

(*a*) *Abstract Operators; The Quantum Condition:* To every dynamical variable A, there corresponds a linear operator $\hat{A}$ in the Hilbert space. The operator has the effect of taking each state vector $|\psi\rangle$ into a new vector denoted by $\hat{A}|\psi\rangle$. Linearity means that

$$\hat{A}(c|\psi\rangle + c'|\chi\rangle) = c\hat{A}|\psi\rangle + c'\hat{A}|\chi\rangle \tag{7.5}$$

The operator $\hat{A}$ is completely defined when its effect on every vector of the space is specified. Actually it is sufficient if its effect on any complete set of vectors $|\phi_1\rangle, |\phi_2\rangle \ldots$ of the space is known, i.e. if $\hat{A}|\phi_1\rangle$, $\hat{A}|\phi_2\rangle, \ldots$ are known. The effect on an arbitrary $|\psi\rangle$ may then be deduced from Eqs. 7.4 and 7.5.

It is to be emphasized that just as the abstract vector $|\phi\rangle$ is a more fundamental object than any of its wave function representations, the operator A is also an abstract object which is to be distinguished from its representations. As a reminder of this distinction we use the carat or 'hat' (^) on the abstract operators while their representations will be written without this symbol. It is not possible to write down any concrete expression for A itself though its representations do have specific forms[2] (such as $-i\hbar\nabla$ for $\hat{\mathbf{p}}$ in the Schrödinger representation). Our main aim in this chapter is to demonstrate how one goes over from the abstract operator to its representations, and incidentally to

[1] This is not strictly true because a Hilbert space admits only vectors of finite norm while states of infinite norm are inescapable in quantum mechanics. Nevertheless, the term Hilbert space is commonly applied to the space of quantum states, as implying just the properties (7.3).

[2] This is not peculiar to quantum mechanics. It is true of operators in any vector space. For example, it is well known that in an anisotropic medium, the electric displacement $\mathbf{D}$ does not have the same direction as the electric field $\mathbf{E}$. They are related by a linear operator $\varkappa$ (the dielectric tensor): $\mathbf{D} = \varkappa\mathbf{E}$. If one represents the vectors by their components with respect to a coordinate system, then and only then does $\varkappa$ acquire a concrete operator form (as a 3×3 matrix).

bring out the fact that different representations of the same abstract operator may look completely different (such as a differential operator and a matrix operator).

In this abstract formalism, the commutation rule

$$[\hat{x},\hat{p}] = i\hbar \tag{7.6}$$

between any position variable and its canonically conjugate momentum is not deduced from anything, but is a fundamental postulate. This is *the quantum condition* which is the basis of quantum mechanics. We shall see below (Sec. 7.5) that the specific forms **x** and $-i\hbar\nabla$ for $\hat{\mathbf{x}}$ and $\hat{\mathbf{p}}$ in the Schrödinger representation can indeed be derived from (7.6).

(*b*) *The Adjoint; Self-adjointness:* To any linear operator $\hat{A}$ there corresponds an adjoint $\hat{A}^\dagger$. It is defined by the statement that for arbitrary $|\psi\rangle$ and $|\phi\rangle$, the scalar product of $|\psi\rangle$ with $\hat{A}^\dagger|\phi\rangle$ is equal to the complex conjugate of the scalar product of $|\phi\rangle$ with $\hat{A}|\phi\rangle$.

$$\langle\psi|\hat{A}^\dagger|\phi\rangle = \langle\phi|\hat{A}|\psi\rangle^* \tag{7.7}$$

Since $|\phi\rangle$, $|\psi\rangle$ are arbitrary, the complex conjugate of this equation states that

$$(\hat{A}^\dagger)^\dagger = \hat{A} \tag{7.8}$$

Eq. 7.7 also defines the action of operators *to the left* on bra vectors. For, if we denote $\hat{A}^\dagger|\phi\rangle$ by $|\chi\rangle$, the left hand side of Eq. 7.7 is $\langle\psi|\chi\rangle$ which is equal to $\langle\chi|\psi\rangle^*$. On comparing this with the right hand side of Eq. 7.7 we obtain $\langle\phi|\hat{A} = \langle\chi| = \overline{|\chi\rangle}$, i.e.

$$\langle\phi|\hat{A} = \overline{\hat{A}^\dagger|\phi\rangle} \tag{7.9}$$

This equation may also be used as an alternative definition of the adjoint. The reader may verify from this, by considering some $|\phi\rangle$ of the form $\hat{B}^\dagger|\phi'\rangle$ that

$$(\hat{B}\hat{A})^\dagger = \hat{A}^\dagger\hat{B}^\dagger \tag{7.10}$$

If a particular operator $\hat{A}$ is such that $\hat{A}^\dagger = \hat{A}$, i.e.

$$\langle\phi|\hat{A}|\psi\rangle = \langle\psi|\hat{A}|\phi\rangle^* \tag{7.11}$$

for all $|\phi\rangle$, $|\psi\rangle$, then $\hat{A}$ is a self-adjoint operator.

(*c*) *Eigenvalues and Eigenvectors:* An eigenvector of any linear operator $\hat{A}$ is a vector whose 'direction' is not changed by the action of the operator, i.e. one which merely gets converted into a multiple of itself. The (real or complex) factor by which it gets multiplied is the eigenvalue of $\hat{A}$ to which the eigenvector belongs. Following Dirac we denote typical eigenvalues by $a, a', a'', \ldots$ and the corresponding eigenvectors by[3] $|a\rangle, |a'\rangle, |a''\rangle, \ldots$. Thus

$$\hat{A}|a\rangle = a|a\rangle,\quad \hat{A}|a'\rangle = a'|a'\rangle, \ldots \tag{7.12}$$

As an illustration of the way one works with abstract vectors and operators, we present now, in Dirac notation, the proof of the orthogonality theorem for eigenvectors of a self-adjoint operator. Suppose $\hat{A}$ in Eq. 7.12 is self-adjoint. Then as a particular case of Eq. 7.11 we have

$$\langle a'|\hat{A}|a\rangle = \langle a|\hat{A}|a'\rangle^* \tag{7.13}$$

[3] Clearly this notation identifies the eigenvectors uniquely only if the eigenvalues are non-degenerate. Further labels required to distinguish between different eigenvectors belonging to the same eigenspace (in case of degeneracy) will be ignored for the present.

On using Eqs. 7.12 this reduces to

$$a \langle a' \mid a \rangle = (a' \langle a \mid a' \rangle)^* = a'^* \langle a' \mid a \rangle \tag{7.14}$$

where Eq. 7.3b has been used in the last step. Thus we have $\langle a' \mid a \rangle (a - a'^*) = 0$. In particular, if $\mid a \rangle = \mid a' \rangle$, then $a = a'$ and this equation reduces to $\langle a \mid a \rangle (a - a^*) = 0$. The norm $\langle a \mid a \rangle \neq 0$. Hence $a = a^*$, i.e. every eigenvalue is real. On using this result, Eq. 7.14 becomes

$$(a - a') \langle a' \mid a \rangle = 0 \tag{7.15}$$

showing that the scalar product between $\mid a' \rangle$ and $\mid a \rangle$ vanishes whenever they belong to distinct eigenvalues $(a \neq a')$. If the vectors $\mid a \rangle, \mid a' \rangle$, etc., are normalized, we can then write

$$\langle a' \mid a \rangle = \delta_{a'a} \tag{7.16}$$

(*d*) *Expansion of the Identity; Projection Operators:* Suppose now that $\hat{A}$ is an observable so that its eigenvectors form a complete set. Then we can write any $|\psi\rangle$ as[4]

$$|\psi\rangle = \sum_a c_a \mid a \rangle \tag{7.17}$$

On taking the scalar product of $\mid a' \rangle$ with this equation and using the orthonormality condition (7.16), we obtain

$$\langle a' |\psi\rangle = \sum_a c_a \, \delta_{aa'} = c_{a'} \tag{7.18}$$

Thus $c_a = \langle a |\psi\rangle$. On substituting this back into Eq. 7.17 we get

$$|\psi\rangle = \sum_a \mid a \rangle \langle a |\psi\rangle \tag{7.19}$$

Now a typical term on the right hand side is a vector whose length is the scalar product of $\mid a \rangle$ and $|\psi\rangle$ and whose direction is that of $\mid a \rangle$; so it is the *projection of* $|\psi\rangle$ on $\mid a \rangle$. The quantity

$$\hat{P}_a \equiv \mid a \rangle \langle a \mid \tag{7.20}$$

is the operator which performs this projection. It is called a *projection operator.* Eq. 7.19 says that the sum of all the projection operators $\hat{P}_a$ leaves any vector $|\psi\rangle$ unchanged. Therefore it is the unit operator.

$$\sum_a \mid a \rangle \langle a \mid = \hat{1} \tag{7.21}$$

The closure relation (3.47) is the Schrödinger representation of Eq. 7.21. This equation, viewed as an expansion of the unit operator, will be found most valuable in the following. As an immediate example we note that it enables us to expand an arbitrary operator $\hat{F}$ as

$$\hat{F} = \hat{1}\hat{F}\hat{1} = \sum_a \mid a \rangle \langle a \mid \hat{F} \sum_{a'} \mid a' \rangle \langle a' \mid = \sum_{a,a'} \mid a \rangle \langle a' \mid F_{aa'},$$

where $F_{aa'} = \langle a \mid \hat{F} \mid a' \rangle$. Thus $\hat{F}$ is expressed as a linear combination (with coefficients $F_{aa'}$) of the operators $\mid a \rangle \langle a' \mid$ which are more general than the projection operators $\hat{P}_a$. If $\hat{F}$ is $\hat{A}$ itself, $\langle a \mid \hat{A} \mid a' \rangle = a\delta_{aa'}$, and the above expansion simplifies to $\hat{A} = \sum_a \mid a \rangle \langle a \mid a = \sum a\hat{P}_a$, showing that

[4] It will be noticed that we use the symbol a (or a') sometimes for a specific eigenvalue and sometimes as a parameter which runs over all the eigenvalues. If a is a continuous parameter (i.e. the eigenvalue spectrum of $\hat{A}$ is continuous) it shall be understood that as in Chapter 3, suitable modifications in the notation are to be made as necessary; e.g., replace sums by integrals, Kronecker delta functions by Dirac delta functions etc.

an observable can be written as a weighted sum of projection operators to its own eigenstates, the weighting factor being the eigenvalue itself. More generally, for any function of $\hat{A}$, we have

$$f(\hat{A}) = \sum_a f(a) \mid a \rangle \langle a \mid = \sum_a f(a) P_a.$$

(*e*) *Unitary Operators:* The simplest definition of unitary operators is through the equation $\hat{U}^\dagger\hat{U} = 1$, or $\hat{U}^\dagger = \hat{U}^{-1}$. The action of such an operator on any state $|\psi\rangle$ leaves the norm unchanged, since the norm of $\hat{U}\,|\psi\rangle$ is $\langle\psi|\,\hat{U}^\dagger.\hat{U}\,|\psi\rangle = \langle\psi|\psi\rangle$. More generally, the scalar product of $\hat{U} \mid \phi\rangle$ and $\hat{U}$ $|\psi\rangle$ reduces to $\langle \phi\,|\psi\rangle$ for arbitrary $\mid \phi \rangle$ and $|\psi\rangle$.

If $\hat{A}$ is any *Hermitian* operator ($\hat{A}^\dagger = \hat{A}$) and α is a *real* number, then $\exp(i\alpha\hat{A})$ is unitary, since $[\exp(i\alpha\hat{A})]^\dagger = \exp[(i\alpha\hat{A})^\dagger] = \exp(-i\alpha\hat{A}) = [\exp(i\alpha\hat{A})]^{-1}$. Conversely, any unitary operator $\hat{U}$ can be written in the form $\hat{U} = \exp(i\alpha\hat{A})$ with α real and $\hat{A}$ Hermitian.

7.4 Representations

(*a*) *Representation of State Vectors; The Wave Function:* It is a familiar fact that the expansion $\mathbf{V} = V_1\mathbf{e}_1 + V_2\mathbf{e}_2 + V_3\mathbf{e}_3$ provides a set of components V_1, V_2, V_3 which *represent* the vector $\mathbf{V}$ with respect to the given basis $(\mathbf{e}_1, \mathbf{e}_2, \mathbf{e}_3)$. In an exactly analogous manner, the coefficients c_a in the expansion (7.17) represent the 'components' of the Hilbert space vector $|\psi\rangle$ with respect to a basis consisting of the complete orthonormal system of vectors $\mid a \rangle$. Eq. 7.18 which gives the value of any component as $c_a = \langle a\,|\psi\rangle$ is the counterpart of the relation $V_i = \mathbf{e}_i.\mathbf{V}$. The set of components $c_a \equiv \langle a\,|\psi\rangle$, considered as a function of the eigenvalue parameter a, represents the state vector $|\psi\rangle$ completely and uniquely. *This function is the wave function representing* $|\psi\rangle$ *in the A representation.* It is instructive to note that the wave function of any of the *basis* vectors, say $\mid a' \rangle$, is a delta function:

$$\text{If } |\psi\rangle = \mid a' \rangle,\ c_a = \langle a \mid a' \rangle = \delta_{aa'} \tag{7.22}$$

We can readily express the normalization condition $\langle\psi|\psi\rangle = 1$ in terms of the c_a with the aid of Eq. 7.21:

$$\begin{aligned}\langle\psi|\psi\rangle &= \langle\psi|\,\hat{1}\,|\psi\rangle = \langle\psi|\,(\sum_a \mid a \rangle \langle a \mid)\,|\psi\rangle \\ &= \sum_a \langle\psi|\,a \rangle \langle a\,|\psi\rangle = \sum_a c_a^* c_a = 1\end{aligned} \tag{7.23}$$

(*b*) *Dynamical Variables as Matrix Operators:* Consider now a linear operator $\hat{F}$ which corresponds to some dynamical variable. Suppose $\hat{F}$ takes a vector $|\psi\rangle$ into $\mid \chi \rangle$:

$$\hat{F}\,|\psi\rangle = \mid \chi \rangle \tag{7.24}$$

How will this operator relation appear in the A-representation? In other words, how will the wave function $\langle a \mid \chi \rangle$ of $\mid \chi \rangle$ be related to that of $|\psi\rangle$? Clearly

$$\begin{aligned}\langle a \mid \chi \rangle &= \langle a \mid \hat{F}\,|\psi\rangle = \langle a \mid \hat{F}\,(\sum_{a'} \mid a' \rangle \langle a' \mid)\,|\psi\rangle \\ &= \sum_{a'} \langle a \mid \hat{F} \mid a' \rangle \langle a'\,|\psi\rangle\end{aligned} \tag{7.25}$$

It will be noted that by introducing the unit operator in the form (7.21), we have been able to bring in the wave function $\langle a' | \psi \rangle$ of $|\psi\rangle$. Eq. 7.25 may be viewed as a matrix equation:

$$\begin{bmatrix} \langle a | \chi \rangle \\ \langle a' | \chi \rangle \\ \langle a'' | \chi \rangle \\ \cdot \\ \cdot \\ \cdot \end{bmatrix} = \begin{bmatrix} \langle a | \hat{F} | a \rangle, & \langle a | \hat{F} | a' \rangle, & \langle a | \hat{F} | a'' \rangle, & \ldots \\ \langle a' | \hat{F} | a \rangle, & \langle a' | \hat{F} | a' \rangle, & \langle a' | \hat{F} | a'' \rangle, & \ldots \\ \langle a'' | \hat{F} | a \rangle, & \langle a'' | \hat{F} | a' \rangle, & \langle a'' | \hat{F} | a'' \rangle, & \ldots \\ - & - & - & \ldots \\ - & - & - & \ldots \\ - & - & - & \ldots \end{bmatrix} \begin{bmatrix} \langle a | \psi \rangle \\ \langle a' | \psi \rangle \\ \langle a'' | \psi \rangle \\ \cdot \\ \cdot \\ \cdot \end{bmatrix} \tag{7.26}$$

Thus if the wave functions representing $|\chi\rangle$ and $|\psi\rangle$ are visualized as column 'vectors' (one-column matrices) the representative of the operator $\hat{F}$ is seen to be a square *matrix* with elements as shown. For this reason, quantities of the form

$$\langle a^{(i)} | \hat{F} | a^{(j)} \rangle \tag{7.27}$$

are referred to as *matrix elements* of $\hat{F}$ in the A-representation. Adopting the notation $[F]_A$ for the representative of $\hat{F}$ in the A-representation, and $(\psi)_A$, $(\chi)_A$ for the columns representing the states $|\psi\rangle$ and $|\chi\rangle$, we rewrite (7.26) succinctly as

$$(\chi)_A = [F]_A (\psi)_A \tag{7.28}$$

When one is already accustomed to the Schrödinger representation from the beginning, the appearance of the wave function as a column matrix and of dynamical variables as square matrices (rather than differential operators) may seem strange. However, the difference in appearance is only superficial. It arises from the fact that in the Schrödinger representation (wherein $\hat{A} = \hat{x}$) the eigenvalue parameter $a = x$ is continuous, whereas in the derivation above, we have tacitly assumed that a is a discrete parameter. We shall see explicitly in the next section that the familiar Schrödinger forms can be recovered from the results obtained above. It may also be mentioned here that in Heisenberg's form of quantum mechanics, dynamical variables (including position and momentum) are represented by matrices.

(*c*) *Products of Operators; The Quantum Condition:* Finally, we observe that any product of operators is represented by the product of the matrices representing the individual operators:

$$[FG]_A = [F]_A\, [G]_A \tag{7.29}$$

The student is urged to verify this by evaluating $\langle a | \hat{F}\hat{G} | a' \rangle$ by 'sandwiching' the expansion $\hat{1} = \Sigma\, | a'' \rangle \langle a'' |$ between $\hat{F}$ and $\hat{G}$.

The quantum condition (7.6) now goes over into the matrix equation $[x]_A\, [p]_A - [p]_A\, [x]_A = i\hbar$. It is interesting that this equation can be satisfied only by infinite-dimensional matrices. To see this, consider the equality of diagonal matrix elements on the two sides:

$$\sum_{a'} \langle a \,|\, \hat{x} \,|\, a' \rangle \langle a' \,|\, \hat{p} \,|\, a \rangle - \sum_{a'} \langle a \,|\, \hat{p} \,|\, a' \rangle \langle a' \,|\, \hat{x} \,|\, a \rangle = i\hbar$$

If we now sum over a also (thus taking the trace of the original equation), the right hand side becomes $i\hbar n$ where n is the dimension of the matrix. The left hand side can be written as

$$\sum_{a} \sum_{a'} \langle a \,|\, \hat{x} \,|\, a' \rangle \langle a' \,|\, \hat{p} \,|\, a \rangle - \sum_{a'} \sum_{a} \langle a' \,|\, \hat{p} \,|\, a \rangle \langle a \,|\, \hat{x} \,|\, a' \rangle$$

by exchanging the labels a,a' in the second term. (This does not affect anything because these are just dummy summation indices). Now, if the order of summation in either term can be reversed—which can always be done in the case of finite sums—the two terms cancel each other and we are left with a contradiction, namely $0 = i\hbar n$. But if the summations are over an infinite number of terms, reversal of the order of summation is not always permissible, and this is what makes it possible to satisfy the quantum condition with infinite-dimensional matrices.

(*d*) *Self-adjointness and Hermiticity:* An important observation is that *if $\hat{F}$ is a self-adjoint operator, then $[F]_A$ is a Hermitian matrix*: for, its matrix element in the $(a^{(i)}, a^{(j)})$ position, namely $\langle a^{(i)} \,|\, \hat{F} \,|\, a^{(j)} \rangle$, is equal to the corresponding element of the transposed complex conjugate (i.e. Hermitian conjugate) matrix, namely $\langle a^{(j)} \,|\, \hat{F} \,|\, a^{(i)} \rangle^*$. This equality is merely a special case of the definition (7.11) of self-adjointness. The reason for the use of the term 'Hermitian' as synonymous with 'self-adjoint' in the literature of quantum mechanics will now be clear.

(*e*) *Diagonalization:* What is the matrix $[A]_A$ for $\hat{A}$ in its own representation? In view of (7.27), (7.12), and (7.16), its matrix elements are

$$\langle a \,|\, \hat{A} \,|\, a' \rangle = a' \delta_{aa'} \tag{7.30}$$

Thus any observable is always diagonal in its own representation, the diagonal elements being just the eigenvalues. The process of solving the eigenvalue problem for an operator is, therefore, often referred to as diagonalization of the operator.

7.5 Continuous Basis—The Schrödinger Representation

The above treatment, with suitable change of notation, applies also to the case when the eigenvalue spectrum of $\hat{A}$ is continuous. We consider here one such case, namely the Schrödinger representation in which $\hat{A}$ is the position operator. It may not appear obvious how the matrix representations of operators, considered in the last section, go over into the familiar differential operator forms. In this section we show explicitly how this happens.

Let $\hat{A}$ be the position operator $\hat{x}$ in one dimension. The basis states of the $\hat{A}$ representation are then the position eigenstates $|\, x \rangle$ defined by

$$\hat{x} \,|\, x \rangle = x \,|\, x \rangle, \quad -\infty < x < \infty \tag{7.31}$$

The counterpart of Eq. 7.16 here is

$$\langle x \,|\, x' \rangle = \delta\,(x - x') \tag{7.32}$$

and the wave function of an arbitrary state $|\psi\rangle$ is given by $\langle x|\psi\rangle$, considered as a function of x. This is the Schrödinger wave function $\psi(x)$:

$$\psi(x) \equiv \langle x | \psi \rangle \tag{7.33}$$

The norm of $|\psi\rangle$ is obtained now as a special case of Eq. 7.23, with the summation sign replaced by an integral over x:

$$\langle\psi|\psi\rangle = \int dx \, |\langle x | \psi \rangle|^2 = \int dx \, |\psi(x)|^2 \tag{7.34}$$

This coincides of course with the definition of norm given in Chapter 2.

The operator $\hat{x}$ is diagonal in its own representation, i.e. its 'matrix elements' $\langle x | \hat{x} | x' \rangle$ vanish for $x \neq x'$. In fact

$$\langle x | \hat{x} | x' \rangle = x\delta(x - x') \tag{7.35}$$

Consequently the wave function of $\hat{x}|\psi\rangle$ is

$$\begin{aligned} \langle x | \hat{x} |\psi\rangle &= \int dx' \langle x | \hat{x} | x' \rangle \langle x' | \psi \rangle \\ &= \int dx' \, x \, \delta(x - x') \, \psi(x') = x\psi(x) \end{aligned} \tag{7.36}$$

It will be noted that the crucial step in this reduction is the introduction of the expansion $\int dx' | x' \rangle \langle x' | = \hat{1}$ of the unit operator, between $\hat{x}$ and $|\psi\rangle$.

To obtain the representation of $\hat{p}$, we make use of the commutation rule $[\hat{x}, \hat{p}] = i\hbar$. Taking the matrix element of both sides, we obtain

$$\langle x | \hat{x}\hat{p} - \hat{p}\hat{x} | x' \rangle = i\hbar \langle x | x' \rangle = i\hbar \, \delta(x - x') \tag{7.37}$$

We reduce the two terms on the left hand side by noting that $\langle x | \hat{x} = x \langle x |$ and $\hat{x} | x' \rangle = x' | x' \rangle$. Then Eq. 7.37 becomes

$$(x - x') \langle x | \hat{p} | x' \rangle = i\hbar \, \delta(x - x'),$$

or

$$\langle x | \hat{p} | x' \rangle = i\hbar \frac{\delta(x - x')}{(x - x')} = i\hbar \frac{\partial}{\partial x'} \delta(x - x') \tag{7.38}$$

The last step uses the identity (C.7) of Appendix C and is crucial in obtaining the differential operator representation of $\hat{p}$. Eq. 7.38 enables us to conclude that the wave function of $\hat{p}|\psi\rangle$ is

$$\begin{aligned} \langle x | \hat{p} |\psi\rangle &= \int dx' \langle x | \hat{p} | x' \rangle \langle x' |\psi\rangle \\ &= i\hbar \int dx' \left[\frac{\partial}{\partial x'} \delta(x - x')\right] \psi(x') \\ &= i\hbar \left[\delta(x - x') \, \psi(x')\right]_{-\infty}^{+\infty} - i\hbar \int \delta(x - x') \frac{\partial\psi(x')}{\partial x'} dx' \\ &= -i\hbar \frac{\partial\psi(x)}{\partial x} \end{aligned} \tag{7.39}$$

In the penultimate step we have carried out one integration by parts and noted that the integrated term vanishes at both limits in view of the properties of the delta function.

The result (7.39) confirms that the representative of $\hat{p}$ in this (Schrödinger) representation is $-i\hbar(\partial/\partial x)$. The fundamental role of the commutation relation in this derivation may be noted.

7.6 Degeneracy; Labelling by Commuting Observables

Returning to matrix representations, let us consider explicitly the case when some or all of the eigenvalues of $\hat{A}$ are degenerate. For the eigenspace belonging to an r-fold degenerate eigenvalue a, we choose a basis consisting of any r orthonormal states $|\, a i \,\rangle$,

$$\hat{A} \,|\, ai \,\rangle = a \,|\, ai \,\rangle, \quad i = 1,2,\ldots, r, \tag{7.40a}$$

$$\langle\, ai \,|\, a'i' \,\rangle = \delta_{aa'}\delta_{ii'} \tag{7.40b}$$

The completeness relation now reads, instead of (7.21), as

$$\sum_{a,i} |\, ai \,\rangle \langle\, ai \,| = \hat{1} \tag{7.40c}$$

Now, let $\hat{B}$ be an observable which *commutes* with $\hat{A}$. Consider the state $\hat{B} \,|\, ai \,\rangle$. Operating on it with $\hat{A}$, we find that

$$\hat{A}\hat{B} \,|\, ai \,\rangle = \hat{B}\hat{A} \,|\, ai \,\rangle = a \,\hat{B} \,|\, ai \,\rangle \tag{7.41}$$

Thus $\hat{B} \,|\, ai \,\rangle$ is in the eigenspace belonging to the eigenvalue a of $\hat{A}$. Taking the scalar product of $|\, a'i' \,\rangle$ with both sides of (7.41) and noting that $\langle\, a'i' \,|\, \hat{A} = a' \,\langle\, a'i' \,|$, we obtain

$$a' \,\langle\, a'i' \,|\, \hat{B} \,|\, ai \,\rangle = a \,\langle\, a'i' \,|\, \hat{B} \,|\, ai \,\rangle \tag{7.42}$$

It follows immediately that all matrix elements of $\hat{B}$ for which $a \neq a'$ vanish. All nonvanishing matrix elements lie within square blocks along the diagonal, each block being associated with a particular eigenvalue a of $\hat{A}$ and of dimension equal to the degeneracy of this eigenvalue. Denoting this block by $B^{(a)}$ with elements $B^{(a)}{}_{i'i}$, we may summarize these statements into the equation:

$$\langle\, a'i' \,|\, \hat{B} \,|\, ai \,\rangle = B^{(a)}{}_{i'i}\, \delta_{a'a} \tag{7.43}$$

or, in explicit matrix form,

$$[B]_A = \left(\begin{array}{c|c|c|c} B^{(a)} & 0 & 0 & \cdot\cdot \\ \hline 0 & B^{(a')} & 0 & \cdot\cdot \\ \hline 0 & 0 & B^{(a'')} & \cdot\cdot \\ \hline \vdots & \vdots & \vdots & \vdots\vdots \end{array}\right), \quad B^{(a)} = \begin{pmatrix} B^{(a)}{}_{11} & B^{(a)}{}_{12} \ldots B^{(a)}{}_{1r} \\ B^{(a)}{}_{21} & B^{(a)}{}_{22} \ldots B^{(a)}{}_{2r} \\ \cdot & \cdot \;\ldots\; \cdot \\ B^{(a)}{}_{r1} & B^{(a)}{}_{r2} \ldots B^{(a)}{}_{rr} \end{pmatrix} \tag{7.44}$$

From Eq. 7.43 we have, with the aid of Eq. 7.40c, that

$$\hat{B} \,|\, ai \,\rangle = \sum_{a'i'} |\, a'i' \rangle \langle\, a'i' \,|\, \hat{B} \,|\, ai \,\rangle = \sum_{i'} |\, ai \,\rangle\, B^{(a)}{}_{i'i} \tag{7.45}$$

This is simply a more precise specification of the statement following Eq. 7.41.

Since we have assumed $\hat{B}$ to be an observable, the matrix $[B]_A$ is Hermitian as shown in Sec. 7.4. Hence each of the 'blocks' (submatrices) such as $B^{(a)}$ is Hermitian too, and can be diagonalized by a unitary transformation. Let $U^{(a)}$ be the matrix which diagonalizes $B^{(a)}$, i.e.

$$(U^{(a)})^{-1}\, B^{(a)}\, U^{(a)} = D^{(a)} \tag{7.46}$$

where $D^{(a)}$ is a diagonal matrix. Its diagonal elements are obviously eigenvalues of $B^{(a)}$. We use these to label the rows and columns of $D^{(a)}$, so that its typical matrix element is $D^{(a)}{}_{b'b} = b\delta_{b'b}$. Now, writing Eq. 7.46 in the form $B^{(a)}\, U^{(a)} = U^{(a)}\, D^{(a)}$, and taking the (i, b) matrix element,

$$\sum_j B^{(a)}{}_{ij} U^{(a)}{}_{jb} = \sum_{b'} U^{(a)}{}_{ib'}\, D^{(a)}{}_{b'b} = bU^{(a)}{}_{ib}, \tag{7.47}$$

This equation enables us to define[5] a new set of states $|\, ab \,\rangle$ which, unlike the $|\, ai \,\rangle$, are *eigenstates of $\hat{B}$ as well as of $\hat{A}$*. For, if we define

[5] It was stated without proof in Sec. 3.13 that it is possible to construct such states. The proof is supplied here.

$$| ab \rangle = \sum_i | ai \rangle U^{(a)}{}_{ib} \tag{7.48}$$

$$\hat{B} | ab \rangle = \sum_i \hat{B} | ai \rangle U^{(a)}{}_{ib} = \sum_{i'i} | ai' \rangle B^{(a)}{}_{i'i} U^{(a)}{}_{ib}$$
$$= \sum_{i'} | ai' \rangle U^{(a)}{}_{i'b}\, b = b \,| ab \rangle \tag{7.49}$$

The second step follows from Eq. 7.45 and the next step from Eq. 7.47.

The set of states $| ab \rangle$ is to be preferred to the original set $| ai \rangle$ as a basis because unlike i which is merely a label for the members of an arbitrarily chosen basis within the eigenspace of $\hat{A}$ for given a, the label b has an immediate physical significance as an eigenvalue of $\hat{B}$. Of course, the symbol $| ab \rangle$ suffices to identify simultaneous eigenvectors of $\hat{A}$ and $\hat{B}$ unambiguously only if *within each eigenspace of* $\hat{A}$, every eigenvalue b of $\hat{B}$ is non-degenerate, i.e. if Eq. 7.49 has only one solution for each pair of values a,b. Then $\hat{A}$ and $\hat{B}$ form a *complete set of commuting observables* for the system. If this is not the case, one or more further observables $\hat{C},\ldots,$ commuting among themselves and with $\hat{A}$ and $\hat{B}$ will have to be employed, as discussed already in Sec. 3.13.

7.7 Change of Basis; Unitary Transformations

We now turn to a consideration of the relation between two different representations. The basis for each representation is formed by the simultaneous eigenstates of a complete commuting set of observables. To avoid unnecessary complexity of notation, we shall consider the case in which each complete set consists of just one observable, and refer to the first representation as the $\hat{A}$-representation, and the second as the $\hat{B}$-representation. It is left to be understood that $\hat{A}$ may stand for a *set* of mutually commuting observables and that the symbol a labelling its eigenstates $| a \rangle$ may stand for a corresponding set of quantum members with similar notation holding good for $\hat{B}$.

We know that the wave function of an arbitrary state $|\psi\rangle$ in the $\hat{A}$-representation is given by $\langle a | \psi \rangle$. It can be easily expressed in terms of the wave function $\langle b |\psi\rangle$ of the same state in the B-representation. Using $\sum_b | b \rangle \langle b | = \hat{1}$, we obtain

$$\langle a |\psi\rangle = \langle a | \hat{1} |\psi\rangle = \sum_b \langle a | b \rangle \langle b |\psi\rangle \tag{7.50}$$

Similarly, given any linear operator $\hat{F}$, the relation between its matrix elements in the two representations, namely $\langle a | \hat{F} | a' \rangle$ and $\langle b | \hat{F} | b' \rangle$ can be obtained in the same way:

$$\langle a | \hat{F} | a' \rangle = \langle a | \hat{1} \hat{F} \hat{1} | a' \rangle$$
$$= \sum_{b'b} \langle a | b \rangle \langle b | \hat{F} | b' \rangle \langle b' | a' \rangle \tag{7.51}$$

Eqs. 7.50 and 7.51 can be written as matrix equations:

$$(\psi)_A = [U]\ (\psi)_B \tag{7.52}$$
$$[F]_A = [U]\ [F]_B\ [U]^\dagger \tag{7.53}$$

where the elements of the matrices $[U]$ and $[U]^\dagger$ are

$$U_{ab} = \langle a | b \rangle,\ [U]^\dagger{}_{ba} = U^*{}_{ab} = \langle b | a \rangle \tag{7.54}$$

It is evident that U is a *unitary matrix*, since

$$[UU^\dagger]_{aa'} = \sum_b U_{ab}U^*_{a'b} = \sum_b \langle a \mid b \rangle \langle b \mid a' \rangle \tag{7.55a}$$
$$= \langle a \mid \hat{1} \mid a' \rangle = \delta_{aa'}$$

and similarly

$$[U^\dagger U]_{bb'} = \delta_{bb'} \tag{7.55b}$$

The transformation of dynamical variables in general, and of the position and momentum variables in particular, through Eq. 7.53, is the counterpart in quantum mechanics of the canonical transformations of classical mechanics. Consequently Eqs. 7.52 and 7.53 are often referred to as *canonical transformations* within the quantum mechanical context.

It is necessary to recognize that the unitary matrix U through which the canonical transformation is performed *need not* be a square matrix. This peculiarity arises from the infinite dimensionality of the space, on account of which one has the possibility of using discrete or continuous or partially discrete and partially continuous bases for representations. For example $\hat{A}$ may have a continuous spectrum of eigenvalues and $\hat{B}$ a discrete spectrum, in which case the matrix elements U_{ab} are labelled by a continuous row index a and a discrete column index b. Eqs. 7.55 in this case would read

$$[UU^\dagger]_{aa'} = \sum_b \langle a \mid b \rangle \langle b \mid a' \rangle = \delta(a - a') \tag{7.56a}$$

$$[U^\dagger U]_{bb'} = \int da \langle b \mid a \rangle \langle a \mid b' \rangle = \delta_{bb'} \tag{7.56b}$$

In particular, if $\hat{A} = \hat{\mathbf{x}}$, then $\langle a \mid b \rangle = \langle x \mid b \rangle$ is the coordinate space wave function of the eigenstate $\mid b \rangle$ of $\hat{B}$. If we denote this wave function by $u_b(x)$, then $\langle b \mid a' \rangle = \langle b \mid x' \rangle = u_b^*(x')$, and Eq. 7.56a is seen to reduce to the closure relation (3.47). Similarly Eq. 7.56b becomes the orthonormality relation (3.46) for the set of eigenfunctions $u_b(x)$. Thus the orthonormality and closure properties appear as statements of the unitarity of a canonical transformation matrix U whose matrix elements are the wave functions $u_b(x)$.

7.8 Unitary Transformations Induced by Change of Coordinate System: Translations

The observables whose eigenstates are used as bases for representations are themselves usually defined in relation to a coordinate system OXYZ in ordinary space. Therefore, any change in the coordinate system causes a change in the observable and thereby induces a transformation of the basis in the Hilbert space of quantum states. Basis transformations brought about in this fashion by a rotation of the coordinate system (keeping the origin fixed) or by a translation (shifting the origin without changing the orientation of the axes) have particular physical significance, as we shall now demonstrate.

Let us consider first the case of a simple system, a single particle in one dimension, and investigate the effect of translations on its wave functions in the Schrödinger representation. With reference to Fig. 7.1, let O be the origin for the x-coordinate in one reference frame S, and let S' be another frame identical to S except that its origin is now at O′, shifted by an amount

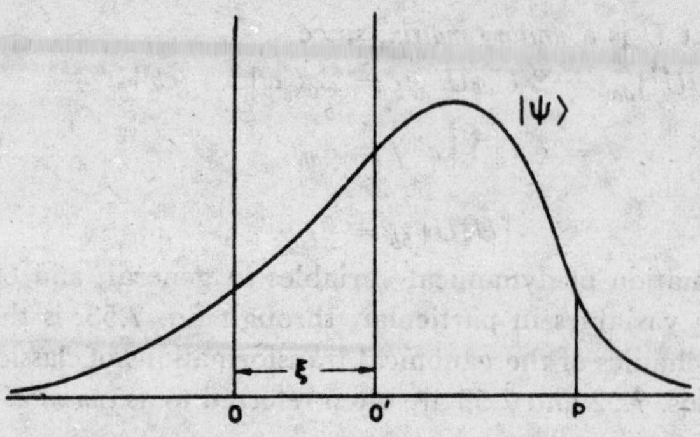

Fig. 7.1 Translation of coordinate frame. The wave function depicted by the curve has the value $\psi(x) \equiv \psi'(x-\xi)$ at P.

ξ in the positive x-direction from O. Let the form of a typical wave function, as depicted in the figure, be represented by functions ψ and ψ' in S and S', respectively. The value of the wave function at a physical point P is given by

$$\psi(x) = \psi'(x'), \; x' = x - \xi \tag{7.57}$$

where x, x' are the coordinates of P measured from O and O' respectively. We can write this as $\psi'(x) = \psi(x+\xi)$. When ξ is infinitesimal we have

$$\psi'(x) \approx \psi(x) + \xi \frac{\partial \psi}{\partial x} = [1 + i\xi\, G_x]\, \psi(x) \tag{7.58a}$$

$$G_x = -\, i \frac{\partial}{\partial x} \tag{7.58b}$$

The change in the *functional form,* i.e. the difference between $\psi'(x)$ and $\psi(x)$ is thus brought about by the action of the operator $i\xi G_x$ on ψ. We call G_x the *generator* of infinitesimal translations along the x-direction. The relation (7.58) is of course not restricted to quantum mechanics; it would hold equally well if ψ were the temperature distribution along a rod or the pressure variation through a column of air. However, in the quantum mechanical context G_x has a special physical significance since the momentum operator here is proportional to $(\partial/\partial x)$. In fact

$$G_x = (p_x/\hbar) \tag{7.59}$$

The two forms $\psi(x)$ and $\psi'(x)$ for the wave function of one and the same physical state $|\psi\rangle$ can be viewed as the result of expanding the state vector in terms of two different bases in Hilbert space. By definition,

$$\psi(x) = \langle x | \psi \rangle, \quad \psi'(x) = {}'\langle x | \psi \rangle \tag{7.60}$$

where $| x \rangle$ is the position eigenstate for a particle at the coordinate x measured from O, and $| x \rangle'$ at coordinate x measured from O' (Fig. 7.2). Therefore, Eqs. 7.58 and 7.59 tell us that

$${}'\langle x | \psi \rangle = [1 + i\xi p_x/\hbar]\, \langle x | \psi \rangle = \langle x \,|\, [1 + i\xi\, \hat{p}_x/\hbar]\, | \psi \rangle \tag{7.61}$$

We have used Eq. 7.39 in the last step. Since Eq. 7.61 is true for arbitrary $|\psi\rangle$, we can conclude that ${}'\langle x | = \langle x \,|\, [1 + i\xi\hat{p}_x/\hbar]$, or taking the conjugate,

$$| x \rangle' = [1 - i\xi\hat{p}_x/\hbar]\, | x \rangle \tag{7.62}$$

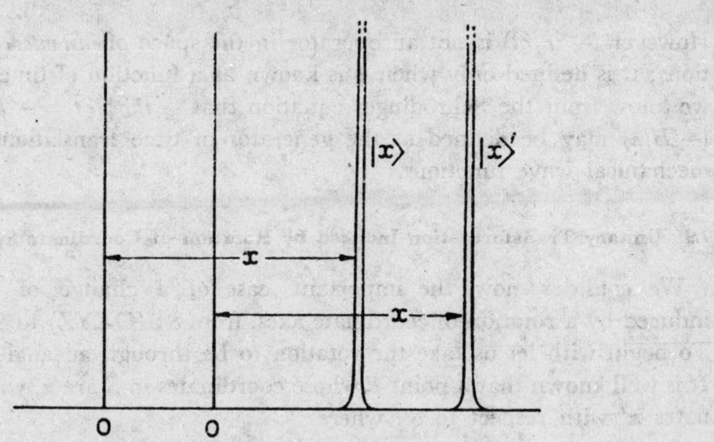

Fig. 7.2 Position eigenstates $|x\rangle$ and $|x\rangle'$ located at points with coordinate x in S and S' respectively.

This is the relation between *the basis states* $|x\rangle$ defined with respect to S and $|x\rangle'$ referred to the translated frame S'.

As regards translation through some finite distance ξ, one can view it as being made up of N infinitesimal transformations through (ξ/N), $N\to\infty$. Eq. 7.62 then generalizes to

$$|x\rangle' = \lim_{N\to\infty}\left[1-\frac{i\xi\hat{p}_x}{\hbar}\right]^N |x\rangle = e^{-i\xi\hat{p}_x/\hbar}|x\rangle \tag{7.63}$$

The corresponding relation between the wavefunctions of a given physical state $|\psi\rangle$ becomes

$$\psi'(x) = {}'\langle x|\psi\rangle = \langle x|e^{i\xi\hat{p}_x/\hbar}|\psi\rangle = e^{i\xi p_x/\hbar}\psi(x) \tag{7.64}$$

The generalization of these to three dimensions is straightforward: one merely replaces x, p_x, ξ, ξp_x, ... by $\mathbf{x}$, $\mathbf{p}$, $\boldsymbol{\xi}$, $\boldsymbol{\xi}.\mathbf{p}$, ...

It is important to recognize that the fact which we have established, namely that *the momentum operator in quantum mechanics is proportional to the generator of infinitesimal translations*, holds quite generally for any arbitrary system. For an n-particle system one merely replaces Eq. 7.57 by

$$\psi(\mathbf{x}_1,\mathbf{x}_2\ldots,\mathbf{x}_n) = \psi'(\mathbf{x}_1-\boldsymbol{\xi},\mathbf{x}_2-\boldsymbol{\xi},\ldots,\mathbf{x}_n-\boldsymbol{\xi})$$

in view of the fact that the coordinates of all the particles are changed by the *same* amount as a result of the translation of origin. This leads to

$$\psi'(\mathbf{x}_1,\mathbf{x}_2,\ldots,\mathbf{x}_n) = \psi(\mathbf{x}) + \boldsymbol{\xi}.\nabla_1\psi(\mathbf{x}) + \boldsymbol{\xi}.\nabla_2\psi(\mathbf{x}) + \ldots + \boldsymbol{\xi}.\nabla_n\psi(\mathbf{x}) = [1 + i\mathbf{p}.\boldsymbol{\xi}/\hbar]\,\psi(\mathbf{x}) \tag{7.65}$$

where $\mathbf{p}$ is now the *total* momentum operator for the system:

$$\mathbf{p} = \mathbf{p}_1 + \mathbf{p}_2 + \ldots + \mathbf{p}_n = -i\hbar\,[\nabla_1 + \nabla_2 + \ldots + \nabla_n] \tag{7.66}$$

For an infinitesimal translation τ of the origin of the *time* coordinate, one finds in a similar fashion that

$$\psi'(\mathbf{x},t) = \psi(\mathbf{x},t+\tau) = \psi(\mathbf{x},t) + i\tau\left[-i\frac{\partial}{\partial t}\psi(\mathbf{x},t)\right]$$

However $(-i\partial/\partial t)$ is not an operator in the space of *instantaneous* wave functions; it is defined only when ψ is known as a function of time. Nevertheless, we know from the Schrödinger equation that $-i\hbar\partial\psi/\partial t = -H\psi$. Therefore, $(-H/\hbar)$ may be defined as the generator of time translations on quantum mechanical wave functions.[6]

7.9 Unitary Transformation Induced by Rotation of Coordinate System

We consider now the important case of a change of representation induced by a rotation of coordinate axes, from S: (OXYZ) to S': (OX′Y′Z′). To begin with let us take the rotation to be through an angle θ about OZ. It is well known that a point P whose coordinates in S are $\mathbf{x}$, will have coordinates $\mathbf{x}'$ with respect to S', where

$$\left.\begin{aligned} x' &= x\cos\theta + y\sin\theta \\ y' &= -x\sin\theta + y\cos\theta \\ z' &= z \end{aligned}\right\} \tag{7.67}$$

For a given state $|\psi\rangle$, the wave function at P has a definite value independent of the orientation of the coordinate axes.[7] Hence the forms ψ and ψ' of the wave function in the two reference frames must be such that $\psi_P = \psi_P'$, or in terms of coordinates,

$$\psi'(\mathbf{x}') = \psi(\mathbf{x}) \tag{7.68}$$

On substituting for $\mathbf{x}'$ from Eq. 7.67 and taking θ to be infinitesimal, we have $\psi'(x + y\theta, y - x\theta, z) = \psi(x,y,z)$. It is more convenient to rewrite this equation (by replacing x by $x - y\theta$ and y by $y + x\theta$ on both sides) as

$$\begin{aligned} \psi'(x,y,z) &= \psi(x - y\theta,\ y + x\theta,\ z) \\ &= \psi(x,y,z) + \theta\left[-y\frac{\partial}{\partial x} + x\frac{\partial}{\partial y}\right]\psi(x,y,z) \end{aligned} \tag{7.69}$$

Here and elsewhere in this chapter (for example in Eq. 7.58) we retain only terms up to the first order in infinitesimal quantities, neglecting second and higher order terms.

We observe that the differential operator appearing in Eq. 7.69 is just the operator for the z-component of the angular momentum, L_z, apart from a numerical factor. In fact, Eq. 7.69 reduces to

$$\psi'(\mathbf{x}) = [1 + i\theta L_z/\hbar]\,\psi(\mathbf{x}) \tag{7.70}$$

Thus $(L_z/\hbar)$ plays the role of the generator of infinitesimal rotations. More generally, if the rotation is about an arbitrary axis along the unit vector $\mathbf{n}$, we have

$$\psi'(\mathbf{x}) = [1 + i\theta\,\mathbf{n}.\mathbf{L}/\hbar]\,\psi(\mathbf{x}) \tag{7.71}$$

For a rotation through a finite angle θ, by considering it as the limit $N \to \infty$

[6] Unlike the space translation generators $\mathbf{p}$ or rotation generators (see next section), $(-H)$ has no universal role. Over and above the limitation to quantum systems, there is also the fact that its explicit form depends on the nature of the particular system, and may vary with time even for a given system. It is only when H is time-independent that one can write the effect of a finite time translation τ as $\psi(\mathbf{x}, t + \tau) = e^{-iH\tau/\hbar}\,\psi(\mathbf{x}, t)$.

[7] This means that the wave function is a scalar function. In the next chapter we shall see that for particles with spin different from zero, one has to use multicomponent (spinor, vector, ...) wave functions. The generators of rotations on vector functions are obtained in Sec. 7.10.

of a combination of N infinitesimal rotations, each through angle (θ/N), one gets

$$\psi'(\mathbf{x}) = e^{i\theta\mathbf{n}.\mathbf{L}/\hbar}\ \psi(\mathbf{x}) \tag{7.72}$$

We can once again view this transformation of the wave function as the result of a change of basis in the Hilbert space from eigenstates $|\,\mathbf{x}\,\rangle$ of position referred to S, to the eigenstates $|\,\mathbf{x}\,\rangle'$ referred to the rotated coordinate system S'. Thus we can rewrite Eq. 7.72 as

$$'\langle\,\mathbf{x}\,|\psi\rangle = \langle\,\mathbf{x}\,|\,e^{i\theta\mathbf{n}.\hat{\mathbf{L}}/\hbar}\,|\psi\rangle \tag{7.73}$$

and hence, in view of the fact that $|\psi\rangle$ can be arbitrary, $'\langle\,\mathbf{x}\,| = \langle\,\mathbf{x}\,|\,e^{i\theta\mathbf{n}.\hat{\mathbf{L}}/\hbar}$ or

$$|\,\mathbf{x}\,\rangle' = e^{-i\theta\mathbf{n}.\hat{\mathbf{L}}/\hbar}\,|\,\mathbf{x}\,\rangle \tag{7.74}$$

Finally, it is not difficult to convince oneself that for an n-particle system, the generators of rotations about the x-, y-, and z-axes are given by $(L_x/\hbar)$, $(L_y/\hbar)$, $(L_z/\hbar)$, where $\mathbf{L}$ is now the sum of the angular momentum operators of the individual particles: $\mathbf{L} = \sum_{\alpha=1}^{n} \mathbf{L}_\alpha$.

7.10 The Algebra of Rotation Generators

We make a brief digression at this point to draw attention to an important fact. The rotation generators have the specific form $(\mathbf{L}/\hbar)$ only when Eq. 7.68 is valid, i.e. ψ is a *scalar field.*[8] However, *the form of the commutation relations* of the generators, namely

$$\left[\frac{L_x}{\hbar}, \frac{L_y}{\hbar}\right] = i\frac{L_z}{\hbar},\ \left[\frac{L_y}{\hbar}, \frac{L_z}{\hbar}\right] = i\frac{L_x}{\hbar},\ \left[\frac{L_z}{\hbar}, \frac{L_x}{\hbar}\right] = i\frac{L_y}{\hbar} \tag{7.75}$$

is very general. We shall illustrate this fact by determining the effect of rotations on the components (V_x, V_y, V_z) of a vector, and more generally, on a vector field.

The transformation of the components of any vector under rotations about the z-axis has exactly the same form as Eq. 7.67. Using matrix notation, we write it as

$$\mathbf{V}' = R_z(\theta)\mathbf{V} \tag{7.76}$$

where $R_z(\theta)$ is the matrix of coefficients of (x,y,z) on the right hand side of Eq. 7.67. For infinitesimal θ we have explicitly,

$$\begin{pmatrix} V_x' \\ V_y' \\ V_z' \end{pmatrix} = \begin{pmatrix} 1 & \theta & 0 \\ -\theta & 1 & 0 \\ 0 & 0 & 1 \end{pmatrix}\begin{pmatrix} V_x \\ V_y \\ V_z \end{pmatrix} = (1 + i\theta\Sigma_z)\begin{pmatrix} V_x \\ V_y \\ V_z \end{pmatrix} \tag{7.77}$$

In the last expression, 1 stands for the unit matrix, and the generator, which is by definition the coefficient of $i\theta$, has been denoted by Σ_z. The form of the matrix Σ_z can be read off from the equation, and the generators Σ_x, Σ_y of rotations about the other axes can be obtained similarly. It is seen that

[8] As long as Eq. 7.68 is valid (even if $\psi(\mathbf{x})$ stands for temperature distribution in space], the form of the generator is identical to $(\mathbf{L}/\hbar)$; but $\mathbf{L}$ has physical interpretation as angular momentum only in the quantum mechanical context.

$$\Sigma_x = \begin{pmatrix} 0 & 0 & 0 \\ 0 & 0 & -i \\ 0 & i & 0 \end{pmatrix}, \; \Sigma_y = \begin{pmatrix} 0 & 0 & i \\ 0 & 0 & 0 \\ -i & 0 & 0 \end{pmatrix}, \; \Sigma_z = \begin{pmatrix} 0 & -i & 0 \\ i & 0 & 0 \\ 0 & 0 & 0 \end{pmatrix} \tag{7.78}$$

The commutators of these matrices may be easily calculated. One finds

$$[\Sigma_x, \Sigma_y] = i\Sigma_z, \;\; [\Sigma_y, \Sigma_z] = i\Sigma_x, \;\; [\Sigma_z, \Sigma_x] = i\Sigma_y \tag{7.79}$$

These are precisely of the form (7.75), with $\mathbf{S} \equiv \hbar\Sigma$ instead of $\mathbf{L}$. This confirms our assertion.

Consider now a vector field, wherein a vector quantity is defined at every point P of some region of space. If, as viewed from S and S', the components of this quantity at P are $\mathbf{V}_P$ and $\mathbf{V}_P'$ respectively, then according to Eq. 7.76,

$$\mathbf{V}_P' = R_z(\theta)\mathbf{V}_P \text{ or } \mathbf{V}'(\mathbf{x}') = R_z(\theta)\mathbf{V}(\mathbf{x}) \tag{7.80}$$

Eq. 7.80 defines a vector field. Its form differs from Eq. (7.68) of the scalar field in having the matrix operator $R_z(\theta)$ on the right hand side. But the relation between $\mathbf{x}'$ and $\mathbf{x}$ is the same in both cases. Hence, taking θ to be infinitesimal, (when $R_z(\theta) \sim 1 + i\theta\Sigma_z$) we can easily obtain the following as the counterpart of Eq. 7.69,

$$\begin{aligned} \mathbf{V}'(x,y,z) &= (1 + i\theta\Sigma_z)\, \mathbf{V}(x - y\theta,\, y + x\theta,\, z) \\ &= (1 + i\theta\Sigma_z)\, [\mathbf{V}(x,y,z) + i\theta(L_z/\hbar)\, \mathbf{V}(x,y,z)] \\ &= \left[1 + \frac{i\theta}{\hbar}(L_z + S_z)\right] \mathbf{V}(x,y,z) \end{aligned} \tag{7.81}$$

As usual, we have retained only terms to the first order in θ. The differential operator L_z acts on each component of $\mathbf{V}$ independently: $L_z\mathbf{V}$ has components L_zV_x, L_zV_y, L_zV_z. Since $S_z \equiv \hbar\Sigma_z$ is a matrix with elements independent of $\mathbf{x}$, it commutes with L_z; it acts on the column $\mathbf{V}$. More generally all components S_x, S_y, S_z of $\mathbf{S}$ commute with those of $\mathbf{L}$.

The generator of rotations (about the z-axis) on vector fields is identified from Eq. 7.81 to be $(L_z + S_z)/\hbar$, and one has similar expressions referring to the other axes. Since $\mathbf{L}$ commutes with $\mathbf{S}$ as already noted, it is easy to show, using Eqs. 7.75 and 7.79, that the components of

$$\mathbf{J} \equiv \mathbf{L} + \mathbf{S} \tag{7.82}$$

also have commutation relations of identical form.

We conclude, therefore, that while the actual expressions for the generators depend on what object is considered, *the form of the commutation relations* (which define what is called a *Lie algebra*) *is purely a characteristic of the operations considered, namely, infinitesimal rotations.*

We shall see in the next chapter (Sec. 8.3b) that the description of particles with one unit of spin angular momentum involves vector wave functions, and that the matrices $\hbar\Sigma$ of Eq. 7.79 appear as spin operators for such particles. Therefore, $\mathbf{J}$ is the total angular momentum operator, which is once again seen to be $\hbar$ times the generator of rotations.

7.11 Transformation of Dynamical Variables

We have shown in the foregoing sections that when the coordinate system is translated or rotated, any state $|\,\phi\,\rangle$ which is imagined to be 'tied' to the coordinate system (and rigidly translated or rotated along with it) goes over into $|\phi\rangle'$ where

$$|\phi\rangle' = e^{-i\alpha\hat{G}}|\phi\rangle \equiv \hat{U}^{-1}|\phi\rangle \tag{7.83}$$

Here $\hat{G}$ is the *generator* which characterizes the nature of the coordinate transformation ($\hat{p}_x/\hbar$ or $\mathbf{n}.\hat{\mathbf{L}}/\hbar$ in the examples considered), and α is a real parameter (ξ or θ in our examples) which specifies the extent or magnitude of the change of coordinate axes. It is to be noted that $\hat{G}$ is a Hermitian operator and hence $\hat{U} = e^{i\alpha\hat{G}}$ is *unitary*.

When a complete set of states $|a\rangle$ 'tied' to the coordinate system is used as a basis for a representation, the change of coordinate axes causes a change of the basis in Hilbert space from $|a\rangle$ to $|a\rangle'$ and correspondingly the *wave function of a given physical state* $|\psi\rangle$ *defined independently of the coordinate system* changes from $\psi(a) = \langle a|\psi\rangle$ to

$$\psi'(a) = {}'\langle a|\psi\rangle = \langle a|\, e^{i\alpha\hat{G}}|\psi\rangle = (U\psi)(a) \tag{7.84}$$

where $(U\psi)(a)$ stands for the wave function of $\hat{U}\,|\psi\rangle$ in the A-representation.

Suppose now that $\hat{F}$ is a dynamical variable defined in relation to S. Let

$$|\chi\rangle = \hat{F}|\phi\rangle \tag{7.85}$$

for an arbitrary state $|\phi\rangle$. If we subject the coordinate system *together with these states* to some transformation characterized by a unitary operator $\hat{U}^{-1}$, the transformed state $|\chi\rangle'$ is given by

$$|\chi\rangle' = \hat{U}^{-1}|\chi\rangle = \hat{U}^{-1}\hat{F}|\phi\rangle = \hat{U}^{-1}\hat{F}\hat{U}|\phi\rangle' \tag{7.86}$$

Thus the dynamical variable defined with respect to the new coordinate system, which relates $|\chi\rangle'$ to $|\phi\rangle'$, is given by

$$\hat{F}' = \hat{U}^{-1}\,\hat{F}\,\hat{U} \tag{7.87}$$

For example, let $\hat{F} = \hat{x}$ and $\hat{U}^{-1} = 1 - i\hat{p}_x\xi/\hbar$, i.e. an infinitesimal translation ξ in the x-direction. Then

$$\begin{aligned}\hat{F}' \equiv \hat{x}' &= [1 - i\hat{p}_x\xi/\hbar]\;\hat{x}\;[1 + i\hat{p}_x\xi/\hbar]\\ &= \hat{x} - \frac{i\xi}{\hbar}\,[\hat{p}_x,\hat{x}] = \hat{x} - \xi\end{aligned} \tag{7.88}$$

This is exactly what we need for Eq. 7.57. Similarly, if an infinitesimal rotation θ about the z-axis is performed, the operator for the momentum component parallel to the new x-axis (OX') is given by

$$\begin{aligned}\hat{p}_x' &= [1 - i\hat{L}_z\theta/\hbar]\;\hat{p}_x\;[1 + i\hat{L}_z\theta/\hbar]\\ &= \hat{p}_x - \frac{i\theta}{\hbar}\,[\hat{L}_z,\hat{p}_x] = \hat{p}_x + \theta\hat{p}_y\end{aligned} \tag{7.89}$$

This is in accordance with the transformation property of $\hat{\mathbf{p}}$ as a vector.

7.12 Symmetries and Conservation Laws

It may happen that certain dynamical variables $\hat{F}$ commute with the generator $\hat{G}$ associated with some particular transformation of coordinate axes. Such dynamical variables, for which

$$[\hat{F},\hat{G}] = 0 \text{ or } \hat{F}' = \hat{U}^{-1}\hat{F}\,\hat{U} = \hat{F} \tag{7.90}$$

are said to be *invariant* under the transformations concerned. For example, z, p_z, L_z are invariant under rotations about the z-axis, but other components of these vectors are *not* invariant, as exemplified by Eq. 7.89 for p_x.

Of special importance is the case when the Hamiltonian of a system is invariant. For, when $[\hat{H},\hat{G}] = 0$ or

$$\hat{U}^{-1} \hat{H} \hat{U} = \hat{H} \tag{7.91}$$

$\hat{G}$ is a constant of the motion; and the proportionality of $\hat{G}$ to observables (such as momentum and angular momentum components in the examples we have considered) implies then that such observables are conserved quantities for the system. Now, the invariance of $\hat{H}$ reflects a *symmetry* of the system. Invariance under translations (or rotations) means that the Hamiltonian has the same form whether expressed in terms of the original or transformed coordinate system; or in more physical terms, that the energy of the system is unaffected by shifting the system as a whole from one place to another (or rotating the system rigidly into a new orientation). The result which we have obtained may therefore be stated in the following general form: *The existence of a symmetry of a quantum system implies the existence of an associated constant of the motion,* i.e. conservation of a certain observable associated with the symmetry. Invariance with respect to translations in some direction implies conservation of the component of momentum in that direction; invariance with respect to rotations about a particular axis means that the angular momentum component parallel to this axis is conserved, and so on.

Closed or isolated systems do possess both translation and rotation invariance — a fact which is reflected in the laws of conservation of energy, momentum and angular momentum of such systems. These laws are believed to be of rigorous and universal validity, without any exceptions.

7.13 Space Inversion

(*a*) *Intrinsic Parity:* Till recently, it had been believed that invariance under space inversion also is a general property of physical systems. By space inversion we mean here the change in description of physical systems resulting from a reversal of the directions of all three coordinate axes. Clearly the coordinates $\mathbf{x}$ and $\mathbf{x}'$ of a point P referred to the original coordinate frame S and the space-inverted frame S' are related by

$$\mathbf{x}' = -\mathbf{x} \tag{7.92}$$

How does space inversion affect the value of a wave function at a specific point P? Before attempting to answer this question it is useful to recall that the values of the physical quantities of classical physics either remain unaltered or change sign under space inversion. These two types of quantities may be said to have, respectively, even and odd intrinsic parities. Simple examples of the even type are the ordinary scalars (temperature, electric charge, energy). Examples of the odd type are the components of ordinary vectors (force, electric field $\mathcal{E}$, momentum). Pseudovectors (magnetic field $\mathcal{H}$, angular momentum) behave as vectors under rotations, but have even intrinsic parity with respect to inversions. Similarly pseudoscalars have the opposite parity (odd) to that of ordinary scalars. They may be constructed by taking the dot product of any ordinary (polar) vector with a pseudovector. It is to be emphasized that the parity referred to here is an intrinsic geometrical property of the entity concerned: the odd intrinsic parity of the electric field $\mathcal{E}$ has nothing to do with whether $\mathcal{E}$ as a function of $\mathbf{x}$ is an odd function or even function or neither. It is simply a statement that $\mathcal{E}$ at any point is an *ordinary vector*. In fact the intrinsic parity is defined even if the quantity in question exists only at a single point, e.g. the charge of a classical particle, which is concentrated in a single point.

Returning now to the wave function, let us suppose that its value ψ_P' at a physical point P, as seen from S' is related to the value ψ_P in S by $\psi_P' = \eta_P \psi_P$, where η is a constant.[9] On performing one more space inversion we get ψ_P'' in S'' to be $\psi_P'' = \eta\psi_P' = \eta^2\, \psi_P$. But two inversions restore the reference frame to its original form. It would seem obvious then that the wave function also returns to its original value, i.e. $\psi_P'' = \psi_P$ and hence $\eta^2 = 1$. In fact, we can take this to be the case for spinless particles (and more generally, for integer spins). However, a subtle point arises in the case of particles of half-integral spin. As we shall see in Example 8.2 (p. 246) for spin $\frac{1}{2}$, a rotation of the coordinate frame through 2π about any axis (say from S to $S_{(2\pi)}$) makes the wave function change sign though there is no physical distinction between S and $S_{(2\pi)}$. This means that we may take ψ_P'' to be $+\psi_P$ or $-\psi_P$ when the spin is half-integral, depending on whether we identify S'' with S itself or with $S_{(2\pi)}$. Consequently, we can have $\eta^2 = +1$ or -1. However, we shall ignore the latter possibility ($\eta = \pm i$) in the following, since none of the numerous particles discovered so far seems to call for such values of η. Therefore we are left with $\eta = \pm 1$. All states ψ of a given particle must be characterized by one and the same value of η. It is called the *intrinsic parity* of the particle. Its defining equation

$$\psi_P' = \eta\psi_P \text{ or } \psi'(-\mathbf{x}) = \eta\psi(\mathbf{x}) \tag{7.93}$$

is to be carefully distinguished from Eq. 4.95 which defines what may be called orbital parity. In the latter case, the values of ψ at *different* physical points (with coordinates $\mathbf{x}', -\mathbf{x}$ in a *given* reference frame) are compared, unlike in Eq. 7.93 where both sides refer to the *same* physical point P. The intrinsic parity of a system of particles is the product of the intrinsic parities of the individual particles constituting the system.

(*b*) *The Unitary Operator of Space Inversion:* As in the case of other coordinate transformations, the wave functions of a state $|\psi\rangle$ in S and S', namely $\langle \mathbf{x}|\psi\rangle \equiv \psi(\mathbf{x})$ and $'\langle \mathbf{x}|\psi\rangle \equiv \psi'(\mathbf{x})$, must be related by a unitary operator, say U_I. (The subscript I stands for 'inversion'):

$$\psi'(\mathbf{x}) = U_I\, \psi(\mathbf{x}), \text{ or } \eta\psi\,(-\mathbf{x}) = U_I\psi(\mathbf{x}) \tag{7.94}$$

This equation defines U_I through its effect on arbitrary wave functions. This

[9] It will be made explicit in Chapter 8 that the wave function for any particle with non-zero spin is a column with a number of components, each of which is a function of $\mathbf{x}$. (We have already seen an example: the vector wave function for spin 1, referred to in Sec. 7.10). In such cases the relation between ψ_P' and ψ_P may involve a constant matrix acting on ψ_P besides the factor η. Such a matrix is in fact necessary in relativistic theories such as the Dirac theory of the electron (Chapter 10).

definition clearly implies that if a particular wave function $\psi(\mathbf{x})$ is to be an eigenfunction of U_I then $\psi(-\mathbf{x})$ must be a multiple of $\psi(\mathbf{x})$, i.e. $\psi(\mathbf{x})$ must have definite orbital or *spatial* parity $\eta_s = \pm 1$. For such a wave function, $U_I\psi(\mathbf{x}) = \eta\eta_s\, \psi(\mathbf{x})$, so that the eigenvalue is $\eta\eta_s$.

The effect of U_I on dynamical variables can be readily found from the definition (7.94). For example, $U_I\,[\mathbf{x}\psi(\mathbf{x})] = \eta[-\mathbf{x}\psi(-\mathbf{x})] = -\mathbf{x}U_I\,\psi(\mathbf{x})$, so that $U_I\mathbf{x} = -\mathbf{x}U_I$, with similar considerations holding good for $\mathbf{p}$. Thus

$$U_I\,\mathbf{x}U_I^{\dagger} = -\mathbf{x},\quad U_I\,\mathbf{p}U_I^{\dagger} = -\mathbf{p},\quad U_I\mathbf{L}U_I^{\dagger} = \mathbf{L} \tag{7.95}$$

etc. If U_I leaves the Hamiltonian H invariant, ($U_IHU_I^{\dagger} = H$ or $U_IH = HU_I$), then we say that the system has *space inversion symmetry.* A consequence of such symmetry is that U_I is a constant of the motion, i.e. *parity is conserved.* In stating above that a definite intrinsic parity is associated with each particle, we have implicitly assumed parity conservation.

As long as parity conservation holds, any reaction among particles must proceed in such a way that the overall parity $\eta\eta_s$ has the same value in the initial and final states i and f.

$$(\eta\eta_s)_i = (\eta\eta_s)_f \tag{7.96}$$

If both the states contain the same set of particles, the intrinsic parities cancel out between the two sides of Eq. 7.96 which then reduces to a statement of conservation of spatial parity. It is only when a reaction involves production or destruction of particles that intrinsic parity comes into play. The ratio of the spatial parities in the initial and final states is then fixed by the ratio of the intrinsic parities. Since the first of these ratios is susceptible to direct experimental measurement, the latter can be inferred therefrom. By examining suitable particle production or decay processes one can thus determine the intrinsic parities of various particles. For example, from the characteristics of the two-photon state resulting from the decay of π° mesons at rest, it has been determined that these mesons have odd intrinsic parity ($\eta = -1$). This information, together with the fact that the spin of π° is zero, is succinctly expressed by saying that π° is a pseudoscalar particle. (Particles with spin 0 and $\eta = +1$ would be called scalar, those with spin 1 and $\eta = -1$ would be vector, etc.).

(*c*) *Parity Nonconservation:* It is now known that not all kinds of interactions of elementary particles conserve parity. The 'strong interactions' (of which the nuclear forces are a manifestation) do possess space-inversion invariance. This is true of electromagnetic interactions too. These are the interactions responsible for the decay of the π° meson, which is therefore a parity-conserving process. But parity is not conserved in processes such as beta decay of nuclei (or of the free neutron), decay of the $\pi^\pm$ mesons etc., caused by the so-called weak interactions.[10] Because of the weakness of the parity-violating interaction,

[10] Elementary particles are thought of as quanta of corresponding quantum fields (like photons for the electromagnetic field, see Chapter 9). Interactions among the fields result in the creation or destruction of quanta (i.e. particles). Various kinds of interactions characterized by typical strengths have been recognized. In the order of decreasing strength are the strong, electromagnetic, weak and gravitational interactions. The suggestion that parity violation occurs in weak interactions is due to T. D. Lee and C. N. Yang, *Phys. Rev.*, **104**, 254, 1956.

it is still true to a very good approximation that each particle has a definite parity. Parity violation reveals itself through the occurrence of processes leading from an initial state of definite overall parity to a final state of mixed parity. Experimental tests for parity violation are based on the fact that if parity is conserved in a process, the cross-section for the process should remain unaltered on replacing the values of the dynamical variables of the particles in the initial and final states by the corresponding space-inverted values; i.e. in two different configurations which are related to each other by space inversion, the process should proceed at the same rate. Any violation of this requirement shows a breakdown of space-inversion invariance, or equivalently, nonconservation of parity. As a simple example, consider a system characterized by a vector (e.g. momentum) and a pseudovector (e.g. angular momentum). The former is represented in Fig. 7.3 by a single arrow, and the latter by a

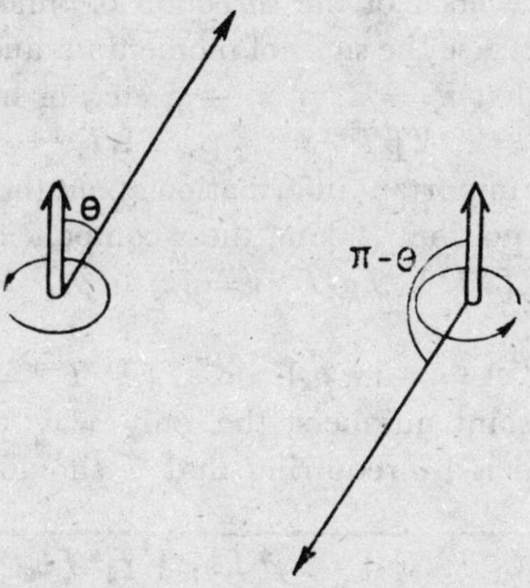

Fig. 7.3 Space-inversion of a system characterized by a vector and a pseudo-vector.

double-line arrow (and also by a directed circle depicting rotation). Space inversion reverses the direction of the vector, but not that of the pseudovector. Therefore, the angle between these two is changed from θ to $\pi - \theta$ by space inversion; and invariance under space inversion would require the cross-sections at both these angles to be the same. The first observation of parity nonconservation came from an experiment set up to test precisely this prediction. In the actual experiment,[11] the pseudovector was the angular momentum of Co^{60} nuclei, oriented in a definite direction with the help of a magnetic field; the vector was the momentum of the electron resulting from the β-decay of the nucleus. The observation was that there is a forward-backward asymmetry in the distribution of decay electrons with respect to the direction of the spin of the Co^{60} nucleus. In other words, the cross-sections for appearance of the electron at θ and $(\pi - \theta)$ were found to be not equal, thereby establishing that parity is not conserved in this process.

[11] C. S Wu, E. Ambler, R. H. Hayward, D. D. Hoppes and R. P. Hudson, *Phys. Rev.*, **105**, 1413, 1957.

7.14 Time Reversal[12]

Suppose we were to reverse the direction of the time axis, i.e. label instants of time by a variable $t' \equiv - t$ which increases towards the past (as in counting years B.C.). How would this affect the description of a quantum system? It is reasonable to expect that the effect on the wave function can be expressed through an operator, say T:

$$\psi'(\mathbf{x},t') = T\psi(\mathbf{x},t), \quad t' = -t \tag{7.97}$$

This relation implies that the effect of time reversal on any (time-independent) operator A representing a dynamical variable is to change it to

$$A' = TAT^{-1} \tag{7.98}$$

This follows from the definition of A': if $\chi = A\psi$, then A' is the operator which relates χ' to ψ'. Since $\chi' = TA\psi = TAT^{-1}(T\psi) = TAT^{-1}\psi'$, we immediately get (7.98).

Now we know that reversal of the direction of time would not affect the position $\mathbf{x}$, but should reverse the signs of momentum and angular momentum. Thus we must require that $\mathbf{x}' = \mathbf{x}$, $\mathbf{p}' = -\mathbf{p}$ etc., or in view of Eq. 7.98,

$$T\mathbf{x}T^{-1} = \mathbf{x}, \quad T\mathbf{p}T^{-1} = -\mathbf{p}, \quad T\mathbf{L}T^{-1} = -\mathbf{L} \tag{7.99}$$

These equations contain important information about the nature of the operator T. From the first two equations (taking the x-components only) it follows that

$$[TxT^{-1}, Tp_xT^{-1}] = [x, -p_x] \tag{7.100}$$

and hence

$$T[x, p_x]\, T^{-1} = -[x, p_x], \text{ i.e. } T(i\hbar)\, T^{-1} = -i\hbar \tag{7.101}$$

Since i, $\hbar$ are just constant numbers, the only way the equation $T(i\hbar) = -(i\hbar)T$ can be satisfied is by requiring that T should *change any number into its complex conjugate.*

$$T(c_1\psi_1 + c_2\psi_2) = c_1^*T\psi_1 + c_2^*T\psi_2 \tag{7.102}$$

This means that T is *not* a linear operator. Any operator having the property (7.102) is said to be *antilinear.*

The antilinearity of T has the following happy consequence. It ensures that if the Hamiltonian H is time-reversal-invariant by virtue of Eqs. (7.99), then $\psi'(\mathbf{x},t')$ satisfies the Schrödinger equation with respect to the reversed time variable.

$$i\hbar\frac{\partial\psi'(\mathbf{x}, t')}{\partial t'} = H\psi'(\mathbf{x}, t') \tag{7.103}$$

This result follows on applying T to both sides of the Schrödinger equation for ψ, and using $T(i\hbar) = -(i\hbar)\,T$ and $TH = HT$ together with Eqs. 7.97.

Since an antilinear operator differs from a linear one only in effecting an extra complex conjugation, it is evidently possible to consider it as a linear operator U times a complex conjugation operator K. So we can write

$$T = UK, \quad T\psi(\mathbf{x}, t) = UK\psi(\mathbf{x}, t) = U\psi^*(\mathbf{x}, t) \tag{7.104}$$

The notation U for the linear part of T is deliberately chosen, to indicate that it is unitary. It has to be unitary, in order that the norm of ψ' be the same as that of ψ. In fact, $(\psi', \psi') = (UK\psi,\ UK\psi) = (K\psi,\ U^\dagger UK\psi) = (\psi^*,\ \psi^*)$

[12] E. P. Wigner, *Göttinger Nachr.*, **31**, 546, 1932

provided $U^\dagger U = 1$, and this reduces to (ψ, ψ) since the norm is real. The same argument shows that any scalar product is changed into its complex conjugate under time reversal:

$$(T\phi, T\psi) = (\phi,\psi)^* \equiv (\psi, \phi) \tag{7.105}$$

This property of T is completely equivalent to its decomposability into the form UK, U being unitary. Any operator which is characterized by either (and hence both) of these properties, is said to be *antiunitary*. The antiunitarity of T implies that if A is any operator,

$$(\phi, A\psi) = (T\phi, TA\psi)^* = (T\phi, A'\,T\psi)^* \tag{7.106}$$

where $A' = TAT^{-1}$. In particular, if A is time-reversal invariant, $A' = A$, the matrix element of A between any two states is equal to the complex conjugate of that between time-reversed states.

The explicit form of U necessarily depends on the system considered and also on the representation used. For a spinless particle, in the coordinate representation (where $\mathbf{x}$ is a real operator and $\mathbf{p} = -i\hbar\nabla$ is imaginary), it is clear that all the requirements (7.99) are satisfied with

$$U = 1, \quad T = K \tag{7.107}$$

However, in the momentum representation ($\mathbf{p}$ real, $\mathbf{x} = -i\hbar\nabla_{\mathbf{p}}$ imaginary) U cannot be unity. To satisfy Eqs. 7.99, it has to be so defined that $U\mathbf{p}U^{-1} = -\mathbf{p}$ and hence

$$T\psi(\mathbf{p}, t) = U\psi^*(\mathbf{p}, t) = \psi^*(-\mathbf{p}, t). \tag{7.108}$$

In the case of particles with nonzero spin, we require, in addition to Eqs. 7.99, that the *spin* angular momentum operator $\mathbf{S}$ should change sign under time reversal:

$$T\mathbf{S}T^{-1} = -\mathbf{S} \tag{7.109}$$

The operators S_x, S_y, S_z are matrices which act on the multicomponent column wave function appropriate to the spin being considered. Such examples, for spin $\frac{1}{2}$ and 1, are given in Eqs. 8.21 and 8.22 (where they are denoted by J_x, J_y, J_z). The representation used there, which is usually the most convenient, is one in which S_x and S_z are real and S_y is imaginary so that $KS_x = S_xK$, $KS_y = -S_yK$, $KS_z = S_zK$. On putting $T = UK$ in Eq. 7.109 and using these results, we get

$$US_xU^{-1} = -S_x, \quad US_yU^{-1} = S_y, \quad US_zU^{-1} = -S_z \tag{7.110}$$

Thus the unitary part of T should reverse the signs of S_x and S_z, but not of S_y. This can obviously be accomplished by a rotation through π about the y-axis. From Sec. 7.9 we know that the operator for such a rotation is $e^{i\pi S_y/\hbar}$ ($S_y/\hbar$ being the generator of rotations about the y-axis). Hence

$$T = e^{i\pi S_y/\hbar}\,K \tag{7.111}$$

Since (iS_y) is real, K commutes with it; using this and the obvious fact that $K^2 = 1$, we obtain

$$T^2 = e^{2\pi i S_y/\hbar} = (-)^{2s} \tag{7.112}$$

The last step is a statement of the fact already referred to in Sec. 7.13a, that a 2π rotation leaves wave functions of integer-spin particles ($2s$ even) unchanged, while those of half-integer-spin particles ($2s$ odd) change sign. Thus $T^2 = +1$

for particles of integral spin and $T^2 = -1$ for half-integral spin. The same result holds for systems of particles since each particle contributes its own factor $(-)^{2s}$ to T^2.

An interesting corollary is that for a particle system with half-integral spin, any state ψ and its time-reversed state $T\psi$ are orthogonal. This may be seen by taking $\phi = T\psi$ in (7.105), whence

$$(\psi, T\psi) = (T^2\psi, T\psi) = -(\psi, T\psi), \ (2s \text{ odd}) \tag{7.113}$$

and hence $(\psi, T\psi)$ vanishes. One consequence is a degeneracy in the eigenvalues of time-reversal-invariant operators belonging to such systems. The invariance implies that if ψ belongs to a certain eigenvalue, $T\psi$ also belongs to the same eigenvalue; and since ψ and $T\psi$ are independent states (being orthogonal) there is obviously a degeneracy.

PROBLEMS

1. If $|\phi\rangle$, $|\psi\rangle$ are unnormalized vectors of a Hilbert space, show that
$$|\langle\phi|\psi\rangle|^2 \leqslant \langle\phi|\phi\rangle\langle\psi|\psi\rangle.$$
This is *Schwarz's inequality*. [Hint: Minimize the norm of $|\phi\rangle + \lambda|\psi\rangle$ with respect to the real and the imaginary parts of the complex parameter λ, and apply Eqs. 7.3].

2. If $H = \mathbf{p}^2/2m + V(\mathbf{x})$, verify that $[x, [x, H]] = -(\hbar^2/m)$. Use this to prove the sum rule
$$\sum_k (E_k - E_n)\,|x_{kn}|^2 = (\hbar^2/2m),$$
where $x_{kn} = \langle k|x|n\rangle$ and $H|n\rangle = E_n|n\rangle$.

3. Write down the matrices for the ladder operators a, $a^\dagger$ of the harmonic oscillator in the representation which diagonalizes the Hamiltonian.

4. Obtain matrices for the operators a and $a^\dagger$ of Prob. 13 of Chapter 4, in the representation in which N is diagonal. Can the matrix of a be diagonalized?

5. Show that the projection operator to the eigenspace belonging to the eigenvalue a of A can be written as $P_a = \prod_{a' \neq a}\left(\dfrac{A-a'}{a-a'}\right)$, where the product is over all eigenvalues $a' \neq a$ of A. (Hint: Show $P_a|a''\rangle = \delta_{a'',a}|a\rangle$)

6. Verify that for a vector field $\mathbf{V}$, $(\Sigma.\mathbf{n})\,\mathbf{V} = i\mathbf{n}\times\mathbf{V}$.

7. If $\mathbf{n}$ is a unit vector, show that the matrices Σ of (7.78) satisfy the equation $(\Sigma.\mathbf{n})^3 = (\Sigma.\mathbf{n})$, and hence that $\exp(i\Sigma.\mathbf{n}\theta) = 1 + i(\Sigma.\mathbf{n})\sin\theta + (\Sigma.\mathbf{n})^2(\cos\theta - 1)$. Verify that the effect of a rotation through θ about the axis $\mathbf{n}$ on the components of a vector is given by
$$\mathbf{V}' = e^{i\Sigma.\mathbf{n}\theta}\,\mathbf{V} = \mathbf{V}\cos\theta + (\mathbf{V}.\mathbf{n})\,\mathbf{n}\,(1-\cos\theta) - (\mathbf{n}\times\mathbf{V})\sin\theta.$$

8. From the form of $\exp(i\Sigma.\mathbf{n}\theta)$ given in Prob. 7, show that the antisymmetric part of the rotation matrix determines the axis of rotation, and the trace determines the angle.

9. Let j_x, j_y, j_z be a set of operators obeying the same commutation rules (7.75) as the components of $(L_x/\hbar)$. Setting $X = \mathbf{j}.\mathbf{n}$ and $Y = \mathbf{j}$ in Prob. 10, Chapter 3, show that all the multiple commutators of X and Y can be evaluated. Hence show that
$$e^{i\theta\mathbf{j}.\mathbf{n}}\,\mathbf{j}\,e^{-i\theta\mathbf{j}.\mathbf{n}} = \mathbf{j}\cos\theta + (\mathbf{j}.\mathbf{n})\,\mathbf{n}\,(1-\cos\theta) + (\mathbf{n}\times\mathbf{j})\sin\theta.$$
Compare with the transformation of vector components in Prob. 7 and comment on the result in the light of the results of Sec. 7.11.

8

Angular Momentum

In this chapter we study some of the most basic aspects of the quantum theory of angular momentum. A study which is general enough to cover spin angular momentum cannot be based on the definition of angular momentum as $\mathbf{L} = \mathbf{x} \times \mathbf{p}$, since the treatment of Chapter 4 based on this definition did not admit $\frac{1}{2}\hbar$ as a possible value. (As will be recalled, experiment demands that spin of this magnitude be attributed to the electron). To generalize the definition, we take the clue from the observation of Sec. 7.9 that the components of $\mathbf{L}$ are proportional to the generators of infinitesimal rotations. We observed there that the commutation relations of the generators are very general, including those of L_x, L_y, L_z as a special case. We postulate, therefore, that the fundamental definition of angular momentum is given by the commutation relations[1]

$$[J_x, J_y] = i\hbar J_z, \ [J_y, J_z] = i\hbar J_x, \ [J_z, J_x] = i\hbar J_y \tag{8.1}$$

The symbol $\mathbf{J}$ is introduced here for a general angular momentum vector, reserving the symbol $\mathbf{L}$ for the particular form $\mathbf{x} \times \mathbf{p}$. We shall see immediately that Eqs. 8.1 are, by themselves, sufficient to determine the eigenvalue spectrum of $\mathbf{J}^2$ as well as that of any component of $\mathbf{J}$, and that the spectrum includes half-integral values. They also lead to a variety of important results which are of great utility in a wide range of applications. On account of space restrictions we will be able to consider only a select few of these.

[1] Commutation relations of exactly the same form have been postulated for another kind of quantity called isotopic spin (or just isospin), which is physically quite different from angular momentum. See Appendix F.

8.1 The Eigenvalue Spectrum

In the derivation to be presented now, the operators

$$J_+ = J_x + iJ_y \text{ and } J_- = J_x - iJ_y \qquad (8.2)$$

will play a central role. The commutation rules (8.1) may readily be rewritten in terms of J_+, J_- and J_z as

$$[J_z, J_+] = \hbar J_+ \qquad (8.3a)$$
$$[J_z, J_-] = -\hbar J_- \qquad (8.3b)$$
$$[J_+, J_-] = 2\hbar J_z \qquad (8.3c)$$

Note that J_+ and J_- are *not* hermitian. Instead,

$$J_+^\dagger = J_- \text{ and } J_-^\dagger = J_+ \qquad (8.4)$$

The square of the angular momentum can now be expressed in various forms:

$$\mathbf{J}^2 = J_x^2 + J_y^2 + J_z^2 \qquad (8.5a)$$
$$= J_- J_+ + \hbar J_z + J_z^2 \qquad (8.5b)$$
$$= J_+ J_- - \hbar J_z + J_z^2 \qquad (8.5c)$$

It is a simple matter to see that these forms are all equal. For instance, $J_- J_+ = J_x^2 + J_y^2 + iJ_x J_y - iJ_y J_x = J_x^2 + J_y^2 - \hbar J_z$, which shows that (8.5b) reduces to (8.5a).

We proceed now to use these facts to arrive at the eigenvalue spectrum. Observe first of all that by virtue of the commutation rules (8.1) or (8.3),

$$[J_x, \mathbf{J}^2] = [J_y, \mathbf{J}^2] = [J_z, \mathbf{J}^2] = 0 \qquad (8.6)$$

Therefore, it is possible (see Sec. 3.13) to find a complete set of simultaneous eigenstates of $\mathbf{J}^2$ and J_z (of course any other component of $\mathbf{J}$ could be used instead of J_z). Consider one such state, belonging to the eigenvalue $\lambda\hbar^2$ of $\mathbf{J}^2$ and $m\hbar$ of J_z. In this state, the expectation values of $\mathbf{J}^2$ and J_z^2 are evidently $\langle \mathbf{J}^2 \rangle = \lambda\hbar^2$ and $\langle J_z^2 \rangle = m^2\hbar^2$. Thus we get

$$\lambda\hbar^2 = \langle J_x^2 \rangle + \langle J_y^2 \rangle + m^2\hbar^2 \qquad (8.7)$$

Now, since J_x and J_y are Hermitian, J_x^2 and J_y^2 are positive operators, and their expectation values are necessarily non-negative (see Eq. 3.26). It follows then from Eq. 8.7 that $\lambda \geqslant m^2$. We do not know the values which λ and m are permitted to take—our immediate objective in fact is to determine what they are—but we know now that if we want to have a state with some (positive) value j for m, the value of λ cannot be less than j^2. Let the smallest value of λ consistent with this condition be λ_j. Given this value of λ, evidently j is the maximum value that m can take. Thus

$$j = m_{\max}, \quad \lambda \geqslant j^2 \geqslant m^2 \qquad (8.8)$$

We now ask the question: given j and hence λ_j (as a function of j which is as yet unknown), what are the permissible values for m?

Let us denote the state vector belonging to the eigenvalues $\lambda_j\hbar^2$ of $\mathbf{J}^2$ and $m\hbar$ of J_z by $|jm\rangle$.

$$\mathbf{J}^2 |jm\rangle = \lambda_j\hbar^2 |jm\rangle \qquad (8.9a)$$
$$J_z |jm\rangle = m\hbar |jm\rangle \qquad (8.9b)$$

We take these states to be normalized; and of course any two states $|jm\rangle$ and $|j'm'\rangle$ with $j \neq j'$ or $m \neq m'$ are orthogonal. Thus

$$\langle j'm' | jm \rangle = \delta_{jj'} . \delta_{mm'} \qquad (8.9c)$$

The crucial observation now is that J_+ and J_- act as 'ladder' operators, which respectively raise and lower the eigenvalue of J_z by one unit, i.e. $J_+ |jm\rangle$ and $J_- |jm\rangle$ belong to the eigenvalues $(m + 1)\hbar$ and $(m - 1)\hbar$ respectively of J_z. This statement is easily verified with the help of Eqs. 8.3. For example, Eq. 8.3a states that $J_z J_+ = J_+ J_z + \hbar J_+$, so that

$$J_z . J_+ |jm\rangle = J_+ J_z |jm\rangle + \hbar J_+ |jm\rangle$$
$$= (m + 1)\hbar . J_+ |jm\rangle \qquad (8.10a)$$

Of course,

$$\mathbf{J}^2 . J_+ |jm\rangle = J_+ . \mathbf{J}^2 |jm\rangle = \lambda_j \hbar^2 J_+ |jm\rangle \qquad (8.10b)$$

Comparing Eqs. 8.10 with 8.9 we conclude that $J_+ |jm\rangle$ is nothing but $|j,m + 1\rangle$ apart from a possible normalization constant. In a similar way we can show that $J_- |jm\rangle \propto |j,m - 1\rangle$. Thus

$$J_+ |jm\rangle = c^+_{jm} \hbar |j,m + 1\rangle \qquad (8.11a)$$
$$J_- |jm\rangle = c^-_{jm} \hbar |j,m - 1\rangle \qquad (8.11b)$$

Here c^+_{jm} and c^-_{jm} are normalization constants, to be determined.

Eq. 8.11a shows that given a state $|jm\rangle$, the state $|j,m + 1\rangle$ must exist *unless* c^+_{jm} vanishes for that particular m. We know that since j is the maximum value of m by definition, there cannot be a state $|j,j + 1\rangle$, i.e. c^+_{jj} must vanish. Hence

$$J_+ |jj\rangle = 0 \qquad (8.12)$$

This equation determines λ_j as a function of j. To see this, consider the equation $\mathbf{J}^2 |jj\rangle = \lambda_j \hbar^2 |jj\rangle$ and substitute the expression (8.5b) for $\mathbf{J}^2$. We get $(J_- J_+ + \hbar J_z + J_z^2) |jj\rangle = \lambda_j \hbar^2 |jj\rangle$. The first term on the left vanishes on account of Eq. 8.12 and since $J_z |jj\rangle = j\hbar |jj\rangle$, we are immediately led to

$$\lambda_j = j(j + 1) \qquad (8.13)$$

We can proceed in an exactly similar way to show that if m' is the minimum value possible for m, then $\lambda_j = m'(m' - 1)$ and hence, by equating this to (8.13), that $m' = -j$. (If m' is the minimum value, a state $|j,m' - 1\rangle$ does not exist and hence $J_- |jm'\rangle = 0$. Considering $\mathbf{J}^2 |jm'\rangle$ with the form (8.5c) for $\mathbf{J}^2$, one immediately gets the stated result). Thus, if we start with the state $|jj\rangle$ having the maximum value j for m, and then reduce the value of m one unit at a time by repeated application of J_- we should, in the end, arrive at the minimum value $-j$. Hence, for given j, the possible values of m are

$$m = j, j - 1, \ldots, -j + 1, -j \qquad (8.14a)$$

Observe that this procedure tells us that the difference between the maximum and minimum values, namely $2j$, must be a non-negative integer. Thus the possible values of j are determined to be

$$j = 0, \tfrac{1}{2}, 1, \tfrac{3}{2}, \ldots \qquad (8.14b)$$

The eigenvalue spectrum has thus been obtained in a remarkably simple fashion, starting from the commutation rules. Moreover, half-integral values of j have automatically emerged from this general treatment. As we already know, half-integral values are possible only when spin is involved.

8.2 Matrix Representation of J in the $| jm \rangle$ Basis

The complete orthonormal set of states $| jm \rangle$ can be used as a basis for a representation (the *angular momentum representation*) in which any function A of the angular momentum components will be represented by a matrix with elements $\langle j'm' | A | jm \rangle$. The rows of the matrix will be labelled by various values of j' and m', and the columns by j and m. Since the basis states are eigenstates of $\mathbf{J}^2$ and J_z, we know from the results of Sec. 7.4 that the matrices for these operators must be diagonal. In fact, we see directly from Eqs. 8.9 and 8.13 that

$$\langle j'm' | \mathbf{J}^2 | jm \rangle = j(j+1)\hbar^2 \delta_{jj'} \delta_{mm'} \tag{8.15a}$$

$$\langle j'm' | J_z | jm \rangle = m\hbar \, \delta_{jj'} \delta_{mm'} \tag{8.15b}$$

As far as J_+ and J_- are concerned, we find by taking the scalar product of $| j'm' \rangle$ with Eqs. 8.11 and using Eq. 8.9c that

$$\langle j'm' | J_+ | jm \rangle = c^+_{jm} \hbar \, \delta_{jj'} \delta_{m',m+1} \tag{8.16a}$$

$$\langle j'm' | J_- | jm \rangle = c^-_{jm} \hbar \, \delta_{jj'} \delta_{m',m-1} \tag{8.16b}$$

The values of c^+_{jm} and c^-_{jm} will now be determined by equating the norms of the two sides in each of Eqs. 8.11.
From Eq. 8.11a we have

$$\langle jm | J_- J_+ | jm \rangle = (c^+_{jm})^* c^+_{jm} . \hbar^2 \langle j,m+1 | j,m+1 \rangle$$
$$= | c^+_{jm} |^2 . \hbar^2 \tag{8.17}$$

In writing the left hand side of this equation, we have recalled the definition (7.3a) of the norm and observed that the bra conjugate to $J_+ | jm \rangle$ is $\langle jm | J_-$ since $J_+^\dagger = J_-$. The left hand side of Eq. 8.17 can be evaluated by expressing $J_- J_+$ in terms of $\mathbf{J}^2$ using Eq. 8.5b. Then it becomes

$$\langle jm | (\mathbf{J}^2 - \hbar J_z - J_z^2) | jm \rangle = [j(j+1) - m - m^2] \hbar^2 \tag{8.18}$$

From the equality of the right hand sides of Eqs. 8.17 and 8.18 we obtain[2]

$$c^+_{jm} = [j(j+1) - m(m+1)]^{1/2} = [(j-m)(j+m+1)]^{1/2} \tag{8.19a}$$

Now, since $J_+ J_- | jm \rangle = (\mathbf{J}^2 + \hbar J_z - J_z^2) | jm \rangle$, we find on evaluating both sides using Eqs. 8.11 and 8.9 that $c^+_{j,m-1} c^-_{jm} = [j(j+1) - m(m-1)]$. Substituting from Eq. 8.19a for $c^+_{j,m-1}$ we obtain

$$c^-_{jm} = [j(j+1) - m(m-1)]^{1/2} = [(j+m)(j-m+1)]^{1/2} \tag{8.19b}$$

This completes the determination of the matrix elements of J_+ and J_-.

Returning now to Eqs. 8.15 and 8.16, we observe that *all* of them contain the factor $\delta_{jj'}$. Therefore, when the full (infinite dimensional) matrix for any of the operators $J_+, J_-, \ldots$ is written out, all nonvanishing elements occur inside 'blocks' appearing along the diagonal, within each of which $j' = j$ (This is a particular case of the general theory of Sec. 7.6). In taking sums, products etc. of two such materices, only those diagonal blocks having the same value of j in the two matrices will combine, and the result is again block-diagonal. Thus instead of dealing with the infinite dimensional matrices we can use the blocks corresponding to each j separately. Such representation

[2] Actually we could include a phase factor $e^{i\alpha}$ also since only $| c^+_{jm} |$ appears in Eq. 8.12 but this is an avoidable complication. We know that phases of states can be arbitrarily chosen, and in making c^+_{jm} real we are merely making the simplest choice.

matrices for specific values of j are given below. The matrix elements are calculated from Eqs. 8.16 and 8.19, using the fact that $J_x = \frac{1}{2}(J_+ + J_-)$ and $J_y = -\frac{1}{2}i(J_+ - J_-)$.

For $j = 0$

$$J_x = J_y = J_z = \mathbf{J}^2 = 0 \tag{8.20}$$

For $j = \frac{1}{2}$

$$J_x = \tfrac{1}{2}\hbar \begin{pmatrix} 0 & 1 \\ 1 & 0 \end{pmatrix}, \; J_y = \tfrac{1}{2}\hbar \begin{pmatrix} 0 & -i \\ i & 0 \end{pmatrix}, \; J_z = \tfrac{1}{2}\hbar \begin{pmatrix} 1 & 0 \\ 0 & -1 \end{pmatrix},$$

$$\mathbf{J}^2 = \frac{3}{4}\hbar^2 \begin{pmatrix} 1 & 0 \\ 0 & 1 \end{pmatrix} \tag{8.21}$$

For $j = 1$

$$J_x = \frac{\hbar}{\sqrt{2}} \begin{pmatrix} 0 & 1 & 0 \\ 1 & 0 & 1 \\ 0 & 1 & 0 \end{pmatrix}, \; J_y = \frac{\hbar}{\sqrt{2}} \begin{pmatrix} 0 & -i & 0 \\ i & 0 & -i \\ 0 & i & 0 \end{pmatrix}, \; J_z = \hbar \begin{pmatrix} 1 & 0 & 0 \\ 0 & 0 & 0 \\ 0 & 0 & -1 \end{pmatrix}$$

$$\mathbf{J}^2 = 2\hbar^2 \begin{pmatrix} 1 & 0 & 0 \\ 0 & 1 & 0 \\ 0 & 0 & 1 \end{pmatrix} \tag{8.22}$$

Note that the rows are labelled by values of m' (and the columns by m), starting with j and going down to $-j$ in unit steps. For example, in the matrix for J_y in (8.22), the elements in the first row are $\langle 11 | J_y | 11 \rangle = 0$, $\langle 11 | J_y | 10 \rangle = -i$, $\langle 11 | J_y | 1, -1 \rangle = 0$, and so on.

To complete the picture, we should also take note of the kind of entities on which these matrices act. The representative of an arbitrary state $|\psi\rangle$ with respect to the $|jm\rangle$ basis is a column whose elements are given by $\langle jm | \psi \rangle$. The matrices of Eqs. 8.21, for example would act to the right on the particular segment of this infinite column, for which $j = \frac{1}{2}$. (It has two elements, labelled by $m = \frac{1}{2}$ and $m = -\frac{1}{2}$ respectively). Similarly the matrices of Eq. 8.22 would operate on the part with $j = 1$, consisting of three elements labelled by $m = 1, 0 - 1$; and so on. The matrices can also act *to the left* on the representative of any bra vector $\langle \psi |$, which is a row with elements $\langle \psi | j'm' \rangle$. The relevant segments of this infinite row would be used for the particular cases $j = 0$, $j = \frac{1}{2}$, $j = 1$ considered explicitly here.

Finally we emphasize once again that the commutation relations are independent of whether $\mathbf{J}$ is orbital or spin angular momentum (or a sum of both), whether it is the angular momentum of a single particle or a many-particle system. If $\mathbf{J}$ is specified to be orbital in nature, $\mathbf{J} = \mathbf{L}$, then there exists a coordinate representation. The eigenstates (which will be denoted by $| lm \rangle$ in this case) then have coordinate space wave functions which, for a single particle, may be denoted by $\langle \theta\varphi | lm \rangle$. We know already that these are spherical harmonics: $\langle \theta\varphi | lm \rangle = Y_{lm}(\theta, \varphi)$. The operators themselves (in the configuration representation, in polar coordinates) are known to have the forms (4.51), from which we deduce

$$L_+ = L_x + i\,L_y = \hbar\, e^{i\varphi} \left(\frac{\partial}{\partial\theta} + i \cot\theta \frac{\partial}{\partial\varphi} \right)$$

$$L_- = L_x - i L_y = \hbar e^{-i\varphi}\left(-\frac{\partial}{\partial\theta} + i \cot\theta \frac{\partial}{\partial\varphi}\right) \tag{8.23}$$

Using the properties of spherical harmonics (Appendix D) one can then show that

$$\begin{aligned} L_+ Y_{lm}(\theta,\varphi) &= [l(l+1) - m(m+1)]^{1/2}\,\hbar Y_{l,m+1}(\theta,\varphi) \\ L_- Y_{lm}(\theta,\varphi) &= [l(l+1) - m(m-1)]^{1/2}\,\hbar Y_{l,m-1}(\theta,\varphi) \end{aligned} \tag{8.24}$$

Eqs. 8.24 are just what are required by Eqs. 8.11 and 8.19. From these it follows, for instance, that

$$\begin{aligned} \int (Y_{l'm'})^* L_\pm Y_{lm} d\Omega &= [l(l+1) - m(m \pm 1)]^{1/2} \hbar\, \delta_{l'l}\, \delta_{m',m\pm1} \\ &= \langle l'm' \mid L_\pm \mid lm \rangle \end{aligned} \tag{8.25}$$

This is a direct verification (from the known nature of the angular momentum eigenfunctions in the Schrödinger representation) that the orbital angular momentum operators do have the matrix representation (8.16) deduced from the commutation rules. To be doubly convinced, the student may evaluate the matrix elements (8.25) for $l = l' = 1$ using the explicit forms (4.90) of Y_{1m} and check that the resulting matrices are just those given in Eqs. 8.22.

8.3 Spin Angular Momentum

Consider now a particle having an intrinsic (spin) angular momentum, which is unrelated to its orbital motion. The operators S_x, S_y, S_z which describe the components of spin cannot be functions of the orbital variables **x**,**p** and must, therefore, commute with any functions of these quantities. In particular,

$$[\mathbf{S},\mathbf{L}] = 0 \tag{8.26}$$

Let us then ignore the orbital degrees of freedom altogether and focus attention on spin alone. We expect that S_x, S_y, S_z must obey the general commutation relations of the form (8.1), and hence that there exist simultaneous eigenstates $| s\, m_s \rangle$ of $\mathbf{S}^2$ and S_z.

$$\begin{aligned} \mathbf{S}^2 | s\, m_s \rangle = s(s+1)\hbar^2 | s\, m_s \rangle, \quad S_z | s\, m_s \rangle = m_s \hbar | s\, m_s \rangle \\ m_s = s,\ s-1, \ldots, -s. \end{aligned} \tag{8.27}$$

At this point we may note one important difference between spin and orbital angular momenta: while any particle can have arbitrary values $l\hbar (l=0,1,2,\ldots)$ for the latter, the spins of any *given kind* of particle have a fixed value. Electrons, protons, neutrons, neutrinos, μ-mesons etc. have $s = \frac{1}{2}$; π-mesons have $s = 0$ and other stable and short-lived particles have other definite values ($s = 1$, $s = \frac{3}{2}$ etc.). Therefore, for a particular particle, only m_s can vary: it can take any of the $(2s + 1)$ values $s, s-1, \ldots, -s$. Thus the $(2s + 1)$ states $| sm_s \rangle$ form a complete set for the spin states of the particle; any arbitrary spin state $| \chi \rangle$ can be written as

$$| \chi \rangle = \sum_{m_s=-s}^{s} | s\, m_s \rangle \langle s\, m_s | \chi \rangle \tag{8.28}$$

The totality of such states constitute what is called the *spin space* of a particle of spin s. The set of $(2s + 1)$ coefficients $\langle sm_s | \chi \rangle$, considered as a function of m_s, is the *spin wave function* representing the spin state $| \chi \rangle$. It is written, in the now-familiar way, as a column whose elements are labelled by the values of m_s in decreasing order starting with $m_s = s$ at the top. If $| \chi \rangle$ itself is one of the basis states, say $| s, m_s' \rangle$, the only non-vanishing element of the wave function (column) is that with $m_s = m_s'$.

(*a*) *Spin-*$\frac{1}{2}$: In the case of the electron ($s = \frac{1}{2}$), the wave functions for the basis states $| \frac{1}{2}, \frac{1}{2} \rangle$ and $| \frac{1}{2}, -\frac{1}{2} \rangle$ are

$$\alpha \equiv \begin{pmatrix} 1 \\ 0 \end{pmatrix} \text{ and } \beta \equiv \begin{pmatrix} 0 \\ 1 \end{pmatrix} \tag{8.29}$$

respectively. The symbols α, β are standard abbreviations for the two columns as indicated, i.e. α is the wave function for the $m_s = \frac{1}{2}$ or *spin up*[3] state while β stands for the $m_s = -\frac{1}{2}$ or *spin down* wave function. The wave function for a general spin state $| \chi \rangle$ — which we will simply denote by χ — is given by

$$\chi = \begin{pmatrix} a \\ b \end{pmatrix} = a \begin{pmatrix} 1 \\ 0 \end{pmatrix} + b \begin{pmatrix} 0 \\ 1 \end{pmatrix} = a\alpha + b\beta,$$

with

$$a = \langle \tfrac{1}{2}\tfrac{1}{2} | \chi \rangle, \; b = \langle \tfrac{1}{2}, -\tfrac{1}{2} | \chi \rangle$$

$$| a |^2 + | b |^2 = (a^* b^*) \begin{pmatrix} a \\ b \end{pmatrix} \equiv \chi^\dagger \chi = 1 \tag{8.30}$$

In accordance with the usual interpretation of wave functions, a is the probability amplitude and $| a |^2$ the probability that the particle has spin up, and $| b |^2$, for spin down. The total probability must of course be unity; this is ensured by normalization as in the last of Eqs. 8.30. The matrices representing S_x, S_y, S_z, which act on the spin wave functions χ of Eq. 8.30 for $s = \frac{1}{2}$, are just the matrices (8.21). Renaming them as s_x, s_y, s_z, we have

$$\mathbf{s} = \tfrac{1}{2}\hbar\boldsymbol{\sigma}$$

$$\sigma_x = \begin{pmatrix} 0 & 1 \\ 1 & 0 \end{pmatrix}, \; \sigma_y = \begin{pmatrix} 0 & -i \\ i & 0 \end{pmatrix}, \; \sigma_z = \begin{pmatrix} 1 & 0 \\ 0 & -1 \end{pmatrix} \tag{8.31}$$

The matrices $\boldsymbol{\sigma}$ are the *Pauli matrices*. Note that α, β are indeed eigenvectors of the matrix σ_z. Further, the roles of $s_+ = s_x + is_y$ and $s_- = s_x - is_y$ as raising and lowering operators are readily apparent

$$s_+ = \tfrac{1}{2}\hbar\sigma_+ = \hbar \begin{pmatrix} 0 & 1 \\ 0 & 0 \end{pmatrix}, \; s_- = \tfrac{1}{2}\hbar\sigma_- = \hbar \begin{pmatrix} 0 & 0 \\ 1 & 0 \end{pmatrix} \tag{8.32a}$$

$$s_+\beta = \hbar\alpha, \; s_-\alpha = \hbar\beta \tag{8.32b}$$

$$s_+\alpha = s_-\beta = 0 \tag{8.32c}$$

Thus s_+ flips the spin from 'down' to 'up' and s_- does the opposite; and they annihilate the states with highest and lowest values of m_s respectively, as expected from the general theory.

The Pauli matrices have some elegant properties which are also very useful

$$\sigma_x^2 = \sigma_y^2 = \sigma_z^2 = 1 \to \sigma^2 = 3, \tag{8.33a}$$

$$\sigma_x\sigma_y = -\sigma_y\sigma_x = i\sigma_z; \; \sigma_y\sigma_z = -\sigma_z\sigma_y = i\sigma_x; \; \sigma_z\sigma_x = -\sigma_x\sigma_z = i\sigma_y \tag{8.33b}$$

$$\sigma_+^2 = \sigma_-^2 = 0; \; \sigma_+\sigma_- = 2(1 + \sigma_z), \; \sigma_-\sigma_+ = 2(1 - \sigma_z) \tag{8.33c}$$

From these it follows (Example 8.2, p. 246) that

[3] When m_s (the z-constant of spin) is positive, the spin vector is visualized as pointing in the positive z-direction which is pictured as 'up'.

$$(\boldsymbol{\sigma}.\mathbf{A})\,(\boldsymbol{\sigma}.\mathbf{B}) = (\mathbf{A}.\mathbf{B}) + i\boldsymbol{\sigma}.(\mathbf{A}\times\mathbf{B}) \tag{8.33d}$$

if the components of **A** and **B** commute with those of $\boldsymbol{\sigma}$.

(*b*) *Spin*-1: The above analysis may be directly extended to higher spins. We briefly consider $s = 1$. In this case the spin wave functions are three-component columns:

$$\chi = \begin{pmatrix} \chi_1 \\ \chi_0 \\ \chi_{-1} \end{pmatrix} = \begin{pmatrix} \langle\, 1 \,|\, \chi \,\rangle \\ \langle\, 0 \,|\, \chi \,\rangle \\ \langle -1 \,|\, \chi \,\rangle \end{pmatrix} \tag{8.34}$$

where we have used the abbreviated notation $|\, m\,\rangle$ for $|\, sm\,\rangle$, $(m = 1, 0, -1)$, since $s = 1$ is fixed. The spin operators S_x, S_y, S_z are now represented by the matrices of Eq. 8.22, which we shall denote by $\mathcal{S}_x$, $\mathcal{S}_y$, $\mathcal{S}_z$ in the present context.[4]

If, instead of the states $|\, m\,\rangle$ which are eigenstates of S_z, we had used a different set of spin states as basis, we would have obtained a different matrix representation for **S**. The following basis is of special interest, for reasons which will become apparent shortly:

$$|\, a\,\rangle = \frac{-1}{\sqrt{2}}\,(|\,1\,\rangle - |\,-1\,\rangle), |\, b\,\rangle = \frac{i}{\sqrt{2}}\,(|\,1\,\rangle + |\,-1\,\rangle), |\, c\,\rangle = |\,0\,\rangle \tag{8.35}$$

According to the general theory of Sec. 7.7, on making this transformation of basis, the matrix representation must undergo a unitary transformation. The new matrices representing spin will be $\mathcal{S}'_x$, $\mathcal{S}'_y$, $\mathcal{S}'_z$, where

$$\boldsymbol{\mathcal{S}}' = U\boldsymbol{\mathcal{S}}U^\dagger \tag{8.36}$$

where the three rows of U are the scalar products of $|\, a\,\rangle$, $|\, b\,\rangle$ and $|\, c\,\rangle$ respectively with the old basis states $|\, m\,\rangle$. Thus

$$U = \begin{pmatrix} -\dfrac{1}{\sqrt{2}} & 0 & \dfrac{1}{\sqrt{2}} \\ -\dfrac{i}{\sqrt{2}} & 0 & -\dfrac{i}{\sqrt{2}} \\ 0 & 1 & 0 \end{pmatrix} \tag{8.37}$$

On evaluating $\boldsymbol{\mathcal{S}}'$ [with $\boldsymbol{\mathcal{S}}$ given by the matrices of (8.22) as already stated], one finds that $\boldsymbol{\mathcal{S}}' = \hbar\boldsymbol{\Sigma}$, where Σ_x, Σ_y, Σ_z are just the matrices (7.78) representing the infinitesimal generators of rotations on the components of *vectors*. We conclude, therefore, that the *proportionality between angular momentum operators and infinitesimal generators of rotations*, which was established in Sec. 7.9 for orbital angular momentum, *holds for spin too*. And we observe further that the components of spin wave functions [with respect to the basis (8.35)] for $s = 1$ behave just like the *Cartesian* components of a vector. The components with respect to the $|\, m\,\rangle$ basis, given in Eq. 8.34 transform like vector components defined with respect to the so-called *spherical basis* defined in Eq. 8.92.

Thus spin-1 particles are described by *vector* wave functions. The two-component wave functions of spin-$\frac{1}{2}$ particles are called *spinors*.

[4] It will be recalled that these same matrices represent also the orbital angular momentum operator in the space of orbital states with $l = 1$. See the closing remarks in Sec. 8.2. This is further reiteration of the fact that the matrices for *integer* j do not convey any information as to what kind of angular momentum is involved. This information has to be supplied from outside. If j is half-integral, we can conclude that the angular momentum arises wholly or partly from spin.

(*c*) *The Total Wave Function:* Let us now relax the exclusive preoccupation with spin and ask: What is the general form of the wave function including the orbital degrees of freedom? Clearly, it still must have two components (when $s = \frac{1}{2}$), but instead of the components being constants like a and b in χ, Eq. 8.30, they may now be functions of $\mathbf{x}$. Thus we may write ψ for a spin-$\frac{1}{2}$ particle as

$$\psi(\mathbf{x}) = \begin{pmatrix} \psi_{1/2}(\mathbf{x}) \\ \psi_{-1/2}(\mathbf{x}) \end{pmatrix} = \psi_{1/2}(\mathbf{x})\alpha + \psi_{-1/2}(\mathbf{x})\beta \qquad (8.38)$$

The probability interpretation is as follows:

$|\psi_{1/2}(\mathbf{x})|^2$ = Probability density for the particle to be in the neighbourhood of $\mathbf{x}$, with spin up.

$\int |\psi_{1/2}(\mathbf{x})|^2 d^3x$ = Probability that the particle is anywhere at all, with spin up.

For spin 'down' replace $\frac{1}{2}$ by $-\frac{1}{2}$ in the above.

$|\psi_{1/2}(\mathbf{x})|^2 + |\psi_{-1/2}(\mathbf{x})|^2 = \psi^\dagger(\mathbf{x})\psi(\mathbf{x})$ = Probability density that the particle is in the neighbourhood of $\mathbf{x}$ with any orientation of spin and

$$\int \psi^\dagger(\mathbf{x})\psi(\mathbf{x})d^3x = 1.$$

Both orbital and spin operators can act on the two-component wave function ψ. The operators $\mathbf{x}$ and $\mathbf{p}$ act on $\psi_{1/2}$ and $\psi_{-1/2}$ individually, without causing any admixture of the two; the spin operators do mix up $\psi_{1/2}$ and $\psi_{-1/2}$ but have no effect on their $\mathbf{x}$-dependence. To illustrate this, we consider the effect of J_x on ψ, where $\mathbf{J}$ is the total (orbital + spin) angular momentum operator: $\mathbf{J} = \mathbf{L} + \mathbf{S} = \mathbf{L} + \frac{1}{2}\hbar\boldsymbol{\sigma}$.

$$J_x\psi = L_x\begin{pmatrix} \psi_{1/2} \\ \psi_{-1/2} \end{pmatrix} + \tfrac{1}{2}\hbar\begin{pmatrix} 0 & 1 \\ 1 & 0 \end{pmatrix}\begin{pmatrix} \psi_{1/2} \\ \psi_{-1/2} \end{pmatrix}$$

$$= \begin{pmatrix} L_x\psi_{1/2} \\ L_x\psi_{-1/2} \end{pmatrix} + \tfrac{1}{2}\hbar\begin{pmatrix} \psi_{-1/2} \\ \psi_{1/2} \end{pmatrix} = \begin{pmatrix} L_x\psi_{1/2} + \frac{1}{2}\hbar\psi_{-1/2} \\ L_x\psi_{-1/2} + \frac{1}{2}\hbar\psi_{1/2} \end{pmatrix} \qquad (8.39)$$

Since electrons have spin-$\frac{1}{2}$, they must be described by two-component wave functions of the form (8.38). (At a more fundamental level, when the requirements of relativity also are to be met, a four-component wave function becomes necessary. See Chapter 10). However, in most atomic problems the effect of spin is very small. That is to say, it makes very little difference (to the energy levels, for example) whether the spin part of the wave function is α or β. Under such circumstances it is enough to use a simple (single-component) wave function, as we have done up to this stage. But much of the so-called fine structure of energy levels is caused by the presence of spin, and in studying this it is essential to use two-component wave functions. The generalization of the above concepts to spin $s > \frac{1}{2}$ is self-evident and will not be pursued here.

Example 8.1—What is the spin wave function (for $s = \frac{1}{2}$) if the spin component in the direction of the unit vector $\mathbf{n}$ has the value $\frac{1}{2}\hbar$? Evidently it is defined by $(\mathbf{s}.\mathbf{n})\chi = \frac{1}{2}\hbar\chi$ or $(\boldsymbol{\sigma}.\mathbf{n})\chi = \chi$. On using (8.31) this becomes

$$\begin{pmatrix} n_3 & n_1 - in_2 \\ n_1 + in_2 & -n_3 \end{pmatrix}\begin{pmatrix} a \\ b \end{pmatrix} = \begin{pmatrix} a \\ b \end{pmatrix}$$

which immediately gives $(a/b) = (n_1 - in_2)/(1 - n_3) = (1 + n_3)/(n_1 + in_2)$.

From this and the normalization $a^*a + b^*b = 1$, we obtain $a = (n_1 - in_2)/[2(1 - n_3)]^{1/2}$, $b = [\frac{1}{2}(1 - n_3)]^{1/2}$. Thus

$$\chi = \frac{n_1 - in_2}{[2(1 - n_3)]^{1/2}} \alpha + [\tfrac{1}{2}(1 - n_3)]^{1/2}\beta.$$

EXAMPLE 8.2—If **A**,**B** are operators which commute with σ but not necessarily with each other (e.g. $A = \mathbf{x}$, $\mathbf{B} = \mathbf{p}$), then $(\boldsymbol{\sigma}.\mathrm{A})(\boldsymbol{\sigma}.\mathrm{B}) = \mathbf{A}.\mathbf{B} + i\boldsymbol{\sigma}.(\mathbf{A} \times \mathbf{B})$. The first term comes from products like $\sigma_x A_x.\sigma_x B_x = A_x B_x$ (since $\sigma_x^2 = 1$). The second term arises from cross-products: e.g. $\sigma_x A_x \sigma_y B_y + \sigma_y A_y \sigma_x B_x = \sigma_x \sigma_y A_x B_y + \sigma_y \sigma_x A_y B_x = i\sigma_z(A_x B_y - A_y B_x)$.

If, in particular, the components of **A** commute among themselves, then $(\boldsymbol{\sigma}.\mathrm{A})^2 = \mathbf{A}^2$. Using this result and the expansion of the exponential function, the student may verify that

$$e^{\frac{1}{2}i(\boldsymbol{\sigma}.\mathbf{n})\theta} = \cos\tfrac{1}{2}\theta + i\,(\boldsymbol{\sigma}.\mathbf{n})\sin\tfrac{1}{2}\theta.$$

Here **n** is a unit vector. This matrix operator is of interest because it represents the effect of a rotation of the coordinate frame (through an angle θ about an axis along **n**) on the spin wave functions of a spin-$\frac{1}{2}$ particle. As we have seen in Secs. 7.9 and 7.10, the generator of such rotations is $(1/\hbar)$ times the component of the angular momentum operator along **n**. Since we are considering spin wave functions, it is the spin angular momentum $\frac{1}{2}\hbar\boldsymbol{\sigma}$ which is involved here. Note that when $\theta = 2\pi$, the transformation $\chi \to \chi' = e^{\frac{1}{2}i(\boldsymbol{\sigma}.\mathbf{n})\theta}\chi$ reduces to $\chi' = -\chi$ (and not $\chi' = \chi$) despite the fact that a rotation through 2π returns the coordinate system to its original position. So the wave functions for spin-$\frac{1}{2}$ (and also $\frac{3}{2}, \frac{5}{2}, \ldots$) are *double-valued*. However, all physical quantities remain single-valued (as they should) since they involve both $\chi^\dagger$ and χ as factors.

8.4 Non-relativistic Hamiltonian Including Spin

Let us consider briefly how the spin of the electron reveals itself in experiments. The electron is a charged particle, and associated with its intrinsic rotation or spin, there is a magnetic moment. This magnetic moment $\boldsymbol{\mu}_s$, which is proportional to the spin **s**, interacts with the magnetic field created by its orbital motion (proportional to the orbital angular momentum **L**). The energy associated with this *spin-orbit interaction*[5] is proportional to **L.s**. When the electron is in the (electric) field of an atom and has a potential energy $V(r)$ — assumed to be spherically symmetric — due to the averaged effects of the other electrons and the nucleus, the above mentioned proportionality factor (including a contribution from relativistic effects) can be shown to be $(2m^2c^2r)^{-1}\,(dV/dr)$. Thus the so-called spin orbit energy appears as a contribution

$$H_{s\cdot o} = \xi(r)\mathbf{L}.\mathbf{s}, \quad \xi(r) = \frac{1}{2m^2c^2} \cdot \frac{1}{r}\frac{dV}{dr} \tag{8.40}$$

to the Hamiltonian of the electron. It has the effect of splitting otherwise

[5] L. H. Thomas, *Nature*, **117**, 514, 1926.

degenerate atomic levels, thereby giving rise to the *fine structure* in atomic spectra.

Secondly, the spin magnetic moment couples to any external electromagnetic field in which the atom may be placed. We know that the effect of such fields on the orbital motion is taken into account by the addition of $e\phi$ to the unperturbed Hamiltonian together with the replacement $\mathbf{p} \to \mathbf{p} - e\mathbf{A}/c$, resulting in

$$H_{\text{orb}} = \frac{1}{2m}\left(\mathbf{p} - \frac{e\mathbf{A}}{c}\right)^2 + V(r) + e\phi \tag{8.41}$$

This simplifies to[6]

$$H_{\text{orb}} = \frac{\mathbf{p}^2}{2m} - \frac{e}{mc}\,\mathbf{A}\cdot\mathbf{p} + V(r) + \frac{e^2\mathbf{A}^2}{2mc^2} \tag{8.42}$$

whenever the scalar and vector potentials can be chosen such that $\phi = \operatorname{div}\mathbf{A} = 0$. This is the case, for instance, if ϕ and $\mathbf{A}$ are due to electromagnetic radiation; or if only constant magnetic (and no electric) fields are present. In the latter case, if the magnetic field is homogeneous (independent of position) we can choose

$$\mathbf{A} = \tfrac{1}{2}\,(\mathcal{H} \times \mathbf{r}) \tag{8.43}$$

Then $\mathbf{A}.\mathbf{p}$ becomes $\frac{1}{2}(\mathcal{H} \times \mathbf{r})\cdot\mathbf{p} = \frac{1}{2}\mathcal{H}.(\mathbf{r}\times\mathbf{p}) = \frac{1}{2}\mathcal{H}.\mathbf{L}$. The corresponding term in (8.42) becomes $-(e/2mc)\mathbf{L}.\mathcal{H}$ which has the form of the interaction energy of $\mathcal{H}$ with a magnetic moment

$$\boldsymbol{\mu}_L = \frac{e}{2mc}\,\mathbf{L} \tag{8.44}$$

This is the magnetic moment associated with orbital motion. When spin is taken into account, the energy of interaction $H_{s\cdot m}$ of the spin magnetic moment $\boldsymbol{\mu}_s$ with $\mathcal{H}$ also has to be added to the Hamiltonian:

$$H_{s\cdot m} = -\boldsymbol{\mu}_s.\mathcal{H},\quad \boldsymbol{\mu}_s = \frac{e}{mc}\,\mathbf{s} = \frac{e\hbar}{2mc}\,\boldsymbol{\sigma} \tag{8.45}$$

Note that the *gyromagnetic ratio* (e/mc) associated with spin has double the value of the ratio in the orbital case. This was first deduced empirically (from atomic spectra), but comes out as an automatic consequence from Dirac's relativistic theory of the electron (Sec. 10.15).

Thus when the atom is in a magnetic field $\mathcal{H}$ (whose direction may be taken to be the z-axis), the total Hamiltonian including spin is

$$H = H_0 + H_{\text{mag}} + H_{s\cdot o}$$

where $H_{s\cdot o}$ is given by Eqs. 8.40 and

$$H_0 = \frac{p^2}{2m} + V,\quad H_{\text{mag}} = -\frac{e\mathcal{H}}{2mc}\,(L_z + 2s_z) \tag{8.46}$$

The term H_{mag} gives rise to the *Zeeman effect.* If $\mathcal{H}$ is so large that the spin-orbit energy can be neglected in comparison to H_{mag}, then $\mathbf{L}^2$, L_z and s_z commute with H in this approximation. So the wave functions $u_{nlm}\,\alpha$ and $u_{nlm}\,\beta$ are

[6] The last term is unimportant in most applications to atomic physics (except when very intense fields—from laser radiation, for example—are involved). But in problems like the motion of a particle in a pure magnetic field ($V(r) = 0$) it is all-important.

eigenfunctions of H (where the u_{nlm} are the simultaneous eigenfunctions of H_0, $\mathbf{L}^2$ and L_z). The corresponding energy eigenvalues are given by

$$E_{nlm} - \frac{e\mathcal{H}\hbar}{2mc}(m_l + 2m_s) \tag{8.47}$$

where $m_s = \frac{1}{2}$ and $-\frac{1}{2}$ for the spin functions α and β respectively. The presence of the spin-orbit term complicates matters very much when $\mathcal{H}$ is weak. (See Example 8.7, p. 266).

8.5 Addition of Angular Momenta

Let $\mathbf{J}_1$ and $\mathbf{J}_2$ be two independent (i.e. commuting) angular momentum vectors, i.e.

$$[\mathbf{J}_1, \mathbf{J}_2] = 0 \tag{8.48}$$

For example, $\mathbf{J}_1$ may be the orbital angular momentum $\mathbf{L}$ and $\mathbf{J}_2$ the spin $\mathbf{S}$ of the same particle; or $\mathbf{J}_1$ and $\mathbf{J}_2$ may be the respective angular momenta of two particles (or for that matter, of two separate systems of particles). Basis states for the system constituted by the two angular momenta may be chosen in different ways. An obvious choice is the set of simultaneous eigenstates $|j_1m_1; j_2m_2\rangle$ of the four commuting operators $\mathbf{J}_1^2$, J_{1z}, $\mathbf{J}_2^2$, J_{2z} which form a complete commuting set of observables. In the following we will be interested in considering the set of $(2j_1 + 1)(2j_2 + 1)$ states for *fixed* j_1 and j_2 ($m_1 = j_1, j_1 - 1, \ldots -j_1$; $m_2 = j_2, j_2 - 1, \ldots, -j_2$). So we suppress the (fixed) values of j_1 and j_2 from the notation for the basis states and write simply $|m_1; m_2\rangle$ for $|j_1, m_1; j_2, m_2\rangle$, leaving it to be understood that

$$\begin{aligned} \mathbf{J}_1^2 |m_1; m_2\rangle &= j_1(j_1 + 1)\hbar^2 |m_1; m_2\rangle \\ \mathbf{J}_2^2 |m_1; m_2\rangle &= j_2(j_2 + 1)\hbar^2 |m_1; m_2\rangle \end{aligned} \tag{8.49}$$

Of course, the explicitly shown quantum numbers m_1, m_2 mean that

$$J_{1z}|m_1;m_2\rangle = m_1\hbar|m_1;m_2\rangle, \; J_{2z}|m_1;m_2\rangle = m_2\hbar|m_1;m_2\rangle \tag{8.50}$$

Noting that $J_{1\pm}$ are ladder operators on m_1 and $J_{2\pm}$ on m_2, we have, from Eqs. 8.11 taken together with Eqs. 8.19, that

$$\begin{aligned} J_{1\pm}|m_1; m_2\rangle &= [(j_1 \mp m_1)(j_1 \pm m_1 + 1)]^{1/2}|m_1 \pm 1; m_2\rangle \\ J_{2\pm}|m_1; m_2\rangle &= [(j_2 \mp m_2)(j_2 \pm m_2 + 1)]^{1/2}|m_1; m_2 \pm 1\rangle \end{aligned} \tag{8.51}$$

The use of the above basis is most appropriate when the two systems characterized by $\mathbf{J}_1$ and $\mathbf{J}_2$ are non-interacting or very nearly so. Under such circumstances m_1, m_2 are (at least approximately) good quantum numbers, and can be used, for instance, to label the unperturbed states in a perturbation treatment of the interaction. However, if the interaction is such as to couple $\mathbf{J}_1$ and $\mathbf{J}_2$ (for example, through a term $\mathbf{J}_1 \cdot \mathbf{J}_2$ in the Hamiltonian) then J_{1z} and J_{2z} are no longer constants of the motion, and it would be more advantageous to use other operators which commute with such an interaction, for defining a basis. Since it is known that the *internal* interactions of a composite system do not affect its total angular momentum (which remains conserved), it is natural to use for this purpose $\mathbf{J}^2$ and J_z (instead of J_{1z} and J_{2z}) where $\mathbf{J}$ is the total angular momentum operator for the *composite* system made up of $\mathbf{J}_1$ and $\mathbf{J}_2$.

$$\mathbf{J} = \mathbf{J}_1 + \mathbf{J}_2, \quad \mathbf{J}^2 = \mathbf{J}_1^2 + \mathbf{J}_2^2 + 2\mathbf{J}_1\cdot\mathbf{J}_2. \tag{8.52}$$

Since $\mathbf{J}_1^2$ and $\mathbf{J}_2^2$ do commute with $\mathbf{J}^2$ and $\mathcal{J}_z$, the simultaneous eigenstates of these four operators will be taken as the new basis. We denote these by $|jm\rangle$, which is to be understood as an abbreviation[7] for $|j_1 j_2 jm\rangle$. The quantum numbers j, m are of course defined by

$$\mathbf{J}^2|jm\rangle = j(j+1)\hbar^2|jm\rangle, \quad \mathcal{J}_z|jm\rangle = m\hbar|jm\rangle \tag{8.53}$$

The immediate question is: given the angular momenta j_1, j_2 of the individual systems, what are the possible values of j for the composite system? The answer was known already from the Old Quantum Theory, and stated in terms of the *vector model:* If $\mathbf{J}_1$, $\mathbf{J}_2$ are thought of as two ordinary vectors of length j_1 and j_2 respectively, these vectors together with $\mathbf{J}_1 + \mathbf{J}_2$ form a triangle (Fig. 8.1). The values of j_1, j_2, j must then be such that a triangle with these sides

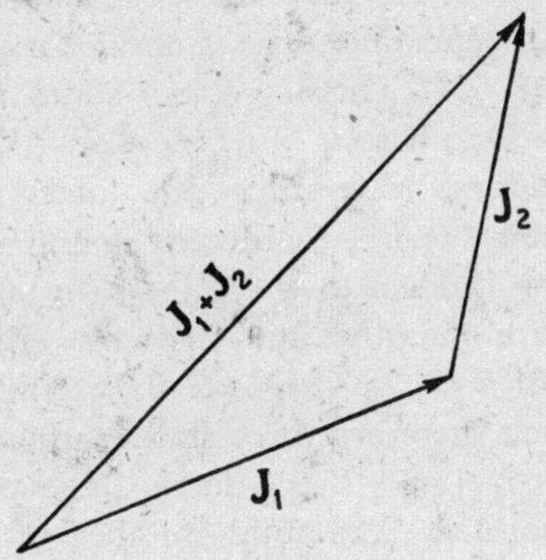

Fig. 8.1 Addition of angular momenta.

can be constructed. This *triangle rule* or *triangle condition* immediately sets the maximum and minimum limits for j as $j_1 + j_2$ and $|j_1 - j_2|$ (corresponding to $\mathbf{J}_1$ and $\mathbf{J}_2$ being parallel and antiparallel respectively). Since angular momentum is quantized, the only allowed values between these limits are $j_1 + j_2 - 1, j_1 + j_2 - 2, \ldots, |j_1 - j_2| + 1$. We shall now see that the same conclusions follow from quantum mechanics.

Observe first of all that the set of states $|jm\rangle$ and the set $|m_1; m_2\rangle$ are simply different bases for the same space — the space of states wherein the two individual angular momenta have values j_1, j_2 respectively. So any $|jm\rangle$ can be written as a linear combination of the states $|m_1; m_2\rangle$:

$$|jm\rangle = \sum_{m_1 m_2} |m_1; m_2\rangle\langle m_1; m_2|jm\rangle \tag{8.54}$$

We have seen in Sec. 7.7, that the coefficients relating two orthonormal bases

[7] The abbreviations $|m_1; m_2\rangle$ for $|j_1 m_1; j_2 m_2\rangle$ and $|jm\rangle$ for $|j_1 j_2 jm\rangle$ create some risk of confusion when particular values are used for m_1, m_2 or j, m. To minimize this we use a semicolon between m_1 and m_2 in $|m_1; m_2\rangle$ while $|jm\rangle$ will be written without semicolon (though sometimes a comma may be necessary). Thus, $|j_1; j - j_1\rangle$ means the state $|m_1; m_2\rangle$ with $m_1 = j_1$, $m_2 = j - j_1$, while $|j_1\, j - j_1\rangle$ or $|j_1, j - j_1\rangle$ would mean $|jm\rangle$ with $j = j_1$, $m = j - j_1$. Attention to the semicolon would resolve doubts in the few places where the possibility of confusion exists.

form a unitary matrix. So the matrix formed by the elements $\langle m_1; m_2 | jm \rangle$, wherein m_1, m_2 together label the rows, and j, m label the columns, must be unitary.

Since $J_z = J_{1z} + J_{2z}$ it is evident from Eqs. 8.50 and the second of Eqs. 8.53 that

$$m = m_1 + m_2 \tag{8.55a}$$

i.e. $$\langle m_1; m_2 | jm \rangle = 0 \text{ unless } m_2 = m - m_1. \tag{8.55b}$$

Hence the double sum in Eq. 8.54 reduces to a single sum:

$$|jm\rangle = \sum_{m_1} | m_1; m - m_1 \rangle \langle m_1; m - m_1 | jm \rangle \tag{8.56}$$

Since the maximum values of m_1 and m_2 are j_1 and j_2 respectively, Eq. 8.55a tells us that the maximum value of m is $j_1 + j_2$, which is then necessarily the maximum value of j too. Since this value for m occurs only once, (when $m_1 = j_1$ and $m_2 = j_2$), $j = j_1 + j_2$ occurs only once. Consider now the next lower value of m, namely $m = j_1 + j_2 - 1$. There are two states with this m (provided neither j_1 nor j_2 is zero): One has $m_1 = j_1$, $m_2 = j_2 - 1$, and the other, $m_1 = j_1 - 1$, $m_2 = j_2$. One of these (or rather, one linear combination of these) must belong to the value $j = j_1 + j_2$ already found, since one state with each value of m from $j_1 + j_2$ down to $-(j_1 + j_2)$ (in unit steps) must go with this j. We are still left with one other state with $m = j_1 + j_2 - 1$, which must belong to a new value of j, $j = j_1 + j_2 - 1$. By extending this procedure to $m = j_1 + j_2 - 2$ (for which there are three states, provided $j_1, j_2 > \frac{1}{2}$), and so on, we find new values $j_1 + j_2 - 2, \ldots$ for j. The process ends when all available states are exhausted. The total number of independent states is $(2j_1 + 1)\ (2j_2 + 1)$, as observed at the beginning of this section, and on the other hand, each value of j has $(2j + 1)$ states associated with it. So the minimum value j_{min} of j is reached when

$$\sum_{j=j_{min}}^{j_1+j_2} (2j + 1) = (2j_1 + 1)\ (2j_2 + 1) \tag{8.57}$$

The left hand side of Eq. 8.57 can be easily summed and one can then solve for j_{min} The result is $j_{min} = |j_1 - j_2|$ as expected.

To sum up, the possible values j of the total angular momentum, resulting from the addition of two given angular momenta j_1, j_2, are

$$j_1 + j_2,\ (j_1 + j_2 - 1), \ldots, |j_1 - j_2| \tag{8.58}$$

8.6 Clebsch-Gordan Coefficients

In order to construct the states $|jm\rangle$ for any of the above values of j, we need to know the coefficients $\langle m_1; m_2 | jm \rangle$ in Eq. 8.54. They are called the *Clebsch-Gordan, Wigner, or vector coupling coefficients*. They depend on j_1 and j_2 also and should, strictly speaking, be written as $\langle j_1 m_1; j_2 m_2 | j_1 j_2 jm \rangle$.

The unitarity of the matrix of Clebsch-Gordan coefficients is expressed through the equations

$$\sum_{jm} \langle m_1;m_2 \mid jm \rangle \langle jm \mid m_1';m_2' \rangle = \langle m_1m_2 \mid m_1'm_2' \rangle = \delta_{m_1m_1'}\delta_{m_2m_2'} \qquad (8.59a)$$

$$\sum_{m_1m_2} \langle jm \mid m_1;m_2 \rangle \langle m_1;m_2 \mid j'm' \rangle = \langle jm \mid j'm' \rangle = \delta_{jj'}\delta_{mm'} \qquad (8.59b)$$

where $\langle jm \mid m_1;m_2 \rangle = \langle m_1;m_2 \mid jm \rangle^*$. These latter quantities appear directly in the expansion

$$\mid m_1;m_2 \rangle = \sum_{jm} \mid jm \rangle \langle jm \mid m_1; m_2 \rangle \qquad (8.60)$$

which is the inverse of Eq. 8.54. Note that in view of Eq. 8.55, only terms with $m_1 + m_2 = m$ contribute to the sums in Eqs. 8.59 and 8.60.

(*a*) *Phase Convention.* The explicit expression for the C-G (Clebsch-Gordan) coefficients is rather complicated. Before we can proceed with its determination, it is necessary to establish certain conventions regarding the choice of phases of the states $\mid jm \rangle$. (It will be recalled that the physical state determines the normalized state vector only to within an arbitrary phase factor). Following standard practice we adopt the convention that for each j, the phase of the state $\mid jj \rangle$ is to be chosen so that the C-G coefficient $\langle j_1j_1;\ j_2, j - j_1 \mid jj \rangle$ which involves the highest value j_1 of m_1, is *positive.* In our abbreviated notation this reads

$$\langle j_1;\ j - j_1 \mid jj \rangle > 0 \qquad (8.61)$$

Once the phase of $\mid jj \rangle$ is fixed in this fashion, the phases of all $\mid jm \rangle$ for this j can be fixed by constructing $\mid jm \rangle$ by repeated application ($j - m$ times) of the lowering operator J_- to $\mid jj \rangle$. From Eqs. 8.11b and 8.19b it is readily seen that if J_- is applied n times to some state $\mid j'm' \rangle$,

$$(J_-)^n \mid j'm' \rangle = c^-(j'm'; n) \mid j', m' - n \rangle \qquad (8.62a)$$

where $c^-(j'm';n) = c^-_{j'm'}\, c^-_{j',m'-1} \cdots c^-_{j',m'-n+1}$ or explicitly,

$$c^-(j'm';n) = \left[\frac{(j' + m')!\,(j' - m' + n)!}{(j' - m')!\,(j' + m' - n)!}\right]^{1/2} \qquad (8.62b)$$

Note that since the factorial function becomes infinite if its argument is a negative integer,

$$c^-(j'm'; n) = 0 \text{ unless } n - j' \leqslant m' \leqslant j' \qquad (8.62c)$$

(It should be remembered that m' and j' are both integers or both half integers, and hence $j' + m'$ and $j' - m'$ are always integers). The definition of $\mid jm \rangle$ in terms of $(\mathbf{J}_-)^{j-m} \mid jj \rangle$ now reads

$$\mid jm \rangle = \frac{(J_-)^{j-m}}{c^-(jj; j - m)} \mid jj \rangle \qquad (8.63)$$

This is obtained by setting $j' = m' = j$ and $n = j - m$ in Eq. 8.62a.

(*b*) *Expression for* $\langle \times;\times \mid j \times \rangle$ *in Terms of*[8] $\langle \times;\times \mid jj \rangle$: Having specified the phases of all the states $\mid jm \rangle$ through Eqs. 8.61 and 8.63, we can now proceed with the determination of the C-G coefficients in stages. Consider Eq. 8.54 for $m = j$, i.e.

[8] To focus attention on those parameters which are assigned specific values, we have replaced all other parameters by small crosses.

$$|jj\rangle = \sum_{\mu_1} |\mu_1, j-\mu_1\rangle\langle \mu_1, j-\mu_1 | jj\rangle \tag{8.64}$$

Now let us operate with $(\mathcal{J}_-)^{j-m}$ on the left hand side, and with same quantity expressed in the form $(\mathcal{J}_{1-} + \mathcal{J}_{2-})^{j-m}$ on the right hand side. Since $\mathcal{J}_{1-}$ and $\mathcal{J}_{2-}$ commute, we can make an ordinary binomial expansion of the latter. Thus,

$$(\mathcal{J}_-)^{j-m}|jj\rangle = \sum_{r=0}^{j-m}\sum_{\mu_1}\langle \mu_1; j-\mu_1|jj\rangle\binom{j-m}{r}\times (\mathcal{J}_{1-})^r(\mathcal{J}_{2-})^{j-m-r}|\mu_1; j-\mu_1\rangle \tag{8.65}$$

where the binomial coefficients have been denoted by

$$\binom{n}{r} \equiv \frac{n!}{r!\,(n-r)!} \tag{8.66}$$

Eqs. 8.62 enable both sides of Eq. 8.65 to be simplified. (Remember that acting on $|m_1; m_2\rangle$, $\mathcal{J}_{1-}$ lowers m_1 and $\mathcal{J}_{2-}$ lowers m_2 in just the same way as $\mathcal{J}_-$ lowers m). Therefore, Eq. 8.65 becomes

$$\begin{aligned} c^-(jj; j-m)\,|jm\rangle &= \sum_r\sum_{\mu_1}\langle \mu_1; j-\mu_1|jj\rangle\binom{j-m}{r}c^-(j_1\mu_1; r).\\ &\qquad c^-(j_2, j-\mu_1; j-m-r)\,|\mu_1 - r;\, m+r-\mu_1\rangle\\ &= \sum_r\sum_{m_1}\langle m_1+r; j-m_1-r|jj\rangle\binom{j-m}{r}c^-(j_1, m_1+r; r).\\ &\qquad c^-(j_2, j-m_1-r; j-m-r)\,|m_1; m-m_1\rangle \end{aligned} \tag{8.67}$$

In the last step we have merely changed the summation index μ_1 to $m_1 = \mu_1 - r$. On comparing Eq. 8.67 with Eq. 8.56, we immediately obtain

$$\langle m_1; m-m_1|jm\rangle = \frac{1}{c^-(jj; j-m)}\sum_r\langle m_1+r; j-m_1-r|jj\rangle\binom{j-m}{r}.\; c^-(j_1, m_1+r; r)\,c^-(j_2, j-m_1-r; j-m-r) \tag{8.68}$$

In the important special case when $m_1 = j_1$, this reduces to

$$\langle j_1; m-j_1|jm\rangle = \frac{c^-(j_2, j-j_1; j-m)}{c^-(jj; j-m)}\langle j_1; j-j_1|jj\rangle \tag{8.69}$$

i.e. only the term with $r=0$ in Eq. 8.68 survives. This is so because, on account of the last inequality in (8.62*c*), $c^-(j_1, m_1+r; r)$ vanishes for all $r > 0$ if $m_1 = j_1$ (and r cannot be negative in view of the way it arises from the binomial expansion).

(*c*) *Expression for* $\langle \times;\times|jj\rangle$ *in Terms of* $\langle j_1; j-j_1|jj\rangle$: Returning to Eq. 8.68, we observe that all the C-G coefficients become known if we know the special set wherein m has its maximum value j. The latter can be readily expressed in terms of $\langle j_1; j-j_1|jj\rangle$ in which m_1 also takes its maximum value. In fact,

$$\langle m_1; j-m_1|jj\rangle = \frac{c^-(j_2, j-m_1; j_1-m_1)}{c^-(j_1 j_1; j_1-m_1)}\langle j_1; j-j_1|jj\rangle \tag{8.70}$$

To prove this we exploit the fact that $\mathcal{J}_+|jj\rangle = 0$ and hence $\langle jj|\mathcal{J}_- = \langle jj|(\mathcal{J}_{1-}+\mathcal{J}_{2-}) = 0$. On taking the scalar product with $|m_1+1; m_2\rangle \equiv |j_1, m_1+1; j_2 m_2\rangle$, we obtain

$$0 = \langle jj|(\mathcal{J}_{1-}+\mathcal{J}_{2-})|m_1+1; m_2\rangle$$

$$= c^-(j_1, m_1+1; 1) \langle jj \mid m_1; m_2 \rangle + c^-(j_2, m_2; 1) \langle jj \mid m_1+1; m_2-1 \rangle,$$

which is nontrivial only for $m_1 + m_2 = j$. Thus, we have

$$\langle jj \mid m_1; j-m_1 \rangle = -\frac{c^-(j_2, j-m_1; 1)}{c^-(j_1, m_1+1; 1)} \langle jj \mid m_1+1; j-m_1-1 \rangle$$

$$= \frac{c^-(j_2, j-m_1; 1)\, c^-(j_2, j-m_1-1; 1) \ldots c^-(j_2, j-j_1+1; 1)}{c^-(j_1, m_1+1; 1)\, c^-(j_1, m_1+2; 1) \ldots c^-(j_1, j_1; 1)} \times$$

$$\times \langle jj \mid j_1; j-j_1 \rangle \qquad (8.71)$$

The product of the $(j_1 - m)$ factors of c^- in the numerator can be written as $c^-(j_2, j-m_1; j_1-m_1)$ and similarly for the denominator. Further, all the quantities on the right hand side of Eq. 8.71 are real in view of (8.61) and (8.62). Hence $\langle jj \mid m_1; j-m_1 \rangle = \langle m_1; j-m_1 \mid jj \rangle$. This completes the proof of Eq. 8.70. Incidentally, the reality of these coefficients implies, by virtue of Eq. 8.68, that *all* C-G coefficients are real.

(*d*) *Determination of* $\langle j_1; j-j_1 \mid jj \rangle$: All that remains now is to determine $\langle j_1; j-j_1 \mid jj \rangle$. This can be readily accomplished by using the unitarity condition (8.59b). Taking j', m, m' to be all equal to j and noting that then $m_2 = j - m_1$, we obtain

$$1 = \sum_{m_1} \left| \langle m_1; j-m_1 \mid jj \rangle \right|^2$$

$$= \langle j_1; j-j_1 \mid jj \rangle^2 . \sum_{m_1} \left[\frac{c^-(j_2, j-m_1; j_1-m_1)}{c^-(j_1, j_1; j_1-m_1)} \right]^2 \qquad (8.72)$$

where we have used Eq. 8.70 in the last step. It so happens that when the expressions for the c^- are introduced from Eq. 8.62b, the series in Eq. 8.72 can be summed using the formula

$$\sum_r \frac{(a+r)!\,(b-r)!}{(c+r)!\,(d-r)!} = \frac{(a+b+1)!\,(a-c)!\,(b-d)!}{(c+d)!\,(a+b-c-d+1)!} \qquad (8.73)$$

Then we obtain

$$\langle j_1; j-j_1 \mid jj \rangle = \left[\frac{(2j_1)!\,(2j+1)!}{(j_1+j_2+j+1)!\,(j_1-j_2+j)!} \right]^{1/2} \qquad (8.74)$$

On combining this with Eq. 8.70 and finally Eq. 8.68, we obtain a general formula for the C-G coefficients.

(*e*) *Table of C-G Coefficients; Symmetry Properties:* Tables 8.1, 8.2 present the

Table 8.1 The C-G coefficients* $\langle j_1\, m-m_2; \tfrac{1}{2}\, m_2 \mid jm \rangle$

j \ m_2	$\frac{1}{2}$	$-\frac{1}{2}$
$j_1 + \frac{1}{2}$	$\left[\frac{j_1+m+\frac{1}{2}}{2j_1+1}\right]^{1/2}$	$\left[\frac{j_1-m+\frac{1}{2}}{2j_1+1}\right]^{1/2}$
$j_1 - \frac{1}{2}$	$-\left[\frac{j_1-m+\frac{1}{2}}{2j_1+1}\right]^{1/2}$	$\left[\frac{j_1+m+\frac{1}{2}}{2j_1+1}\right]^{1/2}$

* More extensive tables of C-G coefficients can be found, e.g. in E. U. Condon and G. H. Shortley, *The Theory of Atomic Spectra*, Cambridge University Press, 1935; M. Rotenberg, R. Bivins, N. Metropolis and John K. Wooten (Jr.), *The 3 j and 6 j Symbols*, M.I.T. Publications, Massachusetts, U.S.A., 1959.

Table 8.2 The C-G coefficients $\langle j_1\ m-m_2;\ 1\ m_2 \mid jm \rangle$

$j \backslash m_2$	1	0	-1
j_1+1	$\left[\frac{(j_1+m)\ (j_1+m+1)}{(2j_1+1)\ (2j_1+2)}\right]^{1/2}$	$\left[\frac{(j_1-m+1)\ (j_1+m+1)}{(2j_1+1)\ (j_1+1)}\right]^{1/2}$	$\left[\frac{(j_1-m)\ (j_1-m+1)}{(2j_1+1)\ (2j_1+2)}\right]^{1/2}$
j_1	$-\left[\frac{(j_1+m)\ (j_1-m+1)}{2j_1(j_1+1)}\right]^{1/2}$	$\frac{m}{[j_1(j_1+1)]^{1/2}}$	$\left[\frac{(j_1-m)\ (j_1+m+1)}{2j_1(j_1+1)}\right]^{1/2}$
j_1-1	$\left[\frac{(j_1-m)\ (j_1-m+1)}{2j_1\ (2j_1+1)}\right]^{1/2}$	$-\left[\frac{(j_1-m)\ (j_1+m)}{j_1\ (2j_1+1)}\right]^{1/2}$	$\left[\frac{(j_1+m+1)\ (j_1+m)}{2j_1\ (2j_1+1)}\right]^{1/2}$

coefficients for the cases where one of the angular momenta (which is taken to be j_2) is $\frac{1}{2}$ or 1. These are the cases which occur most frequently in practice. The same tables can be used to find the coefficients when either of the other angular momenta (j_1 or j) takes one of the above values (and j_2 is arbitrary). One makes use of certain symmetry properties of the C-G coefficients, which will be stated here without proof.[9]

$$\langle j_1m_1; j_2m_2 \,|\, j_3m_3 \rangle = (-)^{j_1+j_2-j_3} \langle j_2m_2; j_1m_1 \,|\, j_3m_3 \rangle \tag{8.75}$$

$$\langle j_1,m_1; j_2m_2 \,|\, j_3m_3 \rangle = (-)^{j_1-m_1} \left(\frac{2j_3+1}{2j_2+1}\right)^{1/2} \langle j_1m_1; j_3, -m_3 \,|\, j_2, -m_2 \rangle \tag{8.76a}$$

$$= (-)^{j_2+m_2} \left(\frac{2j_3+1}{2j_1+1}\right)^{1/2} \langle j_3, -m_3; j_2m_2 \,|\, j_1, -m_1 \rangle \tag{8.76b}$$

$$\langle j_1m_1; j_2m_2 \,|\, j_3m_3 \rangle = (-)^{j_1+j_2-j_3} \langle j_1, -m_1; j_2, -m_2 \,|\, j_3, -m_3 \rangle \tag{8.77}$$

From these basic formulae one can derive others giving the effect of other interchanges of angular momenta. According to Eq. 8.75, interchange of the two angular momenta which are being added can make a difference to the sign of the C-G coefficient. Lack of complete symmetry between the two may appear surprising, but the reason is very simple: we have given preferential treatment to the first angular momentum in fixing the phase convention through (8.61). Note that the interchange of any two j's is to be accompanied by that of the corresponding m's. Further, when one of the j's belongs to the total system, the signs of the two m's which are interchanged should also be changed. This is required in order that the sum of the first two m's in each bracket should remain equal to the third *on both sides* of the equation, without mutual conflict; e.g. in Eq. 8.76, the two sides have $m_1 + m_2 = m_3$ and $m_1 - m_3 = -m_2$, which are mutually consistent.

8.7 Spin Wave Functions for a System of Two Spin-½ Particles

The simplest example of the coupling of two angular momenta is when $j_1 = j_2 = \frac{1}{2}$. As a concrete example we take j_1, j_2 to be the spins s_1 and s_2 of two electrons. The total spin s of the two-electron system can take the values 1 and 0, according to (8.58). The three wave functions for $s = 1$ ($m = 1, 0, -1$) are given in terms of the spin functions $\alpha(1)$, $\beta(1)$ of the first electron and $\alpha(2)$, $\beta(2)$ of the second electron, by

$$\left.\begin{aligned} \chi(1,1) &= \alpha(1)\,\alpha(2) \\ \chi(1,0) &= \frac{1}{\sqrt{2}}\,[\alpha(1)\,\beta(2) + \beta(1)\,\alpha(2)] \\ \chi(1,-1) &= \beta(1)\,\beta(2) \end{aligned}\right\} \tag{8.78}$$

The first and last of these are obtained trivially, since the only way we can get $m = 1$ is by having spin 'up' for both the electrons ($m_1 = m_2 = \frac{1}{2}$), and

[9] See, for instance, A. R. Edmonds, *Angular Momentum in Quantum Mechanics*, Princeton University Press, 1957. It is to be emphasized that in these formulae, the system to which a particular angular momentum (say j_1) pertains is indicated by its *position* in the bracket (and *not* by the subscript). Thus in Eq. 8.75, on the left hand side j_1 is the angular momentum of the first system, and on the right, it belongs to the second system. On the right hand side of Eq. 8.76a, j_2 is the angular momentum of the composite system.

similarly for $m = -1$. Both these wave functions are symmetric under interchange of the two electrons (i.e. the labels 1 and 2), and the same property is expected for χ (1, 0) too, since it differs from the other two only in the 'direction' of the angular momentum vector. Therefore, a symmetric combination of states with $m_1 = \frac{1}{2}$, $m_2 = -\frac{1}{2}$ and $m_1 = -\frac{1}{2}$, $m_2 = \frac{1}{2}$ is chosen for $\chi(1,0)$ with the factor $(1/\sqrt{2})$ introduced for normalization. To verify that this identification is correct, one must show that

$$\mathbf{S}^2\, \chi(1,0) = 1(1+1)\hbar^2\, \chi(1,0);\; S_z\, \chi(1,0) = 0 \tag{8.79}$$

where $\mathbf{S} = \frac{1}{2}\hbar\,(\boldsymbol{\sigma}_1 + \boldsymbol{\sigma}_2)$. Here $\boldsymbol{\sigma}_1$ stands for the three Pauli matrices $\sigma_{1x}, \sigma_{1y}, \sigma_{1z}$ which act on $\alpha(1)$ and $\beta(1)$ only and $\boldsymbol{\sigma}_2$ stands for matrices with the same numerical form as $\boldsymbol{\sigma}_1$, but acting on $\alpha(2)$ and $\beta(2)$, only. Thus $\sigma_{1x}\,\alpha(1)\,\beta(2) = [\sigma_{1x}\,\alpha(1)]\beta(2) = \beta(1)\,\beta(2)$, while $\sigma_{2x}\,\alpha(1)\,\beta(2) = \alpha(1)\,\sigma_{2x}\,\beta(2) = \alpha(1)\,\alpha(2)$ and so on. The verification of Eqs. 8.79 is facilitated by writing

$$\mathbf{S}^2 = \tfrac{1}{4}\hbar^2\,(\boldsymbol{\sigma}_1 + \boldsymbol{\sigma}_2)^2 = \tfrac{1}{4}\hbar^2\,(\boldsymbol{\sigma}_1^2 + \boldsymbol{\sigma}_2^2 + 2\boldsymbol{\sigma}_1\cdot\boldsymbol{\sigma}_2) = \tfrac{3}{2}\hbar^2 + \tfrac{1}{2}\hbar^2\,\boldsymbol{\sigma}_1\cdot\boldsymbol{\sigma}_2 \tag{8.80a}$$

$$= \tfrac{3}{2}\hbar^2 + \tfrac{1}{2}\hbar^2\,(\tfrac{1}{2}\sigma_{1+}\,\sigma_{2-} + \tfrac{1}{2}\,\sigma_{1-}\,\sigma_{2+} + \sigma_{1z}\,\sigma_{2z}) \tag{8.80b}$$

and then using Eqs. 8.32. For example

$$\begin{aligned}\mathbf{S}^2\,\alpha(1)\,\beta(2) &= \tfrac{3}{2}\hbar^2\,\alpha(1)\,\beta(2) + \tfrac{1}{2}\hbar^2\,[\tfrac{1}{2}\,\sigma_{1+}\,\alpha(1).\,\sigma_{2-}\,\beta(2) \\ &\quad + \tfrac{1}{2}\,\sigma_{1-}\,\alpha(1)\,\sigma_{2+}\,\beta(2) + \sigma_{1z}\,\alpha(1)\,\sigma_{2z}\,\beta(2)] \\ &= \tfrac{3}{2}\hbar^2\,\alpha(1)\,\beta(2) + \tfrac{1}{2}\hbar^2\,[0 + 2\beta(1)\,\alpha(2) - \alpha(1)\,\beta(2)] \\ &= \hbar^2\,[\alpha(1)\,\beta(2) + \alpha(2)\,\beta(1)]\end{aligned} \tag{8.81}$$

The student may complete the verification of Eqs. 8.79 in a similar fashion, and also verify that

$$\chi(0,0) = \frac{1}{\sqrt{2}}\,[\alpha(1)\,\beta(2) - \beta(1)\,\alpha(2)] \tag{8.82}$$

This wave function may be deduced directly from the fact that it has to be orthogonal to $\chi(1,0)$ besides having $m = m_1 + m_2 = 0$.

The set of wave functions (8.78), for which the total spin is 1, is collectively referred to as the *triplet* states of the two-electron system and denoted by $^3\chi$ and (8.82) is described as the *singlet* state $^1\chi$. One often talks of the two spins being *parallel* in the former and *antiparallel* in the latter. This terminology is convenient for physical visualization, but is not strictly accurate in view of the nature of the wave function $\chi(1,0)$. It may be noted that on account of Eqs. 8.80a, the triplet and singlet states are eigenstates of $\boldsymbol{\sigma}_1\cdot\boldsymbol{\sigma}_2$ with the eigenvalues 1 and -3 respectively (since $\mathbf{S}^2$ has the values $2\hbar^2$ and 0 in these states). Thus

$$P_t \equiv \tfrac{1}{4}\,(3 + \boldsymbol{\sigma}_1\cdot\boldsymbol{\sigma}_2) \text{ and } P_s \equiv \tfrac{1}{4}\,(1 - \boldsymbol{\sigma}_1\cdot\boldsymbol{\sigma}_2) \tag{8.83}$$

are projection operators to the triplet and singlet states respectively: $P_t\,{}^3\chi = 1$, $P_t\,{}^1\chi = 0$ etc.

We have chosen to construct the states of the two electron system from first principles, though we could have directly made use of the construction using C-G coefficients. It is left to the student to verify that the coefficients appearing in Eqs. 8.78 and 8.82 are really the C-G coefficients: for example, $(1/\sqrt{2})$ in Eq. 8.82 is just $\langle\, \frac{1}{2}, -\frac{1}{2};\, \frac{1}{2}\, \frac{1}{2} \mid 00\,\rangle$. It is to be remembered that the above considerations hold equally well for any two spin-$\frac{1}{2}$ particles (e.g. an electron and a proton, or a proton and a neutron).

EXAMPLE 8.3—Systems of two spin $\frac{1}{2}$ particles as seen in their centre of mass frame are of great interest: e.g. the deuteron (neutron-proton), positronium (electron-positron); particles in collision (electron-proton, nucleon-nucleon, etc.). The system does not make any transitions betwen triplet ($s = 1$) and singlet ($s = 0$) spin states if their mutual interaction H_{int} commutes with $\mathbf{S}^2$, where $\mathbf{S} = \mathbf{s}_1 + \mathbf{s}_2$. One can then deal with triplet and singlet spin states quite independently of each other. In the latter, spin is quite passive: except for the appearance of $\chi(0,0)$ as a factor in wave functions, spin plays no role at all. In the triplet case this happens only if the interaction happens to be spin-independent (e.g. Coulomb interaction). As long as H_{int} commutes with S_z, the triplet eigenfunctions can be obtained as a product of a function of the relative coordinate $\mathbf{x}$ and a spin function $\chi(1; m_s)$. However, there are systems in which H_{int} couples the orbital and spin degrees of freedom. (In the nucleon-nucleon system such coupling is brought about by spin-orbit and tensor interactions, proportional to $\mathbf{L.S}$ and $[(\mathbf{x.s}_1)\ (\mathbf{x.s}_2) - \frac{1}{3}r^2\ (\mathbf{s}_1.\mathbf{s}_2)]$ respectively). Then the eigenfunctions are of the form $u(\mathbf{x})\chi(1;1) + v(\mathbf{x})\chi(1;0) + w(\mathbf{x})$ $\chi(1; -1)$, where there are interrelations among $u(\mathbf{x})$, $v(\mathbf{x})$, $w(\mathbf{x})$, depending on H_{int}. Expressed somewhat differently, m_s is not a good quantum number. This means, in the context of a collision problem, that if the spin function of the system before collision is $\chi(1; m_s)$, the wave function after collision will contain also $\chi(1; m_s')$, $m_s' \neq m_s$. So one has a set of scattering amplitudes $f_{m_s', m_s}(\theta,\varphi)$, m_s, $m_s' = 1, 0, -1$. Those in which $m_s = m_s'$ and $m_s \neq m_s'$ are called 'non-spinflip' and 'spinflip' amplitudes respectively. Of course, if H_{int} commutes with S_z (or in particular, is independent of spin), there is no spinflip, and one has $f_{m_s', m_s}(\theta,\varphi) = \delta_{m_s', m_s} f(\theta,\varphi)$.

8.8 Identical Particles With Spin

(*a*) *Spin-$\frac{1}{2}$ particles; Antisymmetrization of wave functions:* We are now in a position to see in concrete terms how one implements the process of anti-symmetrization which is required of the wave functions of any system of *identical* particles of spin $\frac{1}{2}$ (or $\frac{3}{2}$, $\frac{5}{2}$, . . .). Let us take the simplest case of two electrons (or two protons . . .). As stated in Sec. 3.16, any wave function $\phi(1,2)$ of such a system must necessarily change sign under interchange of the symbols 1 and 2 which stand for values of complete sets of dynamical variables (including spin) of particles 1 and 2 respectively. (This is the Pauli principle.) Thus, if $u_S\ (\mathbf{x}_1, \mathbf{x}_2)$, $u_A\ (\mathbf{x}_1, \mathbf{x}_2)$ are symmetric and antisymmetric *orbital* wave functions, i.e. $u_S\ (\mathbf{x}_1, \mathbf{x}_2) = u_S\ (\mathbf{x}_2, \mathbf{x}_1)$ and $u_A\ (\mathbf{x}_1, \mathbf{x}_2) = -u_A\ (\mathbf{x}_2, \mathbf{x}_1)$, then

$$u_S(\mathbf{x}_1,\mathbf{x}_2)\ {}^1\chi \text{ and } u_A(\mathbf{x}_1,\mathbf{x}_2)\ {}^3\chi \tag{8.84}$$

are possible wave functions for the identical particle system: the antisymmetry of ${}^1\chi$, Eq. 8.82, and the symmetry of the ${}^3\chi$, Eq. 8.78, make both the wave functions (8.84) overall antisymmetric as required. Any linear combination of the above two types of functions is also admissible (but *not* for instance, $u_S{}^3\chi$ which is symmetric, nor $u_S\,\alpha(1)\ \beta(2)$ which is neither symmetric nor antisymmetric).

We have considered in Chapter 5 two examples of two-electron systems: the hydrogen molecule (Sec. 5.9) and the helium atom (Sec. 5.5). In both the cases, the ground state wave function was symmetric with respect to the orbital variables $\mathbf{x}_1$, $\mathbf{x}_2$. The considerations of the last paragraph show then that the spin wave function is necessarily antisymmetric (singlet) in each case. We find thus that the spins of the two electrons are forced to be antiparallel (on account of the Pauli principle) even though our Hamiltonians did not involve the spins at all. If the orbital wave function had been antisymmetric, the spin wave functions would have had to be the triplet, i.e. the system would behave as if the spins were coupled in such a way as to align themselves parallel to each other. This effect has been invoked by Heisenberg in his theory of ferromagnetism.

Another example of interest is that of two spin-$\frac{1}{2}$ particles moving under their mutual interactions only. In the centre of mass frame, where the coordinate variables appear only in the combination $\mathbf{x} = \mathbf{x}_1 - \mathbf{x}_2$, the interchange of $\mathbf{x}_1 \longleftrightarrow \mathbf{x}_2$ leads to $\mathbf{x} \rightarrow -\mathbf{x}$ which is the parity transformation on $\mathbf{x}$. With reference to Example 8.3 (p. 257) we see now that if the particles are identical, antisymmetry of the total wave function requires that *in singlet spin states* the orbital wave function should have *even parity* (s,d ... states allowed but not $p, f, \ldots$). *In triplet spin states, only odd parity* orbital wave functions are permissible. An interesting consequence is that any diatomic molecule in which the atoms are identical and have spin $\frac{1}{2}$ (e.g. hydrogen) can have rotational states with *either* $l = 0, 2, 4, \ldots$ *or* $l = 1,3,5 \ldots$ depending on whether the spin state is singlet (para) or triplet (ortho). In either case, the spacing Δl of l values is an even integer. (This constraint does not exist when the two atoms are different, e.g. HCl.) In the infrared (rotational) spectra of such molecules, lines corresponding to odd Δl are indeed missing.

In a different context, that of scattering, antisymmetrization requires that the scattering amplitude in singlet spin states should remain unchanged under the interchange $(\theta,\varphi) \longleftrightarrow (\pi - \theta,\ \varphi + \pi)$ induced by $\mathbf{x} \longleftrightarrow -\mathbf{x}$, i.e. we must choose $f_+(\theta)$ of (6.115). In triplet states one must take the scattering amplitudes to be the antisymmetric combinations $[f_{m'm}(\theta,\varphi) - f_{m'm}(\pi - \theta,\ \varphi + \pi)]$ of those defined in Example 8.3(p. 257). If the interaction is spin-independent, this reduces to $\delta_{m'm} f_-(\theta,\varphi)$

(*b*) *Spin s:* We have already seen that the spin wave function of a spin-s particle is a $(2s + 1)$-component column. Let the spin wave functions belonging to various values $m\hbar$ of S_z, $(m = s,\ s - 1, \ldots, -s)$ be denoted by $\alpha^{(m)}$. [The correspondence with the earlier notation of the spin-$\frac{1}{2}$ case is $\alpha^{(1/2)} = \alpha$, $\alpha^{(-1/2)} = \beta$.] For a pair of such particles, labelled by 1 and 2, we have $(2s + 1)^2$ wave functions $\alpha^{(m)}(1)\ \alpha^{(m')}(2)$. When the particles are identical, it is more advantageous to use, instead of these, the symmetric and antisymmetric combinations

$$\chi_S^{(m_1,m_2)}(1,2) = \chi^{(m_1)}(1)\ \chi^{(m_2)}(2) + \chi^{(m_2)}(1)\chi^{(m_1)}(2)$$
$$\chi_A^{(m_1,m_2)}(1,2) = \chi^{(m_1)}(1)\ \chi^{(m_2)}(2) - \chi^{(m_2)}(1)\chi^{(m_1)}(2).$$

These can be combined with orbital wave functions of even or odd parity (with respect to the relative coordinate $\mathbf{x}$, in the centre of mass frame) to form complete wave functions with overall symmetry under interchange of particles (for $2s$ even) or antisymmetry (for $2s$ odd). The number of spin functions χ_S and χ_A, and the parity of the orbital functions to be associated with these, are summarized below.

Spin function	Number	Parity of orbital function	Scattering amplitude
χ_S	$(s+1)(2s+1)$	$(-)^{2s}$	$f(\theta,\varphi) + (-)^{2s} f(\pi-\theta, \varphi+\pi)$
χ_A	$s(2s+1)$	$(-)^{2s+1}$	$f(\theta,\varphi) + (-)^{2s+1} f(\pi-\theta, \varphi+\pi)$

The last column shows the scattering amplitude consistent with the stated parity in collisions in which the spin wave functions belong to the set χ_S or χ_A. Now, there is no reason *a priori* for the system to prefer any particular one of the spin states. Therefore, the chance of its having a symmetric spin function is $(s+1)(2s+1)/(2s+1)^2 = (s+1)/(2s+1)$, and for having an antisymmetric one, $s/(2s+1)$. The scattering cross-section in an experiment in which the identical colliding particles are unpolarized should then be

$$\frac{d\sigma}{d\Omega} = \frac{(s+1)}{(2s+1)} \left| f(\theta,\varphi) + (-)^{2s} f(\pi-\theta, \varphi+\pi) \right|^2 + \frac{s}{2s+1} \left| f(\theta,\varphi) + (-)^{2s+1} f(\pi-\theta, \varphi+\pi) \right|^2$$

We have seen an example of this ($s = \frac{1}{2}$) at the end of Sec. 6.19.

In diatomic molecules made up of identical atoms whose nuclei have spin I, the inequality of the number of available spin functions χ_S and χ_A leads to a corresponding inequality in the numbers of molecules formed in such states. Since the χ_S and χ_A go with orbital parity $(-)^{2I}$ and $(-)^{2I+1}$ respectively, this means that the number of molecules with the parity $(-)^{2I}$ for the orbital (rotational) states outnumber those with parity $(-)^{2I+1}$ by a factor $(I+1)/I$. Evidence for this comes from the observation of alternating intensities of lines in the rotational band spectra of such molecules. The lines originate alternately from odd parity and even parity levels, and the above-mentioned factor is reflected in the ratio of intensities of alternate lines.

8.9 Addition of Spin and Orbital Angular Momenta

We go on now to construct wave functions with definite j and m for a single electron of orbital angular momentum l, starting from the orbital wave functions Y_{l,m_l} and spin wave functions χ_{m_s}. This can be done with the aid of the C-G coefficients:

$$u_{l\frac{1}{2};jm} = \sum_{m_l m_s} Y_{lm_l} \chi_{m_s} \langle lm_l; \tfrac{1}{2} m_s \,|\, jm \rangle \tag{8.85a}$$

Since $m_s = \pm \frac{1}{2}$ only, and $m_l = m - m_s$, there are only two terms in this sum. Further, j can only take the values $l \pm \frac{1}{2}$. For these two cases we get, on introducing the C-G coefficients $\langle l, m - m_s; \frac{1}{2}, m_s \,|\, l + \frac{1}{2}, m \rangle$ and $\langle l, m - m_s; \frac{1}{2}, m_s \,|\, l - \frac{1}{2}, m \rangle$ from Table 8.1,

$$u_{l,\frac{1}{2};l+\frac{1}{2},m} = \left(\frac{l+m+\frac{1}{2}}{2l+1}\right)^{1/2} Y_{l,m-\frac{1}{2}}\,\alpha + \left(\frac{l-m+\frac{1}{2}}{2l+1}\right)^{1/2} Y_{l,m+\frac{1}{2}}\,\beta \tag{8.85b}$$

$$u_{l,\frac{1}{2};l-\frac{1}{2},m} = -\left(\frac{l-m+\frac{1}{2}}{2l+1}\right)^{1/2} Y_{l,m-\frac{1}{2}}\,\alpha + \left(\frac{l+m+\frac{1}{2}}{2l+1}\right)^{1/2} Y_{l,m+\frac{1}{2}}\,\beta \tag{8.85c}$$

where $\alpha = \chi_{1/2}$ and $\beta = \chi_{-1/2}$. Using the forms (8.29) for these, the above wave functions can be written explicitly in two-component form as

$$\begin{pmatrix} \left(\frac{l+m+\frac{1}{2}}{2l+1}\right)^{1/2} Y_{l,m-\frac{1}{2}} \\ \left(\frac{l-m+\frac{1}{2}}{2l+1}\right)^{1/2} Y_{l,m+\frac{1}{2}} \end{pmatrix} \text{ and } \begin{pmatrix} -\left(\frac{l-m+\frac{1}{2}}{2l+1}\right)^{1/2} Y_{l,m-\frac{1}{2}} \\ \left(\frac{l+m+\frac{1}{2}}{2l+1}\right)^{1/2} Y_{l,m+\frac{1}{2}} \end{pmatrix} \tag{8.86}$$

respectively. The student is urged to construct these directly by seeking two-component wave functions whose components are made up of the Y_{lm_l} with given l, and which are eigenfunctions of $J_z = L_z + s_z$ and $\mathbf{J}^2 = \mathbf{L}^2 + \mathbf{s}^2 + 2\mathbf{L}.\mathbf{s}$ with $\mathbf{s} = \frac{1}{2}\hbar\sigma$, i.e.

$$J_z = \begin{pmatrix} L_z + \frac{1}{2}\hbar & 0 \\ 0 & L_z - \frac{1}{2}\hbar \end{pmatrix}, \quad \mathbf{J}^2 = \begin{pmatrix} \mathbf{L}^2 + \frac{3}{4}\hbar^2 + \hbar L_z & \hbar(L_x - iL_y) \\ \hbar\,(L_x + iL_y) & \mathbf{L}^2 + \frac{3}{4}\hbar^2 - \hbar L_z \end{pmatrix} \tag{8.87}$$

This form of J_z shows clearly that since J_z is to have the value $m\hbar$ for both the components of u, m_l should be $(m - \frac{1}{2})$ for the upper and $(m + \frac{1}{2})$ for the lower components, as in (8.86). The numerical coefficients are then fixed (up to a normalization constant) by requiring u to be an eigenfunction of $\mathbf{J}^2$.

Another example is provided by a system of two spin-$\frac{1}{2}$ particles as seen in the centre of mass frame of reference. If the orbital motion in this frame is characterized by a definite value of l, the angular parts of the corresponding wave functions are $Y_{lm_l}(\theta, \varphi)$, where θ, φ are the angular coordinates of the relative position vector $\mathbf{x} \equiv \mathbf{x}_1 - \mathbf{x}_2$. For the spin wave functions, we could choose as basis either the set $\alpha(1)\alpha(2)$, $\alpha(1)\beta(2)$, $\beta(1)\,\alpha(2)$, $\beta(1)\,\beta(2)$ or the total-spin eigenfunctions (8.78) and (8.82). Let us choose the latter, and then add the orbital angular momentum $\mathbf{L}$ and the total spin $\mathbf{S}$. When $s = 0$ (singlet state) the addition is trivial since only $j = l$ is then possible. Thus the angular parts of the singlet wave functions with definite j and m are

$$u(l0;\, jm) = Y_{lm}\;\chi(0,0)\,\delta_{jl} \tag{8.88}$$

with $\chi(00)$ given by Eq. 8.82.

For the triplet states on the other hand, we have

$$u\,(l1;\, jm) = \sum_{m_s} Y_{l,m-m_s}\,\chi(1, m_s)\,\langle\, l, m - m_s;\; 1, m_s \,|\, jm \,\rangle \tag{8.89}$$

The values of the C-G coefficients occurring here may be taken from Table 8.2. The $\chi\,(1, m_s)$ are the spin wave functions of Eq. 8.78. Eq. 8.89 gives triplets with $j = l + 1, l$ and $l - 1$, for any given l (except $l = 0$, when $j = 1$ alone is possible).

The use of simultaneous eigenfunctions of $\mathbf{J}^2$ and J_z (rather than L_z and S_z) in considering states with given l and s is particularly advantageous in problems involving spin-orbit coupling. The reason is that the $|\, lsjm \,\rangle$ states, unlike the $|\, lm_l;\, sm_s \,\rangle$ are eigenstates of $\mathbf{L}.\mathbf{S}$. In fact, since $\mathbf{J}^2 = \mathbf{L}^2 + \mathbf{S}^2 + 2\mathbf{L}.\mathbf{S}$, we have quite generally for any system of particles[10]

[10] Capital letters are used conventionally for the angular momentum quantum numbers of many-particle systems.

$$(\mathbf{L.S})\,|LSJM\rangle = \tfrac{1}{2}(\mathbf{J}^2 - \mathbf{L}^2 - \mathbf{S}^2)\,|LSJM\rangle$$
$$= \tfrac{1}{2}\hbar^2\,[J(J+1) - L(L+1) - S(S+1)]\,|LSJM\rangle \qquad (8.90)$$

In the particular example which we were considering, the value of **L.S** is readily seen to be zero in the singlet state, while in triplet states, $\mathbf{L.S} = l\hbar^2$, $-\hbar^2$ or $-(l+1)\hbar^2$ according as $j = l+1$, l or $(l-1)$.

EXAMPLE 8.4.–*Fine structure of the energy levels of alkali atoms:* In the ground state of an alkali atom, all the electrons go into closed shells, except for a single (valence) electron which is an s-state. The closed shells form a highly stable core which has $L = S = 0$. So the ground state and all (moderately) excited states of the atom can be identified by the quantum numbers[11] $(n, l\, m_l m_s)$ or better, $(n, l\, j m)$, of the valence electron which moves in an effective central potential $V(r)$ due to the nucleus and the core electrons. When the spin-orbit interaction is ignored, the energy depends on n and l only. The change $(\Delta E)_{s.o}$ in this energy E_{nl}, produced by $H_{s.o}$ of Eq. 8.40, can be readily calculated from first order perturbation theory as

$$(\Delta E)_{s.o} = \langle n, ljm \,|\, \xi(r)\, \mathbf{L.S} \,|\, n, ljm \rangle$$

This factors into a radial part and an angular part $\langle ljm \,|\, \mathbf{L.S} \,|\, ljm \rangle$. The value of the latter is obtained from Eqs. 8.90, with $S = \frac{1}{2}$, as $\frac{1}{2}\hbar^2\,[j(j+1) - l(l+1) - \frac{3}{4}] = \frac{1}{2}l\,\hbar^2$ or $-\frac{1}{2}(l+1)\hbar^2$ according as $j = l + \frac{1}{2}$ or $l - \frac{1}{2}$. Thus

$$(\Delta E)_{s.o} = l I_{nl}, \qquad (j = l + \tfrac{1}{2})$$
$$= -(l+1)\, I_{nl}, \qquad (j = l - \tfrac{1}{2}),$$

with
$$I_{nl} = \tfrac{1}{2}\hbar^2 \int |R_{nl}|^2\, \xi(r)\, r^2\, dr.$$

Thus each level gets split into a *doublet* (except for $l = 0$, when only $j = l + \frac{1}{2} = \frac{1}{2}$ is possible). To get some idea about the magnitude of the splitting, consider the special case of the hydrogen-like atom, for which $V(r) = -Ze^2/r$ and $\xi(r) = (2m^2c^2r)^{-1}\,(dV/dr) = (Ze^2/2m^2c^2r^3)$. The value of I_{nl} can then be read off from Table 4.2, and one finds

$$(\Delta E)_{s.o} = \frac{A_{nl}}{l+1}, \quad (j = l + \tfrac{1}{2}); \quad \frac{-A_{nl}}{l}, \quad (j = l - \tfrac{1}{2})$$

$$A_{nl} = \frac{e^2\hbar^2 Z^4}{4m^2c^2a^3n^3(l+\frac{1}{2})} = \frac{Z^4}{32\pi^3 n^3 (l+\frac{1}{2})}\left(\frac{\lambda_c}{a}\right)^3 \alpha\, mc^2$$

where the last expression gives A_{nl} in terms of the electron rest energy mc^2 and non-dimensional factors (λ_c is the Compton wavelength, h/mc). The proportionality of this splitting to $\alpha \equiv (e^2/\hbar c)$ is the reason why α is called the *fine structure constant.*

It is to be remembered that to get the actual fine structure for the hydrogen atom, a further effect due to the relativistic variation of electron mass is to be included. In the alkali atoms $V(r)$ does not have the pure Coulomb form, but the above expression gives a general idea of the dependence of the fine structure on n and l.

[11] Here n is the principal quantum number. The spin has the fixed value $\frac{1}{2}$, and is not shown explicitly.

8.10 Spherical Tensors; Tensor Operators

We have seen in Secs. 7.9 and 7.11 that if $|\psi\rangle$ is any state, and F, an operator representing some dynamical variable of a quantum system, then rotation through an infinitesimal angle θ about the direction $\mathbf{n}$ takes $|\psi\rangle$ and F into $|\psi\rangle'$ and F' where $|\psi\rangle' \approx (1 - i\theta\, \mathbf{n}.\mathbf{J})\, |\psi\rangle$ and $F' \approx (1 - i\theta\mathbf{n}.\mathbf{J})\, F\, (1 + i\theta\mathbf{n}.\mathbf{J}) \approx F - i\theta\mathbf{n}.\, [\mathbf{J},F]$. Thus the changes in states $|\psi\rangle$ and dynamical variables F, brought about by infinitesimal rotations, are directly expressed in terms of

$$\mathbf{J}\, |\psi\rangle \text{ and } [\mathbf{J},F] \tag{8.91}$$

respectively. In our consideration of these changes, we shall find it to be advantageous to employ the spherical components of the vector $\mathbf{J}$. The spherical components V_μ ($\mu = 1,0,-1$) of any vector $\mathbf{V}$ is defined through the decomposition

$$\mathbf{V} = \sum_{\mu=-1}^{1} \mathbf{e}_\mu{}^* V_\mu = \Sigma\, (-)^\mu\, \mathbf{e}_{-\mu} V_\mu,$$

$$\mathbf{e}_1 = -\frac{1}{\sqrt{2}}\,(\mathbf{e}_x + i\mathbf{e}_y), \quad \mathbf{e}_0 = \mathbf{e}_z, \quad \mathbf{e}_{-1} = \frac{1}{\sqrt{2}}\,(\mathbf{e}_x - i\mathbf{e}_y) \tag{8.92}$$

where $\mathbf{e}_x$, $\mathbf{e}_y$, $\mathbf{e}_z$ are the orthonormal basis vectors of Cartesian coordinates. Then

$$V_1 = -\frac{1}{\sqrt{2}}\,(V_x + iV_y), \quad V_0 = V_z, \quad V_{-1} = \frac{1}{\sqrt{2}}\,(V_x - iV_y) \tag{8.93}$$

The spherical components of $\mathbf{J}$ are thus

$$J_1 = -\frac{1}{\sqrt{2}}\, J_+, \quad J_0 = J_z, \quad J_{-1} = \frac{1}{\sqrt{2}}\, J_- \tag{8.94}$$

The commutation relations of these operators can be read off from Eqs. 8.3. It is left as an exercise for the student to verify that these can be written as

$$[J_\mu, J_\nu] = \sum_{\mu'} J_{\mu'}\, \langle\, 1\mu' \,|\, J_\mu \,|\, 1\nu\, \rangle \tag{8.95}$$

using the fact that according to Eqs. 8.16 and 8.19,

$$\langle 1\mu' \,|\, J_0 \,|\, 1\mu \rangle = \mu\delta_{\mu\mu'}, \; \langle\, 1\mu' \,|\, J_{\pm 1} \,|\, 1\mu\, \rangle = \mp\, [1 - \tfrac{1}{2}\,(\mu^2 \pm \mu)]^{1/2}\, \delta_{\mu',\mu\pm 1} \tag{8.95a}$$

Returning to Eq. 8.91 we observe that if $|\psi\rangle$ is any of the states $|\, jm\, \rangle$, then the state $J_\mu\, |\psi\rangle = J_\mu |\, jm\, \rangle$ is a linear combination of states having the *same* j and various m'. Specifically,

$$J_\mu \,|\, jm\, \rangle = \sum_{m'} |\, jm'\, \rangle\, \langle\, jm' \,|\, J_\mu \,|\, jm\, \rangle \tag{8.96}$$

Thus the $(2j + 1)$ states $|\, jm\, \rangle$ transform linearly among themselves when subjected to a rotation. Now it is well known that the fundamental characterization of a tensor is through the linear transformation of its components under the effect of a specified class of operations (such as rotations) on the coordinate system or alternatively on the tensor itself. Thus the set of states $|\, jm\, \rangle$ for given j is said to form a *spherical tensor*[12] of rank j, with $(2j + 1)$ components labelled by $m = j, j - 1, \ldots, -j$. Since there is no subset of this set which transforms into itself, the tensor is *irreducible*.

[12] The qualification 'spherical' refers to the fact that the only operations allowed here are rotations about the origin.

Consider now the changes $[J_\mu, F]$ in *operators* F: Suppose we have a set of $(2j+1)$ operators (which we will denote by T_{jm}, $m = j, j-1, \ldots, -j$), such that under commutation with the J_μ the set undergoes the *same* linear transformation as the states $|jm\rangle$, i.e.

$$[J_\mu, T_{jm}] = \sum_{m'} T_{jm'} \langle jm' | J_\mu | jm \rangle \tag{8.97}$$

Such a set is said to constitute an *irreducible spherical tensor operator* of rank j. The essential point which cannot be emphasized too strongly, is that identical coefficients $\langle jm' | J_\mu | jm \rangle$ appear in both (8.96) and (8.97).

The first example of a tensor operator is angular momentum itself. Comparison of Eq. 8.95 with Eq. 8.97 shows that the J_μ constitute a tensor operator of rank 1. In fact any vector operator, such as **x**, **p**, etc. is a tensor operator of rank 1. This is easy to verify, starting from the commutation relations $[J_x, x] = 0$, $[J_x, y] = i\hbar z$, $[J_x, z] = -i\hbar y$, etc. of the Cartesian components.

EXAMPLE 8.5.—If $T^{(1)}{}_{j_1m_1}$, $T^{(2)}{}_{j_2m_2}$ are tensor operators, the quantities T_{jm} defined by

$$T_{jm} \equiv \sum_{m_1m_2} T^{(1)}{}_{j_1m_1} T^{(2)}{}_{j_2m_2} \langle j_1m_1; j_2m_2 | jm \rangle$$

(for given j) do constitute a tensor of rank j as indicated by our choice of notation. The construction of the operators T_{jm} is in exact analogy with that of the states $|jm\rangle$ from states $|j_1m_1; j_2m_2\rangle$ in Eqs. 8.54. Further, the effect of taking the commutators of J_μ with $T^{(1)}{}_{j_1m_1}$ and $T^{(2)}{}_{j_2m_2}$ is also just the same as the effect of $J_{1\mu}$ and $J_{2\mu}$ on $|j_1m_1; j_2m_2\rangle$ by virtue of the definition of tensor operators. Therefore it is evident that the commutator $[J_\mu, T_{jm}]$ will be the same linear combination of the $T_{jm'}$ as $J_\mu |jm\rangle = (J_{1\mu} + J_{2\mu}) |jm\rangle$ will be of the $|jm'\rangle$. Hence the T_{jm} constitute a tensor operator as asserted.

In particular, if $j_1 = j_2$ and $j = 0$, T_{00} is a scalar constructed from the two tensors. In the special case $j_1 = j_2 = 1$, the $T^{(1)}{}_{1m_1}$ and $T^{(2)}{}_{1m_2}$ are the spherical components of vectors $\mathbf{T}^{(1)}$ and $\mathbf{T}^{(2)}$. It may be verified that in this case

$$T_{00} = \Sigma\, (-)^\mu\, T^{(1)}{}_{1\mu}\, T^{(2)}{}_{1,-\mu} = \mathbf{T}^{(1)}.\mathbf{T}^{(2)}$$

The last step follows directly from the definition (8.93) of spherical components.

8.11 The Wigner-Eckart Theorem

This theorem states that the matrix element[13] $\langle \alpha' j' m' | T_{LM} | \alpha jm \rangle$ of any component of a tensor operator decomposes into two factors: the C-G coefficient $\langle jm; LM | j'm' \rangle$ which is quite independent of the physical nature of the tensor operator, and a reduced matrix element[14] $(\alpha' j' \| T_L \| \alpha j)$ which has no dependence at all on the magnetic quantum numbers m, M, m':

[13] The tensor operator may involve degrees of freedom other than the orientational ones: for example **x** contains the radial variable r. The parameter α (or α'), appearing in the matrix element of T_{LM} here stands for one or more quantum numbers required for a complete specification of basis states when such extra degrees of freedom are present. In the context of the hydrogen atom problem, for instance, α would be the principal quantum number n (See Example 8.4, p. 261).

[14] Also called double-bar element. Different authors use different definitions and notations. A comparison of these may be found in A. R. Edmonds, *Angular Momentum in Quantum Mechanics*, Princeton University Press, 1957. Our definition coincides with that of M. E. Rose, *Elementary Theory of Angular Momentum*, John Wiley and Sons, New York, 1957.

$$\langle \alpha' j' m' \mid T_{LM} \mid \alpha j m \rangle = \langle jm; LM \mid j'm' \rangle \langle \alpha' j' \| T_L \| \alpha j \rangle \tag{8.98}$$

This complete separation of the geometric (orientational) and physical aspects is of immense practical utility. The proof of the theorem follows almost trivially from the basic definition of tensor operators. On taking the matrix element of Eq. 8.97 one obtains,

$$\begin{aligned}
&\sum_{M'} \langle \alpha' j' m' \mid T_{LM'} \mid \alpha j m \rangle \langle LM' \mid \mathcal{J}_\mu \mid LM \rangle \\
&= \langle \alpha' j' m' \mid [\mathcal{J}_\mu, T_{LM}] \mid \alpha j m \rangle \\
&= \sum_{m''} \{ \langle j'm' \mid \mathcal{J}_\mu \mid j'm'' \rangle \langle \alpha' j' m'' \mid T_{LM} \mid \alpha j m \rangle \\
&\qquad - \langle \alpha' j' m' \mid T_{LM} \mid \alpha j m'' \rangle \langle jm'' \mid \mathcal{J}_\mu \mid jm \rangle \}
\end{aligned} \tag{8.99}$$

Here we have introduced the expansion

$$1 = \sum_{\alpha'' j'' m''} \mid \alpha'' j'' m'' \rangle \langle \alpha'' j'' m'' \mid \tag{8.100}$$

between $\mathcal{J}_\mu$ and T_{LM}, and used the obvious fact that $\langle \alpha' j' m' \mid \mathcal{J}_\mu \mid \alpha'' j'' m'' \rangle = \langle j'm' \mid \mathcal{J}_\mu \mid j''m'' \rangle \delta_{\alpha'\alpha''} \delta_{j'j''}$. On the other hand one can obtain an equation of identical form with *C-G* coefficients **in the place** of the matrix elements. Specifically,

$$\begin{aligned}
&\sum_{M'} \langle j'm' \mid jm; LM' \rangle \langle LM' \mid \mathcal{J}_\mu \mid LM \rangle \\
&= \sum_{m''} \{ \langle j'm' \mid \mathcal{J}_\mu \mid j'm'' \rangle \langle j'm'' \mid jm; LM \rangle \\
&\qquad - \langle j'm' \mid jm''; LM \rangle \langle jm'' \mid \mathcal{J}_\mu \mid jm \rangle \}
\end{aligned} \tag{8.101}$$

This equation follows from the identity

$$\langle j'm' \mid \mathcal{J}_\mu \mid jm; LM \rangle \equiv \sum_{m''} \langle j'm' \mid \mathcal{J}_\mu \mid j'm'' \rangle \langle j'm'' \mid jm; LM \rangle \tag{8.102}$$

on reexpressing the left hand side as

$$\sum_{m'', M'} \langle j'm' \mid jm''; LM' \rangle \langle jm''; LM' \mid \mathcal{J}_\mu \mid jm; LM \rangle \tag{8.103}$$

and writing $\mathcal{J}_\mu = \mathcal{J}_{1\mu} + \mathcal{J}_{2\mu}$ where $\mathcal{J}_{1\mu}$ acts on the jm variables only and $\mathcal{J}_{2\mu}$ on the LM. Then the last factor in (8.103) reduces to

$$\langle jm'' \mid \mathcal{J}_{1\mu} \mid jm \rangle \delta_{M',M} + \langle LM' \mid \mathcal{J}_{2\mu} \mid LM \rangle \delta_{m'',m}$$

and (8.103) or what is the same thing, the left hand side of Eq. 8.102, becomes[15]

$$\begin{aligned}
&\sum_{m''} \langle j'm' \mid jm''; LM \rangle \langle jm'' \mid \mathcal{J}_\mu \mid jm \rangle \\
&\qquad + \sum_{M'} \langle j'm' \mid jm; LM' \rangle \langle LM' \mid \mathcal{J}_\mu \mid LM \rangle
\end{aligned} \tag{8.104}$$

With this, Eq. 8.102 reduces to Eq. 8.101.

Now, let us compare Eq. 8.99, ignoring the middle member, with Eq. 8.101. The two are identical except that $\langle \alpha' j' m' \mid T_{LM} \mid \alpha j m \rangle$ in one is replaced by $\langle j'm' \mid jm; LM \rangle$ in the other. Evidently the two equations can be mutually consistent if and only if the above two quantities are proportional to each other, with a proportionality factor *independent* of the summation indices occurring in the equations (i.e. the magnetic quantum numbers). This is precisely what is stated by the Wigner-Eckart theorem (Eq. 8.98).

[15] We drop the subscripts 1 and 2 on $\mathcal{J}_{1\mu}$ and $\mathcal{J}_{2\mu}$, with the understanding that $\mathcal{J}_\mu$ stands for whichever angular momentum operator that pertains to the system concerned.

As an immediate corollary we have $\langle j'm' \mid \mathcal{J}_\mu \mid jm \rangle = \langle jm;\ 1\mu \mid j'm' \rangle$ $(j' \| \mathcal{J} \| j)$. To determine the double bar element, we evaluate the matrix element for any convenient value of μ, m, m'. On taking $m = m'$, $\mu = 0$, the value of the left hand side is $m\hbar\,\delta_{jj'}$; so $(j' \| \mathcal{J} \| j)$ must be proportional to $\delta_{jj'}$. The value of the *C-G* coefficient with $j = j'$ is found from Table 8.2 to be $m/[j(j+1)]^{1/2}$. Hence we have

$$(j' \| \mathcal{J} \| j) = \hbar\, [j(j+1)]^{1/2}\, \delta_{jj'},$$
$$\langle j'm' \mid \mathcal{J}_\mu \mid jm \rangle = \langle jm;\ 1\mu \mid j'm' \rangle\, [j(j+1)]^{1/2}\, \delta_{jj'}\, \hbar \tag{8.105}$$

EXAMPLE 8.6. *The quadrupole moment.* Classically, the quadrupole moment of a charge distribution $\rho(\mathbf{x})$ which has axial symmetry is defined as $Q = \int (3z^2 - r^2)\, \rho(\mathbf{x})\, d^3x$, where the axis of symmetry is chosen as the z-axis and the origin is chosen as the centre of charge, i.e. such that $\int \mathbf{x} \mid \rho(\mathbf{x}) \mid d^3x = 0$. In quantum mechanics, a particle of charge e, with the wave function $\psi(\mathbf{x})$, produces a charge distribution $\rho(\mathbf{x}) = e\psi^*(\mathbf{x})\ \psi(\mathbf{x})$ and in this case

$$Q = e \int \psi^*(\mathbf{x})\, (3z^2 - r^2)\, \psi(\mathbf{x}) d^3x.$$

Now, $(3z^2 - r^2)$ is a component $Q_{2,0}$ of a second rank tensor operator. [The student may verify this as a special case of the result of Example 8.5, taking both $T^{(1)}{}_{j_1m_1}$ and $T^{(2)}{}_{j_2m_2}$ $(j_1 = j_2 = 1)$ to be the spherical components of the position operator $\mathbf{x}$, and $j = 2$, $m = 0$. Alternatively, observe that in terms of spherical harmonics, $Q_{2,0} \equiv (3z^2 - r^2) = (16\pi/5)^{1/2}\, r^2\, Y_{2,0}$]. Therefore, if ψ is the angular momentum state $\mid lm \rangle$, we have, from the Wigner-Eckart theorem, that $\langle lm \mid Q_{2,0} \mid lm \rangle = \langle lm;\ 20 \mid lm \rangle\, (l \| Q_2 \| l)$. It is the value of this quantity when $m = l$ that is usually referred to as the quadrupole moment Q in a quantum state of angular momentum l. With this definition,

$$\langle lm \mid Q_{2,0} \mid lm \rangle = Q\, \frac{\langle lm;\ 20 \mid lm \rangle}{\langle ll;\ 20 \mid ll \rangle} = Q \cdot \frac{3m^2 - l(l+1)}{l(2l-1)}.$$

8.12 Projection Theorem for a First Rank Tensor

If $\mathbf{V}$ is any vector (i.e. first rank tensor) operator with spherical components V_μ, then

$$\langle jm' \mid V_\mu \mid jm \rangle = \frac{\langle jm' \mid \mathcal{J}_\mu(\mathbf{J}.\mathbf{V}) \mid jm \rangle}{j(j+1)} \tag{8.106}$$

The operator in the matrix element on the right hand side may be viewed as the projection of $\mathbf{V}$ on to the direction of $\mathbf{J}$, characterized by the 'unit vector' $\mathbf{J}/[j(j+1)]^{1/2}$. So the theorem states essentially that the part of $\mathbf{V}$ which is perpendicular to $\mathbf{J}$ does not contribute to matrix elements between angular momentum states of given j.

As a preliminary to the proof of the theorem, let us evaluate the matrix elements of $\mathbf{J}.\mathbf{V}$:

$$\langle j''m'' \mid \mathbf{J}.\mathbf{V} \mid jm \rangle = \sum_{\nu=-1}^{+1} (-)^\nu \langle j''m'' \mid \mathcal{J}_\nu V_{-\nu} \mid jm \rangle$$
$$= \Sigma\, (-)^\nu \langle j''m'' \mid \mathcal{J}_\nu \mid j'', m''-\nu \rangle \langle j'', m''-\nu \mid V_{-\nu} \mid jm \rangle \tag{8.107}$$

where we have introduced between $\mathcal{J}_\nu$ and $V_{-\nu}$ the expansion (8.100) of the identity and noted that the first matrix element is non-vanishing only for the particular intermediate state $|j'', m''-\nu\rangle$. Now we reduce the above expresssion, by applying the Wigner-Eckart theorem, to

$$(j''\|\mathcal{J}\|j'')\,(j''\|V\|j)\sum_{\nu}(-)^{\nu}\langle j'', m-\nu; 1\nu|j''m''\rangle\,.$$

$$.\langle jm; 1,-\nu|j'', m''-\nu\rangle \qquad (8.108)$$

Finally we bring the sum to the form of the orthonormality relation (8.59b) by writing the last C-G coefficient as

$$\langle jm; 1,-\nu|j'', m''-\nu\rangle = (-)^{j+1-j''}\langle j,-m; 1\nu|j'',-m''+\nu\rangle$$
$$= [(2j''+1)/(2j+1)]^{1/2}.(-)^{j-j''+\nu}\langle j'', m''-\nu; 1\nu|jm\rangle,$$

with the aid of the symmetry properties (8.77) and (8.76b). On using this and noting the fact that the C-G coefficients are real and vanish unless $m''=m$, the sum in (8.108) becomes

$$\delta_{mm''}(-)^{j-j''}\sum_{\nu}\langle j''m|j'', m-\nu; 1\nu\rangle\langle j'', m-\nu; 1\nu|jm\rangle = \delta_{mm''}\delta_{jj''}$$

by virtue of (8.59b). On feeding (8.108) with this simplification into (8.107) we get the desired result:

$$\langle j''m''|(\mathbf{J}.\mathbf{V})|jm\rangle = (j\|\mathcal{J}\|j)\,(j\|V\|j)\,\delta_{mm''}\delta_{jj''} \qquad (8.109)$$

The delta functions here simply reflect the fact that $(\mathbf{J}.\mathbf{V})$ is a scalar, i.e. a tensor operator of rank zero (see Example 8.5, p. 263).

Now, since

$$\langle j'm'|\mathcal{J}_{\mu}(\mathbf{J}.\mathbf{V})|jm\rangle = \langle j'm'|\mathcal{J}_{\mu}|j', m'-\mu\rangle\langle j', m'-\mu|(\mathbf{J}.\mathbf{V})|jm\rangle,$$
$$= (j'\|\mathcal{J}\|j')\langle j', m'-\mu; 1\mu|j'm'\rangle\langle j', m'-\mu|(\mathbf{J}.\mathbf{V})|jm\rangle \qquad (8.110)$$

it follows immediately on using Eq. 8.109 and the expression (8.105) for the reduced matrix element of $\mathbf{J}$ that

$$\langle j'm'|\mathcal{J}_{\mu}(\mathbf{J}.\mathbf{V})|jm\rangle = \delta_{jj'}j(j+1)\langle jm; 1\mu|jm'\rangle\langle(j\|V\|j) \qquad (8.111)$$

We have dropped $\delta_{m'-\mu,m}$, since it is already implied by the C-G coefficient. Finally, on using the Wigner-Eckart theorem in reverse, the right hand side of Eq. 8.111 becomes $\delta_{jj'}j(j+1)\langle jm'|V_{\mu}|jm\rangle$. This completes the proof of Eq. 8.106.

Example 8.7.—*Weak-field Zeeman effect:* An atom has a magnetic moment μ, arising partly from the orbital motion of the electrons ($\boldsymbol{\mu}_L = \mu_0\mathbf{L}$) and partly from the spin ($\boldsymbol{\mu}_s = 2\mu_0\mathbf{S}$). Here $\mu_0 \equiv e\hbar/2mc$ is the Bohr magneton. The energy of its interaction with a magnetic field $\mathcal{H}$ in which it is placed is then

$$H_{\text{mag}} = -\boldsymbol{\mu}.\mathcal{H} = -\mu_0(\mathbf{L}+2\mathbf{S}).\mathcal{H}$$

When the magnetic field is weak, this can be treated as a perturbation, where the unperturbed states are eigenstates of the Hamiltonian which includes all other forces (electrostatic and spin-orbit) on the electrons. Such states, according to the L-S coupling scheme (see Appendix E) are approximate eigenstates of $\mathbf{L}^2$, $\mathbf{S}^2$, $\mathbf{J}^2$ and $\mathcal{J}_z$ and can, therefore, be denoted by $|aLS\mathcal{J}M\rangle$. The corresponding unperturbed energy levels which are independent of M [and hence $(2\mathcal{J}+1)$-fold degenerate] are split by the magnetic field (Zeeman splitting). This is determined by the M-dependent change in energy levels due to H_{mag}, which can be calculated using the first order perturbation theory of a degenerate level (Sec. 5.3). For this we need the matrix elements $\langle aLS\mathcal{J}M'|H_{\text{mag}}|aLS\mathcal{J}M\rangle$. We take $\mathcal{H}$ to be along the z-axis, so that $H_{\text{mag}} = -\mu_z\mathcal{H}$, and then use (8.106). We get

$$(\mu_z)_{M'M} \equiv \langle aLSJM' \mid \mu_z \mid aLSJM \rangle = \frac{\langle aLSJM' \mid J_z(\mathbf{J}.\boldsymbol{\mu}) \mid aLSJM \rangle}{J(J+1)}$$

Since $(\mathbf{J}.\boldsymbol{\mu}) = \mu_0 (\mathbf{L} + \mathbf{S}).(\mathbf{L} + 2\mathbf{S}) = \mu_0 (\mathbf{L}^2 + 2\mathbf{S}^2 + 3\mathbf{L}.\mathbf{S}) = \frac{1}{2}\mu_0 (3\mathbf{J}^2 - \mathbf{L}^2 + \mathbf{S}^2)$ the above expression immediately reduces to

$$(\mu_z)_{M'M} = \frac{3J(J+1) - L(L+1) + S(S+1)}{2J(J+1)} \mu_0 M \delta_{M'M}.$$

Therefore, the secular determinant is already diagonal, and the diagonal elements of H_{mag} directly give the energy change produced by the magnetic field:

$$(\triangle E)_{mag} = \langle aLSJM \mid H_{mag} \mid aLSJM \rangle = - \mathcal{H} (\mu_z)_{MM} = - \mathcal{H} g \mu_0 M,$$

where $g = [3J(J+1) - L(L+1) + S(S+1)]/[2J(J+1)]$ is the *Landé splitting factor*. The splitting is thus into $(2J+1)$ levels with constant spacing $(g\mu_0\mathcal{H})$. Typical values of g are $g = 2$ for $^2S_{1/2}$ states, $g = \frac{2}{3}$ for $^2P_{1/2}$, $g = \frac{4}{3}$ for $^2P_{3/2}$ etc.

PROBLEMS

1. If a spinless particle at rest ($l = 0$) decays into a pair of spin-$\frac{1}{2}$ particles, show that the final state must be a superposition of ^1S and ^3P states.
2. If $s_r = \mathbf{s}.\mathbf{x}/r$, show that $[\mathbf{L}, s_r] = - [\mathbf{s}, s_r] = i\, \mathbf{s} \times \mathbf{x}/r$. Hence $\mathbf{J}$ commutes with s_r as it should, s_r being a scalar operator.
3. Considering an atom in a magnetic field which is so strong that H_{mag} dominates over $H_{S.O}$, determine the pattern of splitting of the levels (8.47) due to the perturbation $H_{S.O}$.
4. Obtain an estimate of the magnetic field strength for which the effects of the terms H_{mag} and $H_{S.O}$ (Sec 8.4) become equal for an alkali atom.
5. A Cartesian coordinate frame S is rotated through an angle θ about an axis $\mathbf{n}$ to yield a new frame S'. If a spin-$\frac{1}{2}$ particle has $m_s = \frac{1}{2}$ with respect to S, what is the probability that it has $m_s = \frac{1}{2}$ with respect to S'?
6. Considering a system of two spin-$\frac{1}{2}$ particles, show that the operator $S_{ij} \equiv [\frac{1}{2} (\sigma_{1i} \sigma_{2j} + \sigma_{2i}\sigma_{1j}) - \frac{1}{3} \delta_{ij} \boldsymbol{\sigma}_1.\boldsymbol{\sigma}_2]$ annihilates the singlet spin state. The tensor interaction Hamiltonian (Example 8.3, p. 257) has this property since it is proportional to the scalar $\sum_{i,j} S_{ij} Q_{ij}$ where $Q_{ij} \equiv (x_i x_j - \frac{1}{3} r^2 \delta_{ij})$.
7. Construct the $^3D_1$ angular momentum eigenfunctions for a system of two spin-$\frac{1}{2}$ particles in its centre of mass frame, from the relevant orbital and spin wave functions.
8. If u_i, u_f are wave functions of the type (8.85) characterized by (l_i, j_i, m_i) and (l_f, j_f, m_f) show that the matrix element $\int u_i^\dagger\, \mathbf{x}\, u_f\, d\tau$ can be nonvanishing only when $\Delta l = \pm 1$, $\Delta j = 0, \pm 1$ (excluding $j_i = j_f = 0$).
9. With P_t, P_s defined by (8.83), show that $(P_t - P_s)$ has the same effect on $^3\chi$ and $^1\chi$ as the interchange of the two particles. This operator, $\frac{1}{2} (1 + \boldsymbol{\sigma}_1.\boldsymbol{\sigma}_2)$, is called the 'spin-exchange' operator.
10. By writing $\boldsymbol{\sigma}_1.\boldsymbol{\sigma}_2$ in terms of the singlet and triplet projection operators, show that $(\boldsymbol{\sigma}_1.\boldsymbol{\sigma}_2)^n = \frac{1}{4} [3 + (-3)^n] + \frac{1}{4} [1 - (-3)^n] (\boldsymbol{\sigma}_1.\boldsymbol{\sigma}_2)$.
11. Construct a simultaneous eigenstate of J_x^2, J_y^2 and J_z^2 from states with $j = 1$.
12. Calculate the matrix elements of $\mathbf{S}.\mathcal{H}$ between triplet spin states of a two-electron system (a) using the Wigner-Eckart theorem and (b) using the explicit structure of the spin

functions ($\mathbf{S} = \mathbf{s}_1 + \mathbf{s}_2$ is the total spin operator and $\mathcal{H}$ stands for a uniform magnetic field).

13. Construct spin wave functions with definite j and m for a three-electron system, carrying out the addition of angular momenta in two alternative ways: $\mathbf{s}_1 + \mathbf{s}_2 = \mathbf{S}'$, $\mathbf{S}' + \mathbf{s}_3 = \mathbf{S}$ and $\mathbf{s}_2 + \mathbf{s}_3 = \mathbf{S}''$, $\mathbf{s}_1 + \mathbf{S}'' = \mathbf{S}$. Comment on the relation between the functions obtained in the two ways.

14. Use the construction of Example 8.5, p. 263 to form a tensor operator T_{2m} from the spherical components of two commuting vector operators $\mathbf{U}$, $\mathbf{V}$. Show that the T_{2m} are linear combinations of components $\frac{1}{2}(U_i V_j + U_j V_i) - \frac{1}{3}\delta_{ij}(\mathbf{U}.\mathbf{V})$ of the symmetric traceless tensor formed from $\mathbf{U}$ and $\mathbf{V}$.

15. Show that the matrix elements of $\mathbf{x}$ between states $u_{l\frac{1}{2};jm}$ and $u_{l'\frac{1}{2};j'm'}$ defined by (8.85a) can be nonvanishing only when $\Delta m = 0, \pm 1$, $\Delta l = \pm 1$, $\Delta j = 0, \pm 1$ (but not if $j = j' = 0$). Here $\Delta l = l - l'$ etc.

9 Evolution with Time

In this chapter we study the dynamics of quantum systems. We first introduce, in Part A, the concept of propagators, which arises naturally from the solution of the Schrödinger equation expressed as an expansion in terms of the eigenfunctions of H. Part B deals with the perturbation method for solving time evolution problems. This theory is quite elementary but has extensive applications. One of the most important applications is to the theory of emission and absorption of radiation by atoms, which we treat briefly. Consideration of various alternative ways of picturing the time evolution of a quantum system is taken up in Part C. In this context, the essential concepts of the quantization of electromagnetic waves (leading to the emergence of photons) are outlined. The final part is devoted to the description of the states of ensembles of systems (and their time evolution) through density matrices.

A. EXACT FORMAL SOLUTIONS: PROPAGATORS

9.1 The Schrödinger Equation; General Solution

As we have noted in Chapter 2 and stated more formally in Sec. 3.14, it is one of the basic postulates of quantum mechanics that the development of the wave function ψ with time is governed by the Schrödinger equation[1]

[1] According to the convention adopted in Chapter 3, $\mathbf{x}$ will denote the set of coordinate variables of all the particles in the system. If the system contains particles with spin, ψ will be a multicomponent wave function with the appropriate number of components, and H will, in general, be a function of the relevant spin matrices as well as $\mathbf{x}$ and $\mathbf{p}$. For simplicity we consider the spinless case initially.

$$i\hbar \frac{\partial}{\partial t} \psi(\mathbf{x},t) = H\psi(\mathbf{x},t) \tag{9.1}$$

where H is the Hamiltonian of the system. If H is independent of time, a general solution of this equation can be readily obtained by using the fact that the eigenfunctions $u_n(\mathbf{x})$ of H form a complete set in configuration space. For this permits us to expand $\psi(\mathbf{x},t)$ at any time t as

$$\psi(\mathbf{x},t) = \Sigma\ c_n(t)\ u_n\ (\mathbf{x}) \tag{9.2}$$

On substituting this in Eq. 9.1, writing $Hu_n = E_n u_n$, and equating the coefficients of the linearly independent functions u_n on both sides, we obtain

$$i\hbar \frac{dc_n}{dt} = E_n c_n, \text{ i.e. } c_n(t) = c_n(0)e^{-iE_n t/\hbar} \tag{9.3}$$

This is a restatement of the fact which we already know, that the time evolution of any stationary state wave function is contained in the simple factor $e^{-iE_n t/\hbar}$. Introducing (9.3) in (9.2), we obtain the general solution of the Schrödinger equation as

$$\psi\ (\mathbf{x},t) = \sum_n c_n(0)e^{-iE_n t/\hbar}\ u_n(\mathbf{x}) \tag{9.4}$$

The constants $c_n\ (0)$ are arbitrary, but if the initial wave function $\psi(\mathbf{x},0)$ at $t = 0$ is specified, it determines the $c_n\ (0)$ through Eq. 3.48.

$$c_n(0) = \int \psi\ (\mathbf{x},0)\ u_n^*(\mathbf{x})\ d\tau \tag{9.5}$$

(This is a consequence of the orthonormality of the set of functions $u_n(\mathbf{x})$, which will be assumed throughout).

9.2 Propagators

Substitution of (9.5) into (9.4) leads to a very interesting result:

$$\psi(\mathbf{x},t) = \Sigma\ [\int \psi(\mathbf{x}',0)\ u_n^*(\mathbf{x}')d\tau']\ e^{-iE_n t/\hbar}\ u_n(\mathbf{x}) = \int G\ (\mathbf{x},\mathbf{x}';t)\ \psi(\mathbf{x}',0)\ d\tau' \tag{9.6}$$

where

$$G(\mathbf{x},\mathbf{x}';t) = \sum_n u_n(\mathbf{x})\ u_n^*\ (\mathbf{x}')\ e^{-iE_n t/\hbar} \tag{9.7}$$

The last step in Eq. 9.6 is an interchange of the order of summation and integration, which we shall assume to be permissible. The interesting point now is that unlike the Schrödinger equation, which is a differential equation stating what happens to ψ in an infinitesimal time interval, Eq. 9.6 directly gives the relation between the wave functions at two instants separated by a *finite* time interval t. It expresses $\psi(\mathbf{x},t)$ for any given $\mathbf{x}$ as a sum of contributions from the initial wave function at various points $\mathbf{x}'$. The contribution from a particular point $\mathbf{x}'$ is seen to be proportional to $G(\mathbf{x},\mathbf{x}';t)$, which may therefore be called a *propagator*. It gives the *amplitude for propagation* of the system from $\mathbf{x}'$ to $\mathbf{x}$ in time t. The amplitudes arising from all the $\mathbf{x}'$ (multiplied by the respective initial values of ψ) are added up to obtain the final amplitude $\psi(\mathbf{x},t)$ for the system to be at $\mathbf{x}$ as stated in Eq. 9.6.

It is obvious that the propagator G of Eq. 9.7 satisfies

$$\left(i\hbar \frac{\partial}{\partial t} - H\right) G(\mathbf{x},\mathbf{x}';t) = 0, \quad G(\mathbf{x},\mathbf{x}';0) = \delta(\mathbf{x} - \mathbf{x}') \tag{9.8}$$

The former follows from the fact that $(i\hbar\ \partial/\partial t)$ introduces a factor E_n in the sum in Eq. 9.7, and so does $H(\mathbf{x},\mathbf{p})$ (which acts on $u_n(\mathbf{x})$ only). The second of Eqs. 9.8 is a consequence of the closure property of the u_n.

If one is interested in the time development for positive times only, i.e. if Eq. 9.6 is to be used only for $t > 0$, then the values of G for negative times do not enter the picture at all. One could then take $G(\mathbf{x},\mathbf{x}';t)$ for $t < 0$ to be anything one likes. Under such circumstances, the use of a function G_R, defined as follows, has been found to be often of advantage:

$$G_R(\mathbf{x},\mathbf{x}';t) = G(\mathbf{x},\mathbf{x}';t)\,\theta(t) \tag{9.9}$$

where $\theta(t)$ is the Heaviside unit function,

$$\theta(t) = 1 \text{ for } t > 0; \; \theta(t) = 0 \text{ for } t < 0 \tag{9.10}$$

Thus G_R coincides with G for positive times and is zero for negative times. Note that G_R makes a discontinuous jump from 0 to $G(\mathbf{x},\mathbf{x}';0) \equiv \delta(\mathbf{x}-\mathbf{x}')$ as t passes through 0 from negative to positive times. Consequently $\partial G_R/\partial t$ involves the Dirac delta function $\delta(t)$. In fact, from Eqs. 9.9 and 9.8 we have [since $d\theta(t)/dt = \delta(t)$],

$$\begin{aligned} i\hbar\frac{\partial}{\partial t}G_R(\mathbf{x},\mathbf{x}';t) &= \theta(t)\; i\hbar\frac{\partial}{\partial t}G(\mathbf{x},\mathbf{x}';t) + i\hbar\, G(\mathbf{x},\mathbf{x}';t)\delta(t) \\ &= \theta(t)\; HG(\mathbf{x},\mathbf{x}';t) + i\hbar\; G(\mathbf{x},\mathbf{x}';0)\delta(t) \end{aligned}$$

Since H does not act on the time variable, the first term can be written as HG_R. Hence

$$\left(i\hbar\frac{\partial}{\partial t} - H\right) G_R(\mathbf{x},\mathbf{x}';t) = i\hbar\;\delta(\mathbf{x}-\mathbf{x}')\delta(t) \tag{9.11}$$

Thus G_R is a *Green's function* corresponding to the linear operator $(i\hbar\,\partial/\partial t - H)$ and Eq. 9.11 may be treated as its defining equation. In this connection, the definition of Green's functions in the context of the time-independent problems in Chapter 6 may be recalled. As was noted there, Green's functions are not uniquely determined by the defining equation. In fact Eq. 9.11 would not be affected if, to a given solution G_R, one adds any solution of the homogeneous equation $(i\hbar\partial/\partial t - H)\; F = 0$. The sum $(G_R + F)$ would again be a Green's function. But if we also insist on the initial condition[2]

$$\lim_{t \to 0_+} G_R(\mathbf{x},\mathbf{x}';t) = \delta(\mathbf{x}-\mathbf{x}') \tag{9.12}$$

then G_R is uniquely given by Eq. 9.9.

The function G_R satisfying Eqs. 9.11 and 9.12 is called the *retarded Green's function* (the subscript R standing for 'retarded'), because it permits contributions to $\psi(\mathbf{x},t)$ only through retarded signals, i.e. those starting at *earlier* times. One could define G_R also for arbitrary initial times t' (instead of 0 in the above discussion). In this case one should write G_R as $G_R(\mathbf{x},\mathbf{x}';t,t')$, but it is easily verified that it depends only on the interval $(t-t')$ and not on t and t' individually. In order to accommodate arbitrary initial times, all one has to do is to replace $\psi(\mathbf{x},0)$ in Eq. 9.6 by $\psi(\mathbf{x},t')$ and replace t by $(t-t')$ in all the subsequent equations. All this is on the supposition that the Hamiltonian H does not depend on time explicitly. If it does, then G_R depends on t and t'

[2] $t \to 0_+$ means that the time 0 is to be approached from the positive time side.

separately. One can nevertheless *define* $G_R(\mathbf{x},\mathbf{x}';t,t')$ through a generalization of Eqs. 9.11 and 9.12:

$$\left[i\hbar\frac{\partial}{\partial t} - H(t)\right]G_R(\mathbf{x},\mathbf{x}';t,t') = \delta(\mathbf{x}-\mathbf{x}')\delta(t-t'),$$

$$\lim_{t\to t'_+} G_R(\mathbf{x},\mathbf{x}';t,t') = \delta(\mathbf{x}-\mathbf{x}') \tag{9.13}$$

It is left to the student to prove that with this definition, $\psi(\mathbf{x},t) \equiv \int G(\mathbf{x},\mathbf{x}';t,t')\psi(\mathbf{x}',t')d\tau'$ does satisfy the Schrödinger equation with the Hamiltonian $H(t)$.

Extensive use of retarded (and other) propagators in quantum physics was pioneered by Feynman.

9.3 Relation of Retarded Propagator to the Green's Function of the Time-Independent Schrödinger Equation

For a system with a time-independent Hamiltonian H, consider the Fourier transform $\int G_R e^{i\omega t}\,dt$ of the retarded propagator $G_R \equiv G\theta(t)$, with G given by Eq. 9.7. It involves integrals of the form $\int e^{-iE_n t/\hbar}\, e^{i\omega t}\,dt = \int e^{i(W-E_n)t/\hbar}\,dt$ where $W = \hbar\omega$. The limits of integration are from 0 to ∞ (since $G_R = 0$ by definition for $t<0$). It may be noticed that the integral does not converge at the upper limit but oscillates infinitely fast like $e^{i(W-E_n)T/\hbar}$ as $T\to\infty$. Therefore, it is customary to introduce an extra factor $e^{-\varepsilon t/\hbar}$ (with $\varepsilon > 0$) in the integrand (which makes the integrals converge) and then take the limit as $\varepsilon\to 0$. With this prescription, the Fourier transform becomes

$$\mathsf{G}^+(\mathbf{x},\mathbf{x}';W) = \int_0^\infty G_R(\mathbf{x},\mathbf{x}';t)\; e^{i(W-E+i\varepsilon)t/\hbar}\,dt$$

$$= -i\hbar\sum_n \frac{u_n(\mathbf{x})\,u_n^*(\mathbf{x}')}{W-E_n+i\varepsilon} \tag{9.14}$$

The superscript $+$ in G^+ is a reminder of the addition of $+i\varepsilon$ to W. It is understood that the limit $\varepsilon\to 0_+$ is to be taken. The point which we now wish to emphasize is that $(\mathsf{G}^+/i\hbar)$ is a Green's function for the time-independent operator $(H-W-i\varepsilon)$:

$$(H-W-i\varepsilon)\;\mathsf{G}^+(\mathbf{x},\mathbf{x}';W) = i\hbar\,\delta(\mathbf{x}-\mathbf{x}') \tag{9.15}$$

This follows from the closure property of the u_n's, after the denominator in Eq. 9.14 is cancelled against $(H-W-i\varepsilon)\,u_n(\mathbf{x}) = (E_n - W - i\varepsilon)\;u_n(\mathbf{x})$ in the numerator. When the system under consideration is a single free particle $(H = -\hbar^2\nabla^2/2m)$, $\mathsf{G}^+(\mathbf{x},\mathbf{x}';W)$ for positive W reduces in fact to the *outgoing* wave Green's function G^+ of Sec. 6.3, apart from constant factors (see Example 9.2, p. 274).

Eq. 9.15 is the configuration space version of the *Lippmann-Schwinger equation*[3]

$$(\hat{H}-W-i\varepsilon)\;\hat{\mathsf{G}}^+(W) = i\hbar\,\hat{1} \tag{9.16}$$

where the carat symbols are used to denote abstract operators. Note that $\mathsf{G}^+(\mathbf{x},\mathbf{x}';W) = \langle\,\mathbf{x}\,|\,\hat{\mathsf{G}}^+\,|\,\mathbf{x}'\,\rangle$ and $\delta(\mathbf{x}-\mathbf{x}') = \langle\,\mathbf{x}'\,|\,\hat{1}\,|\,\mathbf{x}\,\rangle$, $\hat{1}$ being the unit operator.

[3] B. A. Lippmann and J. Schwinger, *Phys. Rev.*, **79**, 469, 1950

It is interesting to observe that the presence of $i\varepsilon$ in the above expressions ensures that the inverse Fourier transform of $\mathbf{G}^+$, which is $(2\pi\hbar)^{-1}\int \mathbf{G}^+ (\mathbf{x},\mathbf{x}';W)\, e^{-iWt/\hbar}\, dW$ is indeed the *retarded* propagator. On substituting (9.14) in this integral, one encounters integrals of the form

$$I_n = \int_{-\infty}^{\infty} \frac{e^{-iWt/\hbar}}{W - E_n + i\varepsilon}\, dW \tag{9.17}$$

We can easily evaluate I_n by a contour integral method. If W is generalized to a complex variable, the above integral is along the real axis in the complex W-plane. We can think of it as part of a closed contour consisting of the real axis and *either* the infinite semicircle Γ enclosing the lower half plane, *or* the

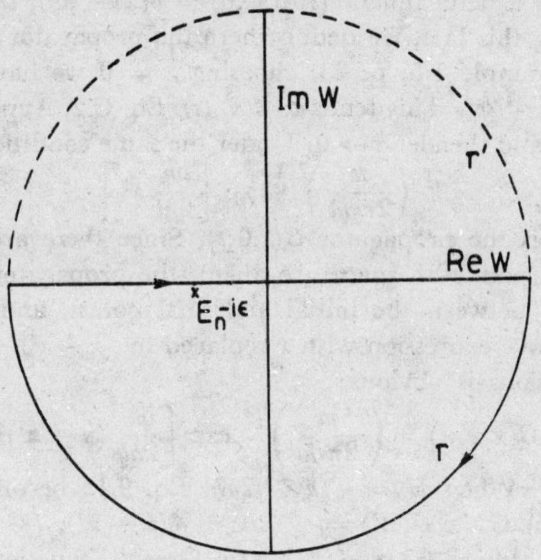

Fig. 9.1 Contours for the evaluation of I_n (Eq. 9.17). The pole of the integrand is just below the real axis, and is indicated by a cross in the figure.

infinite semicircle Γ' over the upper half plane (see Fig. 9.1). As the radius of the semicircle tends to ∞, the real part of $(-\,iWt/\hbar)$ behaves as follows:

$$\begin{aligned} &\textit{For } t > 0: \text{Re}\,(-\,iWt/\hbar) \to -\,\infty \text{ on } \Gamma, +\,\infty \text{ on } \Gamma' \\ &\textit{For } t < 0: \text{Re}\,(-\,iWt/\hbar) \to +\,\infty \text{ on } \Gamma, -\,\infty \text{ on } \Gamma' \end{aligned} \tag{9.18}$$

Thus it is clear that the sensible thing to do is to choose Γ for $t > 0$ and Γ' for $t < 0$, thereby reducing to zero the contribution of the semicircle to the contour integral (through the vanishing of the exponential factor in the integrand). With this choice, the integral (9.17) becomes equal to the integral over the relevant *closed* contour, and its value is then obtained trivially from Cauchy's theorem. We get

$$I_n = 2\pi i\; e^{-i(E_n - i\varepsilon)t/\hbar}\; (t > 0); \quad I_n = 0 \;\; (t < 0) \tag{9.19}$$

The first of these is $2\pi i$ times the residue of the integrand at its pole (at $W = E_n - i\varepsilon$), which lies *within* the closed contour of the $t > 0$ case (remember this contour encloses the *lower* half plane). For $t < 0$ the contour encloses the upper half plane where there are no poles, and so $I_n = 0$. With these values for I_n, one gets (as $\varepsilon \to 0$)

$$\frac{1}{2\pi\hbar}\int_{-\infty}^{\infty} \mathsf{G}^+(\mathbf{x},\mathbf{x}';W)\, e^{-iWt/\hbar}\, dW = \frac{1}{2\pi i}\sum_n u_n(\mathbf{x})u_n^*(\mathbf{x}')\; I_n = G_R(\mathbf{x},\mathbf{x}';t) \quad (9.20)$$

If we had $\mathsf{G}^-(\mathbf{x},\mathbf{x}';W)$ with $W - i\varepsilon$ instead of $W + i\varepsilon$ in Eq. 9.17 this integral would reduce to the *advanced propagator* $G_A(\mathbf{x},\mathbf{x}';t) \equiv G(\mathbf{x},\mathbf{x}';t)\,\theta(-t)$. This is evident from the above derivation: the poles are now in the *upper* half plane (at $W = E_n + i\varepsilon$) and so I_n would vanish for $t > 0$ and be equal to $2\pi i \exp[-i(W + i\varepsilon)t/\hbar]$ for $t < 0$.

EXAMPLE 9.1.—*Propagator for a Free Particle.* Eq. 9.6 shows that if the initial wave function is a delta function: $\psi(\mathbf{x}';0) = \delta(\mathbf{x}' - \mathbf{x}'')$, then $\psi(\mathbf{x},t)$ becomes $G(\mathbf{x},\mathbf{x}'';t)$. Using this fact, we deduce here the propagator for a free particle. Referring to Example 1.5, p. 25, choosing $\bar{k} = 0$ we have at $t = 0$, $\psi(\mathbf{x},0) = a_0 (2\pi/\sigma)^{1/2} e^{-x^2/2\sigma^2}$. This tends to $\delta(x)$ (*cf.* Eq. C.2, Appendix C) if we put $a_0 = 1/(2\pi\sqrt{\sigma})$ and then let $\sigma \to 0$. Under the same conditions, $\psi(x,t)$ becomes

$$\left(\frac{m}{2\pi i\hbar t}\right)^{1/2} \exp\left[\frac{im}{2\hbar t}\, x^2\right]$$

Therefore, this is the propagator $G(x,0;t)$. Since there are no forces on the particle (i.e. all points of space are alike), the propagator can depend only on the distance between the initial and final points, and hence $G(x,x';t)$ is given by the above expression with x replaced by $(x - x')$. The generalization to three dimensions is obvious:

$$G(\mathbf{x},\mathbf{x}';t) = \left(\frac{m}{2\pi i\hbar t}\right)^{3/2} \exp\left[\frac{im}{2\hbar t}\,(\mathbf{x} - \mathbf{x}')^2\right]$$

EXAMPLE 9.2—When $H = -(\hbar^2\nabla^2/2m)$, Eq. 9.15 becomes

$$(\nabla^2 \pm \varkappa^2 + i\varepsilon)\; \mathsf{G}^+(\mathbf{x},\mathbf{x}';W) = -(2im/\hbar)\;\delta(\mathbf{x} - \mathbf{x}'),\; (\varkappa^2 = 2m\,|\,W\,|\,/\hbar^2)$$

showing that $(-\hbar/2im)\; \mathsf{G}^+(\mathbf{x},\mathbf{x}';W)$ is a Green's function of the operator $\nabla^2 \pm \varkappa^2$ in the limit $\varepsilon \to 0_+$. Let us evaluate $\mathsf{G}^+(\mathbf{x},\mathbf{x}';W)$ from the expansion (9.14). In the present case the $u_n(\mathbf{x})$ are the free particle wave functions $(2\pi)^{-3/2} e^{i\mathbf{k}.\mathbf{x}}$, and the sum over n becomes integration over $\mathbf{k}$. Further, the energy eigenvalues $E_n \to E(\mathbf{k}) = \hbar^2 k^2/2m$. Hence we have

$$\mathsf{G}^+(\mathbf{x},\mathbf{x}';W) = (2m/i\hbar)\; I,$$

$$I = \frac{1}{(2\pi)^3}\int \frac{e^{i\mathbf{k}.(\mathbf{x}-\mathbf{x}')}}{\pm\varkappa^2 - \mathbf{k}^2 + i\varepsilon}\, d^3k \qquad \left(\text{for } W = \pm\frac{\hbar^2\varkappa^2}{2m}\right)$$

$$= \frac{1}{(2\pi)^3}\int_0^{\infty}\int_0^{2\pi}\int_0^{\pi} \frac{\exp\,[ik\,|\,\mathbf{x} - \mathbf{x}'\,|\,\cos\alpha]}{\pm\varkappa^2 - \mathbf{k}^2 + i\varepsilon}\,\sin\alpha\; d\alpha\; d\beta\; k^2 dk$$

$$= \frac{1}{(2\pi)^2 i\,|\,\mathbf{x} - \mathbf{x}'\,|}\int_0^{\infty} \frac{e^{ik|\mathbf{x}-\mathbf{x}'|} - e^{-ik|\mathbf{x}-\mathbf{x}'|}}{\pm\varkappa^2 - k^2 + i\varepsilon}\, k\, dk$$

$$= \frac{1}{8\pi^2 i\,|\,\mathbf{x} - \mathbf{x}'\,|}\int_{-\infty}^{\infty} \frac{e^{ik|\mathbf{x}-\mathbf{x}'|} - e^{-ik|\mathbf{x}-\mathbf{x}'|}}{\pm\varkappa^2 - k^2 + i\varepsilon}\, k\; dk$$

In the last step we have used the fact that the integrand is an even function, to extend the region of integration down to $-\infty$. We observe further that the two terms in the integral give the same contribution, as may be seen by setting $k = -k'$ in the first term. Thus we have

$$I = \frac{1}{4\pi^2 i\,|\mathbf{x}-\mathbf{x}'|}\int_{-\infty}^{\infty}\frac{e^{-ik|\mathbf{x}-\mathbf{x}'|}}{\pm\varkappa^2 - k^2 + i\varepsilon}\,k\,dk$$

This integral can be evaluated in the same way as I_n of Eq. 9.17. Here the place of t is taken by $|\mathbf{x}-\mathbf{x}'|$ which is necessarily positive, and hence a contour enclosing the *lower* half plane is to be used. From inspection of the denominator, it is clear that the integrand has two poles, of which only one is in the lower half plane. It is situated at $k_0 \approx -\varkappa - i(\varepsilon/2\varkappa)$, if W is positive, and at $k_0 \approx -i\varkappa + \varepsilon/2\varkappa$, if W is negative. The residue of the integrand at k_0 is easily seen to be $\frac{1}{2}e^{-ik_0|\mathbf{x}-\mathbf{x}'|}$. Hence, substituting the appropriate value of k_0 and finally letting $\varepsilon \to 0$ we obtain

$$\frac{i\hbar}{2m}G^{+}(\mathbf{x},\mathbf{x}';W) \equiv I = -\frac{1}{4\pi}\frac{e^{i\varkappa|\mathbf{x}-\mathbf{x}'|}}{|\mathbf{x}-\mathbf{x}'|},\quad (W>0)$$

$$= -\frac{1}{4\pi}\frac{e^{-\varkappa|\mathbf{x}-\mathbf{x}'|}}{|\mathbf{x}-\mathbf{x}'|},\quad (W<0)$$

Note that what we have obtained for $W > 0$ is just the *outgoing* wave Green's function G^{+} of Eq. 6.19. If we had $-i\varepsilon$ instead of $+i\varepsilon$ in the denominator of Eq. 9.14 we would be automatically led to the *incoming* wave Green's function because in this case the pole in the lower half plane (for $W > 0$) is at $+\varkappa - i(\varepsilon/2\varkappa)$. However, the expression in the case $W < 0$ would remain unaltered.

9.4 Alteration of Hamiltonian: Transitions; Sudden Approximation

As an application of the above theory, let us consider the following question. Suppose the Hamiltonian of a system is H_0 upto time 0 as well as beyond time T, but for $0 < t < T$ it is $H \neq H_0$. (For example H_0 may be the Hamiltonian of an isolated atom, and H its Hamiltonian in an external field which is switched on at time 0 and off at T). The eigenstates and eigenvalues of H_0 are supposed to be known:

$$H_0 u_n = E_n u_n \tag{9.21}$$

Then we ask: What is the probability that the system will be found in an eigenstate u_f of H_0 at time T if it was in u_i at time 0?

Firstly we observe that even if the change from H_0 to H at $t = 0$ is a sudden discontinuous change, the wave function ψ remains a continuous function of time, though $(\partial\psi/\partial t)$ will then have a discontinuity at $t = 0$. (This is evident from the (time-dependent) Schrödinger equation). Therefore, there is no ambiguity in talking of the wave function $\psi(\mathbf{x},0)$ at $t = 0$ and similarly of $\psi(\mathbf{x},T)$. Since the Hamiltonian during $(0,T)$ is H, any arbitrary initial state $\psi(x,0)$ evolves into

$$\psi(\mathbf{x},T) = \int G(\mathbf{x},\mathbf{x}';T,0)\,\psi(\mathbf{x}',0)\,d\tau' \tag{9.22}$$

where G is the propagator associated with H (*not* H_0). If the Hamiltonian had remained H_0 throughout, any eigenstate u_n would have simply evolved into $u_n\, e^{-iE_nT/\hbar}$ in time T. The amplitude for finding such a state in $\psi(\mathbf{x},T)$ is given by the expansion coefficient c_n in

$$\psi(\mathbf{x},T) = \Sigma\, c_n u_n(\mathbf{x})\, e^{-iE_nT/\hbar}; \quad c_n = e^{iE_nT/\hbar} \int u_n^*\psi d\tau \tag{9.23}$$

(cf. Eq. 3.48). The last integral may be evaluated in terms of the initial state, by introducing (9.22) for $\psi(\mathbf{x},T)$. For the particular case we are interested in, namely that when $\psi(\mathbf{x},0)$ itself is an eigenstate $u_i(\mathbf{x})$ of H_0, we obtain

$$c_{fi} = e^{iE_fT/\hbar} \iint u_f^*(\mathbf{x})\, G(\mathbf{x},\mathbf{x}';T,0)\, u_i(\mathbf{x}')\, d\tau d\tau' \tag{9.24}$$

where we have written c_{fi} instead of c_f for obvious reasons. Eq. 9.24 may be more simply expressed as $c_{fi} = e^{iE_fT/\hbar}\ (u_f, G u_i)$, the second factor being just the matrix element of the propagator between the states f and i. This is the *transition amplitude* for transitions $i \rightarrow f$ caused by the alteration of the Hamiltonian to H during $(0,T)$.

If H remains *time-independent* in the interval $(0,T)$, we can use an expansion of the form (9.7) for G: with $v_n(\mathbf{x})$, W_n defined by

$$Hv_n(\mathbf{x}) = W_n v_n(\mathbf{x}) \tag{9.25}$$

$$G(\mathbf{x},\mathbf{x}';T,0) = \sum_l v_l(\mathbf{x}) v_l^*(\mathbf{x}')\ e^{-iW_lT/\hbar} \tag{9.26}$$

Substitution of this in 9.24 yields

$$c_{fi} = \iint u_f^*(\mathbf{x})\ [\sum_l v_l(\mathbf{x}) v_l^*(\mathbf{x}')\ e^{i(E_f-W_l)T/\hbar}]\ u_i(\mathbf{x}')\ d\tau d\tau' \tag{9.27}$$

This is the transition amplitude due to an alteration $(H - H_0)$ which is maintained constant for a time interval T.

If T is small enough, i.e. if $(H - H_0)$ corresponds to a sudden, transient, disturbance of constant magnitude, then a simple approximation to c_{fi} is possible. It is called the *sudden approximation.* To obtain this, we observe that the integrals $\int u_f^* v_l d\tau$ and $\int v_l^* u_i d\tau$ which occur in Eq. 9.27 can have appreciable values only for a certain set of states v_l (depending on the particular initial and final states i and f). Their energies W_l lie within some finite range. Consider the case when T is small enough so that $(E_f - W_l)\ T/\hbar \ll 1$ for all W_l within this range. Then we can write $e^{i(E_f-W_l)T/\hbar} \approx 1 + i(E_f - W_l)\, T/\hbar$. The contribution of the first term to c_{fi} is just δ_{fi}, because when the exponential is replaced by unity, the factor in square brackets in Eq. 9.27 becomes $\delta(\mathbf{x} - \mathbf{x}')$ due to the closure property of the v's. So it is only the second term which causes transitions $i \rightarrow f$, $i \neq f$. It gives rise to the sum $\Sigma\ v_l(\mathbf{x}) v_l^*(\mathbf{x}')$ $(E_f - W_l)$ in the integrand, which can be written as $(E_f - H)\ \Sigma\ v_l(\mathbf{x}) v_l^*(\mathbf{x}')$ $= (E_f - H)\ \delta(\mathbf{x} - \mathbf{x}')$. With this, we get (for $f \neq i$)

$$c_{fi} = \frac{iT}{\hbar} \int u_f^*(\mathbf{x})\ (E_f - H)\ u_i(\mathbf{x}) d\tau = \frac{iT}{\hbar} \int u_f^*(H_0 - H)\ u_i\, d\tau \tag{9.28}$$

The hermiticity of H_0 has been used in the last step.

The transition amplitude (in the sudden approximation) is thus seen to be proportional to the matrix element of the change in the Hamiltonian and to the time for which this change persists. If H varies with time during $(0,T)$,

Eq. 9.28 can still be shown to be valid with replacement of $(H_0 - H)T$ by $\int_0^T (H_0 - H)\, dt$.

B. PERTURBATION THEORY FOR TIME EVOLUTION PROBLEMS

9.5 Perturbative Solution for Transition Amplitude

We have been concerned so far with solutions of the time evolution problem which are in principle exact. However, in practice it is difficult to determine the exact propagator except in very simple cases, and it is almost impossible if the Hamiltonian is time-dependent. Therefore, one usually has to depend on approximate methods to answer questions regarding time evolution.

We present here a *perturbation method* which is useful whenever H can be written as

$$H = H_0 + H' \tag{9.29}$$

and the following conditions hold: H_0 constitutes the major part of the Hamiltonian, is time-independent, and all its eigenvalues and orthonormalized eigenfunctions are known[4]

$$H_0 u_n = E_n u_n, \quad \int u_m{}^* u_n d\tau = \delta_{mn} \tag{9.30}$$

The part H' of H is a small *perturbation* on H_0; it may or may not depend on time. The problem is to determine $\psi(\mathbf{x},t)$ at arbitrary times t, given $\psi(\mathbf{x},0)$, the wave function at time 0.

The procedure, which is due to Dirac,[5] is based on an expansion of $\psi(\mathbf{x},t)$ in terms of the eigensolutions $u_n(\mathbf{x})e^{-iE_nt/\hbar}$ of the unperturbed time-dependent wave equation,

$$\psi(\mathbf{x},t) = \Sigma\, a_n(t) u_n(\mathbf{x})\, e^{-iE_nt/\hbar} \tag{9.31}$$

The quantities to be determined here are the coefficients $a_n(t)$ and these are of direct physical interest. According to the theory of Sec. 3.9, the expansion (9.31) implies that $|\, a_f(t)\, |^2$ is the probability[6] with which the system described by $\psi(\mathbf{x},t)$ will be found in the energy eigenstate u_f. We shall now see how the a_f may be determined starting from the (time-dependent) Schrödinger equation of the problem.

The Schrödinger equation is given by

$$i\hbar \frac{\partial \psi}{\partial t} = H\psi = H_0\psi + H'\psi \tag{9.32}$$

On substituting for ψ from Eq. 9.31 we obtain

[4] Note the change in notation from the preceding sections. The u_n here are eigenfunctions of the *unperturbed* Hamiltonian H_0 and not of the total Hamiltonian H.

[5] P. A. M. Dirac, *Proc. Roy. Soc.*, Lond., **A 112**, 661, 1926.

[6] Since probabilities, by definition, can take values only between 0 and 1, it follows that $0 \leq |\, a_n(t)\, |^2 \leq 1$ for all n. This inequality need not hold if n is one of a *continuum* of eigenstates, since $|\, a_n(t)\, |^2$ is then *not* a probability but a probability *density*. In this case the integral of $|\, a_n\, |^2$ over any part of the continuum is a probability (namely the probability that the system may be found with quantum numbers lying within this part of the continuum). The value of every such *integral* must lie between 0 and 1 though the probability density $|\, a_n\, |^2$ can have arbitrary positive values.

$$\sum_n (i\hbar \dot{a}_n + E_n a_n) u_n e^{-iE_nt/\hbar} = \sum_n (H_0 u_n + H' u_n) a_n e^{-iE_nt/\hbar} \tag{9.33}$$

Since $H_0 u_n = E_n u_n$, this term on the right hand side cancels with the corresponding term on the left hand side. On multiplying the remaining terms by u_f^* and integrating, we get, in view of the orthonormality of the eigenfunctions,

$$i\hbar\, \dot{a}_f e^{-iE_ft/\hbar} = \sum_n a_n H'_{fn} e^{-iE_nt/\hbar}$$

or

$$\dot{a}_f = (i\hbar)^{-1} \sum a_n H'_{fn} e^{i\omega_{fn}t} \tag{9.34}$$

Here $H'_{fn} = \int u_f^* H' u_n \, d\tau$ is a matrix element of the perturbation, and ω_{fn} is the Bohr angular frequency pertaining to the pair of unperturbed levels f,n:

$$\hbar\,\omega_{fn} = (E_f - E_n) \tag{9.35}$$

The set of equations (9.34) is completely equivalent to the Schrödinger equation 9.32.

We observe from Eq. 9.34 that if the perturbation H' were absent, all the a_f would be constants.[7] The perturbation H' makes them vary with time. (For this reason this method is often referred to as the *method of variation of constants*). Since H' is supposed to be very small, it is reasonable to try and express $a_f(t)$ as a series

$$a_f = a_f^{(0)} + a_f^{(1)} + a_f^{(2)} + \dots \tag{9.36}$$

wherein the order of each term (denoted by the superscript) gives the number of factors of H' contained in it. On substituting this in Eq. 9.34 and equating terms of the same order on both sides, we get

$$\dot{a}_f^{(0)} = 0, \quad \dot{a}_f^{(r+1)} = (i\hbar)^{-1} \sum a_n^{(r)} H'_{fn} e^{i\omega_{fn}t} \tag{9.37}$$

These equations can, in principle, be integrated successively upto any desired order.

Suppose now that the perturbation H' begins to act at time 0. At $t = 0$, when the perturbation has not yet had a chance to act, $a_f^{(1)}, a_f^{(2)}, \dots$ must be necessarily zero. So $a_f(0) = a_f^{(0)}(0)$. In practice one is most interested in the situation where the system (e.g. an atom) is *in a particular unperturbed state*, say u_i before the perturbation is 'turned on' (e.g. radiation begins arriving) at time 0. This would mean $a_f(0) = \delta_{fi}$. With this choice, our initial conditions are

$$a_f^{(0)}(0) = \delta_{fi}, \quad a_f^{(r)}(0) = 0, \; r = 1,2,\dots \tag{9.38}$$

Eqs. 9.37 are to be solved subject to these initial conditions. The first equation says that $a_f^{(0)}(t)$ is really a constant, so that $a_f^{(0)}(t) = a_f^{(0)}(0) = \delta_{fi}$. On feeding this into the second equation with $r = 0$, we get $\dot{a}_f^{(1)} = (i\hbar)^{-1} H'_{fi} e^{i\omega_{fi}t}$ and hence

$$a_f^{(1)}(t) = (i\hbar)^{-1} \int_0^t H'_{fi}(t') e^{i\omega_{fi}t'} dt' \tag{9.39}$$

in view of the initial condition (9.38). Substituting this into Eq. 9.37 with $r = 1$, we obtain

$$a_f^{(2)} = (i\hbar)^{-2} \sum_l \int_0^t dt'' H'_{fl}(t'') e^{i\omega_{fl}t''} \int_0^{t'} dt' H'_{li}(t') e^{i\omega_{li}t'} \tag{9.40}$$

and so on for all r. (Clearly $a_f^{(r)}$ involves an r-fold integral whose structure is

[7] It is with this in mind that $e^{-iE_nt/\hbar}$ was pulled out as a separate factor in the coefficient of $u_n(\mathbf{x})$ in Eq. 9.31. Compare Eqs. 9.31 and 9.4.

evident from the above examples). The coefficient $a_f(t)$ in Eq. 9.31 is then obtainable as the sum of all the $a_f^{(r)}(t)$.

Now, the physical significance of the quantities $|a_f(t)|^2$ as probabilities (or probability densities) has been already noted below Eq. 9.31. With the initial condition $a_f(0) = \delta_{fi}$ which we have assumed, $|a_f(t)|^2$ is acually a *transition probability*: it is the probability for finding the system in u_f at time t, *given that* it was in u_i at time 0. Correspondingly, $a_f(t)$ is the transition amplitude (apart from the oscillatory exponential factor which, for the time interval in question, is $e^{-iE_ft/\hbar}$). Eq. 9.36 pictures this amplitude as being made up of contributions of various orders. Eqs. 9.39 and 9.40 suggest a very interesting physical interpretation for these. Considering $a_f^{(1)}(t)\ e^{-iE_ft/\hbar}$ from Eq. 9.39, one can separate the integrand into three factors: $e^{-iE_f(t-t')/\hbar}$, $(i\hbar)^{-1} H'_{fi}(t')$ $e^{-iE_it'/\hbar}$, which can be interpreted (from right to left) as the amplitudes for the system to remain in i from 0 to t', that for a jump or transition to take place from i to f at time t', and that for the system to remain in f from t' to t. The integration over t' takes care of the fact that the jump may occur at any t' between 0 and t (Fig. 9.2). A similar interpretation can evidently be given

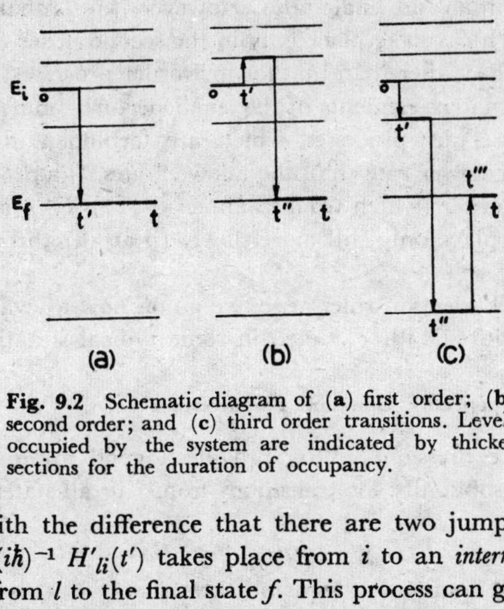

Fig. 9.2 Schematic diagram of (**a**) first order; (**b**) second order; and (**c**) third order transitions. Levels occupied by the system are indicated by thicker sections for the duration of occupancy.

for Eq. 9.40, with the difference that there are two jumps: the first jump, represented by $(i\hbar)^{-1} H'_{li}(t')$ takes place from i to an *intermediate* state l and the next, at t'', from l to the final state f. This process can go through via any of the states of the system as intermediate state. Therefore, there is to be a sum over l, and of course integration over the times t', t'' $(t'' > t')$ at which the jumps can take place. In general, the rth order contribution represents the amplitude for the system to go from i to f in r steps. Eq. 9.36 then gives the total transition amplitude as the sum of an infinite number of such contributions. If all the terms can be evaluated and the sum converges, then the result is exact. But in practice one can rarely go beyond the first two or three orders. The approximation thus introduced is good provided H' is a sufficiently small perturbation.

9.6 Selection Rules

In many cases of interest, $H'(t)$ can be expressed in the form

$$H'(t) = H'^0 f(t) \tag{9.41}$$

where H'^0 itself is time-independent. Then the matrix elements also factor out in a similar fashion, and we can write

$$a_f^{(1)}(t) = (i\hbar)^{-1} H'^0_{fi} \int_0^t f(t')\, e^{i\omega_{fi}t'}\, dt' \tag{9.42}$$

$$a_f^{(2)}(t) = (i\hbar)^{-2} \sum_l H'^0_{fl}\, H'^0_{li} \int_0^t f(t'')\, e^{i\omega_{fl}t''}\, dt'' \int_0^t f(t')\, e^{i\omega_{li}t'}\, dt' \tag{9.43}$$

etc. Eq. 9.42 shows that $a_f^{(1)}(t)$ is nonzero only if H'^0_{fi} is. Thus, transitions in the lowest (first) order can take place only between pairs of states (f,i) for which $H'^0_{fi} \neq 0$. Transitions between any other states are *forbidden* in the *first* order. We have thus obtained a *selection* rule.

Even if H'^0_{fi} vanishes, transitions between i and f may still be possible in higher orders. For instance, $a_f^{(2)}$ can be nonzero if there exists some intermediate state l such that H'^0_{fl} and H'^0_{li} are both nonzero. We say then this intermediate state *connects* the initial and final states. However, the probability $|a_f^{(2)}(t)|^2$ for transitions which can take place only in the second order (being forbidden in the first) is much smaller than that for first order processes, since it involves a product of four matrix elements of the small perturbation H', compared to two in $|a_f^{(1)}(t)|^2$. Thus processes which are forbidden in the first order proceed at a much slower rate than the allowed ones. If a process is forbidden also in the second order (which would happen if $H'^0_{fl}\, H'^0_{li}$ is zero for *all* l) then it could take place only still more slowly (if at all) through the third or higher orders.

Having seen how selection rules arise, we go on now to evaluate the actual transition probabilities $|a_f^{(1)}(t)|^2$ etc., in some typical situations.

9.7 First Order Transitions: Constant Perturbation

The first order expressions $a_f^{(1)}(t)$ provide a good approximation to the $a_f(t)$ if the total probability for transitions from i to all states $f \neq i$ is much less than unity:

$$\Sigma' |a_f^{(1)}(t)|^2 \ll 1 \tag{9.44}$$

(The prime on the summation sign indicates that $f = i$ is to be excluded from the sum over f). In considering the explicit expressions for transition probabilities obtained below, this validity criterion is to be kept in mind.

(*a*) *Transition Probability:* We now consider the specific case of a perturbation H' which lasts from time 0 to t and is constant during this period. In this case H'_{fi} can be pulled outside the integral in Eq. 9.39, which then reduces to

$$a_f^{(1)}(t) = -H'_{fi}\frac{e^{i\omega_{fi}t}-1}{\hbar\omega_{fi}} \tag{9.45}$$

$$|a_f^{(1)}(t)|^2 = \frac{|H'_{fi}|^2}{\hbar^2}\cdot\frac{4\sin^2\frac{1}{2}\omega_{fi}t}{\omega_{fi}^2} \tag{9.46}$$

A striking feature of this expression is that the transition probability from i to f (which we assume for the moment to be a discrete level) does *not* change monotonically with time, but varies simple harmonically (between 0 and the maximum value $4\,|\,H'_{fi}\,|^2/\hbar^2\omega_{fi}^2$) with a frequency equal to the Bohr frequency $\nu_{fi} = (\omega_{fi}/2\pi)$ associated with this pair of levels. The smaller the energy difference between the two levels, the larger is the maximum value of the probability, but so also is the time ($\propto \omega_{fi}^{-1}$) taken to reach this maximum.

Let us now consider $|\,a_f^{(1)}(t)\,|^2$ as a function of the energy difference between the initial and final levels (i and f), for fixed t. The behaviour of the factor $(4/\omega_{fi}^2)\sin^2\frac{1}{2}\omega_{fi}t$ is depicted by the curve in Fig. 9.3. The main peak of

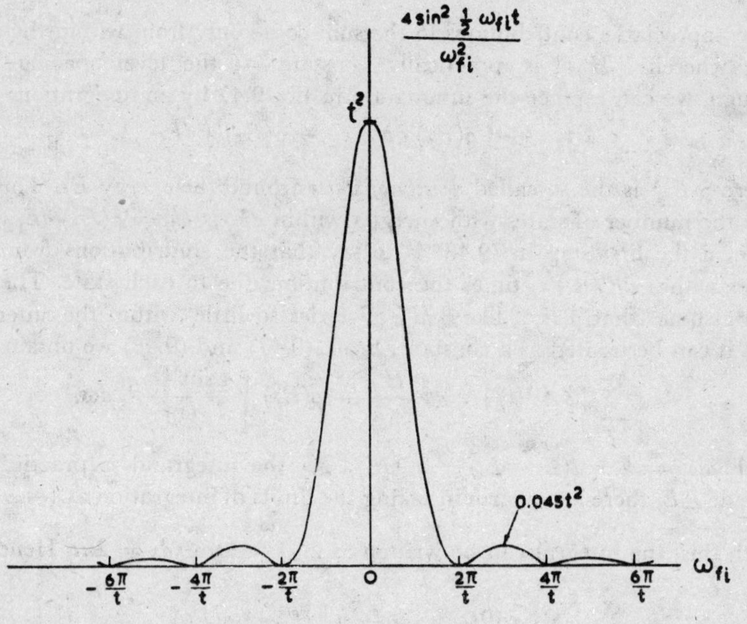

Fig. 9.3 The transition probability as a function of the energy difference between initial and final states, for fixed t.

the curve, which occurs at $\omega_{fi} = 0$ is of height t^2. The subsidiary peaks are very much smaller, the height of the biggest of these (at $\omega_{fi} \approx \pm 3\pi/t$) being less than 5% of that of the main one. Therefore we see that transitions from i take place with appreciable probability only to those levels f such that ω_{fi} falls under the main peak: $|\,\omega_{fi}\,| \leqslant (2\pi/t)$. In other words, the magnitude of the energy difference, $|\,\hbar\omega_{fi}\,|$, between the initial and final states is very unlikely to be significantly higher than $\hbar(2\pi/t) = (h/t)$. This result is generally considered as an expression of an *energy-time uncertainty relation*. The energy changes ΔE caused by a perturbation which lasts for a time Δt (which in the present case is t itself), and which is constant (or at least smoothly varying) within that interval, are $\approx (h/\Delta t)$ so that $(\Delta E)\,(\Delta t) \approx h$. We find thus that energy is conserved, subject to the uncertainty ΔE which is small when t is large

(*b*) *Closely Spaced Levels: Constant Transition Rate:* Returning to Eq. 9.46, let us apply it to a system whose energy levels are very closely spaced. We consider a situation where (i) there is a large number of states f within the energy interval $\triangle E$ covered by the main peak in Fig. 9.3, and (ii) $\triangle E$ is yet small enough that the variation of $|H'_{fi}|$ with f within this interval is negligible. Under these conditions the total transition probability $\Sigma|a_f^{(1)}(t)|^2$ to all states is given by

$$\sum_f |a_f^{(1)}|^2 = \frac{|H'_{fi}|^2}{\hbar^2} \sum_f \frac{4\sin^2 \frac{1}{2}\omega_{fi}t}{\omega_{fi}^2} \tag{9.47}$$

since appreciable contributions to the sum come only from within the interval $\triangle E$, wherein $|H'_{fi}|$ is practically constant. If the level spacing is close enough, we can replace the summation in Eq. 9.47 by an integration:

$$\sum_f \ldots \rightarrow \int \rho(E_f)\, dE_f \ldots \rightarrow \rho(E_f) \int dE_f \ldots \tag{9.48}$$

where $\rho(E_f)$ is the so-called *density of states* around the energy E_f. This means that the number of states with energies within dE_f is $dn_f = \rho(E_f)\, dE_f$. What is done in the first step in (9.48) is to say that the contributions from all the states within dE_f is dn_f times the contribution due to each state. The second step assumes that $\rho(E_f)$, like $|H'_{fi}|$, varies so little within the interval ΔE that it can be treated as a constant. From (9.47) and (9.48) we obtain

$$\sum_f |a_f^{(1)}(t)|^2 = \frac{|H'_{fi}|^2}{\hbar} \rho(E_f) \int \frac{4\sin^2\frac{1}{2}\omega_{fi}t}{\omega_{fi}^2}\, d\omega_{fi} \tag{9.49}$$

since $d\omega_{fi} = \hbar^{-1}\, d(E_f - E_i) = \hbar^{-1} dE_f$. As the integrand is practically zero outside $\triangle E$, there is no harm in taking the limits of integration as $(-\infty, +\infty)$. With this, the integral can be written as $2t \int_{-\infty}^{\infty} x^{-2}\sin^2 x\, dx = 2\pi t$. Hence

$$\sum_f |a_f^{(1)}(t)|^2 = \frac{2\pi}{\hbar} t\, |H'_{fi}|^2 \rho(E_f) \tag{9.50}$$

It is very interesting that this expression is directly proportional to t. This fact enables us to define a transition probability *per unit time.*

$$w = \frac{1}{t} \sum_f |a_f^{(1)}(t)|^2 = \frac{2\pi}{\hbar} |H'_{fi}|^2 \rho(E_f) \tag{9.51}$$

for transitions from i to any of the states f with energy near E_i. The formula (9.51) is what Fermi calls a 'Golden Rule.' Almost all calculations of intensities of spectral lines, and many of cross-sections for various processes, are based on the use of this rule.

The above presentation is incomplete in one important respect: We have paid no attention to the fact that the index labelling any state is, in general, a composite one, standing for a set of quantum numbers. Besides the energy itself, which may be taken as one of the quantum numbers, there may be several others which we shall collectively indicate by the symbol α. [For instance if the system is a free particle in a box (Sec. 2.5) the stationary states are momentum eigenstates. The energy depends only on the magnitude of the momentum, but two more quantum numbers are needed to specify the direction of the momentum and thus identify the state uniquely.] Thus

$$i \rightarrow (E_i, \alpha_i), \quad f \rightarrow (E_f, \alpha_f) \text{ etc.} \tag{9.52}$$

A little reflection shows that the derivation presented above would still go through if α_i and α_f are held fixed throughout and the summations in Eqs. 9.47 – 9.51 are understood to be over final states with all possible energies but with the *same* fixed α_f. Eq. 9.51 should then be modified to read

$$w_{\alpha_f,\alpha_i} = \left[\frac{2\pi}{\hbar}\left| H'_{E_f\alpha_f,E_i\alpha_i}\right|^2 \rho(E_f,\alpha_f)\right]_{E_f=E_i} \tag{9.53}$$

This gives the probability per unit time for transitions from a given initial state to final states with specified α_f. If one is interested only in the result for transitions to states within some range of values of α_f, one can sum the above expression over the α_f within the desired range.

Finally we remind the reader of the condition (9.44) which places an upper limit on the time t for which Eq. 9.50 remains valid. It will be recalled that there is also a lower limit imposed by the requirement that the energy interval $(\Delta E = h/t)$ over which the summation in Eq. 9.47 is effective, should be small enough, so that the matrix elements and $\rho(E_f)$ are effectively constant within ΔE for given α_f. It is the existence of these limitations which determines how large H' can be: A time interval consistent with these will not exist if H' is so large that the upper limit on t imposed by Eqs. 9.50 and 9.44 becomes very low and clashes with the above mentioned lower limit.

9.8 Transitions in the Second Order: Constant Perturbation

As long as the first order transition amplitude $a_f^{(1)}$ is nonzero ($H'_{fi} \neq 0$ for the particular pair of levels i,f), the second and higher order contributions can usually be ignored since H' is supposed to be very small. But if H'_{fi} vanishes, transitions are governed by the second order term $a_f^{(2)}$. With H' constant from 0 to t, we find from Eqs. 9.40 that

$$a_f^{(2)} = (i\hbar)^{-2} \sum_l H'_{fl} H'_{li} \cdot F_{fli}^{(2)}(t) \tag{9.54}$$

where

$$F_{fli}^{(2)}(t) = \int_0^t e^{i\omega_{fl}t''}\, dt'' \int_0^{t''} e^{i\omega_{li}t'}\, dt' \tag{9.55a}$$

$$= -\frac{1}{\omega_{li}}\left[\frac{e^{i\omega_{fi}t} - 1}{\omega_{fi}} - \frac{e^{i\omega_{fl}t} - 1}{\omega_{fl}}\right] \tag{9.55b}$$

It is not difficult to convince oneself, from an examination of this last form, that the value of $|F^{(2)}|$ is largest when $\omega_{li} = \omega_{fi} = \omega_{fl} = 0$, $(E_i = E_f = E_l)$ and that it becomes relatively very small when any of these quantities differs from zero by more than $\sim (2\pi/t)$. This means that when t is large, first of all there is approximate energy conservation (with an 'uncertainty' $\Delta E \approx h/t$), and secondly, in the sum over intermediate states l in Eq. 9.54, only those which have energies E_l close to E_i and E_f (within approximately h/t) contribute appreciably. These properties make it possible to show that $a_f^{(2)}$ reduces effectively to a form which closely resembles the expression for $a_f^{(1)}$. In carrying out this reduction, we shall assume that the states are closely spaced, with a density of states $\rho(E_l)$ around energy E_l, and that H'_{fl}, H'_{li} and $\rho(E_l)$ are practically constant within any energy interval $\Delta E \approx h/t$ (with t chosen sufficiently large). The process is simplified if we go back to the double integral (9.55a) for $F^{(2)}$ and reexpress it in a form wherein the variables of integration are

$$T = \tfrac{1}{2}(t'' + t') \text{ and } \tau = (t'' - t') \tag{9.56}$$

integrating first over τ (for fixed T) and then over T. The limits of integration with respect to these variables may be obtained by inspection of the region of

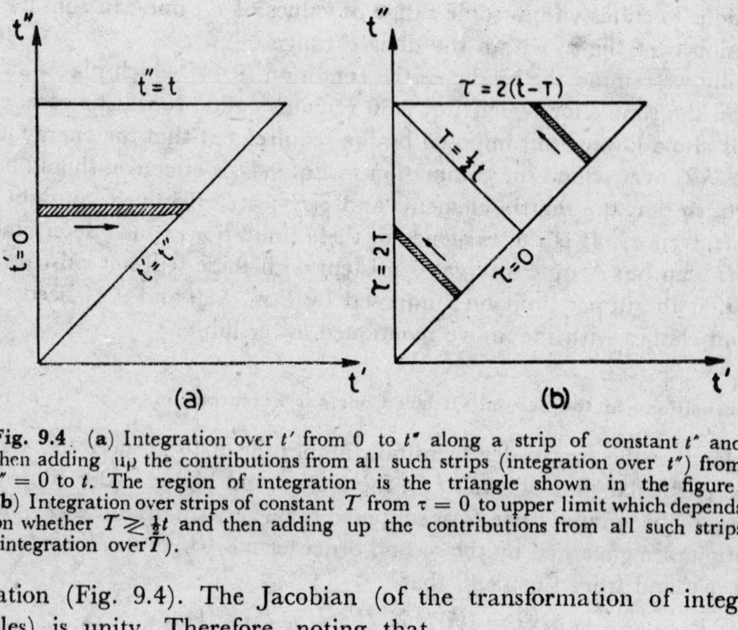

Fig. 9.4 (**a**) Integration over t' from 0 to t'' along a strip of constant t'' and then adding up the contributions from all such strips (integration over t'') from $t'' = 0$ to t. The region of integration is the triangle shown in the figure; (**b**) Integration over strips of constant T from $\tau = 0$ to upper limit which depends on whether $T \gtrless \frac{1}{2}t$ and then adding up the contributions from all such strips (integration over T).

integration (Fig. 9.4). The Jacobian (of the transformation of integration variables) is unity. Therefore, noting that

$$(\omega_{fl}\, t'' + \omega_{li}\, t') = [\omega_{fi}\, T + \tfrac{1}{2}\,(\omega_{fl} - \omega_{li})\tau],$$

we write $F^{(2)}$ as

$$F_{fli}^{(2)} = \int_0^{\frac{1}{2}t} \exp\,(i\omega_{fi}T)\, dT \int_0^{2T} \exp\,[\tfrac{1}{2}i(\omega_{fl} - \omega_{li})\tau]\, d\tau$$

$$+ \int_{\frac{1}{2}t}^{t} \exp\,(i\omega_{fi}T)\, dT \int_0^{2(t-T)} \exp\,[\tfrac{1}{2}i(\omega_{fl} - \omega_{li})\tau]\, d\tau \tag{9.57}$$

If this is substituted in Eq. 9.54, the summation over l will involve just

$$\Sigma H'_{fl}\, H'_{li} \exp\,(-\,iE_l\tau/\hbar) \to \int H'_{fl}\, H'_{li} \exp\,(-iE_l\tau/\hbar)\, \rho(E_l)\, dE_l \tag{9.58}$$

Recall now that the structure of $F_{fli}^{(2)}$ is such that only a small range ΔE of

E_l would contribute appreciably to this sum, and that $\rho(E_l)$ and the matrix elements vary very little in this range. Thus the value of $a_f^{(2)}$ will be unaffected by whatever we might assume about the value of these quantities outside $\triangle E$; in particular, we can consider them (as far as the evaluation of $a_f^{(2)}$ is concerned) to be practically independent of E_l. But then the integral in (9.58), which appears as the Fourier transform of $H'_{fl} H'_{li} \rho(E_l)$, would vanish except for $\tau \approx 0$. So, if we go back to the expression (9.54) for $a_f^{(2)}$, with $F_{fli}^{(2)}$ given by Eq. 9.57, we see that only the part $\tau \approx 0$ of the integral over τ in the latter contributes appreciably to $a_f^{(2)}$. There is no harm then in taking the upper limits of the τ-integrations in both the terms of Eq. 9.57 to be ∞, and finally we can even add $i\varepsilon$ ($\varepsilon \to 0_+$) to $\frac{1}{2}(\omega_{fl} - \omega_{li})$ in the exponent, to make these integrals convergent. (Remember that this step does not affect the integrand in the region which matters, $\tau \approx 0$ since we make $\varepsilon \to 0$). These steps convert the τ-integrals in both terms of Eq. 9.57 to

$$\int_0^\infty \exp\left[\tfrac{1}{2}i(\omega_{fl} - \omega_{li})\tau - \varepsilon\tau\right] d\tau = \frac{1}{i[\frac{1}{2}(\omega_{fl} - \omega_{li}) + i\varepsilon]} \tag{9.59}$$

and $F_{fli}^{(2)}$ then reduces to

$$\frac{e^{i\omega_{fi}t} - 1}{i\omega_{fi}} \cdot \frac{1}{i\left[\frac{1}{2}(\omega_{fl} - \omega_{li}) + i\varepsilon\right]} \tag{9.60}$$

On substituting this in Eq. 9.54 and replacing the sum over l by an integral, we finally obtain[8]

$$a_f^{(2)} = \frac{e^{i\omega_{fi}t} - 1}{\hbar\omega_{fi}} \int \frac{H'_{fl} H'_{li}}{E_l - \frac{1}{2}(E_f + E_i) - i\varepsilon} \rho(E_l)\, dE_l \tag{9.61}$$

It is very interesting that this expression has just the same form as Eq. 9.45 for $a_f^{(1)}$ except for the replacement of $(-H'_{fi})$ by the integral in Eq. 9.61. This integral is often called the *second order matrix element*. Discussion of the transition *probabilities* will follow exactly the same lines as in the first order case, except that the second order matrix element must now be used. Since only the matrix element with $E_f = E_i$ appears in the final expression for the transition probability per unit time, the denominator in the integral of Eq. 9.61 may be replaced by $(E_l - E_i - i\varepsilon)$.

9.9 Scattering of a Particle by a Potential

As an illustration of first order perturbation theory we shall rederive the Born approximation scattering cross-section, Eq. 6.25a, using Eq. 9.51. We take H_0 to be the free-particle Hamiltonian and the potential $V(\mathbf{r})$ to be the perturbation. The whole system will be assumed to be in a large cubical box of sides L. The initial and final states, describing the particle incident with momentum $\mathbf{k}_0$ and scattered with momentum $\mathbf{k}$, will be represented by the box-normalized wave functions

[8] We have absorbed a factor $\hbar$ into ε in going from (9.60) to (9.61), but clearly it does not matter whether we write ε or $\varepsilon\hbar$, since the limit $\varepsilon \to 0_+$ is to be taken.

$$u_i(\mathbf{x}) = L^{-3/2}\, e^{i\mathbf{k}_0.\mathbf{x}}, \quad u_f(\mathbf{x}) = L^{-3/2}\, e^{i\mathbf{k}.\mathbf{x}} \tag{9.62}$$

Note that f stands for the magnitude $k = |\mathbf{k}|$ and direction (θ,φ) of the vector. The former defines the energy $E_f = \hbar^2k^2/2m$, and (θ,φ) are the α_f's of this problem. Energy conservation, $E_f = E_i$ is taken care of by choosing $|\mathbf{k}_0| = k = |\mathbf{k}|$. The matrix element H'_{fi} is given by

$$H'_{fi} = \int u_f^* \, V(\mathbf{x})\, u_i \, d\tau = L^{-3} \int V(\mathbf{x})\, e^{-i\mathbf{K}.\mathbf{x}} d\tau \tag{9.63}$$

where $\mathbf{K} = \mathbf{k} - \mathbf{k}_0$.

In determining the density of final states, it is to be remembered that we are interested in transitions to states with momentum directions lying within a solid angle $d\Omega = \sin\theta \, d\theta \, d\varphi$. The number of wave vectors $\mathbf{k}$ (quantized, because of the presence of the box) which lie within $d\Omega$ and whose lengths are between k and $k + dk$ has been calculated already in Sec. 2.5. It is equal to $(L/2\pi)^3.k^2dk\, d\Omega$. This is the number of states with energies between E_f and $E_f + dE_f$, where $E_f = (\hbar^2k^2/2m)$, $dE_f = (\hbar^2k/m)\, dk$. Therefore, the density of final states is given by

$$\rho(E_f) = \left(\frac{L}{2\pi}\right)^3 k^2dkd\Omega . \frac{m}{\hbar^2kdk} = \left(\frac{L}{2\pi}\right)^3 \cdot \frac{mk}{\hbar^2}\, d\Omega \tag{9.64}$$

On introducing Eqs. 9.63 and 9.64 in Eq. 9.51, we get the probability w per unit time for scattering into the direction $d\Omega$ of a particle whose incident wave function is u_i of Eq. 9.62. The particle flux associated with this wave function is evidently $|u_i|^2 v = L^{-3}\,(\hbar k/m)$. Since the differential scattering cross-section is defined to be the number of scattered particles per unit solid angle per *unit* incident flux, we finally obtain

$$\begin{aligned} \frac{d\sigma}{d\Omega}(\theta,\varphi) &= \frac{w}{d\Omega}\cdot\frac{L^3m}{\hbar k} \\ &= \left(\frac{m}{2\pi\hbar^2}\right)^2 \left| \int V(\mathbf{x}) e^{-i\mathbf{K}.\mathbf{x}}\, d\tau \right|^2 \end{aligned} \tag{9.65}$$

This expression coincides with what was obtained from the Born approximation in Sec. 6.4. Observe that the size L of the box — which was artificially introduced to help in determining $\rho(E_f)$ — does not appear in the final result.

9.10 Inelastic Scattering: Exchange Effects

In the scattering by a fixed potential (which is equivalent to an infinitely massive object) there is no transfer of energy (though there is of momentum) between the particle and the potential. But if the scattering centre has a structure and energy levels of its own, there can be transfer of energy (as well as momentum), i.e. the scattering may be *inelastic*.[9] As an example of such a process, we consider the inelastic scattering of an electron by a hydrogen atom. We assume the electron to be sufficiently fast that a first order perturbation calculation would be good enough, yet not so fast as to cause any

[9] A collision is elastic if there is no exchange of energy between the colliding objects as seen in their *centre of mass* frame. If one of the objects is very much heavier than the other and initially at rest in the laboratory, (e.g. a static atom on which an electron is incident) then the centre of mass frame is essentially the laboratory frame itself.

appreciable recoil of the atom as a whole. To begin with, we ignore exchange effects arising from the fact that the incident and atomic electrons are identical particles.

The unperturbed Hamiltonian of this problem is

$$H_0 = -\frac{\hbar^2}{2m}\nabla_1^2 - \frac{\hbar^2}{2m}\nabla_2^2 - \frac{e^2}{r_2} \tag{9.66}$$

where the subscripts 1 and 2 refer to the incident and atomic electrons respectively. The interaction of the incident electron with the atomic electron and the nucleus is taken as the perturbation

$$H' = -\frac{e^2}{r_1} + \frac{e^2}{r_{12}} \tag{9.67}$$

We shall suppose that initially the atom is in its ground (1*s*) state, and that the external electron has a momentum $\hbar\mathbf{k}_0$. We seek the cross-section for excitation of the atom to the 2*s* state as a result of the collision, with the electron going off with momentum $\hbar k$. Then the wave functions of the initial and final states i, f are

$$\begin{aligned} i &: \; L^{-3/2} e^{i\mathbf{k}_0.\mathbf{x}_1} u_{100}(\mathbf{x}_2) \\ f &: \; L^{-3/2} e^{i\mathbf{k}.\mathbf{x}_1} u_{200}(\mathbf{x}_2) \end{aligned} \tag{9.68}$$

where the lengths of $\mathbf{k}_0$ and $\mathbf{k}$ are related by energy conservation:

$$\frac{\hbar^2 k_0^2}{2m} - \frac{e^2}{2a_0} = \frac{\hbar^2 k^2}{2m} - \frac{e^2}{8a_0}, \quad a_0 = \frac{\hbar^2}{me^2} \tag{9.69}$$

From (9.67) and (9.68)

$$H'_{fi} = L^{-3} \iint e^{-i\mathbf{K}.\mathbf{x}_1} u^*_{200}(\mathbf{x}_2) \left(\frac{e^2}{r_{12}} - \frac{e^2}{r_1}\right) u_{100}(\mathbf{x}_2)\, d\tau_2\, d\tau_1 \tag{9.70}$$

The density of states $\rho(E_f)$ is the same as (9.64), but the incident flux now involves k_0 (which is no longer equal to k.) When this is taken into account the differential cross-section is easily seen to be

$$\frac{d\sigma}{d\Omega}(\theta) = \frac{k}{k_0}\left(\frac{m}{2\pi\hbar^2}\right)^2 L^6 \,|H'_{fi}|^2 \tag{9.71}$$

What remains now is to evaluate H'_{fi} from Eq. 9.70.

The term e^2/r_1 in the integrand does not contribute to the integral, because it is free of $\mathbf{x}_2$ and so the $d\tau_2$ integration gives zero since $u_{100}(\mathbf{x}_2)$ and $u_{200}(\mathbf{x}_2)$ are orthogonal. The integral of the other term is easily evaluated by integrating over $\mathbf{x}_1$ first:

$$\begin{aligned} \int \frac{e^{-i\mathbf{K}.\mathbf{x}_1}}{r_{12}}\, d\tau_1 &= e^{-i\mathbf{K}.\mathbf{x}_2} \int \frac{e^{-i\mathbf{K}.\mathbf{x}_{12}}}{r_{12}}\, d\tau_{12} \\ &= \frac{4\pi}{K^2} e^{-i\mathbf{K}.\mathbf{x}_2} \end{aligned} \tag{9.72}$$

The first step here is simply a change of the integration variable from $\mathbf{x}_1$ to $\mathbf{x}_{12} \equiv \mathbf{x}_1 - \mathbf{x}_2$. The value of the resulting integral has been evaluated already in Example 6.1, p. 185. (What we use here is the special case of that result for $\varkappa = 0$). On introducing the above result in Eq. 9.70 we are led to

$$H'_{fi} = \frac{4\pi e^2}{L^3 K^2} \int e^{-i\mathbf{K}.\mathbf{x}_2} u^*_{200}(\mathbf{x}_2)\, u_{100}(\mathbf{x}_2)\, d\tau_2$$

$$= \frac{16\sqrt{2}\,\pi\, a_0^2 e^2}{L^3(K^2a_0^2 + \frac{9}{4})^3} \tag{9.73}$$

This value is obtained by using the explicit forms (Sec. 4.16) of the hydrogen atom wave functions and carrying out the integration. On feeding this into Eq. 9.71 we obtain

$$\frac{d\sigma}{d\Omega}(\theta) = \frac{128\, a_0^2\, k}{k_0(K^2a_0^2 + \frac{9}{4})^6} \tag{9.74}$$

The total cross-section can be readily obtained by using the fact that $K^2 = k_0^2 + k^2 - 2k_0k\cos\theta$, which gives $\sin\theta\, d\theta = (2k_0k)^{-1}\, dK^2$. So $\int (d\sigma/d\Omega)\, d\Omega = 2\pi \int (d\sigma/d\Omega) \sin\theta\, d\theta$ reduces to $(\pi/kk_0) \int (d\sigma/d\Omega)\, dK^2$. The limits of the integration over K^2 are evidently $(k_0 + k)^2$ and $(k_0 - k)^2$. Thus we obtain

$$\sigma = \frac{128\pi}{5\, k_0^2}\left\{\left[(k_0 - k)^2 a_0^2 + \frac{9}{4}\right]^{-5} - \left[(k_0 + k)^2 a_0^2 + \frac{9}{4}\right]^{-5}\right\} \tag{9.75}$$

These expressions for $d\sigma/d\Omega$ and σ, obtained by a perturbation calculation, have their best validity when the incident energy is much higher than the binding energies in the atom: $(\hbar^2 k_0^2/2m) \gg (e^2/2a_0)$ which is to say, $k_0^2a_0^2 \gg 1$. Then $k_0 \approx k$ and under these conditions the total cross-section becomes $\sigma \approx (128\pi/5k_0^2)\ (2/3)^{10}$. It is a small fraction of [approximately $\frac{1}{2}\ (k_0a_0)^{-2}$ times] the 'geometrical' cross-section πa_0^2. As for the differential cross-section, we see that except at small momentum transfers (such that Ka_0 is of the order of unity or less), $d\sigma/d\Omega$ is a very steeply decreasing function of momentum transfer, going down like $(Ka_0)^{-12}$. In the case of high energies referred to above, $K \approx 2k_0 \sin\frac{1}{2}\theta$ and $Ka_0 \approx 1$ means $\theta \approx (1/k_0a_0)$; thus, as θ goes beyond this small value (for some given large momentum), the differential cross-section falls off like $\text{cosec}^{12}\, \frac{1}{2}\theta$. This compares with $\text{cosec}^4\, \frac{1}{2}\theta$ for *elastic* scattering by an atom (e.g. screened Coulomb potential, Example 6.1, p. 185). The very rapid decrease with angle is characteristic of inelastic cross-sections.

It is useful to state the main features of the above derivation in more general terms. If a particle of momentum $\mathbf{k}_i$ collides with a scatterer having its own internal structure, the interaction Hamiltonian depends explicitly on the coordinates of the particles within the scatterer (which we denote collectively by $\mathbf{x}_2$): $H' = H'(\mathbf{x}_1, \mathbf{x}_2)$. The matrix element for scattering of the particle (to a final momentum $\mathbf{k}_f$) accompanied by a change in the internal state of the scatterer from $u_i(\mathbf{x}_2)$ to $u_f(\mathbf{x}_2)$ is

$$L^{-3} \iint e^{-i\mathbf{k}_f.\mathbf{x}_1}\, u_f^*(\mathbf{x}_2)\, H'(\mathbf{x}_1,\mathbf{x}_2)\, u_i(\mathbf{x}_2)\, e^{i\mathbf{k}_i.\mathbf{x}_1}\, d\tau_1 d\tau_2 \tag{9.76}$$

which may be rewritten as

$$L^{-3} \int e^{-i\mathbf{K}.\mathbf{x}_1}\, H'_{\text{eff}}(\mathbf{x}_1) d\tau_1 \tag{9.77}$$

where

$$H'_{\text{eff}}(\mathbf{x}_1) = \int u_f^*(\mathbf{x}_2)\, H'(\mathbf{x}_1,\mathbf{x}_2)\, u_i(\mathbf{x}_2) d\tau_2 \tag{9.78}$$

Eq. 9.77 is of exactly the same form as Eq. 9.63 of potential scattering. The only difference is that in $\mathbf{K} = \mathbf{k}_f - \mathbf{k}_i$, $|\mathbf{k}_f| \neq |\mathbf{k}_i|$ here. Subject to this, *inelastic scattering can be treated simply as scattering by an 'effective potential'* H'_{eff} which is to be computed as in Eq. 9.78. Note that the wave functions of the initial and final states of the scatterer enter in an essential manner in the determination of H'_{eff}, which should strictly be written as $(H'_{eff})_{fi}$. The integration in Eq. 9.78 has the consequence that $H'_{eff}(\mathbf{x}_1)$ is a much smoother function of $\mathbf{x}_1$ than $H'(\mathbf{x}_1, \mathbf{x}_2)$, and correspondingly poorer in Fourier components of high K. That is why the scattering amplitude falls off much faster with K in inelastic scattering than in elastic scattering. For the particular example considered above (inelastic scattering with excitation of a hydrogen atom from the $1s$ to the $2s$ state), we have

$$H'_{eff} = \frac{2\sqrt{2}e^2}{27a_0^2}(3r_1 + 2a_0)\,e^{-3r_1/2a_0}$$

The expressions obtained above for the cross-sections do not take into account the effects of identity of the two electrons, and are, to that extent, incorrect. To get the correct results, we should also include the matrix element for *exchange scattering*, wherein the electron 2 (which was originally in the atom) comes out as the free (scattered) electron in the final state, and electron 1 takes its place in the atom. The simplest way to do this is by symmetrizing or antisymmetrizing (Sec. 3.16) the final state wave function

$$L^{-3/2}\,[e^{i\mathbf{k}.\mathbf{x}_1}u_{200}(\mathbf{x}_2) \pm e^{i\mathbf{k}.\mathbf{x}_2}u_{200}(\mathbf{x}_1)] \tag{9.79}$$

while leaving the initial wave function as it is in (9.68). Such a procedure would correspond to saying that whichever electron is incident can, *by convention*, be labelled 1; while after the interaction it is impossible to say which of the electrons comes out, thus necessitating symmetrization or antisymmetrization of the final wave function.[10] The use of (9.79) as final state would mean that to the matrix element (9.70) for direct (non-exchange) scattering which we shall now denote by $H'^{(d)}_{fi}$, one would have to add

$$H'^{(ex)}_{fi} = \pm L^{-3}\iint e^{-i\mathbf{k}.\mathbf{x}_2}u^*_{200}(\mathbf{x}_1)\left(\frac{e^2}{r_{12}} - \frac{e^2}{r_1}\right)e^{i\mathbf{k}_0.\mathbf{x}_1}u_{100}(\mathbf{x}_2)\,d\tau_1 d\tau_2 \tag{9.80}$$

representing exchange scattering. In the formula (9.71) for the differential cross-section, H'_{fi} should then be replaced by $H'^{(d)}_{fi} + H'^{(ex)}_{fi}$. The evaluation of the integral (9.80) is a rather cumbersome process and we do not present it here.

9.11 Double Scattering by Two Non-Overlapping Scatterers

As an illustration of second order perturbation theory, we consider the following problem. Let A, B be two scatterers centred at $\mathbf{R}_A$, $\mathbf{R}_B$ and having no overlap. By this we mean that there is no point $\mathbf{x}$ where the influence of both A and B can be simultaneously felt. Then the interaction of an incident particle with A and B can be written as

$$H' = H'_A(\mathbf{x}) + H'_B(\mathbf{x}) \tag{9.81}$$

with $H'_A(\mathbf{x})\,H'_B(\mathbf{x}) = 0$ for all $\mathbf{x}$. The scatterers may be fixed potentials of

[10] One could proceed more formally by using properly symmetrized/antisymmetrized *and normalized* wave functions for *both* the initial and final states, i.e. $2^{-1/2}[u_i(1,2) \pm u_i(2,1)]$ and $2^{-1/2}[u_f(1,2) \pm u_f(2,1)]$, where $u_i(1,2)$ and $u_f(1,2)$ are the functions denoted by i and f in Eq. 9.68. The normalization factors are not exactly $2^{-1/2}$ since $e^{i\mathbf{k}.\mathbf{x}}$ is not orthogonal to u_{100} or u_{200} but the difference is proportional to $L^{-3/2}$ and goes to zero as $L \to \infty$. The student may verify that this procedure would also give the matrix element as $H'^{(d)}_{fi} + H'^{(ex)}_{fi}$.

range less than $R = |\mathbf{R}|$ where $\mathbf{R} = \mathbf{R}_A - \mathbf{R}_B$, or they may have internal structure of their own (e.g. two well-separated atoms). The latter case can also be considered as scattering by appropriate effective potentials (Sec. 9.10), and so we shall use the language of potential scattering in the following. We are interested in finding the probability that the particle, incident with momentum $\mathbf{k}_i$, interacts with *both* A and B and then emerges with momentum $\mathbf{k}_f$. More specifically, we wish to demonstrate that if the incident particle is fast ($k_i a \gg 1$, where a is the 'size' of each of the scatterers), then appreciable probability for the above process exists only if $\mathbf{k}_i$, $\mathbf{k}_f$ *and the relative direction of the two scatterers* (defined by $\mathbf{R} = \mathbf{R}_A - \mathbf{R}_B$) are all nearly parallel. This is just what one would expect classically. But in quantum mechanics, where the wave function extends all over space, it is not at all obvious *a priori* that a similar result should hold. The fact that it does is gratifying; it shows that the formation of well-defined tracks by high energy particles (in photographic emulsions, cloud chambers, bubble chambers, etc.), is quite consistent with quantum mechanics.

The double scattering process which we are interested in does not go through in the first order of perturbation theory. It can take place in the second order, and so what we have to consider is the second order matrix element (9.61). From the occurrence of the product $H'_{fl} H'_{li}$ in this quantity, one can see directly that transitions wherein the initial and final momenta are not nearly parallel have only negligible probability. For, H'_{li} is proportional to $\int e^{-i\mathbf{K}_{li}\cdot\mathbf{x}} H' d\tau$ where $\mathbf{K}_{li} = \mathbf{k}_l - \mathbf{k}_i$; and we already know that this integral falls off rapidly with $K_{li} = |\mathbf{K}_{li}|$ when the value of K_{li} exceeds $(1/a)$. Similarly H'_{fl} is very small unless $K_{fl} \lesssim (1/a)$. In the case of large initial momenta, $k_i a \gg 1$, we also know that $k_i \approx k_l \approx k_f$ (even if the transitions are inelastic); using k to denote any of these, we see readily that the conditions $K_{li} \lesssim (1/a)$ and $K_{fl} \lesssim (1/a)$ become $\theta_{li}, \theta_{fl} \lesssim (1/ka)$ where the θ's are the angles between the relevant pairs of vectors. Evidently then, the angular deviation between $\mathbf{k}_i$ and $\mathbf{k}_f$ can only be of the order of $(1/ka)$.

Next, we observe that since $H' = H'_A + H'_B$ here, the product $H'_{fl} H'_{li}$ breaks up into four terms. Only two of these involve *both* H'_A and H'_B, and we are concerned only with these. Their sum is

$$(H'_A)_{fl} (H'_B)_{li} + (H'_B)_{fl} (H'_A)_{li} \tag{9.82}$$

It is enough to examine the contribution of any one of these to the second order matrix element (9.61) since the other can then be obtained simply by interchanging A and B. Thus what we have to deal with is the expression

$$\sum_l \frac{(H_A')_{fl} (H_B')_{li}}{E_l - E_i - i\varepsilon}$$
$$= \frac{1}{L^6} \sum_l \frac{\iint \exp[-i\mathbf{K}_{fl}\cdot\mathbf{x}]\, H'_A(\mathbf{x}) \exp[-i\mathbf{K}_{li}\cdot\mathbf{x}']\, H'_B(\mathbf{x}')\, d\tau\, d\tau'}{(\hbar^2/2m)(k_l^2 - \varkappa^2 - i\varepsilon)} \tag{9.83}$$

where[11] $\varkappa^2 = k_i^2$. The sum over l can be evaluated:

[11] If A has levels ε_l of its own, then $E_l = (\hbar^2/2m)\, k_l^2 + \varepsilon_l$, and $E_i = (\hbar^2/2m)\, k_i^2 + \varepsilon_i$. Then $\varkappa^2 = k_i^2 + (2m/\hbar^2)(\varepsilon_i - \varepsilon_l)$. However, in the high energy approximation the second term is negligible.

$$\frac{1}{L^3}\sum_l \frac{\exp\,[\,i\mathbf{k}_l.(\mathbf{x}-\mathbf{x}')]}{k_l^2-\varkappa^2-i\varepsilon} \to \frac{1}{(2\pi)^3}\int \frac{\exp\,[\,i\mathbf{k}_l.(\mathbf{x}-\mathbf{x}')]}{k_l^2-\varkappa^2-i\varepsilon}\,d^3k_l \qquad (9.84)$$

The last integral is obtained by passing to the limit $L \to \infty$ and observing that within a volume element d^3k_l of momentum space there are $(2\pi/L)^3 d^3k_l$ states which contribute equally to the sum. The value of this integral is known to be $\exp\,(i\varkappa\,|\,\mathbf{x}-\mathbf{x}'\,|\,)/[4\pi\,|\,\mathbf{x}-\mathbf{x}'\,|\,]$ (See Example 9.2, p. 274). On substituting this in Eq. 9.83, it reduces to

$$\frac{2m}{4\pi\hbar^2 L^3}\iint \frac{1}{|\,\mathbf{x}-\mathbf{x}'\,|}\,e^{-i\mathbf{k}_f.\mathbf{x}}\,H'_A(\mathbf{x})\,e^{i\varkappa|\mathbf{x}-\mathbf{x}'|}\,e^{i\mathbf{k}_i.\mathbf{x}'}\,H'_B(\mathbf{x}')\,d\tau d\tau' \qquad (9.85)$$

At this point we invoke the assumption that the distance R between A and B is much larger than the range a of either. So, writing $\mathbf{x}-\mathbf{x}'$ as

$$(\mathbf{x}-\mathbf{R}_A)-(\mathbf{x}'-\mathbf{R}_B)+(\mathbf{R}_A-\mathbf{R}_B)\equiv \mathbf{x}_A-\mathbf{x}_B'+\mathbf{R},$$

we observe that the contributions to the above integral come from regions of $\mathbf{x}_A$, $\mathbf{x}_B$ such that $|\,\mathbf{x}_A\,|$, $|\,\mathbf{x}_B\,| \lessapprox a \ll R$. Using this property we can expand

$$|\,\mathbf{x}-\mathbf{x}'\,|\equiv|\,\mathbf{R}+\mathbf{x}_A-\mathbf{x}'_B\,|\simeq R+\frac{1}{R}\,(\mathbf{x}_A.\mathbf{R}-\mathbf{x}_B.\mathbf{R}) \qquad (9.86)$$

In the denominator only the leading term R is important, and it can be pulled outside the integral. Then, defining the vector of magnitude $\varkappa$ in the direction of R as $\mathbf{k}_R=\varkappa\,(\mathbf{R}/R)$, we get the integral to be

$$\approx \frac{m}{2\pi\hbar^2 R\,L^3}\int H'_A(\mathbf{x})\,e^{-i(\mathbf{k}_f-\mathbf{k}_R).\mathbf{x}}\,d\tau\int H'_B(\mathbf{x}')\,e^{i(\mathbf{k}_i-\mathbf{k}_R).\mathbf{x}'}\,d\tau' \qquad (9.87)$$

The kind of arguments which we used earlier in this section now show that this quantity is very small unless $|\,\mathbf{k}_f-\mathbf{k}_R\,|\lessapprox(1/ka)$ and $|\,\mathbf{k}_i-\mathbf{k}_R\,|\lessapprox(1/ka)$. Thus, for large ka the vectors $\mathbf{k}_i$, $\mathbf{k}_f$, $\mathbf{k}_R$ all must have nearly the same length and direction. Since the direction of $\mathbf{k}$ is that of $\mathbf{R}$, the result we claimed above is proved: if the particle is to interact with both A and B, the line joining A and B must be nearly parallel to $\mathbf{k}_i$.

A further interesting result follows if we consider the second term in (9.82). Interchanging A and B in the above derivation is equivalent to changing $\mathbf{R}$ to $-\mathbf{R}$ and $\mathbf{k}_R$ to $-\mathbf{k}_R$, so that the final result is that $\mathbf{k}_i$, $\mathbf{k}_f$ and $-\mathbf{k}_R$ should be very nearly parallel. The change in sign becomes very significant when we consider the *order* in which the interactions with A and B take place. The second term in (9.82) corresponds to the particle interacting with A first and then with B, because it is H'_A which takes the initial state i to l, and H'_B then takes l to f. (The reverse is the case with the first term). Our result then is that B is excited first if the vector $\mathbf{R}$ (pointing from B to A) is parallel to $\mathbf{k}_i$ while A is excited first if $-\mathbf{R}$ (pointing from A to B) is parallel to $\mathbf{k}_i$. This is again just what one expects from classical mechanics.

9.12 Harmonic Perturbations

(*a*) *Amplitude for Transition with Change of Energy:* So far we have been dealing with a constant perturbation which acts over a time interval t. We shall now consider a harmonic perturbation[12]

[12] This is not Hermitian by itself, and one should add to it its Hermitian conjugate which is also harmonic, with the time-dependent factor $e^{i\omega t}$. See below for discussion of this point.

$$H'^0 e^{-i\omega t}, \quad (\omega > 0) \tag{9.88}$$

supposed to act during the time interval $(0,t)$. Substituting this in Eq. 9.42, we immediately obtain for the first order transition amplitude

$$a_f^{(1)}(t) = -H'^0_{fi} \frac{e^{i(\omega_{fi}-\omega)t} - 1}{\hbar(\omega_{fi} - \omega)} \tag{9.89}$$

This expression differs from Eq. 9.45 only in having $(\omega_{fi} - \omega)$ instead of just ω_{fi}. With this replacement, the entire discussion of Sec. 9.7 remains valid for harmonic perturbations too. In particular, one can conclude that for large t, only those transitions in which

$$\omega_{fi} - \omega = 0 \quad \text{or} \quad E_f - E_i = \hbar\omega \tag{9.90}$$

can take place with appreciable probability. This means that the perturbation (9.88) can induce transitions from E_i only to a level E_f whose energy is higher than E_i by just $\hbar\omega$. Such a transition may be described as *absorption* of energy $\hbar\omega$ by the system (from the perturbing agency).

As for the Hermitian conjugate $(H'^0)^\dagger e^{i\omega t}$ of (9.88), it leads to a transition amplitude which differs from (9.89) in having $(H'^0)^\dagger_{fi} = (H'^0_{if})^*$ as the overall factor, and $(\omega_{fi} + \omega)$ instead of $(\omega_{fi} - \omega)$. It induces only transitions in which

$$(\omega_{fi} + \omega) = 0 \quad \text{or} \quad E_f - E_i = -\hbar\omega \tag{9.91}$$

In this case one finds that E_f is lower, that is, energy $\hbar\omega$ is *given away* by the system to the perturbing agency.

The actual perturbation has to be Hermitian and must, therefore, be

$$H' = H'^0 e^{-i\omega t} + (H'^0)^\dagger e^{i\omega t} \tag{9.92}$$

In the above discussion we have considered the effects of the two terms separately. Strictly speaking one should take the total amplitude (including the contributions of both terms) and take its absolute square to find the transition probability. On doing this one gets two interference terms consisting of cross products of the two parts. However, since one factor effectively vanishes except when $\omega \approx \omega_{fi}$ and the other vanishes unless $\omega \approx -\omega_{fi}$, their product (for a specific pair of initial and final states) is zero for all practical purposes. That is to say, the two parts of H' in effect act independently of each other, and the treatment given above is adequate.

(*b*) *Transitions Induced by Incoherent Spectrum of Perturbing Frequencies:* These notions have obvious applicability to the theory of absorption and emission of radiation by atoms, and we shall take this up in the next section. In this connection the case of a discrete pair of levels (e.g. the levels of an atom) is of interest. As long as the perturbation contains only a single frequency, the transition probability oscillates with time as in (9.46) with $(\omega_{fi} \mp \omega)$ instead of ω_{fi}. However, a transition probability proportional to time can arise under the following conditions: (i) The perturbation involves a whole spectrum of frequencies ω which are so closely spaced that very many such frequencies are contained within the interval $(1/t)$; (ii) these are incoherent, in the sense that the phases of the different frequency components are unrelated to each other, but (iii) the magnitudes of these perturbations and the spacing of the frequencies are smooth functions of ω. Condition (ii) means that to determine the total transition probability induced by the whole spectrum of frequencies, one simply adds up the probabilities (*not* the amplitudes) arising from the different frequency components. Condition (i) then enables this sum to be replaced by an integral. Thus the total transition probability would be

$$\sum_{\omega} \left| a_f^{(1)}(t;\omega) \right|^2 = \sum_{\omega} \frac{|H'^0_{fi}(\omega)|^2}{\hbar^2} \cdot \frac{4\sin^2 \frac{1}{2}(\omega_{fi}-\omega)t}{(\omega_{fi}-\omega)^2}$$

$$\to \int \frac{|H'^0_{fi}(\omega)|^2}{\hbar^2} \frac{4\sin^2\frac{1}{2}(\omega_{fi}-\omega)t}{(\omega_{fi}-\omega)^2} \rho(\omega)\, d\omega \tag{9.93}$$

where we have substituted for $|a_f^{(1)}|^2$ from Eq. 9.46 with the appropriate modifications, and introduced $\rho(\omega)d\omega$ as the number of different frequencies present in the perturbation within the range $d\omega$. The part of the perturbation which pertains to frequency ω has been denoted by $H'^0(\omega)\, e^{-i\omega t}$. Condition (iii) above states that $|H'^0_{fi}(\omega)|^2$ and $\rho(\omega)$ vary smoothly with ω. Hence they can be treated as constants within the small frequency interval of width $\sim 1/t$ around $\omega = \omega_{fi}$, within which alone the remaining factor in the integrand has an appreciable value. We can, therefore, pull them outside the integral, giving them their values at $\omega = \omega_{fi}$. The remaining integral is just the same as in Eq. 9.49, except for a trivial difference in the variable of integration, and its value is $2\pi t$. Hence, the transition probability *per unit time* obtained by dividing (9.93) by t, is found to be

$$\frac{2\pi}{\hbar^2} \left| H'^0_{fi}(\omega_{fi}) \right|^2 \rho(\omega_{fi}) \tag{9.94}$$

In one respect the considerations of the last paragraph need to be amplified. The perturbation associated with any given $\omega > 0$ should be taken as

$$H'(\omega) = H'^0(\omega)\, e^{-i\omega t} + H'^0(-\omega)\, e^{i\omega t}$$

with $H'^0(-\omega) = [H'^0(\omega)]^\dagger$ to ensure Hermiticity. Thus the condition (ii) is to be enforced only for ω's of one sign. Though the two terms of $H'(\omega)$ for given ω are completely correlated, their contributions to the transition probability can be separately computed, as explained earlier in this section. In order to take into account both these contributions, the sum over ω in (9.93) is to be understood as extending over both positive and negative ω, with

$$H'^0_{fi}(-\omega) = [H'^0_{if}(\omega)]^*, \quad \rho(\omega) = \rho(-\omega) \tag{9.95}$$

The latter condition is true because, along with any given $\omega > 0$, $-\omega$ also is present automatically. From these observations one can draw an important inference. If m and n are any two states (with $E_m > E_n$, say, so that $\omega_{mn} > 0$), the probabilities for the upward transition ($i = n, f = m$) and the downward transition ($i = m, f = n$) between this particular pair of states are, according to (9.94),

$$\frac{2\pi}{\hbar^2} |H'^0_{mn}(\omega_{mn})|^2 \rho(\omega_{mn}) \text{ and } \frac{2\pi}{\hbar^2} |H'^0_{nm}(-\omega_{mn})|^2 \rho(-\omega_{mn}) \tag{9.96}$$

respectively (since $\omega_{nm} = -\omega_{mn}$). But these two are equal, by virtue of Eq. 9.95. Therefore, we have the important result that probabilities for upward and downward transitions between a given pair of levels, induced by Hermitian perturbations of the type considered above, are identical.

We shall now apply the foregoing results to the perturbation of an atom by the field of electromagnetic radiation, which under most circumstances satisfies the conditions (i)–(iii).

9.13 Interaction of an Atom with Electromagnetic Radiation[13]

As the model of an atom, we use a (spinless) particle of charge e bound to a field of force in which its potential energy is $V(\mathbf{x})$. We have noted earlier that electromagnetic *radiation* can be characterized by a vector potential $\mathbf{A}$ which is transverse (div $\mathbf{A} = 0$), with the scalar potential $\phi = 0$. The Hamiltonian of the atom in the presence of the radiation is then given by

$$H = H_0 + H',$$
$$H_0 = -\frac{\hbar^2}{2m}\nabla^2 + V(\mathbf{x}); \; H' = \frac{ie\hbar}{mc}\mathbf{A}.\nabla \tag{9.97}$$

where a term proportional to $\mathbf{A}^2$ is neglected (see footnote, p. 247). For monochromatic light with the propagation vector $\mathbf{k}$ and angular frequency ω, one has a real simple harmonic wave, say

$$\mathbf{A}(\mathbf{x},t) = 2\mathbf{A}_0 \cos(\mathbf{k}.\mathbf{x} - \omega t + \alpha) \tag{9.98}$$
$$\mathbf{A}_0.\mathbf{k} = 0, \; \omega = ck \tag{9.99}$$

where $\mathbf{A}_0$ is a constant real vector and α is an arbitrary (real) phase constant. The first of Eqs. 9.99 ensures that div $\mathbf{A} = 0$. The electric and magnetic fields associated with this $\mathbf{A}$ are

$$\mathcal{E} = -\frac{1}{c}\frac{\partial \mathbf{A}}{\partial t} = -2k\mathbf{A}_0 \sin(\mathbf{k}.\mathbf{x} - \omega t + \alpha)$$
$$\mathcal{H} = \text{curl}\,\mathbf{A} = -2\mathbf{k} \times \mathbf{A}_0 \sin(\mathbf{k}.\mathbf{x} - \omega t + \alpha) \tag{9.100}$$

Introducing (9.98) in H' of (9.97), we express the perturbation as

$$H' = H'^0 e^{-i\omega t} + (H'^0)^\dagger e^{i\omega t},$$
$$H'^0 = \frac{ie\hbar}{mc} e^{i(\mathbf{k}.\mathbf{x}+\alpha)} \mathbf{A}_0.\nabla, \; (H'^0)^\dagger = \frac{-ie\hbar}{mc} e^{-i(\mathbf{k}.\mathbf{x}+\alpha)} \mathbf{A}_0.\nabla \tag{9.101}$$

From the results of the last section we can immediately conclude that the first term of H' gives rise to upward transitions ($E_f > E_i$) with a probability proportional to $|H'^0{}_{fi}|^2$ where

$$H'^0{}_{fi} = \frac{ie\hbar}{mc} e^{i\alpha} \int u_f{}^* \, e^{i\mathbf{k}.\mathbf{x}} \, \mathbf{A}_0.\nabla u_i \, d\tau \tag{9.102}$$

provided $\hbar\omega \approx E_f - E_i$. The second term generates downward transitions ($E_f < E_i$) with probability proportional to the absolute square of the matrix element

$$(H'^0)^\dagger{}_{fi} = \frac{ie\hbar}{mc} e^{-i\alpha} \int u_f{}^* \, e^{-i\mathbf{k}.\mathbf{x}} \, \mathbf{A}_0.\nabla u_i \, d\tau \tag{9.103}$$

if $\hbar\omega \approx E_i - E_f$. We shall assume that in making an upward transition the system *absorbs* a quantum $\hbar\omega$ of electromagnetic energy from the radiation field, and that in a downward transition it *emits* such a quantum. This is one of the Bohr postulates. To prove it from the point of view of quantum

[13] For a detailed account of radiation theory, see W. Heitler, *The Quantum Theory of Radiation*, 3rd ed., Oxford University Press, 1954.

mechanics, we need a more sophisticated view of the electromagnetic field than we have used up to now. It will be noticed that in our mathematical description of the electromagnetic field, the notion of a quantum has not entered at all. We lay this question aside for the moment and shall return to it later (Sec. 9.19). Meanwhile we shall freely use the concept of emission and absorption.

We are mostly interested in the situation where the electromagnetic radiation is not strictly monochromatic, but is an incoherent radiation with a spread of frequencies. (This is the case with light from ordinary sources.) Then the perturbation contains a range of (uncorrelated) frequency components, and the theory developed in the last section applies. The *transition probability per unit time* for an upward $(n \rightarrow m)$ or downward $(m \rightarrow n)$ transition between a given pair of levels m, n (with $E_m > E_n$) is given by Eq. 9.96. Since the matrix elements (9.102) and (9.103) depend on the direction $\hat{\mathbf{k}}$ of $\mathbf{k}$ and on the polarization vector $\mathbf{e} \equiv \mathbf{A}_0/|\mathbf{A}_0|$, we shall consider waves with $\hat{\mathbf{k}}$ lying within $d\Omega$ and given polarization $\mathbf{e}$. In denoting the number of such waves within the angular frequency range $\Delta\omega$ by $\rho(\omega)\,\Delta\omega$, it will be understood that $\rho(\omega)$ is proportional to $d\Omega$. The number $\rho(\omega)\,\Delta\omega$ can be expressed in terms of the energy $\varepsilon(\omega)\,\Delta\omega$ per unit volume of the box, contributed by waves characterized by the same $\Delta\omega$, $d\Omega$, $\mathbf{e}$ and $\hat{\mathbf{k}}$ as before (note that $\varepsilon(\omega) \propto d\Omega$). Evidently $\varepsilon(\omega)$ is $\rho(\omega)$ times the energy density contributed by each wave.[14] The contribution of the wave (9.98) to the instantaneous energy density is $(1/8\pi)(\mathcal{E}^2 + \mathcal{H}^2) = (k^2\mathbf{A}_0^2/\pi)\sin^2(\mathbf{k.x} - \omega t + \alpha)$, where we have used the fact that $\mathbf{k}$ is perpendicular to $\mathbf{A}_0$. Its value averaged over one period is $(k^2\mathbf{A}_0^2/2\pi) = (\omega^2\mathbf{A}_0^2/2\pi c^2)$. Hence $\varepsilon(\omega) = (\omega^2\mathbf{A}_0^2/2\pi c^2)\,\rho(\omega)$, or

$$\rho(\omega) = \frac{2\pi c^2}{\omega^2\mathbf{A}_0^2}\,\varepsilon(\omega) \tag{9.104}$$

On substituting this (with $\omega = \omega_{mn}$) into Eq. 9.94, we obtain the probability per unit time for transitions (induced by waves with $\hat{\mathbf{k}}$ within $d\Omega$ and given $\mathbf{e}$) between the pair of levels, $m\ n$, as

$$\left[\frac{4\pi^2c^2}{\hbar^2\omega^2\mathbf{A}_0^2}\,\varepsilon(\omega)\,|H'^0_{mn}|^2\right]_{\omega=\omega_{mn}} \tag{9.105}$$

With the introduction of the matrix element from Eq. 9.103 we obtain the following expression for the probability per unit time for the downward transition $m \rightarrow n$:

$$P_{nm} = \left[\frac{4\pi^2e^2}{m^2\omega^2}\,\varepsilon(\omega)\,|\mathbf{e}.\mathbf{I}_{nm}|^2\right]_{\omega=\omega_{mn}} \tag{9.106}$$

$$\mathbf{I}_{nm} = \int u_n{}^* e^{-i\mathbf{k.x}}\,\nabla u_m\,d\tau \tag{9.107}$$

It may be recalled from the theory of the last section that the probability P_{mn} for the opposite transition $n \rightarrow m$ is also the same.

We observe that the probability P_{nm} per unit time for emission/absorption (according as the system is in the upper/lower level initially) is proportional to

[14] Since the radiation is assumed incoherent, the energy densities due to the different waves simply add up.

the energy density $\varepsilon(\omega)$ in the radiation field at $\omega = \omega_{mn}$. If this result were strictly valid, emission could only be *induced* by radiation already present [at a rate proportional to $\varepsilon(\omega)$], but could not take place *spontaneously* in the absence of any radiation. Actually, spontaneous emission *does* take place. The reason why our theory has failed to account for spontaneous emission is that we have ignored the quantum nature of the electromagnetic field itself. We shall see in Sec. 9.20 that on treating the radiation field as a quantum field, one automatically gets a nonvanishing probability for spontaneous emission besides recovering the formula (9.106) for induced emission.

9.14 The Dipole Approximation: Selection Rules

(*a*) *The Dipole Approximation:* In processes involving emission or absorption of radiation it is most often the case that the wavelength of the radiation is much larger than the size a of the atom, molecule, etc., with which it interacts, i.e. $ka \ll 1$. (For example, $k = 2\pi/\lambda \sim 10^5$ cm^{-1} in the optical region of electromagnetic radiation, while the size of atoms emitting it is $\sim 10^{-8}$ cm giving $ka \sim 10^{-3}$). Then $e^{-i\mathbf{k}.\mathbf{x}} \approx 1$, and $\mathbf{I}_{nm}$ simplifies to

$$\mathbf{I}_{nm} \approx \int u_n^* \nabla \, u_m d\tau \tag{9.108}$$

Now, substituting

$$\nabla = \frac{i}{\hbar}\,\mathbf{p} = \frac{m}{\hbar^2}\,[\mathbf{x}, H_0] \tag{9.109}$$

and using the Hermiticity of H_0 as well as the fact that $H\,u_m = E_m u_m$ etc., we can reduce (9.108) to

$$\mathbf{I}_{nm} = \frac{m}{\hbar^2}\,(E_m - E_n) \int u_n^* \mathbf{x}\, u_m\, d\tau \tag{9.110}$$

This form brings out the fact that in the long wave length approximation, it is the matrix element of the electric dipole operator $e\mathbf{x}$ which determines the transition probabilities and selection rules. The approximation $e^{i\mathbf{k}\mathbf{x}} \approx 1$ is, therefore, called the *dipole approximation*. With the use of Eq. 9.110 in Eq. 9.106, P_{nm} reduces to

$$P_{nm} = \left[\frac{4\pi^2 e^2}{\hbar^2}\,\varepsilon(\omega)\,|\,\mathbf{e}.\mathbf{x}_{nm}\,|^2\right]_{\omega=\omega_{mn}} \tag{9.111}$$

Observe that this quantity does not involve the direction of $\mathbf{k}$. So it is easy to calculate the transition rate due to unpolarized radiation with all directions of propagation (e.g. black body radiation within an isothermal cavity). We merely have to average P_{nm} over all directions of $\mathbf{e}$, i.e. if e_1, e_2, e_3 are the components of $\mathbf{e}$, replace the products $e_i\, e_j$ in

$$\mathbf{e}.|\,\mathbf{x}_{nm}\,|^2 \equiv \sum_{ij} e_i\, e_j\, (\mathbf{x}_{nm})_i\, (\mathbf{x}_{nm}^*)_j \tag{9.112}$$

by their averages. It is evident that the averaging should give $\overline{e_1^2} = \overline{e_2^2} = \overline{e_3^2} = \frac{1}{3}\,(\overline{e_1^2} + \overline{e_2^2} + \overline{e_3^2}) = \frac{1}{3}$, while $\overline{e_1 e_2}$ etc. vanish. Thus $\overline{e_i e_j} = \frac{1}{3}\delta_{ij}$, and the averaged value of $|\,\mathbf{e}.\mathbf{x}_{nm}\,|^2$ becomes $\frac{1}{3}\mathbf{x}_{nm}.\mathbf{x}_{nm}^*$. Using this average value of Eq. 9.111 we obtain $\overline{P}_{nm}$ for the case of unpolarized ambient radiation to be

$$\bar{P}_{nm} = \left[\frac{4\pi^2 e^2}{3\hbar^2}\, \varepsilon(\omega)\, \mathbf{x}_{nm}.\mathbf{x}_{nm}^*\right]_{\omega=\omega_{mn}} \tag{9.113}$$

(*b*) *Selection Rules:* It is evident from Eq. 9.111 that dipole transitions between states f, i are possible only if $\mathbf{e}.\mathbf{x}_{fi} \neq 0$. If, as is almost always the case, the states under consideration have definite angular momenta, this condition gives selection rules governing the change of angular momentum. It may be proved from the properties of spherical harmonics that $\int (Y_{l_f m_f})^* \, \mathbf{x} Y_{l_i m_i} d\Omega$ can be nonvanishing only when $l_f - l_i = \pm 1$. One can infer the same result more generally by observing that $\mathbf{x}$ is a vector operator, i.e. a spherical tensor of rank 1. Hence, its matrix elements between the angular momentum eigenstates are proportional to the C-G coefficient $\langle\, l_i m_i;\, 1m \mid l_f m_f \,\rangle$, by the Wigner-Eckart Theorem (Sec. 8.11). Hence it vanishes if $|\Delta l| \equiv |l_i - l_f|$ exceeds 1. Further, the matrix elements for $\Delta l = 0$ vanish because $\mathbf{x}$ is of odd parity. We are thus left with $\Delta l = \pm 1$. It is also clear from the properties of the C-G coefficient that $\Delta m \equiv m_f - m_i$ is limited to $\Delta m = \pm 1, 0$. The polarization vector $\mathbf{e}$ determines which of these can occur. For example, if $\mathbf{e}$ is along the z-direction, $\mathbf{e}.\mathbf{x}_{fi} = z_{fi}$. Since $z = r\cos\theta$ is independent of φ, this corresponds to $m = 0$ in the C-G coefficient, and only $\Delta m = 0$ is possible in this case. Transitions which respect the dipole selection rules are called *allowed transitions*.[15]

Forbidden transitions, which are forbidden by the selection rules of the dipole approximation, may occur, but with greatly reduced probability. These arise from the higher order terms in the expansion of $e^{i\mathbf{k}.\mathbf{x}}$ as $\Sigma\, (i\mathbf{k}.\mathbf{x})^n/n!$ or better, as $\Sigma\, (2l+1)\, i^l j_l(kr)\, P_l(\cos\theta)$, where θ is the angle between $\mathbf{x}$ and $\mathbf{k}$. In either case, the nth term of the series is of order $(ka)^{n-1}$, since only the region $r \leqslant a$ (where the wave functions do not vanish) is effective in the integral, and for $ka \ll 1$, $j_l(ka) \propto (ka)^l$. Therefore, each term is smaller than the preceding one by a factor of order ka (typically 10^{-3}) and the corresponding probability is smaller by $\sim (ka)^2$. Evidently, the forbidden lines arising from even the $l = 1$ term in the above expansion is very much weaker than those from allowed transitions. Incidentally, the terms with $l = 1, 2, \ldots$ lead to contributions from electric quadrupole, octupole, ... moments to the transitions.

Finally, we observe that certain transitions are *strictly forbidden* in the sense that the exact (first order) transition matrix element vanishes; $\mathbf{e}.\mathbf{I}_{fi}$ in Eq. 9.106 is zero. This is the case if both i and f are s-states. To see this, take the z-axis along the direction of $\mathbf{e}$. Then $\mathbf{e}.\nabla = \partial/\partial z$, and $\mathbf{e}.\mathbf{I}_{fi} = \int u_f^*\, e^{i\mathbf{k}.\mathbf{x}}\, (\partial u_i/\partial z)\, d\tau$. Further, since $\mathbf{k}$ is perpendicular to $\mathbf{e}$, $\mathbf{k}.\mathbf{x}$ reduces to $(k_x x + k_y y)$.

Therefore, the z-part of the integral is just $\int u_f^*\, (\partial u_i/\partial z)\, dz$, which is zero because the integrand is of odd parity if u_f, u_i are both s-states. It may be noted that even these 'strictly forbidden' transitions can take place in higher orders of perturbation theory.

[15] Selection rules for allowed transitions between states (8.85) of spin-$\frac{1}{2}$ particles are $\Delta l = \pm 1$, $\Delta j = 0, \pm 1$ ($j = 0 \to j = 0$ forbidden) $\Delta m = 0, \pm 1$. (Problem 15, Chapter VIII).

9.15 The Einstein Coefficients: Spontaneous Emission

There is a simple argument, due to Einstein,[16] by which the probability for induced emission (or absorption) can be related to that for spontaneous emission. It was put forward before the advent of quantum mechanics, and uses only the general concept of quantized atomic levels. The quantum properties of the electromagnetic field enter the argument indirectly through the use of the Planck distribution law for black body radiation. The basic observation is that the 'black body' radiation present in an isothermal cavity is in dynamic equilibrium with the walls of the cavity: absorption of radiation from the cavity by 'atoms' or other 'oscillators' constituting the walls is going on continually, while on the other hand, emission of radiation into the cavity is also continually taking place. Since the energy density $\varepsilon(\nu)$ of radiation at any frequency ν remains steady, it follows that within every interval $\triangle\nu$, absorption and emission must balance each other. This balancing depends on two types of factors. One is the probability *per atom* per unit time for an upward transition absorbing radiation of energy $h\nu$, or for a downward transition emitting radiation $h\nu$. The latter consists of two parts, that for spontaneous emission, denoted by A, and that for induced emission which is believed proportional to $\varepsilon(\nu)$ and is, therefore, written as $B\varepsilon(\nu)$. For absorption, one has $B'\varepsilon(\nu)$. Focussing our attention on a particular pair of states m, n such that $E_m - E_n = h\nu$, we observe that the rate of emission of radiation into the cavity by transitions $m \rightarrow n$ is equal to the number of atoms N_m whose state is m, times the probability per atom for a downward transition. Similarly, the rate of absorption of radiation is $N_n B'\varepsilon(\nu)$. It is these rates which have to balance:

$$N_m\,[A + B\varepsilon(\nu)] = N_n B'\varepsilon(\nu) \tag{9.114}$$

Now, we know from statistical mechanics that if an ensemble of systems (atoms, oscillators, ...) is in thermal equilibrium at temperature T, the average number of systems occupying some state of energy E is proportional to the Boltzmann factor $e^{-\beta E}$ where $\beta = (1/kT)$, k being the Boltzmann's constant. Thus, in our case $(N_m/N_n) = e^{-\beta E_m}/e^{-\beta E_n}$, and with $E_m - E_n = h\nu$, we have $N_n = N_m\, e^{\beta h\nu}$. Introducing this in Eq. 9.114 and rearranging terms, we can write it as

$$\varepsilon(\nu) = \frac{(A/B')}{(B/B')\, e^{\beta h\nu} - 1} \tag{9.115}$$

Thus the energy distribution within the radiation in the cavity is expressed in terms of the *Einstein coefficients* A, B, B'. But this distribution must coincide with the Planck distribution (1.3). Hence we obtain the following relations among the coefficients:

$$B' = B; \quad A = \frac{8\pi h\nu^3}{c^3}\, B \tag{9.116}$$

Our quantum mechanical calculations in the last section did produce equal values for the probabilities for absorption and induced emission, as demanded by the first of the above equations. From the definitions of B and B',

[16] A. Einstein, *Phys Z.*, **18**, 121 (1917).

these are to be identified with[17] $[P_{nm}/\varepsilon(\nu)] \equiv [P_{nm}/2\pi\varepsilon(\omega)]$, averaged over all directions and polarizations of the radiation, in view of the nature of radiation in a cavity. The second of Eqs. 9.116 tells us now that the probability per unit time for spontaneous emission $P_{nm}^{(\text{spont.})}$ (which is A itself) should be related to P_{nm} through

$$P_{nm}^{(\text{spont.})} = \frac{\hbar\omega^3}{\pi^2 c^3} \frac{P_{nm}}{\varepsilon(\omega_{mn})} = \left[\frac{4e^2\omega^3}{3\hbar c^3} \mathbf{x}_{nm}\cdot\mathbf{x}_{nm}^*\right]_{\omega=\omega_{mn}} \tag{9.117}$$

The last expression is obtained from the averaged value (9.113) for P_{nm} in the dipole approximation.

C. ALTERNATIVE PICTURES OF TIME EVOLUTION

9.16 The Schrödinger Picture: Transformation to other Pictures

We have so far based our treatment of the time evolution of quantum systems on Schrödinger's approach: the basic dynamical variables $\mathbf{x}$, $\mathbf{p}$, spin, etc., are placed in correspondence with operators which are time-independent, while the wave function ψ changes with time in accordance with the Schrödinger equation. For the purpose of this description of temporal development, it does not matter whether we use coordinate space language, or the momentum (or any other) representation. To emphasize this we may well use the abstract vector space language and express what is called the *Schrödinger picture* of the time evolution through the equations

$$i\hbar \frac{\partial}{\partial t} |\psi\rangle = H |\psi\rangle \tag{9.118}$$

$$\frac{d}{dt} A = \frac{\partial A}{\partial t}, \text{ where } A(t) \equiv A(\mathbf{x}, \mathbf{p}; t) \tag{9.119}$$

Eq. 9.119 states simply that linear operators A corresponding to any functions of the basic dynamical variables $\mathbf{x}$, $\mathbf{p}$ (and possibly spin, etc., which we do not show explicitly) can depend on time only through explicit appearance of the time parameter in the functional form.

Now, neither the operators nor the state vectors (or their representatives, the wave functions) can themselves be measured; all results of measurements are expressible in terms of expectation values or matrix elements, $\langle \phi | A |\psi\rangle$, of linear operators. Such quantities are quite unaffected by simultaneous transformations

$$|\psi\rangle \rightarrow U|\psi\rangle, \; A \rightarrow UAU^\dagger \tag{9.120}$$

on states and operators, even if U is time-dependent, provided U is a unitary operator: $UU^\dagger = U^\dagger U = 1$ (Remember that if $|\psi\rangle \rightarrow U|\psi\rangle$, $\langle\psi| \rightarrow \langle\psi| U^\dagger$). However, when U is chosen to be time-dependent, the equations of motion of the transformed states and operators no longer have the forms (9.118) and (9.119). Thus the *picture* of time evolution is altered,

[17] The energy densities expressed in terms of ν or ω are related by $\varepsilon(\nu)\,d\nu = \varepsilon(\omega)d\omega = \varepsilon(\omega)2\pi d\nu$.

though there is no change in the physical content expressed through matrix elements.

There are two special choices of U which lead to interesting new pictures. Their construction is related to the self-evident observation that the solution of Eq. 9.118 can be formally written as

$$|\psi(t)\rangle = V(t, t_0)\,|\psi(t_0)\rangle \tag{9.121a}$$

$$V(t, t_0) = e^{-iH(t-t_0)/\hbar} \tag{9.121b}$$

provided H is independent of time. Clearly,

$$i\hbar\frac{\partial V(t, t_0)}{\partial t} = HV(t, t_0) \text{ and } \lim_{t\to t_0} V(t, t_0) = 1 \tag{9.122}$$

and

$$V^{-1}(t, t_0) = V(t_0, t), \quad V(t, t_1)\,V(t_1, t_0) = V(t, t_0) \tag{9.123}$$

If H is time-dependent, Eq. 9.121a would still remain valid if V were defined through Eqs. 9.122, though V would no longer have the explicit form (9.121b). Since V translates the state at time t_0 into that at time t, it is called the *operator of (finite) time translation* or the *time evolution operator*.

It is interesting to observe that the propagator G defined in Sec. 9.2 is nothing but the coordinate space representation of the time evolution operator. This follows from the fact that if we take the $\langle \mathbf{x} | \ldots | \mathbf{x}' \rangle$ matrix element of Eq. 9.122, it becomes $i\hbar(\partial/\partial t)\langle \mathbf{x} | V | \mathbf{x}' \rangle = H(\mathbf{x}, \mathbf{p})\langle \mathbf{x} | V | \mathbf{x}' \rangle$ with the initial condition $\langle \mathbf{x} | V(t_0, t_0) | \mathbf{x}' \rangle = \langle \mathbf{x} | 1 | \mathbf{x}' \rangle = \delta(\mathbf{x} - \mathbf{x}')$. Since the form of the equation and the initial condition coincide with those in Eqs. 9.8, it follows that the solutions must also coincide, i.e.

$$\langle \mathbf{x} | V(t, t_0) | \mathbf{x}' \rangle = G(\mathbf{x}, \mathbf{x}'; t, t_0) \tag{9.124}$$

When H is time-independent, one can verify explicitly from Eq. 9.121b that $\langle \mathbf{x} | V | \mathbf{x}' \rangle$ reduces to the expression (9.7) for G. If the eigenstates of H are denoted by $|n\rangle$, $H|n\rangle = E_n|n\rangle$, we have

$$\begin{aligned}\langle \mathbf{x} | V(t, t_0) | \mathbf{x}' \rangle &= \sum_{n,n'} \langle \mathbf{x} | n \rangle \langle n | e^{-iH(t-t_0)/\hbar} | n' \rangle \langle n' | \mathbf{x}' \rangle \\ &= \sum_{n} \langle \mathbf{x} | n \rangle\, e^{-iE_n(t-t_0)/\hbar} \langle n | \mathbf{x}' \rangle \end{aligned} \tag{9.125}$$

This coincides with (9.7) as stated, since $\langle \mathbf{x} | n \rangle = u_n(\mathbf{x})$ and $\langle n | \mathbf{x}' \rangle = u_n^*(\mathbf{x}')$.

9.17 The Heisenberg Picture

States $|\psi(t)\rangle_H$ and operators $A_H(t)$ in the Heisenberg picture are obtained from $|\psi(t)\rangle$ and $A(t)$ of the Schrödinger picture by choosing U in Eqs. 9.120 to be $V^{-1}(t, t_0)$:

$$|\psi(t)\rangle_H = V^{-1}(t, t_0)\,|\psi(t)\rangle, \quad A_H(t) = V^{-1}(t, t_0)\,A(t)\,V(t, t_0) \tag{9.126}$$

We then have, by virtue of Eq. 9.121a, $|\psi(t)\rangle_H = |\psi(t_0)\rangle$, which is independent of t. Thus

$$i\hbar\frac{\partial}{\partial t}|\psi(t)\rangle_H = 0 \tag{9.127}$$

Further, differentiating the equality $V(t, t_0)A_H(t) = AV(t, t_0)$ and using Eq. 9.122, we get

$$V.i\hbar \frac{d}{dt} A_H(t) + HVA_H(t) = i\hbar \frac{\partial A}{\partial t} V + A.HV$$

Multiplying by V^{-1} on the left hand side and introducing $1 = VV^{-1}$ where necessary, we obtain

$$i\hbar \frac{d}{dt} A_H(t) = [A_H, H_H] + i\hbar \left(\frac{\partial A}{\partial t}\right)_H \tag{9.128}$$

According to Eqs. 9.127 and 9.128, the state vectors $|\psi\rangle_H$ remain independent of time and the time evolution is manifested entirely through variation of the *linear operators* $A_H(t)$ representing the dynamical variables. Eq. 9.128 and the commutation rule $[x_H(t), p_H(t)] = i\hbar$, between canonically conjugate position-momentum pairs, were the central features of Heisenberg's version of quantum mechanics. Hence this picture of the time evolution is named after him.

It may be observed that the concept of dynamical variables as time-dependent quantities is something which the Heisenberg picture has in common with classical mechanics. In fact, the Heisenberg equation of motion (9.128) closely resembles the classical Hamiltonian equations of motion written in terms of Poisson brackets:

$$\frac{dA}{dt} = \{A, H\} + \frac{\partial A}{\partial t} \tag{9.129}$$

where $\{A, B\}$ denotes the Poisson Bracket of classical dynamical variables A, B (see Appendix A). Comparing Eqs. 9.129 and 9.128, we see that the classical equation of motion goes over into the quantum mechanical (Heisenberg) equation if dynamical variables are made operators, and the replacement

$$\{A, B\} \rightarrow \frac{1}{i\hbar} [A, B] \tag{9.130}$$

is made. The same replacement also gives the *quantum condition* correctly. Classically, we have $\{q_i, p_j\} = \delta_{ij}$. On making the above replacement we obtain

$$[q_i(t), p_j(t)] = i\hbar\, \delta_{ij} \tag{9.131}$$

which is the correct commutation rule for the qs and ps in quantum mechanics. Since the Heisenberg qs and ps vary with time, the commutation rule is required to be true at all times t; we have indicated this explicitly in Eq. 9.131.

9.18 Matrix Mechanics — The Simple Harmonic Oscillator

In formulating quantum mechanics through the quantum condition and the equation of motion (9.128), Heisenberg assumed also that the operators representing dynamical variables are matrix operators. Therefore this formulation is known as *matrix mechanics*. It is instructive to see how the idea of matrices comes as a natural extension of Bohr's postulates regarding the structure of atoms and their interaction with radiation.

It will be recalled that according to Bohr, emission or absorption of radiation is associated with pairs of atomic levels, the frequency of the radiation being related to the energy difference between the levels: $\hbar\omega_{mn} = E_m - E_n$.

This suggests a picture of the atom as a collection of emitters/absorbers, one for each *pair* of energy levels (k, n) with 'coordinate' x_{kn} varying simple harmonically: $x_{kn}(t) \propto e^{i\omega_{kn}t}$. The set of all these quantities forms a coordinate *matrix* $x(t)$, to be associated with the atom as a whole, and in the same way one is led to a momentum matrix $p(t)$. Products of such matrices, formed through the standard matrix multiplication rule, do have the correct time-dependence for their (k, n) elements: for example

$$(x(t), p(t))_{kn} = \Sigma\, x_{kl}(t)\, p_{ln}(t) = \Sigma\, x_{kl}(0)\, p_{ln}(0)\, e^{i(\omega_{kl}+\omega_{ln})t}$$
$$= e^{i\omega_{kn}t}\, \Sigma\, x_{kl}(0)\, p_{ln}(0) = (x(0), p(0))_{kn}\, e^{i\omega_{kn}t}.$$

The operator for the energy [the Hamiltonian, constructed from $x(t)$ and $p(t)$], is to be rather special in that the energy is associated with individual levels (rather than pairs of levels), and it is to be time-independent. Both these requirements are ensured if $H(x, p)$ is a diagonal matrix, $H_{kn} = 0$ for $k \neq n$; then the diagonal values H_{nn}, which are time-independent (since $\omega_{nn} = 0$) can be identified as the energies of the levels n.

Let us see how these ideas, combined with the quantum condition

$$[x(t),\ p(t)] = i\hbar \tag{9.132}$$

and the Heisenberg equation of motion leads to the energy spectrum of a simple system, the harmonic oscillator. Observe first of all that with[18]

$$H = \frac{p^2}{2m} + \tfrac{1}{2} m\omega_c^2\, x^2 \tag{9.133}$$

one finds from Eq. 9.128, with the aid of Eq. 9.132, that

$$\dot{x} = (i\hbar)^{-1}\, [x,H] = (p/m) \tag{9.134}$$

$$\dot{p} = (i\hbar)^{-1}\, [p,H] = -\, m\omega_c^2 x \tag{9.135}$$

Combining the two equations, one gets

$$\ddot{x} = (\dot{p}/m) = -\,\omega_c^2 x \tag{9.136}$$

and so, for each element $x_{kn}(t)$, one must have $\ddot{x}_{kn} = -\,\omega_c^2 x_{kn}$. Since x_{kn} is required to have the time dependence $x_{kn}(t) = x_{kn}(0)\ e^{i\omega_{kn}t}$, we are led to

$$x_{kn}(0)\ (\omega_{kn}^2 - \omega_c^2) = 0 \tag{9.137}$$

Thus the only nonvanishing x_{kn} are those for which $\omega_{kn} = \pm\,\omega_c$. Expressed differently, given any k, there are only two states n, with $E_k - E_n = \pm\,\hbar\omega_c$ for which $x_{kn} \neq 0$. In case n is the ground state, which we label by $n = 0$, only $x_{1,0}$ is nonzero, where 1 stands for the first excited state whose energy is necessarily $E_0 + \hbar\omega_{1,0} = E_0 + \hbar\omega_c$. Starting from these, the above result leads naturally to a series of states labelled by $n = 0, 1, 2, \ldots$ with energies E_0, $E_0 + \hbar\omega_c$, $E_0 + 2\hbar\omega_c$, ... such that

$$x_{kn} = 0 \text{ if } k - n \neq \pm\, 1 \tag{9.138}$$

An identical statement holds for p_{kn} too, since in view of Eq. 9.134

$$p_{kn} = m\dot{x}_{kn} = mi\omega_{kn}\, x_{kn} \tag{9.139}$$

Let us use Eqs. 9.138 and 9.139 to evaluate the diagonal (n, n) elements of the quantum condition (9.132). We get

[18] It should be kept in mind that x, p are matrices varying with time, though we have suppressed the argument t for brevity.

$$i\hbar = (xp)_{nn} - (px)_{nn}$$
$$= (x_{n,n+1}\, p_{n+1,n} + x_{n,n-1}\, p_{n-1,n}) - (p_{n,n+1}\, x_{n+1,n} + p_{n,n-1}\, x_{n-1,n})$$
$$= 2im\omega_c\, (x_{n,n+1}\, x_{n+1,n} - x_{n,n-1}\, x_{n-1,n}) \tag{9.140}$$

except for $n = 0$ when $(n - 1)$ does not exist and one has

$$i\hbar = 2im\omega_c . x_{0,1}\, x_{1,0} \tag{9.141}$$

Eq. 9.140 is evidently a recurrence relation

$$x_{n,n+1}\, x_{n+1,n} = \frac{\hbar}{2m\omega_c} + x_{n-1,n}\, x_{n,n-1} \tag{9.142}$$

which, when iterated, yields

$$x_{n,n+1}\, x_{n+1,n} = \frac{(n + 1)\hbar}{2m\omega_c} \tag{9.143}$$

Then Eq. 9.139 leads to

$$p_{n,n+1}\, p_{n+1,n} = (- mi\omega_c)\, x_{n,n+1} \cdot (mi\omega_c)\, x_{n+1,n}$$
$$= (m\omega_c)^2 \cdot \frac{(n + 1)\hbar}{2m\omega_c} \tag{9.144}$$

The matrix for H is now readily evaluated. One observes that

$$(x^2)_{nn} = (x_{n,n+1}\, x_{n+1,n} + x_{n,n-1}\, x_{n-1,n})$$
$$= \frac{(2n + 1)\hbar}{2m\omega_c} \tag{9.145}$$

where Eq. 9.143 has been used in the last step. Similarly $(p^2)_{nn} = (m\omega_c)^2$ $(2n + 1)\hbar/(2m\omega_c)$. Hence

$$H_{nn} = \frac{1}{2m}\, (p^2)_{nn} + \tfrac{1}{2} m\omega_c^2 (x^2)_{nn} = (n + \tfrac{1}{2})\hbar\omega_c \tag{9.146}$$

Thus we have recovered the energy levels of the quantum harmonic oscillator from Heisenberg's matrix mechanics. The student may verify that $H_{kn} = 0$ if $k \neq n$. It may be noted that the matrix elements of x given in Eq. 4.24 are consistent with (9.143).

9.19 Electromagnetic Wave as Harmonic Oscillator; Quantization: Photons

While matrix mechanics *per se* is primarily of historic interest, the Heisenberg picture of time evolution, expressed through time dependence of operators, is of great utility, especially in the quantum theory of fields. It is through the quantization of the electromagnetic field that the concept of a photon as something which carries a discrete amount of electromagnetic energy (and momentum) is given substance. To accomplish this, what one does, basically, is to recognize that an electromagnetic wave with a given wave vector $\mathbf{k}$ and definite polarization is equivalent to a simple harmonic oscillator and can be quantized as such. When this is done, the possible energy and momentum values of the field are found to be discrete and uniformly spaced, with spacings $\hbar\omega \equiv \hbar ck$ and $\hbar\mathbf{k}$ respectively. So the excitations of the field become equivalent to having a number of photons, each with energy $\hbar\omega$ and momentum $\hbar k$. We shall now present the main aspects of this procedure.

(*a*) *Classical Electromagnetic Wave: Equivalence to Oscillators:* We start with a classical electromagnetic wave described by the vector potential

$$\mathbf{A}(\mathbf{x}, t) = L^{-3/2} [\mathbf{A}(\mathbf{k}, t)\, e^{i\mathbf{k}.\mathbf{x}} + \mathbf{A}^*(\mathbf{k}, t)\, e^{-i\mathbf{k}.\mathbf{x}}] \tag{9.147}$$

Both the terms in this expression have to go together since the electromagnetic fields (and hence **A**) are *real* fields. It is convenient to think of the field as being confined within a box of volume L^3, with periodic boundary conditions. The factor $L^{-3/2}$ serves to normalize the exponential functions within this box.

We have already noted that for free electromagnetic waves (i.e. when charges and currents are absent), the gauge can be so chosen that the scalar potential ϕ vanishes and **A** is solenoidal:

$$\text{div}\,\mathbf{A} = 0 \Rightarrow \mathbf{k}.\mathbf{A}(\mathbf{k}, t) = \mathbf{k}.\mathbf{A}^*(\mathbf{k}, t) = 0 \tag{9.148}$$

The vector **A** is thus transverse to **k**, and can have only two independent components, which correspond to two independent polarizations of the wave. Let us take the case of a wave which is linearly polarized, with **A** in the direction of a (real) unit vector **e**: $\mathbf{A}(\mathbf{x}, t) = \mathbf{e}A(\mathbf{x},t)$. Then

$$\mathbf{A}(\mathbf{k}; t) = \mathbf{e}\, A(\mathbf{k}, t);\ \mathbf{A}^*(\mathbf{k}, t) = \mathbf{e}A^*(\mathbf{k}, t);\ \mathbf{e}.\mathbf{k} = 0 \tag{9.149}$$

$$\mathbf{A}(\mathbf{x}, t) = L^{-3/2}\, \mathbf{e}\, [A(\mathbf{k}, t)\, e^{i\mathbf{k}.\mathbf{x}} + A^*(\mathbf{k}, t)\, e^{-i\mathbf{k}.\mathbf{x}}] \tag{9.150}$$

The electric and magnetic fields associated with this vector potential are[19]

$$\mathcal{E} = -\frac{1}{c}\frac{\partial \mathbf{A}}{\partial t} = -\frac{\mathbf{e}}{cL^{3/2}} [\dot{A}(\mathbf{k})\, e^{i\mathbf{k}.\mathbf{x}} + \dot{A}^*(\mathbf{k})\, e^{-i\mathbf{k}.\mathbf{x}}] \tag{9.151}$$

$$\mathcal{H} = \text{curl}\,\mathbf{A} = \frac{i\mathbf{k} \times \mathbf{e}}{L^{3/2}} [A(\mathbf{k})\, e^{i\mathbf{k}.\mathbf{x}} - A^*(\mathbf{k})\, e^{-i\mathbf{k}.\mathbf{x}}] \tag{9.152}$$

The total energy of this field within the box is given by $H = (8\pi)^{-1} \int (\mathcal{E}^2 + \mathcal{H}^2)\, d\tau$, where the integration is over the volume of the box. This integral is easily evaluated on noting that $\int e^{2i\mathbf{k}.\mathbf{x}}\, d\tau = \int e^{-2i\mathbf{k}.\mathbf{x}}\, d\tau = 0$, on account of the periodic boundary conditions, and that $(\mathbf{k} \times \mathbf{e})^2 = k^2$ since **k** is orthogonal to **e**. We then obtain

$$\begin{aligned} H &= (8\pi)^{-1} \int (\mathcal{E}^2 + \mathcal{H}^2)\, d\tau \\ &= \frac{1}{4\pi}\left[\frac{1}{c^2}\dot{A}(\mathbf{k})\,\dot{A}^*(\mathbf{k}) + k^2 A(\mathbf{k})\, A^*(\mathbf{k})\right] \end{aligned} \tag{9.153}$$

By separating the real and imaginary parts of $A(\mathbf{k})$ and writing

$$A(\mathbf{k}) = (2\pi c^2)^{1/2} [Q_1(\mathbf{k}) + iQ_2(\mathbf{k})],\quad A^*(\mathbf{k}) = (2\pi c^2)^{1/2} [Q_1(\mathbf{k}) - iQ_2(\mathbf{k})] \tag{9.154}$$

we reexpress H as

$$H = \tfrac{1}{2} \sum_{j=1}^{2} \{[\dot{Q}_j(\mathbf{k})]^2 + \omega^2 [Q_j(\mathbf{k})]^2\},\quad \omega = ck \tag{9.155}$$

This expression has just the same form as the sum of the energies of two harmonic oscillators[20] of unit mass, whose coordinate variables are Q_1, Q_2. For such oscillators, the momentum variables would, of course, be

$$P_j = \dot{Q}_j \tag{9.156}$$

(*b*) ***Quantization of Field Oscillators: Time Evolution:*** So far we have been dealing with the *classical* electromagnetic field. To quantize the field, we

[19] For brevity, we suppress the time variable, but it is to be remembered that $A(\mathbf{k})$, $A^*(\mathbf{k})$ etc., are time-dependent.

[20] The appearance of *two* oscillators, despite our confining attention to a single polarization, has to do with the fact that $\mathbf{A}(\mathbf{x}, t)$ contains two waves, one propagating in the direction of **k** and the other opposite to **k**. The 'oscillators' are just the normal modes of the field, wherein the field quantities at all points of space vary harmonically in unison.

simply convert the oscillators into quantum oscillators by requiring that Q_j and P_j be operators satisfying the quantum conditions:

$$[Q_i(t), P_j(t)] = i\hbar\,\delta_{ij} \qquad (i, j = 1, 2)$$
$$[Q_i(t), Q_j(t)] = [P_i(t), P_j(t)] = 0 \tag{9.157}$$

It follows immediately that each of the oscillators can have only the discrete set of energies $(n + \frac{1}{2})\hbar\omega$. In other words, the field energy in the box (associated with each mode $j = 1, 2$ of the field) can only take the values $(n + \frac{1}{2})\hbar\omega$, $n = 0, 1, 2, \ldots$ The fact that the energy can differ from the minimum (zero point) energy only in multiples of the quantum $\hbar\omega$ invites the interpretation that these quanta are photons. We shall return to this point later, after evaluating also the momentum of the field.

Let us now define, for the oscillators $j = 1, 2$, lowering and raising operators a_j and $a_j\dagger$ following the prescription of Eqs. 4.27:

$$a_j = \left(\frac{\omega}{2\hbar}\right)^{1/2} Q_j + i\left(\frac{1}{2\hbar\omega}\right)^{1/2} P_j, \quad a_j\dagger = \left(\frac{\omega}{2\hbar}\right)^{1/2} Q_j - i\left(\frac{1}{2\hbar\omega}\right)^{1/2} P_j \tag{9.158}$$

When expressed in terms of these quantities, the commutation relations (9.157) and the Hamiltonian (9.155) become

$$[a_i, a_j\dagger] = \delta_{ij}, \; [a_i, a_j] = [a_i\dagger, a_j\dagger] = 0 \tag{9.159}$$
$$H = \sum_j (a_j\dagger a_j + \tfrac{1}{2})\,\hbar\omega \tag{9.160}$$

It is easily verified from these that

$$[H, a_j] = -\hbar\omega a_j, \quad [H, a_j\dagger] = \hbar\omega a_j\dagger \tag{9.161}$$

But according to the Heisenberg equation of motion (9.128), the left hand sides of these equations are, respectively, $-i\hbar\, da_j/dt$ and $-i\hbar\, da_j\dagger/dt$. Hence

$$\frac{da_j}{dt} = -i\omega a_j, \quad \frac{da_j\dagger}{dt} = i\omega a_j\dagger \tag{9.162}$$

The solution of these equations can be immediately obtained as:

$$a_j(t) = a_j(0)\, e^{-i\omega t}, \quad a_j\dagger(t) = a_j\dagger(0)\, e^{i\omega t} \tag{9.163}$$

(c) *Separation of Waves Moving in Opposite Directions:* Now that we know the nature of the time-dependence of the operators a_j and $a_j\dagger$, we can use this knowledge to separate $\mathbf{A}(\mathbf{x}, t)$ into two parts: One part in which the phases appearing in the exponent are $\pm(\mathbf{k}.\mathbf{x} - \omega t)$, and the other with phases $\pm(\mathbf{k}.\mathbf{x} + \omega t)$. Evidently, the first part propagates in the direction of $\mathbf{k}$ and the other opposite to $\mathbf{k}$ (i.e. in the direction of $-\mathbf{k}$). By substituting the inverse of Eqs. 9.158, namely

$$Q_j = \left(\frac{\hbar}{2\omega}\right)^{1/2}(a_j + a_j\dagger) \text{ and } P_j = -i\left(\frac{\hbar\omega}{2}\right)^{1/2}(a_j - a_j\dagger) \tag{9.164}$$

in Eq. 9.154, and the resulting expressions for $A(\mathbf{k}, t)$ and[21] $A\dagger(\mathbf{k}, t)$ into Eq. 9.150, we find the part of $\mathbf{A}(\mathbf{x}, t)$ propagating in the direction of $\mathbf{k}$ to be

$$\begin{aligned}\mathbf{A}_\mathbf{k}(\mathbf{x}, t) &= \left(\frac{2\pi\hbar c^2}{\omega L^3}\right)^{1/2} \mathbf{e}\,[a(\mathbf{k}, t)\, e^{i\mathbf{k}.\mathbf{x}} + a\dagger(\mathbf{k}, t)\, e^{-i\mathbf{k}.\mathbf{x}}] \\ &= \left(\frac{2\pi\hbar c^2}{\omega L^3}\right)^{1/2} \mathbf{e}\,[a(\mathbf{k}, 0)\, e^{i(\mathbf{k}.\mathbf{x}-\omega t)} + a\dagger(\mathbf{k}, 0)\, e^{-i(\mathbf{k}.\mathbf{x}-\omega t)}]\end{aligned} \tag{9.165}$$

[21] Note that complex conjugation denoted by * is replaced by Hermitian conjugation (†) when classical quantities are promoted to quantum operators. Thus $A^*(\mathbf{k},t) \to A\dagger(\mathbf{k},t)$.

where

$$a(\mathbf{k}, t) = 2^{-1/2}[a_1(t) + i\, a_2(t)] \tag{9.166}$$

The other part, propagating in the direction $-\mathbf{k}$ has a form similar to Eq. (9.165) except for the replacement of $\mathbf{k}$ everywhere by $-\mathbf{k}$, with $a(-\mathbf{k}, t) \equiv 2^{-1/2}[a_1(t) - ia_2(t)]$. These two parts constitute *independent* degrees of freedom of the field, since the operators $a(\mathbf{k})$, $a^\dagger(\mathbf{k})$ commute with $a(-\mathbf{k})$, $a^\dagger(-\mathbf{k})$. This is easily verified from the definitions of these quantities, using the commutation relations (9.159). It is a simple matter to verify that the operator H for the total energy of the field in the box, Eq. 9.160, also breaks up into two parts, $H = H_\mathbf{k} + H_{-\mathbf{k}}$, where

$$H_\mathbf{k} = [a^\dagger(\mathbf{k})\, a(\mathbf{k}) + \tfrac{1}{2}]\, \hbar\omega \tag{9.167}$$

and $H_{-\mathbf{k}}$ has the same form with $a(-\mathbf{k})$ instead of $a(\mathbf{k})$.

(*d*) *Photons:* Let us hereafter focus attention on $\mathbf{A}_\mathbf{k}(\mathbf{x}, t)$. It is equivalent to a simple quantum harmonic oscillator, because its energy operator $H_\mathbf{k}$ has exactly the same form (4.29b) as that of a simple harmonic oscillator, and so does the commutation relation between $a(\mathbf{k})$ and $a^\dagger(\mathbf{k})$:

$$[a(\mathbf{k}),\, a^\dagger(\mathbf{k})] = 1 \tag{9.168}$$

This follows, of course, from Eqs. 9.166 and 9.159. We shall now see that the total momentum of the field $\mathbf{A}_\mathbf{k}(\mathbf{x}, t)$ in the box can also be expressed in a very suggestive form. The energy flux associated with the propagating wave $\mathbf{A}_\mathbf{k}(\mathbf{x}, t)$ is given by the Poynting vector $(c/4\pi)\, (\mathcal{E} \times \mathcal{H})$; this is c times the energy density, which in turn is c times the momentum per unit volume. Therefore, the total momentum $\mathbf{P}_\mathbf{k}$ in the box is $(1/4\pi c) \int (\mathcal{E} \times \mathcal{H})\, d\tau$. Since $\mathcal{E}$ and $\mathcal{H}$ corresponding to $\mathbf{A}_\mathbf{k}(\mathbf{x}, t)$ of Eq. 9.165 is given by Eqs. 9.151 and 9.152 with $A(\mathbf{k}) = (2\pi\hbar c^2/\omega)^{1/2}\, a(\mathbf{k})$ and $\dot{A}(\mathbf{k}) = (2\pi\hbar c^2/\omega)^{1/2}\, \dot{a}(\mathbf{k}) = -i\,(2\pi\hbar c^2\omega)^{1/2}\, a(\mathbf{k})$, we obtain

$$\begin{aligned}\mathbf{P}_\mathbf{k} &= -\tfrac{1}{4} i\hbar L^{-3}\, [\mathbf{e} \times (\mathbf{k} \times \mathbf{e})] \int [-ia\, e^{i\mathbf{k}.\mathbf{x}} + ia^\dagger\, e^{-i\mathbf{k}.\mathbf{x}}]\, [a\, e^{i\mathbf{k}.\mathbf{x}} - a^\dagger\, e^{-i\mathbf{k}.\mathbf{x}}]\, d\tau \\ &= \tfrac{1}{2}\hbar\mathbf{k}\, (aa^\dagger + a^\dagger a) = \hbar\mathbf{k}\, (a^\dagger a + \tfrac{1}{2})\end{aligned} \tag{9.169}$$

Now, as a consequence of the formal identity of the commutation rule (9.168) with Eq. 4.28 of the harmonic oscillator case, we can conclude from the treatment of Sec. 4.4b that the eigenvalues of the operator

$$\mathcal{N}(\mathbf{k}) = a^\dagger(\mathbf{k})\, a(\mathbf{k}) \tag{9.170}$$

are the natural numbers $n_\mathbf{k} = 0, 1, 2, \ldots$ It follows immediately from Eqs. 9.167 and 9.169 that the energy and momentum of the field for given $\mathbf{k}$ can take only the values $(n_\mathbf{k} + \frac{1}{2})\, \hbar\omega$ and $(n_\mathbf{k} + \frac{1}{2})\, \hbar\mathbf{k}$ respectively. They differ from their minimum values ($\frac{1}{2}\hbar\omega$ and $\frac{1}{2}\hbar\mathbf{k}$) by an integer multiple $n_\mathbf{k}$ of $\hbar\omega$ and $\hbar\mathbf{k}$. It is then natural to say that the quantum state for which $\mathcal{N}(\mathbf{k})$ has the value $n_\mathbf{k}$ is one in which there are $n_\mathbf{k}$ quanta of radiation, each carrying energy $\hbar\omega$ and momentum $\hbar\mathbf{k}$. These quanta are the *photons*.

It is usual to denote the state with $n_\mathbf{k}$ photons by $|\, n_\mathbf{k} \rangle$, in the abstract vector space notation of Sec. 7.3(c). By definition, it is an eigenstate of the operator $\mathcal{N}_\mathbf{k}$ which is called the *number operator* for obvious reasons:

$$\mathcal{N}_\mathbf{k}\, |\, n_\mathbf{k} \rangle = n_\mathbf{k}\, |\, n_\mathbf{k} \rangle,\ n_\mathbf{k} = 0,1,2,\ldots \tag{9.171}$$

Borrowing once again from Sec. 4.4c (where we had u_n instead of what we call $|n_{\mathbf{k}}\rangle$ here), we note that the effect of $a(\mathbf{k})$ and $a^\dagger(\mathbf{k})$ on the $|n_{\mathbf{k}}\rangle$ is given by

$$a(\mathbf{k})\,|n_{\mathbf{k}}\rangle = \sqrt{n_{\mathbf{k}}}\,|n_{\mathbf{k}}-1\rangle;\; a^\dagger(\mathbf{k})\,|n_{\mathbf{k}}\rangle = \sqrt{n_{\mathbf{k}}+1}\,|n_{\mathbf{k}}+1\rangle \quad (9.172)$$

It is assumed here that the states are normalized: $\langle n_{\mathbf{k}}\,|\,n'_{\mathbf{k}}\rangle = \delta_{n_{\mathbf{k}}n'_{\mathbf{k}}}$. Since $a(\mathbf{k})$ reduces the photon number $n_{\mathbf{k}}$ by one, and $a^\dagger(\mathbf{k})$ increases it by one, these operators are respectively called *annihilation* and *creation operators* (of photons). Similar operators play a fundamental role in quantum field theories of other kinds of particles too.

We have been working so far in the Heisenberg picture wherein the basic operators $a(\mathbf{k})$, $a^\dagger(\mathbf{k})$ vary according to Eq. 9.163. This variation is such that $\mathcal{N}_{\mathbf{k}}$ and hence $H_{\mathbf{k}}$ remain constant. The states, of course, are time-independent in this picture. The general state of this field oscillator is some linear combination $|\psi\rangle_H = \Sigma\, c(n_{\mathbf{k}})\,|n_{\mathbf{k}}\rangle$, wherein the sum is over values of the occupation number $n_{\mathbf{k}}$. In such a state the number of photons is not definite: there is a probability $|c(n_{\mathbf{k}})|^2$ for finding $n_{\mathbf{k}}$ photons.

If we go over to the Schrödinger picture, the operators $a(\mathbf{k})$, $a^\dagger(\mathbf{k})$ become time-independent and the states evolve with time. In particular, if $|\psi(0)\rangle = \Sigma\, c(n_{\mathbf{k}})\,|n_{\mathbf{k}}\rangle$, then

$$|\psi(t)\rangle = \sum_{n_{\mathbf{k}}} c(n_{\mathbf{k}})\, e^{-i(n_{\mathbf{k}}+\frac{1}{2})\omega t}\,|n_{\mathbf{k}}\rangle \quad (9.173)$$

Finally we make a few remarks on the electro-magnetic field in general. It is, of course, composed of waves with all possible propagation vectors $\mathbf{k}$ — an infinite number of them. [When confined to a cubical box, the admissible vectors $\mathbf{k}$ are those whose components are integral multiples of $(2\pi/L)$, as we already know]. It can be shown that these different waves or modes of the field independently contribute terms of the same form as (9.167) to the total Hamiltonian of the field, i.e. the total Hamiltonian is simply a sum of terms corresponding to the various modes. The same holds for the momentum. The eigenstates are characterized by the number of photons $n_{\mathbf{k}_1}$ with momentum $\mathbf{k}_1$, $n_{\mathbf{k}_2}$ with momentum $\mathbf{k}_2$ and so on, and are denoted by $|n_{\mathbf{k}_1}, n_{\mathbf{k}_2},\ldots\rangle$; but we shall not pursue the general case any further here.

9.20 Atom Interacting with Quantized Radiation: Spontaneous Emission

In Sec. 9.13 we considered the interaction of an atom with a vector potential $\mathbf{A}$ which was taken to be a *classical field,* i.e.,a numerical-valued function of $\mathbf{x}$. Since $\mathbf{A}$ had a simple harmonic time dependence, its effect on the atom was a harmonic perturbation. Here we treat the same physical problem, but taking the electromagetic wave also as a quantum system. As we have seen in the last section, its dynamical variables are $a, a^\dagger$. The quantum field, whose Hamiltonian is $H_R = (a^\dagger a + \frac{1}{2})$, interacts with another quantum system, the atom, whose dynamical variables are $\mathbf{x},\mathbf{p}$ and which has the Hamiltonian $H_A = (\mathbf{p}^2/2m) + V(\mathbf{x})$. So the unperturbed Hamiltonian H_0 and the unperturbed eigenstates[22] $u_m(\mathbf{x}) \mid n \rangle$ of the total system (atom + radiation) are given by

$$H_0 = H_R + H_A; \quad H_A\, u_m = W_m\, u_m, \quad H_R \mid n \rangle = (n + \tfrac{1}{2})\hbar\omega \mid n \rangle.$$

$$H_0 u_m(\mathbf{x}) \mid n \rangle = [W_m + (n + \tfrac{1}{2})\hbar\omega]\; u_m(\mathbf{x}) \mid n \rangle. \tag{9.174}$$

The interaction between the two is given by the same H' as in Eq. 9.97, but with $\mathbf{A}$ taken now as the *operator* field (9.165) with one difference: a and $a^\dagger$ will be taken as time-independent.

$$H' = \frac{ie\hbar}{mc}\,\mathbf{A}.\nabla = \left(-\frac{2\pi\hbar^3 e^2}{m^2\omega L^3}\right)^{1/2} [a\, e^{i\mathbf{k}.\mathbf{x}} + a^\dagger e^{-i\mathbf{k}.\mathbf{x}}]\; \mathbf{e}.\nabla \tag{9.175}$$

In taking the dynamical variables $a, a^\dagger$ of the radiation as well as $\mathbf{x},\mathbf{p}$ of the atom to be time-independent, we are simply adopting the Schrödinger picture for the entire system. Note that as far as the radiation is concerned, $\mathbf{x}$ is *not* a dynamical variable but merely a label for the spatial point at which the dynamical quantity, the field, is considered. In the present context, $\mathbf{A}(\mathbf{x})$ stands for the radiation field at the position $\mathbf{x}$ of the atomic electron, and so the $\mathbf{x}$ appearing in $\mathbf{A}$ is the position variable of the electron; it acts on the wave functions $u(\mathbf{x})$. The operators $a, a^\dagger$ have no effect on the $u_m(\mathbf{x})$ but operate on the states $\mid n \rangle$ of the radiation.

The perturbation H' has no explicit time dependence. So we can use the theory of Sec. 9.7 to compute probabilities for transitions caused by the interaction. If we are interested in the initial and final states

$$i : u_i(\mathbf{x}) \mid n_i \rangle, \qquad f : u_f(\mathbf{x}) \mid n_f \rangle \tag{9.176}$$

the transition amplitude in the first order is governed by the matrix element of H' between these:

$$H'_{fi} = \left(-\frac{2\pi\hbar^3 e^2}{m^2\omega L^3}\right)^{1/2} \mathbf{e}.\; [\langle n_f \mid a \mid n_i \rangle \int u_f^*\, e^{i\mathbf{k}.\mathbf{x}}\, \nabla u_i\, d\tau$$
$$+ \langle n_f \mid a^\dagger \mid n_i \rangle \int u_f^*\, e^{-i\mathbf{k}.\mathbf{x}}\, \nabla u_i\, d\tau] \tag{9.177}$$

Observe now that in view of Eq. 9.172,

$$\langle n_f \mid a \mid n_i \rangle = \sqrt{n_i}\; \delta_{n_f, n_i - 1}; \; \langle n_f \mid a^\dagger \mid n_i \rangle = \sqrt{n_i + 1}\; \delta_{n_f, n_i + 1} \tag{9.178}$$

So the first term in Eq. 9.177 describes a process in which one quantum is annihilated (*absorbed* by the atom); while the second term governs *emission* processes in which one quantum is created ($n_f = n_i + 1$). Let us consider the latter case. Denoting the integral in the last term by $\mathbf{I}'_{fi}$, we have

$$(H'_{fi})_{\text{emission}} = \left(-\frac{2\pi\hbar^3 e^2}{m^2\omega L^3}\right)^{1/2} \mathbf{e}.\mathbf{I}'_{fi}\; \sqrt{n_i + 1}\;\; \delta_{n_f, n_i + 1} \tag{9.179}$$

The most remarkable fact is the appearance of the factor $\sqrt{n_i + 1}$ *which is nonzero even if there is no radiation present initially* ($n_i = 0$). Thus, quantization of the radiation field has immediately led to the possibility of *spontaneous emission.*

[22] For the present we continue to consider a single mode characterized by the wave vector $\mathbf{k}$; but for brevity of notation, the argument $\mathbf{k}$ will be suppressed. Later in the section we shall take account of the presence of modes with various wave vectors.

We use the notation W_m for the energy of the atom here, reserving the symbol E for the energy of the total system. The use of a mixed notation for states — the ordinary wave function language for the atom and abstract vector notation for the radiation — should not cause any difficulty. There is a possibility of confusion at this stage if the abstract notation is used for both at the same time.

To calculate the actual transition probabilities, we use first order perturbation theory. The probability of a transition from i to f in time t is given by $|a_f^{(1)}(t)|^2$, (Eq. 9.46). Introducing H'_{fi} from Eq. 9.179 we get for the present case,

$$|a_f^{(1)}(t)|^2 = \frac{2\pi\hbar e^2}{m^2\omega L^3} |\mathbf{e.I'}_{fi}|^2 \, n_f \, \delta_{n_f, n_i+1} \frac{4 \sin^2 \frac{1}{2} \omega_{fi} t}{\omega_{fi}^2} \tag{9.180}$$

where $\hbar\omega_{fi}$ is the energy difference between the initial and final states of the atom + electromagnetic field:

$$\hbar\omega_{fi} = [W_f + (n_f + \tfrac{1}{2})\hbar\omega] - [W_i + (n_i + \tfrac{1}{2})\hbar\omega] = W_f - W_i + \hbar\omega \tag{9.181}$$

Upto this point we have confined our attention to transitions involving a single mode of the electromagnetic field. Actually the atom is in interaction with all the field modes characterized by various wave vectors **k** and polarizations **e**. So the operators $a, a^\dagger$ and the numbers n_f, n_i in Eqs. 9.177 to 9.181 could belong to any one of these modes. Eq. 9.180 with n_f, n_i replaced by $n_f(\mathbf{k},\mathbf{e})$ and $n_i(\mathbf{k},\mathbf{e})$ gives the probability for a transition in which the atom goes from u_i to u_f raising the number of photons in the mode $(\mathbf{k},\mathbf{e})$ from $n_i(\mathbf{k},\mathbf{e})$ to $n_f(\mathbf{k},\mathbf{e}) = n_i(\mathbf{k},\mathbf{e}) + 1$ — in other words, emitting a photon of wave vector **k** and polarization **e**. So the total probability for the emission of a photon of given **e** and any **k** within an infinitesimal range of wave vectors is given by

$$\sum |a_f^{(1)}(t)|^2 = \frac{2\pi\hbar e^2}{m^2\omega L^3} \sum_{\mathbf{k}} |\mathbf{e.I'}_{fi}|^2 \, [n_i(\mathbf{k},\mathbf{e}) + 1] \, . \, \frac{4 \sin^2\frac{1}{2}\omega_{fi} t}{\omega_{fi}^2} \tag{9.182}$$

Taking the sum over all **k** with directions within $d\Omega$ and magnitudes near **k**, and assuming that $n_i(\mathbf{k},\mathbf{e})$ and $|\mathbf{e.I'}_{fi}|$ do not vary appreciably within this range, we take these factors out of the sum. The arguments used in the reduction of Eq. 9.47 to 9.51 can now be applied, and they lead to the following transition probability per unit time

$$P_{fi} = \frac{1}{t} \sum |a_f^{(1)}(t)|^2 = \frac{2\pi}{\hbar} \, . \, \hbar^2 \rho(E_f) \, . \, \frac{2\pi\hbar e^2}{m^2\omega L^3} |\mathbf{e.I'}_{fi}|^2 \, [n_i(\mathbf{k},\mathbf{e}) + 1] \tag{9.183}$$

The whole expression is to be evaluated at some typical **k** within the range of interest, for which one has, by energy conservation,

$$\hbar ck \equiv \hbar\omega = W_i - W_f ; \; \omega = \omega_{if} \tag{9.184}$$

The density of final states $\rho(E_f)$ is dN_f/dE_f where dN_f is the number of modes of the field with wave vectors within the volume element $k^2 dk d\Omega$ in **k**-space. We know this to be $k^2 dk \, d\Omega . (L/2\pi)^3$, and further, $dE_f = d(W_f + \hbar\omega) = c\hbar dk$ since we are considering a specific final state of the atom with given energy W_f. Thus, we have

$$\rho(E_f) = \frac{k^2}{c\hbar} d\Omega \left(\frac{L}{2\pi}\right)^3 \tag{9.185}$$

It may also be noted that since each photon carries energy $\hbar\omega$, the radiation energy present *initially* in the box per mode is $n_i(\mathbf{k},\mathbf{e})\hbar\omega$. Since the number of modes between ω and $\omega + \Delta\omega$ is $\rho(\omega)\Delta\omega \equiv \rho(E_f) \, \Delta E_f = \rho(E_f)\hbar\Delta\omega$ the radiation energy per unit angular frequency range per unit volume of the box is $\rho(E_f) \, \hbar . \, n_i(\mathbf{k},\mathbf{e})\hbar\omega/L^3$. This is the initial energy density $\varepsilon(\omega)$. So we can express $\rho(E_f) \, n_i(\mathbf{k},\mathbf{e})$ in terms of $\varepsilon(\omega)$ as

$$\rho(E_f)\ n_i(\mathbf{k},\mathbf{e}) = \frac{\varepsilon(\omega)L^3}{\hbar^2\omega} \tag{9.186}$$

We can now rewrite Eq. 9.183 as

$$P_{fi} = \frac{(2\pi\hbar)^2 e^2}{m^2\omega L^3}\ |\,\mathbf{e}.\mathbf{I}'_{fi}\,|^2\ [\rho(E_f)\, n_i + \rho(E_f)] \tag{9.187}$$

and substitute for the two terms in square brackets from Eqs. 9.185 and 9.186. We obtain

$$P_{fi} = \left[\frac{4\pi^2 e^2}{m^2\omega^2}\,\varepsilon(\omega) + \frac{e^2\hbar\omega}{2\pi m^2 c^3}\, d\Omega\right]\ |\,\mathbf{e}.\mathbf{I}'_{fi}\,|^2 \tag{9.188}$$

It is clear that the first term[proportional to ε (ω)]is the probability for induced emission; indeed it coincides with Eq. 9.106 which we obtained from the 'semiclassical' theory. The second term gives the probability for spontaneous emission. If we take its sum over both the independent directions of polarization for given **k** and then integrate over all directions of **k**, the result will be the total probability for spontaneous emission in any direction with any polarization. This should coincide with Eq. 9.117 when the dipole approximation is used for $\mathbf{I}'_{fi}$. In fact it does. To show this we proceed as follows.

Firstly we note that summation of $|\,\mathbf{e}.\mathbf{I}'_{fi}\,|^2$ over polarizations (for given **k**) yields $|\,\mathbf{e}^{(1)}.\mathbf{I}'_{fi}\,|^2 + |\,\mathbf{e}^{(2)}.\mathbf{I}'_{fi}\,|^2$ where $\mathbf{e}^{(1)}, \mathbf{e}^{(2)}$ are (unit) polarization vectors perpendicular to each other and to the unit vector $\hat{\mathbf{k}} \equiv (\mathbf{k}/k)$. The components of this orthonormal set of vectors have the 'closure' property

$$e_j^{(1)}e_l^{(1)} + e_j^{(2)}e_l^{(2)} + \hat{k}_j\hat{k}_l = \delta_{jl} \tag{9.189}$$

On using this we get

$$\sum_{\alpha=1}^{2} |\,\mathbf{e}^{(\alpha)}.\mathbf{I}'_{fi}\,|^2 = \sum_{\alpha=1}^{2}\sum_{j,l=1}^{3} e_j^{\alpha}e_l^{\alpha}\,(I'_{fi})_j(I'^*_{fi})_l = \sum_{jl}\,(\delta_{jl} - \hat{k}_j\hat{k}_l)(I'_{fi})_j(I'^*_{fi})_l \tag{9.190}$$

If we integrate this over all directions of **k**, assuming I'_{fi} to be independent of **k**, it is clear that $\int \hat{k}_j\hat{k}_l\, d\Omega = 0$ if $j \neq l$ while the integrals of $\hat{k}_1^2, \hat{k}_2^2, \hat{k}_3^2$ are all equal. Each of these will then be equal to one-third the integral of $\hat{\mathbf{k}}^2$ (and $\hat{\mathbf{k}}^2 = 1$). Hence $\int (\delta_{jl} - \hat{k}_j.\hat{k}_l)\, d\Omega = \frac{2}{3}\delta_{jl} \int d\Omega = (8\pi/3)\,\delta_{jl}$. Using this result, we get the total probability for spontaneous emission to be

$$\frac{e^2\hbar\omega}{2\pi m^2 c^3}\int d\Omega \sum_{\alpha=1}^{2}\left|\,\mathbf{e}^{(\alpha)}.\mathbf{I}_{fi}\,\right|^2 = \frac{4e^2\hbar\omega}{3m^2c^3}\,\mathbf{I}'_{fi}.\mathbf{I}'^*_{fi} \tag{9.191}$$

In the dipole approximation $\mathbf{I}'_{fi}$ is given by Eq. 9.110, which satisfies our assumption above that $\mathbf{I}'_{fi}$ is independent of **k**. On substituting it in Eq. 9.191, the latter reduces to Eq. 9.117, confirming our assertion.

9.21 The Interaction Picture

We now turn to a picture of the time evolution of quantum systems which is intermediate between the Schrödinger and Heisenberg pictures. It is based on a splitting of the Hamiltonian into two parts, say $H = H_0 + H'$. (The familiar context for such a splitting is when H' reflects some interaction in which the system is involved, and in whose absence the Hamiltonian would have been just H_0). The unitary operator U of Eq. 9.120 is so chosen in this case that the variation of operators is entirely due to H_0, and the states would not vary but for the existence of H'. As we shall now see, the appropriate choice is $U = V_0^{-1}(t,t_0)$ where V_0 satisfies Eq. 9.122 with H replaced by H_0. So we define quantities in the interaction picture as

$$|\,\psi(t)\,\rangle_{\mathrm{I}} = V_0^{-1}\,(t,t_0)\,|\,\psi(t)\,\rangle, \quad A_{\mathrm{I}}(t) = V_0^{-1}(t,t_0)A(t)V_0(t,t_0) \tag{9.192}$$

$$i\hbar\frac{dV_0(t,t_0)}{dt} = H_0V_0(t,t_0) \tag{9.193}$$

By differentiating the first of Eqs. 9.192 after multiplying by V_0, we obtain

$$H_0V_0\,|\,\psi(t)\,\rangle_{\mathrm{I}} + V_0\,i\hbar\frac{\partial}{\partial t}\,|\,\psi(t)\,\rangle_{\mathrm{I}} = i\hbar\frac{\partial}{\partial t}\,|\,\psi(t)\,\rangle$$
$$= H\,|\,\psi(t)\,\rangle = HV_0\,|\,\psi(t)\,\rangle_{\mathrm{I}}$$

Hence

$$V_0 i\hbar\frac{\partial}{\partial t}\,|\,\psi(t)\rangle_{\mathrm{I}} = H'V_0\,|\,\psi(t)\,\rangle_{\mathrm{I}};$$

$$i\hbar\frac{\partial}{\partial t}\,|\,\psi(t)\,\rangle_{\mathrm{I}} = H_{\mathrm{I}}'\,|\,\psi(t)\,\rangle_{\mathrm{I}},\ H'_{\mathrm{I}} = V_0^{-1}H'V_0 \tag{9.194}$$

Similarly, differentiating $V_0A_{\mathrm{I}}(t) = A(t)V_0$ from the second of Eqs. 9.192 and remembering Eq. 9.119 for the Schrödinger picture operator $A(t)$, we obtain

$$i\hbar\frac{dA_{\mathrm{I}}(t)}{dt} = [(A_{\mathrm{I}}(t),(H_0)_{\mathrm{I}}] + \left(\frac{\partial A}{\partial t}\right)_{\mathrm{I}} \tag{9.195}$$

In practice, H_0 is always chosen time-independent, and then $V_0(t,t_0)$ is simply $\exp\,[iH_0(t-t_0)/\hbar]$, which commutes with H_0; hence $(H_0)_{\mathrm{I}} = H_0$. Nevertheless, we have written $(H_0)_{\mathrm{I}}$ in Eq. 9.195 to emphasize that it is to be thought of as a function of the interaction picture operators $\mathbf{x}_{\mathrm{I}},\mathbf{p}_{\mathrm{I}}$; i.e. $(H_0)_{\mathrm{I}}(t) = H_0(\mathbf{x}_{\mathrm{I}}(t), \mathbf{p}_{\mathrm{I}}(t))$. More generally, the Schrödinger picture operator $A(\mathbf{x},\mathbf{p};t)$ goes over into

$$A_{\mathrm{I}}(t) \equiv A(\mathbf{x}_{\mathrm{I}}(t),\mathbf{p}_{\mathrm{I}}(t);t) \tag{9.196}$$

Eq. 9.195 for the interaction picture operators has exactly the same form as the Heisenberg equation of motion, with one important difference: it is not the complete Hamiltonian H but the unperturbed part H_0 which enters here. The operators $A_{\mathrm{I}}(t)$ evolve as if the 'interaction' H' did not exist: they obey what may be called the 'free' equations of motion (corresponding to $H \to H_0$). The effects of H' are manifested entirely through the time-variation of the *states* $|\,\psi(t)\,\rangle_{\mathrm{I}}$. But for the presence of H', the states would be time-independent. Consequently, if H' is small, so is the change of $|\,\psi(t)\,\rangle_{\mathrm{I}}$ with time. In these circumstances, the use of perturbation theory becomes possible and reasonable. It is, in fact, as a starting point for perturbative solutions that the interaction picture is of the greatest value.

It is a straightforward matter to obtain a perturbative solution of the form

$$|\,\psi(t)\,\rangle_{\mathrm{I}} = \sum_n |\,\psi^{(n)}(t)\,\rangle_{\mathrm{I}} \tag{9.197}$$

where $|\,\psi^{(n)}(t)\,\rangle_{\mathrm{I}}$ is of the nth order in H_{I}', i.e. contains n factors of H_{I}'. Substituting this in (9.194) and equating terms of the same order on the two sides, one gets

$$i\hbar \frac{\partial}{\partial t} \mid \psi^{(0)}(t) \rangle_I = 0,$$

$$i\hbar \frac{\partial}{\partial t} \mid \psi^{(n)}(t) \rangle_I = H_I' \mid \psi^{(n-1)}(t) \rangle_I, \qquad n = 1,2,\ldots \tag{9.198}$$

At the initial instant t_0, when the 'interaction' has had no time to act, we take the whole state to be of the zeroth order:

$$\mid \psi^{(0)}(t_0) \rangle_I = \mid \psi(t_0) \rangle_I, \quad \mid \psi^{(n)}(t_0) \rangle_I = 0, \quad (n = 1,2,\ldots) \tag{9.199}$$

The solutions of Eqs. 9.198 subject to these initial conditions are evidently

$$\mid \psi^{(0)}(t) \rangle_I = \mid \psi^{(0)}(t_0) \rangle_I;$$

$$\mid \psi^{(n)}(t) \rangle_I = (i\hbar)^{-1} \int_{t_0}^{t} H_I'(t') \mid \psi^{(n-1)}(t') \rangle_I \, dt' \tag{9.200}$$

The last equation which relates the terms of order n and $(n-1)$ can be solved recursively. It yields

$$\mid \psi^{(n)}(t) \rangle_I = (i\hbar)^{-n} \int_{t_0}^{t} dt_n \, H_I'(t_n) \int_{t_0}^{t_n} H_I'(t_{n-1}) \, dt_{n-1} \ldots$$

$$\ldots \int_{t_0}^{t_2} H_I'(t_1) dt_1 \mid \psi^{(0)}(t_0) \rangle_I \tag{9.201}$$

Substituting this in Eq. 9.197 we finally have

$$\mid \psi(t) \rangle_I = V_I(t,t_0) \mid \psi(t_0) \rangle_I \tag{9.202}$$

where the operator $V_I(t,t_0)$ relating the interaction picture state at time t to that at time t_0 is given by the infinite series

$$V_I(t,t_0) = 1 + (i\hbar)^{-1} \int_{t_0}^{t} H_I'(t_1) dt_1 + (i\hbar)^{-2} \int_{t_0}^{t} H_I'(t_2) dt_2 \int_{t_0}^{t_2} H_I'(t_1) dt_1 + \ldots \tag{9.203}$$

One could use Eq. 9.202 as the definition of $V_I(t,t_0)$ independently of any perturbation expansion, and by substituting this equation in (9.194) and observing that $\mid \psi(t_0) \rangle_I$ is quite arbitrary, obtain a differential equation for V_I:

$$i\hbar \frac{\partial V_I(t,t_0)}{\partial t} = H_I' V_I(t,t_0); \; V_I(t_0,t_0) = 1 \tag{9.204}$$

The series (9.203) is the perturbative solution of this equation.

Having obtained the perturbative solution of the time evolution problem by working in the interaction picture, we can express the state $\mid \psi(t) \rangle$ in the more familiar Schrödinger picture as

$$\mid \psi(t) \rangle = V_0(t,t_0) \mid \psi(t) \rangle_I = V_0(t,t_0) V_I(t,t_0) \mid \psi(t_0) \rangle_I \tag{9.205}$$

Since the states in both pictures coincide at time t_0, this becomes

$$\mid \psi(t) \rangle = V(t,t_0) \mid \psi(t_0) \rangle, \quad V(t,t_0) = V_0(t,t_0) V_I(t,t_0) \tag{9.206}$$

Thus the time evolution operator $V(t,t_0)$ of the Schrödinger picture is obtained. Using the expression (9.203) for V_I and the fact that $H_I'(t) = V_0^{-1}(t,t_0) H' V_0(t,t_0)$, we see that V itself can be written as

$$V(t,t_0) = \sum_{n=0}^{\infty} V^{(n)}(t,t_0),$$

$$V^{(n)}(t,t_0) = (i\hbar)^{-n} V_0(t,t_0) \int_{t_0}^{t} \int_{t_0}^{t_n} \ldots \int_{t_0}^{t_2} [V_0^{-1}(t_n,t_0) H' V_0(t_n,t_0)$$

$$. V_0^{-1}(t_{n-1},t_0) H' V_0(t_{n-1},t_0) \ldots V_0^{-1}(t_1,t_0) H' V_0(t_1,t_0)] dt_1 dt_2 \ldots dt_n$$

$$= (i\hbar)^{-n} \int_{t_0}^{t} \int_{t_0}^{t_n} \ldots \int_{t_0}^{t_2} [V_0(t,t_n)\, H' V_0(t_n,t_{n-1})\, H' \ldots V_0(t_2,t_1) H'$$

$$. V_0(t_1,t_0)] dt_1 dt_2 \ldots dt_n \qquad (9.207)$$

In the last step we have used the fact that $V_0\,(t_a, t_b)\; V_0\,(t_b,t_c) = V_0(t_a,t_c)$, Eq. 9.123.

Since V_0 is the time evolution operator for the unperturbed system, the picture of the progress of the system as given by the above expression is one of unperturbed propagation from t_0 to t_1, t_1 to $t_2, \ldots, t_n$ to t (represented by the V_0 factors), with 'interruptions' (action of the perturbation) at $t_1,t_2,\ldots,t_n$ manifested through the factors $(H'/i\hbar)$ occurring between pairs of V_0's. The total $V(t,t_0)$ is the sum of contributions from channels with $n = 0, 1, 2, \ldots$ to ∞.

It may be recalled at this point that we had encountered a similar picture while dealing with transition amplitudes in perturbation theory in Sec. 9.5. The transition amplitude from the state $|\, i\, \rangle$ to the state $|\, f \rangle$ is $\langle f\, |\psi(t)\, \rangle$ under the condition that $|\psi(t_0)\, \rangle = |\, i\, \rangle$; this is simply $\langle f |\; V(t,t_0)\; |\, i\, \rangle$. The nth order contribution to it is $\langle f |\; V^{(n)}\,(t,t_0)\; |\, i\, \rangle$. This may be immediately related to $a_f^{(n)}$ of Sec. 9.5 by introducing eigenstates of H_0 as intermediate states between the H's. For example, setting $t_0 = 0$ and making use of the fact that $V_0\,(t_a\,, t_b) = \exp\,[-\,iH_0\,(t_a - t_b)/\hbar]$, we obtain for the second order transition amplitude

$$\langle f |\; V^{(2)}(t,0)\; |\, i\, \rangle = (i\hbar)^{-2} \int_0^t \int_0^{t_2} \sum_m \langle f |\; V_0(t,t_2)\; H' \;|\, m\, \rangle$$

$$\langle\, m\, |\; V_0(t_2,t_1) H' V_0(t_1,0)\; |\, i\, \rangle$$

$$= (i\hbar)^{-2} \sum_m \int_0^t \int_0^{t_2} e^{-iE_f(t-t_2)/\hbar}\; H'_{fm}\; e^{-iE_m(t_2-t_1)/\hbar}\; H'_{mi}\; e^{-iE_it_1/\hbar}\; dt_1 dt_2 \qquad (9.208)$$

which reduces to $a_f^{(2)}$ of Eq. 9.40 apart from a factor $e^{-iE_ft/\hbar}$. Thus the method of variation of constants merely leads to a special representation of the general results of this section, which is pictured diagrammatically as in Fig. 9.2.

Another form which has been much in vogue in recent years is obtained if, instead of using a representation in terms of eigenstates of H_0 as above, we use the coordinate representation. In this case we use the fact, already noted in Sec. 9.16, Eq. 9.124, that in the coordinate representation V_0 is explicitly the propagator G_0 associated with the (unperturbed) Hamiltonian H_0: $\langle\, \mathbf{x}\, |\; V_0(t,t_0)\; |\, \mathbf{x}'\, \rangle = G_0\,(\mathbf{x},\mathbf{x}';t,t_0)$. We also note that since $\mathbf{x},\mathbf{p}$ are 'diagonal' in the sense of Sec. 7.5, $\langle\, \mathbf{x}\, |\, H'\, |\, \mathbf{x}'\, \rangle = \delta(\mathbf{x} - \mathbf{x}')\; H'(\mathbf{x}, -\, i\hbar\nabla)$. So the matrix element of the second order term $V^{(2)}$ for example, becomes (with $t_0 = 0$)

$$\langle\, \mathbf{x}\, |\; V^{(2)}(t,0)\; |\, \mathbf{x}_0\, \rangle = (i\hbar)^{-2} \int_0^t \int_0^{t_2} \{ \int d^3x_1 \int d^3x_2 .$$

$$. \langle\, \mathbf{x}\, |\; V_0(t,t_2)\; |\, \mathbf{x}_2\, \rangle\; H'(2)\; \langle\, \mathbf{x}_2\, |\; V_0(t_2,t_1)\; |\, \mathbf{x}_1\, \rangle\; H'(1) .$$

$$. \langle\, \mathbf{x}_1\, |\; V_0(t_1,0)\; |\, \mathbf{x}_0\, \rangle \}\; dt_1\; dt_2$$

$$= (i\hbar)^{-2} \int_0^t \int_0^{t_2} \int d^3x_1 \int d^3x_2 G_0(\mathbf{x},\mathbf{x}_2;t,t_2) H'(2) G_0(\mathbf{x}_2,\mathbf{x}_1;t_2,t_1) .$$

$$. H'(1) G_0(\mathbf{x}_1,\mathbf{x}_0;t_1,0)\; dt_1\; dt_2$$

Here we have made the obvious abbreviation $H'(1)$ for H' at $\mathbf{x}_1$, etc.

This representation of the time evolution operator leads to the famous Feynman diagrams. For the case of a single particle, with H_0 describing free motion (no forces), the Feynman diagrams corresponding to the first three terms of V are shown in Fig. 9.5. The straight segments depict free propagation

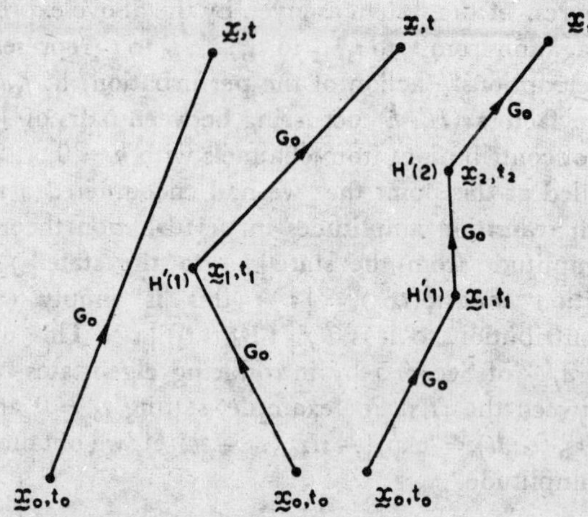

Fig. 9.5 Feynman diagrams for the propagation of a particle from $(\mathbf{x}_0, t_0)$ to $(\mathbf{x}, t)$. The three diagrams shown depict the zeroth, first and second order contributions to the propagator

(governed by the propagator G_0) from one space-time point to another, while the kinks or vertices are the space-time points where the perturbation has acted. The coordinate space matrix element of any $V^{(n)}$ can be written down directly by looking at the corresponding Feynman diagram. Reading from *right to left*, the factors which appear are $G_0(\mathbf{x}_1,\mathbf{x}_0; t_1,t_0),(i\hbar)^{-1}H'(1)$, $G_0(\mathbf{x}_2,\mathbf{x}_1; t_2,t_1)$, $(i\hbar)^{-1}H'(2),\ldots$ and so on. The whole expression is to be integrated over $\mathbf{x}_1,\mathbf{x}_2\ldots,\mathbf{x}_n$ and over $t_1,t_2,\ldots,t_n$ with the ordering $t_0 \leqslant t_1 \leqslant t_2 \leqslant \ldots \leqslant t$.

9.22 The Scattering Operator

In scattering experiments, the time interval between the firing of the projectile and its observation after scattering is extremely large compared to the time during which the colliding particles are actually in interaction. So, for all practical purposes this interval may be considered infinite. For this reason, in the application of the above theory to scattering problems, it is customary to assume that $t_0 \to -\infty$ and $t \to +\infty$. The operator

$$S = \lim_{t_0 \to -\infty,\, t \to +\infty} V(t,t_0) \tag{9.210}$$

is known as the scattering operator. Its matrix elements $\langle f | S | i \rangle$ between the unperturbed (free) states of the system of particles constitute the so-called *scattering matrix* or *S-matrix*.

The nth order term in the perturbation expansion of S is obtained by taking the above limits in Eq. 9.207. It will be noticed that while the lower limits of the time integrations all go to $-\infty$, the upper limit goes to $+\infty$ only in the last integral (over t_n). One can formally extend all the integrations to $+\infty$, provided one replaces $V_0(t_a,t_b)$ everywhere by the retarded operator

$$V_0{}^{R}(t_a,t_b) = V_0(t_a,t_b)\,\theta(t_a - t_b) \tag{9.211}$$

where the θ function vanishes whenever t_b exceeds t_a. This property ensures that the integrand is zero throughout the extended part of the integration region, so that nothing is changed by putting all the upper limits as $+\infty$. The coordinate representation of $V_0{}^{R}$ is the retarded propagator defined in Sec. 9.2. When the coordinate representation is used for the S-operator, it is an advantage to have all the integrals over space *and* time coordinates going uniformly from $-\infty$ to $+\infty$, especially in the context of relativistic problems where space-time symmetry is an important feature.

It is perhaps pertinent to observe that the limiting process (9.210) is not a very well defined mathematical procedure. It is made clear-cut by the $i\varepsilon$ prescription of Sec. 9.3 which is equivalent to replacing $V^{R}(t_a,t_b)$ by the limit $(\varepsilon \to 0_+)$ of $V^{R}(t_a,t_b)\,e^{-\varepsilon(t_a-t_b)}$.

D. TIME EVOLUTION OF ENSEMBLES

9.23 The Density Matrix[23]

When the average properties of an ensemble of identical non-interacting systems are of interest, and information on the individual members of the ensemble is not needed, it is useful to introduce the concept of the density matrix. It is a matrix which describes the *state of the ensemble*, in a sense which will be now made precise.

Consider first a single system in a state $|\psi\rangle$. In some representation whose basis states we denote generally by $|m\rangle$, a column vector with elements $c_m = \langle m|\psi\rangle$ represents the state $|\psi\rangle$. By multiplying this column *from the right* by its transposed complex conjugate row, one gets a square matrix (generally of infinite dimension) whose (m,n) element is $c_m c_n{}^*$. Denoting this matrix by ρ, we have[24]

$$\rho_{mn} \equiv c_m c_n{}^* = \langle m|\psi\rangle\langle\psi|n\rangle \tag{9.212a}$$

or

$$\rho_{mn} = \langle m|\hat{\rho}|n\rangle, \quad \hat{\rho} = |\psi\rangle\langle\psi| \tag{9.212b}$$

The abstract operator $\hat{\rho}$ is called the *density operator* corresponding to the state $|\psi\rangle$ and its matrix representative ρ, with elements ρ_{mn}, is the *density matrix*.

[23] J. von Neumann, *Gottinger Nachr.* 1, 245, 1927. A useful review may be found in D. ter Haar, *Rep. Progr. Phys.*, 24, 304, 1961.
[24] The 'hat' notation for abstract operators is temporarily revived here, to distinguish them from their matrix representatives.

The expectation value of any operator $\hat{F}$ of the system can be computed from a knowledge of the density matrix.

$$\langle F \rangle_\psi = \langle \psi | \hat{F} | \psi \rangle = \sum_{m,n} \langle \psi | m \rangle \langle m | \hat{F} | n \rangle \langle n | \psi \rangle$$

$$= \sum_{m,n} \langle m | \hat{F} | n \rangle \langle n | \psi \rangle \langle \psi | m \rangle = \sum_{n,m} F_{mn} \rho_{nm},$$

i.e.

$$\langle F \rangle = \sum_m (F\rho)_{mm} = \mathrm{Tr}\,(F\rho) \tag{9.213}$$

where TrA means the trace (sum of diagonal elements) of the matrix A.

Suppose now that we have an ensemble of identical systems, the ith member of the ensemble being in the state $| \psi^{(i)} \rangle$ with the associated density matrix $\rho^{(i)}$. The expectation value of $\hat{F}$ for this particular member is $\langle F \rangle^{(i)} = \mathrm{Tr}(F\rho^{(i)})$. What is the average of the $\langle F \rangle^{(i)}$ for the whole ensemble? This is readily computed if we know the numbers $\mathcal{N}_i$ of members which are in the various states i, or equally well if we know the fractions $f_i = (\mathcal{N}_i/\mathcal{N})$ where $\mathcal{N}$ is the total number of members in the ensemble. The answer is evidently

$$\langle \overline{F} \rangle = \sum_i f_i \langle F \rangle^{(i)} = \sum_i f_i \,\mathrm{Tr}\,(F\rho^{(i)})$$

or

$$\langle \overline{F} \rangle = \mathrm{Tr}\,(F\rho) \tag{9.214}$$

where

$$\rho = \sum_i f_i \,\rho^{(i)} \tag{9.215}$$

Thus the *ensemble average* $\langle \overline{F} \rangle$ of F (which is really a double average as indicated by the brackets and the bar) depends only on the matrix ρ defined as in Eq. 9.215. Once the density matrix ρ *of the ensemble* is known, there is no need to go back to the states of the individual systems while computing ensemble averages. The matrix elements of ρ are, of course, given by

$$\rho_{mn} = \sum_i f_i \, c_m^{(i)} c_n^{(i)*} = \langle m | \hat{\rho} | n \rangle, \; \hat{\rho} = \sum_i f_i | \psi^{(i)} \rangle \langle \psi^{(i)} | \tag{9.216}$$

It can hardly be overemphasized that though $\hat{\rho}$ is an operator, it does *not* stand for any dynamical variable but for the state (by definition). The following basic properties of the matrix ρ may be noted: It is Hermitian and positive, and further,

$$\mathrm{Tr}\,\rho = 1, \; \mathrm{Tr}\,\rho^2 \leqslant 1 \tag{9.217}$$

Hermiticity ($\rho_{mn} = \rho_{nm}{}^*$) is evident from the definition. It implies that ρ can be diagonalized by a unitary transformation. If the diagonal form is $\rho^{(d)} = U^\dagger \rho U$, and its elements (the eigenvalues of ρ) are ρ_a, then

$$\rho_a = \sum_{mn} U_{ma}{}^* \rho_{mn} U_{na} = \sum_i f_i \Big(\sum_m U_{ma}{}^* c_m^{(i)}\Big) \Big(\sum_n U_{na} c_n^{(i)}\Big)^*$$

$$= \sum_i f_i \Big| \sum_n U_{na} c_n^{(i)*} \Big|^2 \geqslant 0 \tag{9.218}$$

Thus, all the eigenvalues are non-negative, which is what is meant by saying that ρ is a positive matrix. The fact that $\mathrm{Tr}\,\rho = \sum_m \rho_{mm} = 1$ follows readily from Eq. 9.216 since $\sum_m | c_m^{(i)} |^2 = 1$ and $\sum f_i = 1$. Since the trace is unchanged by the unitary transformation, from the diagonal form we have $\Sigma\rho_a = 1$. Now, if more than one of the ρ_a is nonzero, each is a positive number less than unity and hence $\rho_a^2 < \rho_a$ and $\Sigma\, \rho_a^2 < 1$, which is to say Tr $\rho^2 < 1$. If only one of the ρ_a, say ρ_{a_0}, is nonzero, then it is unity ($\rho_a = \delta_{aa_0}$) and evidently then Tr$\rho^2 = 1$. In such a case, the inverse of the diagonalization process, $\rho = U\rho^{(d)}U^\dagger$, yields

$$\rho_{mn} = \sum_a U_{ma}\rho_a U_{na}^* = \sum_a U_{ma}\delta_{aa_0} U_{na}^* = U_{ma_0}U_{na_0}^* \tag{9.219}$$

This is seen to be a special case of the first of Eqs. 9.216 corresponding to all f_i being zero except one, say $f_i = \delta_{ii_0}$, with $c_m^{(i_0)} = U_{a_0m}$. Thus it describes a state of the ensemble wherein *all* the systems have the *same* state $|\psi^{(i_0)}\rangle$. Such a state of the ensemble is called a *pure* state. It is characterized by Tr$\rho^2 = 1$ while for any 'impure' state Tr $\rho^2 < 1$.

9.24 Spin Density Matrix

The simplest example of a density matrix is one describing the average *spin state* of an ensemble of particles whose orbital wave functions are all the same. Consider, for instance, a beam of electrons all of which have the same momentum but various orientations of spin. Since the spin space in this case is two-dimensional, the spin density matrix is of the order 2×2. Now, it is easy to convince oneself that any Hermitian 2×2 matrix can be expressed as a real linear combination of the unit matrix and the Pauli matrices. So the spin-$\frac{1}{2}$ density matrix can quite generally be written as

$$\rho = \tfrac{1}{2}(I + \boldsymbol{\sigma}.\mathbf{P}) \tag{9.220}$$

where I is the 2×2 unit matrix. The factor half ensures that Tr $\rho = 1$ (the Pauli matrices being traceless). The ensemble average of any component of spin may be obtained using Eq. 9.214. For the x-component, for instance, the operator F is σ_x (apart from a factor $\frac{1}{2}\hbar$) and one gets

$$\langle \overline{\sigma_x} \rangle = \text{Tr}\,(\sigma_x\rho) = \text{Tr}\,\tfrac{1}{2}(\sigma_x + P_x + i\sigma_z P_y - i\sigma_y P_z) = P_x \tag{9.221}$$

Thus the vector $\mathbf{P}$ directly gives the average spin vector of the ensemble. It is called the *polarization vector*. If $\mathbf{P} = 0$, the beam is *unpolarized* and the corresponding density matrix is just $\frac{1}{2}.I$. Further, since

$$\text{Tr}\,\rho^2 = \text{Tr}\,\tfrac{1}{4}(I + \mathbf{P}^2 + 2\boldsymbol{\sigma}.\mathbf{P}) = \tfrac{1}{2}(I + \mathbf{P}^2) \tag{9.222}$$

to the maximum value 1 of Trρ^2 there corresponds the maximum value 1 for $\mathbf{P}^2$. Thus $|\mathbf{P}| = 1$ denotes a pure state (full polarization).

As a numerical example, let 1/3 of all the particles have spin pointing in the negative z-direction, and the remainder in the positive x-direction. The spin wave functions and the corresponding weights are

$$\begin{pmatrix}0\\1\end{pmatrix},\ f_1 = \tfrac{1}{3} \quad \text{and} \quad \frac{1}{\sqrt{2}}\begin{pmatrix}1\\1\end{pmatrix},\ f_2 = \tfrac{2}{3}$$

respectively, and hence

$$\rho = \tfrac{1}{3}\begin{pmatrix}0\\1\end{pmatrix}(0,1) + \tfrac{2}{3}.\frac{1}{\sqrt{2}}\begin{pmatrix}1\\1\end{pmatrix}\cdot\frac{1}{\sqrt{2}}(1,1) = \tfrac{1}{3}\begin{pmatrix}1 & 1\\1 & 2\end{pmatrix}$$

It is left to the reader to verify that this matrix satisfies (9.217) and that $\mathbf{P} = (1/3,\ 0,\ 1/6)$.

It may be noted that in general, if the spin is j, the spin density matrix is $(2j+1)$-dimensional.

9.25 The Quantum Liouville Equation

The development of the states of the individual systems with time endows $\hat{\rho}$ also with a time evolution. From

$$i\hbar\frac{d}{dt}\,|\,\psi^{(i)}\,\rangle = \hat{H}\,|\,\psi^{(i)}\,\rangle \quad \text{and} \quad -i\hbar\frac{d}{dt}\,\langle\,\psi^{(i)}\,| = \langle\,\psi^{(i)}\,|\,\hat{H}$$

we obtain
$$i\hbar\frac{d\hat{\rho}}{dt} = \sum_i f_i\,(\hat{H}\,|\,\psi^{(i)}\,\rangle\,\langle\,\psi^{(i)}\,| - |\,\psi^{(i)}\,\rangle\,\langle\,\psi^{(i)}\,|\,\hat{H}) \tag{9.223}$$

$$= \hat{H}\hat{\rho} - \hat{\rho}\hat{H} = [\hat{H},\hat{\rho}] \tag{9.224}$$

Here we have used the fact that all the systems, being identical, have the same $\hat{H}$. This equation has a superficial resemblance to the Heisenberg equation of motion for dynamical variables, though the sign of the right hand side is wrong. But it is important to keep in mind that here we are working in the Schrödinger picture, and the variation of $\hat{\rho}$ arises precisely from the variation of the states.

In classical mechanics, the properties of ensembles are described in terms of a phase space density function (also denoted by ρ) which is a function defined over the phase space of any typical member of the ensemble. In other words, if the description of any member requires r position coordinates $q \equiv (q_1 q_2 \ldots q_r)$ and corresponding momenta $p \equiv (p_1 p_2 \ldots p_r)$ then ρ is a function $\rho(q,p)$. Its rate of change is governed by the Liouville equation, $d\rho/dt = \{\rho, H\}$ where $\{\rho, H\}$ is the Poisson bracket of ρ and H. Clearly Eq. 9.224 is its quantum analogue. So it is often called the *quantum Liouville equation.*

9.26 Magnetic Resonance

As an illustration of the foregoing theory, consider the spin system of Sec. 9.24 in the presence of a magnetic field $\mathcal{H}$. The Hamiltonian of an electron in such a field is[25]

$$H = \mu_B \sigma.\mathcal{H}, \qquad \mu_B = (e\hbar/2mc) \tag{9.225}$$

With this H, and ρ as in Eq. 9.220, $[\rho, H] = \mu_B[\tfrac{1}{2}\sigma.\mathbf{P},\ \sigma.\mathcal{H}] = i\mu_B\sigma.(\mathbf{P}\times\mathcal{H})$, as may be easily verified from the properties (Sec. 8.3a) of the Pauli matrices. Also $d\rho/dt = \tfrac{1}{2}\sigma.d\mathbf{P}/dt$. Using these we can write the equation of motion (9.224) as an equation for the vector $\mathbf{P}$:

$$\frac{d\mathbf{P}}{dt} = -(2\mu_B/\hbar)\ \mathbf{P}\times\mathcal{H} \tag{9.226}$$

Since $(d\mathbf{P}/dt)$ is perpendicular to both $\mathbf{P}$ and $\mathcal{H}$, the motion of the vector $\mathbf{P}$ is a

[25] We have denoted the charge on the electron by $-e$ with e being positive. For simplicity, we suppose the electrons to be free. The purely orbital part of the Hamiltonian is not relevant here and is omitted.

precession about the instantaneous direction of $\mathcal{H}$. If $\mathcal{H}$ is time-independent (and is in the z-direction, say) this is a steady precession around the z-axis with angular frequency $\omega_0 = (2\mu_B \mathcal{H}_0/\hbar) \equiv 2\omega_L$ with $\omega_L = 2\pi\nu_L$ where $\nu_L = (e\mathcal{H}/4\pi mc)$ is the Larmor frequency. So one may straightaway write down the solution as $P_x(t) = P_x(0) \cos \omega_0 t - P_y(0) \sin \omega_0 t$, $P_y(t) = P_y(0) \cos \omega_0 t + P_x(0) \sin \omega_0 t$, $P_z(t) = P_z(0)$. An alternative way of saying the same thing is that if one goes over to a coordinate system which rotates about the z-axis with angular frequency ω_0, the components of $\mathbf{P}$ in that frame would remain time-independent. They are

$$P_x' \equiv P_x \cos \omega_0 t + P_y \sin \omega_0 t$$
$$P_y' \equiv P_y \cos \omega_0 t - P_x \sin \omega_0 t, \; P_z' = P_z \tag{9.227}$$

Their values in fact reduce to $P_x(0)$, $P_y(0)$, $P_z(0)$.

In magnetic resonance experiments a strong constant field $\mathcal{H}_0$ in the z-direction together with a weak rotating field of magnitude $\mathcal{H}'$ in the perpendicular plane is employed:

$$\mathcal{H} = (\mathcal{H}' \cos \omega t, + \mathcal{H}' \sin \omega t, \mathcal{H}_0) \tag{9.228}$$

The polarization vector as seen in the rotating coordinate system mentioned above would no longer be constant but it should be slowly varying since $\mathcal{H}'$ is small. The equations for P_x', P_y', P_z' are easily obtained from their definitions, knowing $d\mathbf{P}/dt$. It is left to the student to verify that they are:

$$\dot{P}_x' = - P_z' \omega' \sin (\omega_0 - \omega)t, \; \dot{P}_y' = - P_z' \omega' \cos (\omega_0 - \omega)\, t$$
$$\dot{P}_z' = P_x' \omega' \sin (\omega_0 - \omega)t + P_y' \omega' \cos (\omega_0 - \omega)t \tag{9.229}$$

where $\omega' = (2\mu_B \mathcal{H}'/\hbar)$. As expected, all the derivatives are small ($\propto \omega' \propto \mathcal{H}'$). These equations can be solved without difficulty. Differentiating the last equation twice, and using the other two equations to eliminate $\dot{P}_x'$ and $\dot{P}_y'$, one gets

$$\ddot{P}_z' = - \Omega^2 \dot{P}_z', \; \Omega^2 = (\omega_0 - \omega)^2 + \omega'^2 \tag{9.230}$$

Hence

$$P_z = a \cos \Omega\, t + b, \; (a, b \text{ constants}) \tag{9.231}$$

We shall not write down the expressions for the other two components, since the main point of interest is the time dependence of the energy of the system, to which they contribute very little when $\mathcal{H}' \ll \mathcal{H}_0$. The energy is

$$\text{Tr } (\rho H) = \mu_B \text{ Tr } (\rho \sigma . \mathcal{H}) \approx \mu_B \mathcal{H}_0 P_z = \mu_B \mathcal{H}_0 (a \cos \Omega t + b) \tag{9.232}$$

If $\Omega \gtrsim \omega$, within each cycle of the applied alternating field the energy of the ensemble goes through one or more complete cycles, absorbing energy from, and returning it to, the applied field in this process. But if $\omega \approx \omega_0$ (*resonance* between the applied frequency ω and the 'natural' precession frequency ω_0 in the constant field $\mathcal{H}_0$), then $\Omega \approx \omega' \ll \omega$; and it takes very many cycles of the applied field before $\cos \Omega t$ changes sign. Throughout this time the energy change is monotonic. Since the ensemble would have been in a state of minimum possible energy in the constant field $\mathcal{H}_0$ before the alternating field is switched on, this monotonic variation of energy is an increase, i.e., energy is continuously absorbed from the external source. This absorption reveals the occurrence of resonance and enables one to determine the resonance frequency ω_0 with very high precision.

PROBLEMS

1. Show that the propagator for the simple harmonic oscillator is
$$G(x,x';t) = \left[\frac{m\omega}{2\pi i\hbar \sin\omega t}\right]^{1/2} \exp\left[\frac{im\omega}{2\hbar \sin\omega t}\left\{(x^2 + x'^2)\cos\omega t - 2xx'\right\}\right]$$
(Use Eq. 9.7, and evaluate the sum using one of the representations (D.7) of the H_n).

2. Show that the total cross-section for the scattering of an electron by a hydrogen atom with excitation of the latter from the $1s$ to any $2p$ state is (in the high energy approximation, neglecting exchange effects) $\sigma \approx (576\pi/k_0^2)(2/3)^{12} \ln(4k_0a_0)$.

3. Solve the Heisenberg equations of motion for x and p (i) for a free particle, and (ii) for a harmonic oscillator. Use the result of the first case to find the time-dependence of the dispersion in the coordinate.

4. Show that for a simple harmonic oscillator, the operator $A \equiv p(t)\sin\omega t - m\omega\, x(t)\cos\omega t$ is time-independent. Can it be simultaneously diagonalized with the Hamiltonian?

5. The energy of interaction of the spin magnetic moment of a particle with a constant magnetic field in the z-direction has the form $H = -\omega S_z$. Transform the operators S_x and S_y to the Heisenberg representation and show that their variation can be interpreted as due to a precession of the angular momentum about the z-direction.

6. Evaluate the probability per unit time for spontaneous radiative transitions from the $2p$ state of the hydrogen atom and hence show that the mean life time in this state is 1.60×10^{-9} sec.

7. The intensity of a spectral line is proportional to $\omega_{mn}\,\Sigma P_{nm}$, where P_{nm} is given by (9.117) and the sum is over all states n, m belonging to the particular pair of energy levels considered. Show that the ratio of intensities of the first two Lyman lines of hydrogen (Ly α: $2p \rightarrow 1s$; Ly β: $3p \rightarrow 1s$) is ~ 3.2.

10

Relativistic Wave Equations

10.1 Generalization of the Schrödinger Equation

In this chapter we consider two types of wave equations which have been proposed for the description of particles travelling at speeds close to that of light.[1] At these speeds, the Hamiltonian of the (free) particle is no longer given by $(\mathbf{p}^2/2m)$; hence Schrödinger's equation, obtained from such a Hamiltonian by the prescription of Sec. 2.1, is not applicable to relativistic particles. One could try to generalize the equation by using, instead of $(\mathbf{p}^2/2m)$, the relativistic expression for the energy, namely $E = (c^2\mathbf{p}^2 + m^2c^4)^{1/2}$. The operator replacement $E \to i\hbar\partial/\partial t$, $\mathbf{p} \to -i\hbar\nabla$ would then lead to $i\hbar\partial\psi/\partial t = (-\hbar^2c^2\nabla^2 + m^2c^4)^{1/2}\psi$. This equation has some obviously unattractive features. The space and time differential operators enter in it in very different ways. This is in contrast with the quite symmetric role of the space and time coordinates (as different components of a single four-vector) in relativity theory. The meaning of the operator $(-\hbar^2c^2\nabla^2 + m^2c^4)^{1/2}$ is itself unclear. One could get around this by passing to the momentum representation. But even this possibility would disappear if the quantity under the square root were to be modified to include functions of $\mathbf{x}$: for example, through the replacement $\mathbf{p} \to (\mathbf{p} - e\mathbf{A}/c)$ which becomes necessary when electromagnetic fields are present (assuming that the particle has charge e). Schrödinger suggested (immediately after his formulation of nonrelativistic quantum mechanics) that in order to avoid the difficulties arising from the square root, the operator replacement of $\mathbf{p}$ and E be made in the relativistic expression for E^2:

[1] Such particles are said to be *relativistic*, referring to the fact that their behaviour is understood in terms of the theory of relativity.

$$E^2 = c^2\mathbf{p}^2 + m^2c^4 \tag{10.1}$$

The resulting equation is

$$-\hbar^2\frac{\partial^2\psi}{\partial t^2} = -\hbar^2c^2\nabla^2\psi + m^2c^4\psi,$$

or

$$\frac{1}{c^2}\frac{\partial^2\psi}{\partial t^2} - \nabla^2\psi + \left(\frac{mc}{\hbar}\right)^2\psi = 0 \tag{10.2}$$

The implications of this equation in the quantum mechanical context were studied by Klein and Gordon[2] and the equation is now generally known under their names. Meanwhile Dirac succeeded in constructing another equation which is of the *first* order in $(\partial/\partial t)$ and ∇, unlike Eq. 10.2, and yet involves these operators in a fully symmetric way. The Dirac equation constrains the spin of the particle to be $\frac{1}{2}$. Its application to the electron was phenomenally successful, not only in helping to understand its properties in a natural way, but more spectacularly, in the prediction of its anti-particle, the positron (which was later discovered in cosmic rays). The Klein-Gordon equation has nothing to say about the spin of the particle; it is the one to be used for particles of spin 0, like the π- and K-mesons.

A. THE KLEIN-GORDON EQUATION

10.2 Plane Wave Solutions; Charge and Current Densities

Solutions of Eq. 10.2 corresponding to particles of definite momentum $\mathbf{p} = \hbar\mathbf{k}$ may be obtained by substituting $\psi(\mathbf{x},t) = f(t)\,e^{i\mathbf{k.x}}$. This leads to $d^2f/dt^2 = [c^2\mathbf{k}^2 + (mc^2/\hbar)^2]f$. Solving this we obtain (apart from a constant normalization factor), the plane wave solutions:

$$\psi(\mathbf{x},t) = e^{i(\mathbf{k.x}\mp\omega t)} = e^{i(\mathbf{p.x}-Et)/\hbar} \tag{10.3}$$

$$E = \pm\hbar\omega = \pm(c^2\mathbf{p}^2 + m^2c^4)^{1/2} \tag{10.4}$$

In contrast to the nonrelativistic case where the coefficient $(E/\hbar)$ of $(-it)$ in the exponent is the *positive* quantity $(\mathbf{p}^2/2m\hbar)$, here we have solutions with $-\omega$ as well as with $+\omega$. The appearance of the 'negative energy' solutions (characterized by $+\omega t$ in the exponent) is typical of relativistic wave equations; their interpretation will be considered later.

Another difference from the nonrelativistic case is that $\psi^*\psi$ cannot be interpreted as the probability density $P(\mathbf{x},t)$. We expect $P(\mathbf{x},t)$ to satisfy a continuity equation of the form (2.36), namely $\partial P/\partial t + \text{div}\,\mathbf{S} = 0$, which would ensure that $\int P(\mathbf{x},t)\,d^3x$ is time-independent. To obtain such an equation we multiply Eq. 10.2 on the left by ψ^*, its complex conjugate equation by ψ, and subtract. The resulting equation can be written as

$$\frac{1}{c^2}\frac{\partial}{\partial t}\left(\psi^*\frac{\partial\psi}{\partial t} - \psi\frac{\partial\psi^*}{\partial t}\right) - \nabla.(\psi^*\nabla\psi - \psi\nabla\psi^*) = 0$$

[2] O. Klein, *Z. Physik*, **37**, 895, 1926; W. Gordon, *Z. Physik*, **40**, 117, 121, 1926.

This is a continuity equation, with

$$P(\mathbf{x},t) = \frac{i\hbar}{2mc^2}\left(\psi^* \frac{\partial \psi}{\partial t} - \psi \frac{\partial \psi^*}{\partial t}\right) \tag{10.5a}$$

$$\mathbf{S}(\mathbf{x},t) = -\frac{i\hbar}{2m}(\psi^* \nabla\psi - \psi\nabla\psi^*) \tag{10.5b}$$

A convenient choice of a common constant factor in P and $\mathbf{S}$ has been made here. With this choice, $\mathbf{S}$ coincides exactly with the corresponding nonrelativistic expression (2.37b). However, P is quite different. It vanishes identically if ψ is real, and in the case of complex wave functions, P can even be made negative by choosing $\partial\psi/\partial t$ appropriately. (Since Eq. 10.2 is of the second order in time derivatives, both ψ and $\partial\psi/\partial t$ can be arbitrarily prescribed). Clearly, P cannot be a probability density. One could multiply P by a charge e and then interpret it as a charge density (which can be positive or negative), and $\mathbf{S}$ as the corresponding electric current density. However, P can still have different signs at different points, and this is hardly satisfactory in the description of a single particle of given charge. It is now known that Eq. 10.2 can be used to describe a system of arbitrary numbers of particles and their antiparticles (which have the opposite charge) by treating ψ itself as an operator function instead of a numerical-valued function. At that level, the above-mentioned properties of P are no longer objectionable; but here we confine ourselves to the one-particle treatment, overlooking the difficulties.

10.3 Interaction with Electromagnetic Fields; Hydrogen-like Atom

To take account of the presence of electromagnetic fields acting on the particle, we follow the now-familiar procedure[3] of replacing $\mathbf{p}$ by $\mathbf{p} - e A/c$ and E by $E - e\phi$. Eq. 10.2 then goes over into the form

$$\left(i\hbar\frac{\partial}{\partial t} - e\phi\right)^2 \psi - c^2\left(-i\hbar\nabla - \frac{e\mathbf{A}}{c}\right)^2 \psi - m^2c^4\psi = 0 \tag{10.6}$$

or more explicitly,

$$\frac{1}{c^2}\left(\frac{\partial^2\psi}{\partial t^2} + \frac{2ie}{\hbar}\phi\frac{\partial\psi}{\partial t} + \frac{ie}{\hbar}\frac{\partial\phi}{\partial t}\psi - \frac{e^2\phi^2}{\hbar^2}\psi\right)$$
$$-\left(\nabla^2\psi - \frac{2ie}{\hbar c}\mathbf{A}.\nabla\psi - \frac{ie}{\hbar c}\,\mathrm{div}\mathbf{A}\,\psi - \frac{e^2\mathbf{A}^2}{\hbar^2c^2}\psi\right) + \left(\frac{mc}{\hbar}\right)^2\psi = 0 \tag{10.7}$$

If the potentials $\mathbf{A}$, ϕ are independent of t, solutions of the form

$$\psi(\mathbf{x},t) = u(\mathbf{x})\, e^{-iEt/\hbar} \tag{10.8}$$

exist. On substituting this into Eq. 10.6 we find that $u(\mathbf{x})$ must satisfy

$$(E - e\phi)^2 u = [-\hbar^2c^2\nabla^2 + 2ie\hbar c\,\mathbf{A}.\nabla + ie\hbar c\,(\mathrm{div}\,\mathbf{A}) + e^2\mathbf{A}^2 + m^2c^4]\,u \tag{10.9}$$

Let us solve this equation with $\mathbf{A} = 0$ and $e\phi = -Ze^2/r$, which represents a Coulomb field. The resulting equation,

$$\left(E + \frac{Ze^2}{r}\right)^2 u = (-\hbar^2c^2\nabla^2 + m^2c^4)\,u \tag{10.10}$$

[3] This procedure is consistent with relativity theory because $(c\mathbf{p},E)$ transform as a four-dimensional vector, and so do $(\mathbf{A},\phi)$. If a field which is scalar under Lorentz transformations were to be introduced, it would have to be clubbed together with the scalar mc^2 in the equation.

is the relativistic counterpart[4] of Eq. 4.100 with $V(r) = -Ze^2/r$ for a hydrogen-like atom, with a spinless particle replacing the electron. Since the potential is spherically symmetric, we can obtain solutions which are separated into radial and angular parts:

$$u(\mathbf{x}) = R(r)\, Y_{lm}(\theta,\varphi) \tag{10.11}$$

The radial wave equation for $R(r)$ now becomes

$$\left[\frac{1}{r^2}\frac{d}{dr}\left(r^2\frac{dR}{dr}\right) + \frac{E^2 - m^2c^4}{\hbar^2c^2} + \frac{2EZe^2}{\hbar^2c^2r} - \frac{l(l+1) - Z^2e^4/\hbar^2c^2}{r^2}\right]R = 0 \tag{10.12}$$

This equation has precisely the same form as the radial wave equation for the hydrogen atom, except for the values of the constant coefficients. Note that for bound states, E must be *less* than the rest energy mc^2, so that the second term in Eq. 10.12 is negative. (The corresponding term in Eq. 4.130 is $2\mu E/\hbar^2$). The last term in Eq. 10.12 corresponds to the centrifugal term of 4.130; but instead of $[l(l+1)]$ we now have $[l(l+1) - Z^2e^4/\hbar^2c^2]$ which may be written as $s(s+1)$ with

$$s = -\tfrac{1}{2} + \tfrac{1}{2}\,[(2l+1)^2 - 4\gamma^2]^{1/2} \tag{10.13}$$

$$\gamma \equiv Ze^2/\hbar c \equiv Z\alpha \approx (Z/137) \tag{10.14}$$

Now, by performing the scale transformation $\rho = \varkappa r$ and introducing the definitions

$$\varkappa = \frac{2}{\hbar c}\,(m^2c^4 - E^2)^{1/2}, \quad \lambda = \frac{2ZEe^2}{\hbar^2c^2\varkappa} = \frac{2E\gamma}{\hbar c\varkappa} \tag{10.15}$$

we can reduce Eq. 10.12 to

$$\frac{1}{\rho^2}\frac{d}{d\rho}\left(\rho^2\frac{dR}{d\rho}\right) + \left(\frac{\lambda}{\rho} - \frac{1}{4} - \frac{s(s+1)}{\rho^2}\right)R = 0 \tag{10.16}$$

Except for the appearance of s instead of l, this equation is identical to Eq. 4.132 of the hydrogen atom theory, and the solutions obtained there can be taken over directly. In particular, from the requirement that the wave function be well behaved at infinity, we get the condition:

$$\lambda = n' + s + 1, \quad n' = 0,\ 1,\ 2,\ \ldots \tag{10.17}$$

which is the counterpart of Eq. 4.139. In view of the definition (10.15) of λ, this equation states that

$$\gamma E\,(m^2c^4 - E^2)^{-1/2} = n' + \tfrac{1}{2} + [(l+\tfrac{1}{2})^2 - \gamma^2]^{1/2} \tag{10.18}$$

On solving this equation for E, we get the quantized energy levels.

The right hand side of Eq. 10.18 is very nearly $n' + l + 1$, since $\gamma \ll 1$. If the left hand side is to be equal to this finite number despite the appearance of the factor γ, $(m^2c^4 - E^2)$ must clearly be of the order γ^2. So, we can make an expansion of E in powers of γ^2, as $E = mc^2.(1 - a\gamma^2 - b\gamma^4 \ldots)$ where a, b,... are constants. By introducing this expression in (10.18), expanding both sides in powers of γ^2 (keeping terms up to γ^4) and comparing coefficients, we can determine a and b. We find that

$$E = mc^2\left[1 - \frac{\gamma^2}{2n^2} - \frac{\gamma^4}{2n^4}\left(\frac{n}{l+\frac{1}{2}} - \frac{3}{4}\right)\right] \tag{10.19}$$

where n is the principal quantum number, $n = (n' + l + 1)$.

[4] It is to be remembered that E here includes the rest energy mc^2 of the particle.

The first term in $(E - mc^2)$, namely $-(mc^2\gamma^2/2n^2) = -(mZ^2e^4/2\hbar^2n^2)$, gives just the Balmer levels (4.142) of the hydrogen atom. But Eq. 10.19 has an additional term *depending on* l, which gives a fine structure to each of these levels. The total spread of a level due to fine structure, i.e. the change in E as l goes from 0 to $(n - 1)$ for given n, is clearly

$$\frac{mc^2\gamma^4}{2n^3}\left(\frac{1}{\frac{1}{2}} - \frac{1}{n-\frac{1}{2}}\right) = \frac{mc^2\gamma^4}{n^3}\cdot\frac{n-1}{n-\frac{1}{2}} \tag{10.20}$$

This spread is more than twice what the Dirac theory predicts for spin-$\frac{1}{2}$ particles in a Coulomb field (Sec. 10.14), and it is the latter which is in agreement with experimental data on hydrogen-like atoms. This is as it should be, since electrons do have spin $\frac{1}{2}$.

There is one final point which deserves mention. We proved in Sec. 4.12c that the admissible radial wave functions behave like r^l near the origin (cf. Eq. 4.112). The replacement of l by s leads to a behaviour like r^s in the present context. Now, it is easy to see from Eq. 10.13 that with $\gamma \ll 1$, $s \approx l - \gamma^2/(2l + 1)$. This becomes *negative* when $l = 0$ and r^s diverges at the origin in this case. Thus the s-state $(l = 0)$ wave functions are not 'well behaved' at the origin. But the singularity is so mild that the probability integral remains finite, and so these wave functions are admissible.

10.4 Nonrelativistic Limit

The energy E (which here includes the rest energy) is very close to mc^2 in nonrelativistic problems, i.e. $|E - mc^2| \ll mc^2$. Hence if we write the factor $e^{-iEt/\hbar}$ determining the time dependence of any stationary state as $e^{-imc^2t/\hbar}.\chi(t)$ then $|i\hbar\,\partial\chi/\partial t|/mc^2 = |E - mc^2|/mc^2 \ll 1$. In the case of electromagnetic interactions, $|e\phi/mc^2|$ may be taken to be of the same order as the above quantity, since it contributes in the first order to $(E - mc^2)$. On the other hand $|ie\hbar\partial\phi/\partial t|/(mc^2)^2$, $(e\phi)^2/(mc^2)^2$, $|\hbar^2\partial^2\chi/\partial t^2|/(mc^2)^2$, etc. are of the second order of smallness. These statements may be made also if χ is defined for any general wave function through

$$\psi(\mathbf{x},t) = e^{-imc^2t/\hbar}\,\chi(\mathbf{x},t) \tag{10.21}$$

If we now substitute this form into Eq. 10.7 and drop all terms which are of the second order of smallness, it may be verified that it reduces to

$$2mc^2i\hbar\frac{\partial\chi}{\partial t} + [\hbar^2c^2\nabla^2 - 2ie\hbar c\,\mathbf{A}.\nabla - ie\hbar c \operatorname{div}\mathbf{A} - e^2\mathbf{A}^2 - 2mc^2\,e\phi]\,\chi = 0 \tag{10.22}$$

Except for the notation χ instead of ψ, this equation is identical with Eq. 2.24 obtained by the introduction of electromagnetic potentials into the Schrödinger equation. It may be shown that the same limiting process applied to Eq. 10.5a takes $P(\mathbf{x},t)$ into $\psi^*\psi$. It is, indeed, satisfactory that the relativistic equation (10.7) has the correct nonrelativistic limit.

B. THE DIRAC EQUATION

10.5 Dirac's Relativistic Hamiltonian[5]

The Klein-Gordon equation is of the second order in $(\partial/\partial t)$, and so the formalism of quantum mechanics as developed in Chapter 3 does not apply to it. Since $P(\mathbf{x},t)$ is not of the form $\psi^*\psi$, one cannot define expectation values etc. in the usual fashion nor can one even identify operators for all dynamical variables. For instance, a Hamiltonian operator H is not defined; if it were, ψ would satisfy a *first* order equation $i\hbar\, \partial\psi/\partial t = H\psi$. Dirac observed that only a first order equation of this form would be free of the difficulties experienced in connection with the Klein-Gordon equation (Sec. 10.2). He proceeded, therefore, to postulate the existence of a Hamiltonian operator H for relativistic particles, requiring further that it should be *linear* in the momentum components. This requirement was made in order to ensure that (with the replacement $\mathbf{p} \to -i\hbar\nabla$) the wave equation would be linear in the space differential operators as well as in $\partial/\partial t$, thus explicitly preserving the relativistic symmetry between space and time.

The Dirac Hamiltonian is

$$H = c\boldsymbol{\alpha}.\mathbf{p} + \beta mc^2 \tag{10.23}$$

where β and the coefficients α_x, α_y, α_z of the momentum components are as yet undetermined. (The relativistic Hamiltonian must necessarily include the rest energy, and since energy is on the same footing as $c\mathbf{p}$ in relativity theory, H is taken linear in mc^2 too). It is clear that α_x, α_y, α_z and β have to be independent of $\mathbf{p}$ if H is to be linear in $\mathbf{p}$ as postulated. They must be independent of $\mathbf{x}$ and t too; otherwise they would be equivalent to space-time dependent potentials, so that H would no longer describe a *free* particle. Nevertheless α_x, α_y, α_z, β are not just numbers. If they were, the square of the energy (Hamiltonian) operator (10.23) would contain terms proportional to p_xp_y, p_xmc^2, etc. But the relativistic expression (10.1) for the square of the energy contains no such terms. Dirac, therefore, postulated that α_x, α_y, α_z, β *do not commute* among themselves (though, being independent of $\mathbf{x}$ and $\mathbf{p}$, they commute with these quantities). The commutation relations of α_x, α_y, α_z, β were determined by requiring that H^2 should reduce to $(c^2\mathbf{p}^2 + m^2c^4)$. Writing out H^2 as HH to facilitate observance of the order of noncommuting factors during multiplication we have

$$c^2(\alpha_xp_x + \alpha_yp_y + \alpha_zp_z + \beta mc)\,(\alpha_xp_x + \alpha_yp_y + \alpha_zp_z + \beta mc) = c^2(\mathbf{p}^2 + m^2c^2) \tag{10.24}$$

Comparing the coefficients of $(p_x)^2$, $(p_y)^2$, $(p_z)^2$ and $(mc^2)^2$ on the two sides, we obtain

$$\alpha_x^2 = \alpha_y^2 = \alpha_z^2 = \beta^2 = 1 \tag{10.25a}$$

On the right hand side of Eq. 10.24, products like p_xp_y do not appear. Therefore, the coefficient $(\alpha_x\alpha_y + \alpha_y\alpha_x)$ on the left hand side must vanish. Similar statements hold for coefficients of p_yp_z and p_zp_x. Thus,

[5] P.A.M. Dirac, *Proc. Roy. Soc., Lond.*, **A 117**, 610, 1928.

$$\alpha_x\alpha_y + \alpha_y\alpha_x = \alpha_y\alpha_z + \alpha_z\alpha_y = \alpha_z\alpha_x + \alpha_x\alpha_z = 0 \tag{10.25b}$$

For similar reasons, the coefficients of p_x, p_y, p_z on the left hand side must vanish too. Hence

$$\alpha_x\beta + \beta\alpha_x = \alpha_y\beta + \beta\alpha_y = \alpha_z\beta + \beta\alpha_z = 0 \tag{10.25c}$$

These are the essential constraints on the α's and β.

The simplest noncommuting quantities (independent of space-time variables and differential operators) are matrices. Therefore, α_x, α_y, α_z, β are taken to be matrices. According to Eqs. 10.25b and 10.25c, each of these matrices *anticommutes* with every other, i.e. their product changes sign when the order of factors is reversed ($\alpha_x\alpha_y = -\alpha_y\alpha_x$, etc.). Further, each matrix has to be *Hermitian*, in order that H be Hermitian. So, the Dirac equation is

$$i\hbar\frac{\partial\psi}{\partial t} = H\psi = -i\hbar c\boldsymbol{\alpha}.\nabla\psi + \beta mc^2\psi \tag{10.26}$$

where $\alpha_x, \alpha_y, \alpha_z$, β are constant matrices which are Hermitian and satisfy Eqs. 10.25. Their explicit forms are as yet unspecified. Since ψ is multiplied by these matrices from the left, ψ itself must be a *column* with the same number of elements or components as the dimension of the matrix. We shall see below that this number is 4. Each of the components ψ_1, ψ_2, ψ_3, ψ_4 of ψ is still some function of **x** and t.

10.6 Position Probability Density; Expectation Values

For many purposes it is unnecessary to know anything about the matrices α_x, α_y, α_z, β beyond the defining properties (10.25). For instance, we shall now show quite generally that the position probability density $P(\mathbf{x},t)$ for the Dirac particle can be identified with $\psi^\dagger\psi$, where $\psi^\dagger$ is a *row* consisting of the elements ψ_1^*, ψ_2^*, ψ_3^*, ψ_4^*:

$$P(\mathbf{x},t) = \psi^\dagger(\mathbf{x},t)\,\psi(\mathbf{x},t) = \psi_1^*\psi_1 + \psi_2^*\psi_2 + \psi_3^*\psi_3 + \psi_4^*\psi_4 \tag{10.27}$$

To verify that this quantity does obey a continuity equation of the form (2.36), we need the equation satisfied by $\psi^\dagger$. This can be obtained by taking the Hermitian conjugate[6] of Eq. 10.26.

$$-i\hbar\frac{\partial\psi^\dagger}{\partial t} = i\hbar c\nabla\psi^\dagger.\boldsymbol{\alpha} + mc^2\psi^\dagger\beta \tag{10.28}$$

From Eqs. 10.26 and 10.28 we see that

$$\begin{aligned} i\hbar\frac{\partial}{\partial t}(\psi^\dagger\psi) &= i\hbar\frac{\partial\psi^\dagger}{\partial t}.\psi + \psi^\dagger.i\hbar\frac{\partial\psi}{\partial t} \\ &= -(i\hbar c\nabla\psi^\dagger.\boldsymbol{\alpha} + mc^2\psi^\dagger\beta)\,\psi + \psi^\dagger\,(-i\hbar c\boldsymbol{\alpha}.\nabla\psi + \beta mc^2\psi) \\ &= -i\hbar c\nabla.(\psi^\dagger\boldsymbol{\alpha}\psi) = -i\hbar\,\text{div}\,\mathbf{S}(\mathbf{x},t) \end{aligned} \tag{10.29}$$

where

$$\mathbf{S}(\mathbf{x},t) = c\,(\psi^\dagger\boldsymbol{\alpha}\psi) \tag{10.30}$$

Thus P and **S** satisfy the continuity equation and may be interpreted as probability density and probability current density respectively. Since P is evidently positive definite, the interpretational difficulties experienced with the Klein-Gordon equation do not arise here.

[6] Remember that the order of factors in any matrix product is reversed by Hermitian conjugation. We also use the fact that α, β are Hermitian, $\alpha = \alpha^\dagger$, $\beta = \beta^\dagger$.

The normalization of the Dirac wave function is done by requiring that

$$\int P(\mathbf{x},t)\, d^3x \equiv \int \psi^\dagger(\mathbf{x},t)\, \psi(\mathbf{x},t)\, d^3x = 1 \tag{10.31}$$

and the expectation value of any operator A is defined by

$$\langle A \rangle = \int \psi^\dagger A \psi \, d^3x \tag{10.32}$$

It is to be noted that A may be a matrix, *or* an 'orbital' operator formed from $\mathbf{x}$ and $\mathbf{p}$ and not involving matrices, *or* a combination of both (as is the case with the Hamiltonian). Orbital operators act independently on all the components of ψ: e.g. $\mathbf{p}\psi$ is the column with components $\mathbf{p}\psi_1$, $\mathbf{p}\psi_2$, $\mathbf{p}\psi_3$, $\mathbf{p}\psi_4$. Matrix operators mix up the different components according to the ordinary laws of matrix multiplication.

10.7 Dirac Matrices

To get a clearer idea of the nature of the Dirac equation (10.26), we need to determine explicitly a set of matrices obeying Eqs. 10.25. Since it is known that the eigenvalues of a matrix satisfy any algebraic equation obeyed by the matrix, we can conclude, from Eqs. 10.25a, that the only eigenvalues that any of our matrices can have are $+1$ and -1. Secondly, we can write $\alpha_x = \alpha_x\beta^2 = \alpha_x\beta\beta = -\beta\alpha_x\beta$. On taking the trace ($=$ sum of diagonal elements $=$ sum of eigenvalues) of the matrices on both sides, we have $\mathrm{tr}\,\alpha_x = -\mathrm{tr}\,\beta\alpha_x\beta = -\mathrm{tr}\,\alpha_x\beta^2 = -\mathrm{tr}\,\alpha_x$. (The penultimate step follows because the trace of a product of matrices is unaltered by transferring the first matrix to the end). Thus $\mathrm{tr}\,\alpha_x = 0$ which means that α_x must have the same number of $+1$ eigenvalues as -1. The same is true of the other three matrices. One consequence of this result is that the dimension n of the matrices has to be even. The simplest possibility, $n = 2$ is ruled out because there are just three anticommuting matrices of this dimension, while we need four. The three Pauli matrices

$$\sigma_x = \begin{pmatrix} 0 & 1 \\ 1 & 0 \end{pmatrix}, \quad \sigma_y = \begin{pmatrix} 0 & -i \\ i & 0 \end{pmatrix}, \quad \sigma_z = \begin{pmatrix} 1 & 0 \\ 0 & -1 \end{pmatrix} \tag{10.33}$$

do anticommute, and the square of each is the unit matrix:

$$\sigma_x\sigma_y = -\sigma_y\sigma_x = i\sigma_z, \quad \sigma_y\sigma_z = -\sigma_z\sigma_y = i\sigma_x,$$
$$\sigma_z\sigma_x = -\sigma_x\sigma_z = i\sigma_y; \quad \sigma_x^2 = \sigma_y^2 = \sigma_z^2 = 1 \tag{10.34}$$

But if one takes any other 2×2 matrix with arbitrary elements and requires that it should anticommute with these, one finds immediately that all its elements must vanish.

With dimension $n = 4$, one *can* find four anticommuting matrices. In fact the following set of matrices does satisfy all the Eqs. 10.25 and also the condition of Hermiticity:

$$\beta = \left(\begin{array}{cc:cc} 1 & 0 & 0 & 0 \\ 0 & 1 & 0 & 0 \\ \hdashline 0 & 0 & -1 & 0 \\ 0 & 0 & 0 & -1 \end{array}\right), \quad \alpha_x = \left(\begin{array}{cc:cc} 0 & 0 & 0 & 1 \\ 0 & 0 & 1 & 0 \\ \hdashline 0 & 1 & 0 & 0 \\ 1 & 0 & 0 & 0 \end{array}\right)$$

$$\alpha_y = \left(\begin{array}{cc|cc} 0 & 0 & 0 & -i \\ 0 & 0 & i & 0 \\ \hline 0 & -i & 0 & 0 \\ i & 0 & 0 & 0 \end{array}\right), \quad \alpha_z = \left(\begin{array}{cc|cc} 0 & 0 & 1 & 0 \\ 0 & 0 & 0 & -1 \\ \hline 1 & 0 & 0 & 0 \\ 0 & -1 & 0 & 0 \end{array}\right) \tag{10.35}$$

The dotted 'partition' lines serve to bring out that these matrices are made up of 2 × 2 blocks which are Pauli matrices or the unit matrix I. They may be written more compactly in terms of these 2 × 2 *submatrices* as

$$\beta = \begin{pmatrix} I & 0 \\ 0 & -I \end{pmatrix}, \alpha_x = \begin{pmatrix} 0 & \sigma_x \\ \sigma_x & 0 \end{pmatrix}, \alpha_y = \begin{pmatrix} 0 & \sigma_y \\ \sigma_y & 0 \end{pmatrix}, \alpha_z = \begin{pmatrix} 0 & \sigma_z \\ \sigma_z & 0 \end{pmatrix} \tag{10.36}$$

Such partitioned matrices can be multiplied as if the submatrices were just elements of an ordinary matrix, except that care must be taken now to retain the order of the factors unchanged. For example,

$$\alpha_x\alpha_y = \begin{pmatrix} \sigma_x\sigma_y & 0 \\ 0 & \sigma_x\sigma_y \end{pmatrix} = \begin{pmatrix} i\sigma_z & 0 \\ 0 & i\sigma_z \end{pmatrix} \equiv i\,\Sigma_z \tag{10.37}$$

say, while $\alpha_y\alpha_x = -i\,\Sigma_z$ because $\sigma_y\,\sigma_x = -i\sigma_z$ occurs in this product. The use of the partitioned forms (10.36) helps to simplify some of the calculations which follow.

It must be particularly noted that the set (10.35) or (10.36) is by no means a unique solution of Eqs. 10.25. In fact it is evident that any set $\boldsymbol{\alpha}' = S\boldsymbol{\alpha}S^{-1}$, $\beta' = S\beta S^{-1}$ where S is any nonsingular matrix, will also satisfy Eqs. 10.25. It can be shown, conversely, that any set of 4 × 4 matrices obeying Eqs. 10.25 is related to every other such set by a similarity transformation. So, all such sets are equivalent, and the choice of a particular set such as (10.36) is merely a matter of convenience for particular purposes. An alternative set which we shall find useful later, is

$$\alpha_x = \begin{pmatrix} \sigma_x & 0 \\ 0 & -\sigma_x \end{pmatrix}, \alpha_y = \begin{pmatrix} 0 & I \\ I & 0 \end{pmatrix}, \alpha_z = \begin{pmatrix} \sigma_z & 0 \\ 0 & -\sigma_z \end{pmatrix}, \beta = \begin{pmatrix} 0 & iI \\ -iI & 0 \end{pmatrix} \tag{10.38}$$

It is obvious that if we constructed matrices of dimension 8, 12, 16..., in the form

$$\alpha_x = \begin{pmatrix} \alpha_x' & 0 & 0 & . \\ 0 & \alpha_x'' & 0 & . \\ 0 & 0 & \alpha_x''' & . \\ . & . & . & . \end{pmatrix} \text{etc.,} \quad \beta = \begin{pmatrix} \beta' & 0 & 0 & . \\ 0 & \beta'' & 0 & . \\ 0 & 0 & \beta''' & . \\ . & . & . & . \end{pmatrix} \tag{10.39}$$

where $(\alpha_x', \alpha_y', \alpha_z', \beta')$, $(\alpha_x'', \alpha_y'', \alpha_z'', \beta'')$..are individually sets of 4 × 4 matrices satisfying Eqs. 10.25, then the large matrices (10.39) would also obey Eqs. 10.25. It can be shown that *all solutions* of Eqs. 10.25 can be brought to the form (10.39) by a similarity transformation. However, using them in the Dirac equation would be equivalent to writing a number of separate equations with 4 × 4 matrices and this does not give anything new. So, we need consider only the Dirac equation with 4 × 4 matrices α_x, α_y, α_z, β.

10.8 Plane Wave Solutions of the Dirac Equation; Energy Spectrum

We now return to Eq. 10.26 and seek plane wave solutions of the form

$$\psi(\mathbf{x},t) = u(\mathbf{p})\, e^{i(\mathbf{p}.\mathbf{x}-Et)/\hbar} \tag{10.40}$$

which characterize particles of given momentum $\mathbf{p}$. The energy E is to be determined by substituting Eq. 10.40 into the Dirac equation. We get

$$E\,u(\mathbf{p}) = (c\boldsymbol{\alpha}.\mathbf{p} + \beta mc^2)\, u(\mathbf{p}) \tag{10.41}$$

Let us take α_x, α_y, α_z, β to be the 4×4 matrices (10.35), so that ψ and u would now be 4-component columns on which these matrices can operate. To take advantage of the partitioned form (10.36) of α_x, α_y, α_z and β, we shall imagine u also to be partitioned into two two-component parts:

$$u = \begin{pmatrix} u_1 \\ u_2 \\ u_3 \\ u_4 \end{pmatrix} = \begin{pmatrix} v \\ w \end{pmatrix}; \quad v = \begin{pmatrix} u_1 \\ u_2 \end{pmatrix}, \quad w = \begin{pmatrix} u_3 \\ u_4 \end{pmatrix} \tag{10.42}$$

On substituting Eqs. 10.36 and 10.42 into Eq. 10.41 we obtain

$$E\begin{pmatrix} v \\ w \end{pmatrix} = c\begin{pmatrix} 0 & \boldsymbol{\sigma}.\mathbf{p} \\ \boldsymbol{\sigma}.\mathbf{p} & 0 \end{pmatrix}\begin{pmatrix} v \\ w \end{pmatrix} + mc^2\begin{pmatrix} I & 0 \\ 0 & -I \end{pmatrix}\begin{pmatrix} v \\ w \end{pmatrix} \tag{10.43}$$

Carrying out the matrix multiplications on the right hand side and rearranging terms, we write the last equation as a set of two equations for the 2-component quantities v, w:

$$(E - mc^2)\, v = c\,(\boldsymbol{\sigma}.\mathbf{p})\, w \tag{10.44a}$$

$$(E + mc^2)\, w = c\,(\boldsymbol{\sigma}.\mathbf{p})\, v \tag{10.44b}$$

Multiplying the first of these equations by $(E + mc^2)$ and substituting for $(E + mc^2)\, w$ from the second, we get

$$(E^2 - m^2c^4)\, v = c^2\,(\boldsymbol{\sigma}.\mathbf{p})^2\, v = c^2\mathbf{p}^2 v \tag{10.45}$$

The last step follows from Eqs. 10.34. Thus we find that the column v must vanish identically unless the numerical factor $(E^2 - c^2\mathbf{p}^2 - m^2c^4)$ is zero. Hence

$$E = \pm\,(c^2\mathbf{p}^2 + m^2c^4)^{1/2} \equiv E_{\pm} \tag{10.46}$$

The same result would follow if we eliminated v rather than w from Eqs. 10.44. Eq. 10.46 is just what one should expect from the fact that E, according to Eq. 10.41, is an eigenvalue of the Dirac Hamiltonian H, and we have required at the very beginning that $H^2 = (c^2\mathbf{p}^2 + m^2c^4)$.

We note that the energy spectrum of a free Dirac particle consists of *two branches* corresponding to the two signs in (10.46). The positive branch starts at $E = mc^2$ (when $\mathbf{p} = 0$) and extends to $+\infty$ ($|\mathbf{p}| \to \infty$), while the negative energies begin at $E = -mc^2$ and go down to $-\infty$ as $|\mathbf{p}| \to \infty$. There is a 'forbidden gap' of width $2mc^2$ between the two branches, within which no energy levels exist. The presence of levels with infinitely negative energies poses a serious problem. Deferring this problem to Sec. 10.10 we now determine the eigen functions $u(\mathbf{p})$ belonging to any energy eigenvalue E.

We observe first of all that we may take either v or w to be an *arbitrary* two-component column, and then determine the other from *one* of the Eqs. 10.44. (If we combine the two equations, we simply get Eq. 10.46 as already seen; no further restriction on v or w is obtained). We will use Eq. 10.44b for determining positive energy solutions ($E = E_+$) because Eq. 10.44a is indeterminate for the particular (positive) energy $E = mc^2$ corresponding to $\mathbf{p} = 0$. For similar reasons we use Eq. 10.44a when $E = E_-$. Thus we write

$$w = \frac{c\,(\boldsymbol{\sigma}.\mathbf{p})\,v}{E_+ + mc^2}. \quad (E = E_+) \tag{10.47}$$

$$v = -\frac{c\,(\boldsymbol{\sigma}.\mathbf{p})\,w}{E_+ + mc^2}, \quad (E = E_- = -E_+) \tag{10.48}$$

These equations reveal one very useful fact. If the particle is nonrelativistic ($cp \ll mc^2$, $E_+ - mc^2 \ll mc^2$ i.e. $E_+ \approx mc^2$), w is of much smaller magnitude than v for positive energy solutions. According to Eq. 10.47, their ratio is $\approx (cp/2E_+) = \frac{1}{2}v_p/c$, where v_p is the velocity of the particle.[7] One often refers to the upper half (v) of the wave functions as the *large components* and the lower half (w) as the *small components* in nonrelativistic situations. For negative energy solutions, the positions of the large and small components are reversed, as is evident from Eq. 10.48.

We proceed now to solve the above equations explicitly. We note that Eq. 10.47 provides *two* independent solutions, since there are two linearly independent possibilities for the two-component column v. If we take the simplest possibilities $\begin{pmatrix}1\\0\end{pmatrix}$ and $\begin{pmatrix}0\\1\end{pmatrix}$ for the independent two-component objects, and use the fact that

$$\boldsymbol{\sigma}.\mathbf{p} = \sigma_x p_x + \sigma_y p_y + \sigma_z p_z = \begin{pmatrix} p_z & p_x - ip_y \\ p_x + ip_y & -p_z \end{pmatrix} \tag{10.49}$$

$$\boldsymbol{\sigma}.\mathbf{p}\begin{pmatrix}1\\0\end{pmatrix} = \begin{pmatrix} p_z \\ p_x + ip_y \end{pmatrix}, \quad \boldsymbol{\sigma}.\mathbf{p}\begin{pmatrix}0\\1\end{pmatrix} = \begin{pmatrix} p_x - ip_y \\ -p_z \end{pmatrix} \tag{10.50}$$

we get from Eq. 10.47 two independent *positive energy solutions:*

$$v = \begin{pmatrix}1\\0\end{pmatrix}, \quad w = \begin{pmatrix} cp_z/(E_+ + mc^2) \\ cp_+/(E_+ + mc^2) \end{pmatrix}, \quad u = \begin{pmatrix} v \\ w \end{pmatrix} \tag{10.51}$$

$$v = \begin{pmatrix}0\\1\end{pmatrix}, \quad w = \begin{pmatrix} cp_-/(E_+ + mc^2) \\ -cp_z/(E_+ + mc^2) \end{pmatrix}, \quad u = \begin{pmatrix} v \\ w \end{pmatrix} \tag{10.52}$$

Here we have used the abbreviations

$$p_+ = p_x + ip_y, \; p_- = p_x - ip_y \tag{10.53}$$

The last statement in each of Eqs. 10.51 and 10.52 is a reminder that the four-component column u is made up of v and w.

In a similar way, using the same two linearly independent two-component objects now for w in Eq. 10.48, we get, with the aid of Eqs. 10.50, the following *negative energy solutions:*

$$v = \begin{pmatrix} -cp_z/(E_+ + mc^2) \\ -cp_+/(E_+ + mc^2) \end{pmatrix}, \quad w = \begin{pmatrix}1\\0\end{pmatrix}, \quad u = \begin{pmatrix} v \\ w \end{pmatrix} \tag{10.54}$$

$$v = \begin{pmatrix} -cp_-/(E_+ + mc^2) \\ cp_z/(E_+ + mc^2) \end{pmatrix}, \quad w = \begin{pmatrix}0\\1\end{pmatrix}, \quad u = \begin{pmatrix} v \\ w \end{pmatrix} \tag{10.55}$$

For each solution we can arrange to make $u^\dagger u \equiv u_1{}^*u_1 + u_2{}^*u_2 + u_3{}^*u_3 + u_4{}^*u_4 = 1$, by multiplying all components by $\{1 + [c^2\mathbf{p}^2/(E_+ + mc^2)^2]\}^{-/12} = [\frac{1}{2}\,(1 + mc^2/E_+)]^{1/2}$.

[7] We use v_p for 'particle velocity' since the symbol v is already reserved for the upper half of the Dirac wave function.

We have now obtained *four* independent solutions of the Dirac equation *for a given momentum* **p**. It remains to see how the negative energy solutions are to be understood, and what the reason is for the existence of *two* positive energy (or negative energy) solutions. The answer to the second question will now be provided, by showing that the Dirac particle has spin-$\frac{1}{2}$ so that there are two independent spin orientations for given momentum and energy.

10.9 The Spin of the Dirac Particle

It is an elementary fact of mechanics that the angular momentum of any free particle (indeed, of any isolated system) should be conserved. Let us now consider any component of the angular momentum operator $\mathbf{L} = \mathbf{x} \times \mathbf{p}$, say L_z, and see whether it is conserved for a particle whose Hamiltonian is given by Eq. 10.23. In view of (3.87) we have[8]

$$\begin{aligned} i\hbar \frac{dL_z}{dt} &= [L_z, H] = [xp_y - yp_x, c\boldsymbol{\alpha}.\mathbf{p} + \beta mc^2] \\ &= [xp_y, c\alpha_x p_x] - [yp_x, c\alpha_y p_y] \\ &= i\hbar c\,(\alpha_x p_y - \alpha_y p_x) \end{aligned} \qquad (10.56)$$

In evaluating the commutator we have used the fact that x and y from L_z do not commute with p_x and p_y respectively from H, but all other quantities commute. Eq. 10.56 shows that L_z is *not* conserved. This can only mean that **L** is not the complete angular momentum; there must be another part (spin angular momentum **S**) such that $\mathbf{L} + \mathbf{S}$ is conserved. We shall now verify that if S_z is taken to be $\frac{1}{2}\hbar\Sigma_z$ where Σ_z is defined by Eq. 10.37, then $(L_z + S_z)$ does commute with H. We note first that $\Sigma_z \equiv -i\alpha_x\alpha_y$ commutes with α_z and β by virtue of the anticommutation rules (10.25b). For example $\alpha_x\alpha_y\alpha_z = -\alpha_x\alpha_z\alpha_y = \alpha_z\alpha_x\alpha_y$. But $[\alpha_x\alpha_y, \alpha_x] = \alpha_x\alpha_y\alpha_x - \alpha_x\alpha_x\alpha_y = -\alpha_y\alpha_x^2 - \alpha_x^2\alpha_y = -2\alpha_y$ and similarly $[\alpha_x\alpha_y, \alpha_y] = 2\alpha_x$. On using these results we obtain[8]

$$\begin{aligned} i\hbar \frac{d}{dt}(\tfrac{1}{2}\hbar\Sigma_z) = \tfrac{1}{2}\hbar[\Sigma_z, H] &= -\tfrac{1}{2}i\hbar\,[\alpha_x\alpha_y, c\,\alpha_x p_x + c\alpha_y p_y] \\ &= -i\hbar\,c\,[\alpha_x p_y - \alpha_y p_x] \end{aligned} \qquad (10.57)$$

From Eqs. 10.56 and 10.57 we see that $(L_z + \frac{1}{2}\hbar\Sigma_z)$ is conserved, and therefore $S_z = \frac{1}{2}\hbar\Sigma_z$ may be taken as the z-component of the spin angular momentum operator. More generally,

$$\mathbf{S} = \tfrac{1}{2}\hbar\boldsymbol{\Sigma};\ \Sigma_x = -i\alpha_y\alpha_z,\ \Sigma_y = -i\alpha_z\alpha_x,\ \Sigma_z = -i\alpha_x\alpha_y \qquad (10.58a)$$

[8] Since we are using the Schrödinger picture of time evolution (ψ varying with time in accordance with the Dirac equation, and dynamical variables A represented by time-independent operators A_{op}), dL_z/dt in Eq. 10.56 is to be understood as an abbreviation for $(dL_z/dt)_{op}$ — and not $d(L_z)_{op}/dt$ which is zero since $(L_z)_{op} = -i\hbar(x\partial/\partial y - y\partial/\partial x)$ is time-independent. (See Sec. 3.14). Similarly what one has in Eq. 10.57 is $(d\Sigma z/dt)_{op}$, the operator representing the rate of change of the physical quantity whose representative is the matrix Σ_z.

If we switch over to the Heisenberg picture of time evolution (Sec. 9.17) there will be no more Dirac *equation* (since ψ is time-independent in this picture). But all operators will change with time, and the Dirac *Hamiltonian* will control this time-variation through the Heisenberg equation of motion. For L_z and Σ_z (which are time-dependent operators in this picture) we get precisely Eqs. 10.56 and 10.57. The Heisenberg picture is explicitly made use of in Example 10.1, p. 333.

Though these are 4×4 matrices, they have precisely the same algebraic properties as the Pauli matrices:

$$\Sigma_x^2 = \Sigma_y^2 = \Sigma_z^2 = 1, \ \Sigma_x\Sigma_y = i\Sigma_z, \ \Sigma_y\Sigma_z = i\Sigma_x, \ \Sigma_z\Sigma_x = i\Sigma_y \quad (10.58b)$$

With the representation (10.36) for $\boldsymbol{\alpha}$, the explicit form of Σ is

$$\Sigma = \begin{pmatrix} \boldsymbol{\sigma} & 0 \\ 0 & \boldsymbol{\sigma} \end{pmatrix} \quad (10.59)$$

Now that we have the spin operators, it is easy to determine what the value of the spin is. Since $\Sigma_z^2 = 1$ the eigenvalues of Σ_z are $+1$ and -1, and hence those of S_z are $\frac{1}{2}\hbar$ and $-\frac{1}{2}\hbar$. Further,

$$\mathbf{S}^2 = \tfrac{1}{4}\hbar^2 (\Sigma_x^2 + \Sigma_y^2 + \Sigma_z^2) = \tfrac{3}{4}\hbar^2 \quad (10.60)$$

From these facts it is evident that the Dirac equation describes particles of spin $\frac{1}{2}$, such as the electron. The theory based on this equation is commonly referred to as the *Dirac theory of the electron.*

EXAMPLE 10.1 *Zitterbewegung.* The fact that the orbital motion of the electron is coupled to the spin through the term $c(\boldsymbol{\alpha}.\mathbf{p})$ in the Dirac Hamiltonian has rather peculiar consequences. The use of the Heisenberg picture of time evolution facilitates the understanding of these. So let $\mathbf{x}$, $\mathbf{p}$, $\boldsymbol{\alpha}$, $\boldsymbol{\Sigma}$, ... be time-dependent operators which take the familiar forms (e.g. Eq. 10.59 for Σ, $-i\hbar\nabla$ for $\mathbf{p}$) at one instant of time (say $t = 0$). Thereafter they vary according to the Heisenberg equation of motion:

$$(dA/dt) = (i\hbar)^{-1} [A,H], \text{ or } A(t) = e^{iHt/\hbar} A(0) e^{-iHt/\hbar}.$$

Consider the operator for any component of the velocity of the Dirac particle. For example, $(dx/dt) = (i\hbar)^{-1} [x,H] = c\alpha_x$. The eigenvalues of this operator are $\pm c$ which would imply, according to the theory of Sec. 3.9, that the particle can only travel with the velocity of light! As for the acceleration $(d^2x/dt^2) = (d/dt)\, c\alpha_x$, we obtain

$$\frac{d}{dt}(c\alpha_x) = \frac{1}{i\hbar}[c\alpha_x,H] = -\frac{2H}{i\hbar}\,c\alpha_x + \frac{2}{i\hbar}\,c^2 p_x.$$

To get the last form we have used the fact that $H\alpha_x + \alpha_x H = 2cp_x$, by Eqs. 10.25. Since H and p_x remain time-independent (they commute with H) the above equation can be written as

$$\frac{d}{dt}\left(e^{-2iHt/\hbar}\, c\alpha_x\right) = e^{-2iHt/\hbar}\,\frac{2c^2 p_x}{i\hbar},$$

and immediately integrated, yielding

$$e^{-2iHt/\hbar}\, c\alpha_x(t) - c\alpha_x(0) = (e^{-2iHt/\hbar} - 1)\, H^{-1} c^2 p_x,$$

$$\Rightarrow c\alpha_x(t) = H^{-1}c^2 p_x + e^{2iHt/\hbar}\,[c\alpha_x(0) - H^{-1} c^2 p_x]$$

The first term, which in a state of definite energy E becomes $(c^2 p_x/E)$, is just what we would have intuitively expected for the x-component of the velocity, since the classical expression for the velocity of a relativistic particle is $(c^2\mathbf{p}/E)$. The actual velocity operator $c\alpha_x$ differs from this through the second term which oscillates extremely rapidly, the frequency $2E/\hbar$ being $> 2mc^2/\hbar$. This 'jittery motion' or *Zitterbewegung* of the particle, due to the spin-orbit coupling, is what makes the acceleration formally nonzero (despite the particle being free) and leads to $\pm c$ as eigenvalues for velocity. But the Zitterbewegung is much too fast to be observed, and for all practical purposes, the velocity operator is effectively $H^{-1} c^2 p_x$.

10.10 Significance of Negative Energy States; Dirac Particle in Electromagnetic Fields

We now turn to the resolution of the difficulties arising from the existence of solutions with negative energies $E_- = -(c^2\mathbf{p}^2 + m^2c^4)^{1/2}$ which extend down to $-\infty$. The fundamental difficulty is that even a weak electromagnetic field or other perturbation could cause a particle which is in a state of finite positive energy to undergo a quantum transition to a state of energy $E \to -\infty$, resulting in the release of an infinite amount of energy. No such thing takes place in Nature, of course. To get around this problem, which is a serious one, Dirac postulated that *all negative energy states are ordinarily occupied* by electrons and that this 'sea' of negative energy electrons would have no physically observable effects whatever. Since electrons obey Fermi-Dirac statistics (the Pauli exclusion principle) these occupied states cannot accommodate any more electrons. Thus transitions to negative energy states are prevented. One *could* make the occupant of one of the negative energy states jump to a (vacant) positive energy state and thus create a 'hole' in the negative energy sea by supplying enough energy to it. (The amount of energy required for this purpose must necessarily exceed the width $2mc^2$ of the 'gap' between the negative and positive energy branches). The 'hole' represents a deviation from normal; its creation means the *removal* of a certain amount of *negative* energy and *negative* charge from the 'sea', which is equivalent to the *creation* of an equal amount of *positive* energy and charge. Thus the hole manifests itself as a particle of charge $|e|$ and positive energy. The electron whose expulsion created the hole becomes observable too, since it goes to a positive energy state. The whole process may, therefore, be described as the disappearance of the quantum of energy supplied (say a photon) with creation of a pair of observable particles: a positive energy electron and another particle differing from the electron only in the sign of its charge. No such particle was known at the time Dirac's theory was formulated, but it was discovered soon afterwards in cosmic rays, and named *positron*. The discovery of the conversion of high energy gamma rays into electron-positron pairs was a real triumph of the Dirac theory. This led to the conjecture that other particles like the proton would also have their own *antiparticles*, and these have since been discovered.

The association between negative energy electron states and positron states of positive energy, which is the basic idea in the above interpretation, is not a purely *ad hoc* postulate. It can be actually demonstrated by considering the Dirac equation with electromagnetic interaction included. By following the usual prescription $\mathbf{p} \to \mathbf{p} - e\mathbf{A}/c$, $E \to E - e\phi$, we obtain from Eq. 10.26 the following equation for a Dirac particle in electromagnetic fields:

$$i\hbar\frac{\partial\psi}{\partial t} = \boldsymbol{\alpha}.\ (-i\hbar c\nabla - e\mathbf{A})\psi + \beta mc^2\psi + e\phi\psi \tag{10.61}$$

Let $\mathbf{A},\phi$ be time-independent. Then there exist stationary states $\psi(\mathbf{x},t) = u(\mathbf{x})$ $e^{-iEt/\hbar}$ determined by the eigenvalue equation

$$Eu(\mathbf{x}) = \boldsymbol{\alpha}.\ (-i\hbar c\nabla - e\mathbf{A})\ u(\mathbf{x}) + \beta mc^2\ u(\mathbf{x}) + e\phi u(\mathbf{x}) \tag{10.62}$$

We assume the fields to be weak enough so that the energy spectrum does not differ very much from that of the free particle, and in particular, separate positive- and negative-energy branches exist. We make a specific choice of the matrices $\boldsymbol{\alpha},\beta$ such that all the elements of α_x, α_y, α_z are real, while those of β are pure imaginary, e.g. the set (10.38).

Let us now consider a particular state $u(\mathbf{x})$ for which E is negative, i.e. $E = -\ |E|$ in Eq. 10.62. If we now take the complex conjugate of Eq. 10.62, and recall that we have chosen the reality properties of $\boldsymbol{\alpha},\beta$ such that $\alpha^* = \alpha$ and $\beta^* = -\beta$, we get

$$-\ |E|\ u^*(\mathbf{x}) = \boldsymbol{\alpha}.\ (+i\hbar c\nabla - e\mathbf{A})\ u^*(\mathbf{x}) - \beta mc^2 u^*(\mathbf{x}) + e\phi u^*(\mathbf{x}) \tag{10.63}$$

(Note that $\mathbf{A}$ and ϕ, being real fields, are unaffected by the complex conjugation). On multiplying Eq. 10.63 by -1, we discover that $u^*(\mathbf{x})$ satisfies an equation of precisely the same form as Eq. 10.62 *except* that e is replaced by $-e$ throughout, and the eigenvalue is the positive quantity $|E|$. The association between negative energy states of a particle of charge $+e$ and positive energy states of a particle of charge $-e$ is thus made evident. (The fact that it is the *absence* of the negative energy electron which constitutes a positron, is reflected in the *complex conjugate* relationship between the two wave functions).

10.11 Relativistic Electron in a Central Potential: Total Angular Momentum

We have seen in Sec. 10.9 that even for a free electron it is only the total angular momentum $\mathbf{J}$ that is conserved (and not $\mathbf{L}$ and $\mathbf{S}$ individually). If the electron is in a central field characterized by $\mathbf{A} = 0$, $e\phi = V(r)$, the Hamiltonian is

$$H = c\,\boldsymbol{\alpha}.\mathbf{p} + \beta mc^2 + V(r) \tag{10.64}$$

and $\mathbf{J}$ is still conserved since it evidently commutes with the extra term $V(r)$. Therefore, in attempting to solve the eigenvalue problem for this Hamiltonian it would be natural to seek simultaneous eigenstates of H and $\mathbf{J}^2$. However, unlike in the nonrelativistic case where the Schrödinger Hamiltonian could be directly expressed in terms of $\mathbf{L}^2$ and radial operators, we cannot expect to write H of Eq. 10.64 in terms of $\mathbf{J}^2$. This is because H is linear in $\mathbf{p}$ while $\mathbf{J}^2$ involves $\mathbf{p}$ quadratically through its orbital part. It turns out, fortunately, that there exists an operator K which is something like a 'square root' of $\mathbf{J}^2$ (in the same sense that the free Dirac Hamiltonian is a square root of $c^2\mathbf{p}^2 + m^2c^4$). To show this we write

$$\mathbf{J}^2 = \mathbf{L}^2 + 2\mathbf{L}.\mathbf{S} + \mathbf{S}^2 = \mathbf{L}^2 + \hbar\mathbf{L}.\boldsymbol{\Sigma} + \tfrac{3}{4}\hbar^2 \tag{10.65}$$

and use the identity[9]

[9] The proof of this result is the same as that of a similar identity involving Pauli matrices, given in Example 8.2, p. 246.

$$(\mathbf{\Sigma.A})\,(\mathbf{\Sigma.B}) = \mathbf{A.B} + i\mathbf{\Sigma}.(\mathbf{A} \times \mathbf{B}) \tag{10.66}$$

to express $\mathbf{L}^2$ also in terms of $\mathbf{\Sigma.L}$. Taking $\mathbf{A} = \mathbf{B} = \mathbf{L}$ in Eq. 10.66 we get $\mathbf{L}^2 = (\mathbf{\Sigma.L})^2 - i\mathbf{\Sigma}.(\mathbf{L} \times \mathbf{L}) = (\mathbf{\Sigma.L})^2 + \hbar(\mathbf{\Sigma.L})$ and hence from Eq. 10.65,

$$\mathbf{J}^2 = (\mathbf{\Sigma.L})^2 + 2\hbar\mathbf{\Sigma.L} + \tfrac{3}{4}\hbar^2 = (\mathbf{\Sigma.L} + \hbar)^2 - \tfrac{1}{4}\hbar^2 \tag{10.67}$$

where we have used the fact that $\mathbf{L} \times \mathbf{L} = i\hbar\mathbf{L}$. Thus $\mathbf{J}^2 + \frac{1}{4}\hbar^2$ can be expressed as the square of the operator $(\mathbf{\Sigma.L} + \hbar)$ or equally well as

$$\mathbf{J}^2 + \tfrac{1}{4}\hbar^2 = K^2, \quad K = \beta(\mathbf{\Sigma.L} + \hbar) \tag{10.68}$$

The inclusion of the factor β has the valuable consequence that K commutes with H, as may be easily verified.

Now, we know from vector addition of the orbital angular momentum—which can take values $l\hbar$ ($l = 0,1,2\ldots$)—with the spin ($\frac{1}{2}\hbar$) that the possible values of the total angular momentum are given by $j\hbar$ with

$$j = \tfrac{1}{2}, \tfrac{3}{2}, \tfrac{5}{2}, \ldots \tag{10.69}$$

(Note that any value of j may arise either from $l = j - \frac{1}{2}$ or $l = j + \frac{1}{2}$). The first of Eqs. 10.68 tells us that if the eigenvalues of K are denoted by[10]

$k\hbar$, then $k^2 = j(j+1) + \frac{1}{4} = (j + \frac{1}{2})^2$. Hence

$$k = \pm\,(j + \tfrac{1}{2}) = \pm\,1, \pm\,2, \pm\,3, \ldots \tag{10.70}$$

Let us consider the eigenvalue equation $Ku = \hbar ku$. On using the definition (10.68) of K and with the representations (10.36) and (10.59) for β and Σ respectively, this becomes

$$\begin{pmatrix} \boldsymbol{\sigma}.\mathbf{L} + \hbar & 0 \\ 0 & -\boldsymbol{\sigma}.\mathbf{L} - \hbar \end{pmatrix} \begin{pmatrix} v \\ w \end{pmatrix} = k\hbar \begin{pmatrix} v \\ w \end{pmatrix} \tag{10.71}$$

Since an eigenfunction of K is automatically an eigenfunction of $\mathbf{J}^2$ we also have,

$$j(j+1)\hbar^2 \begin{pmatrix} v \\ w \end{pmatrix} = \mathbf{J}^2 \begin{pmatrix} v \\ w \end{pmatrix} = \begin{pmatrix} [\mathbf{L}^2 + \hbar\boldsymbol{\sigma}.\mathbf{L} + \frac{3}{4}\hbar^2]\, v \\ [\mathbf{L}^2 + \hbar\boldsymbol{\sigma}.\mathbf{L} + \frac{3}{4}\hbar^2]\, w \end{pmatrix} \tag{10.72}$$

where we have used Eq. 10.65. Substituting for $\boldsymbol{\sigma}.\mathbf{L}v$ and $\boldsymbol{\sigma}.\mathbf{L}w$ from Eq. 10.71 we obtain

$$\begin{aligned} j(j+1)\,\hbar^2 v &= [\mathbf{L}^2 + (k\hbar - \hbar)\,\hbar + \tfrac{3}{4}\hbar^2]\, v \\ j(j+1)\hbar^2\, w &= [\mathbf{L}^2 - (k\hbar + \hbar)\,\hbar + \tfrac{3}{4}\hbar^2]\, w \end{aligned} \tag{10.73}$$

It is clear now that v and w are eigenstates of $\mathbf{L}^2$, but with *different* values of l. For v, we have $l\,(l+1) = j(j+1) - (k-1) - \frac{3}{4}$, which reduces to $(l + \frac{1}{2})^2 = j^2$ if $k = +\,(j + \frac{1}{2})$ and $(l + \frac{1}{2})^2 = (j+1)^2$ if $k = -\,(j + \frac{1}{2})$. Since j and l are both positive by definition, we are thus led to the following identifications:

$$\text{For } v\text{:}\quad \begin{aligned} k &= j + \tfrac{1}{2} \rightarrow l = j - \tfrac{1}{2} \\ k &= -\,(j + \tfrac{1}{2}) \rightarrow l = j + \tfrac{1}{2} \end{aligned} \tag{10.74a}$$

A similar reduction shows that

$$\text{For } w\text{:}\quad \begin{aligned} k &= j + \tfrac{1}{2} \rightarrow l = j + \tfrac{1}{2} \\ k &= -\,(j + \tfrac{1}{2}) \rightarrow l = j - \tfrac{1}{2} \end{aligned} \tag{10.74b}$$

[10] The quantum number k introduced here is not to be confused with the propagation constant k used in earlier chapters.

We shall find these results to be particularly useful in understanding the outcome of the relativistic theory of the hydrogen atom.

10.12 Radial Wave Equations

We return now to the problem of separating the equation

$$[c\boldsymbol{\alpha}.\mathbf{p} + \beta mc^2 + V(r)]\, u = Eu \tag{10.75}$$

into radial and angular parts. This will be achieved by expressing $(\boldsymbol{\alpha}.\mathbf{p})$ and hence H in terms of K and other quantities which commute with K. To do this we need the following identity:

$$(\boldsymbol{\alpha}.\mathbf{A})\ (\boldsymbol{\alpha}.\mathbf{B}) = \mathbf{A}.\mathbf{B} + i\boldsymbol{\Sigma}.\ (\mathbf{A} \times \mathbf{B}) \tag{10.76}$$

This identity is similar to (10.66) and may be proved readily by using Eqs. 10.25 and the definition of $\boldsymbol{\Sigma}$. It holds whenever $\mathbf{A}$ and $\mathbf{B}$ commute with $\boldsymbol{\alpha}$, whether or not they commute with each other.

Putting $\mathbf{A} = \mathbf{x}$ and $\mathbf{B} = \mathbf{p}$ in Eq. 10.76 we obtain

$$\begin{aligned}(\boldsymbol{\alpha}.\mathbf{x})(\boldsymbol{\alpha}.\mathbf{p}) &= \mathbf{x}.\mathbf{p} + i\boldsymbol{\Sigma}.(\mathbf{x} \times \mathbf{p}) \\ &= \mathbf{x}.\mathbf{p} + i\boldsymbol{\Sigma}.\mathbf{L} = \mathbf{x}.\mathbf{p} + i(\beta K - \hbar)\end{aligned} \tag{10.77}$$

Multiplying the left hand side of Eq. 10.77 with $(\boldsymbol{\alpha}.\mathbf{x})^{-1} = r^{-2}\,(\boldsymbol{\alpha}.\mathbf{x})$, we get

$$\boldsymbol{\alpha}.\mathbf{p} = r^{-2}\,(\boldsymbol{\alpha}.\mathbf{x})\ [\mathbf{x}.\mathbf{p} + i(\beta K - \hbar)] \tag{10.78}$$

Defining the *radial components* α_r and p_r of $\boldsymbol{\alpha}$ and $\mathbf{p}$ as

$$\alpha_r = r^{-1}\ (\boldsymbol{\alpha}.\mathbf{x}) \tag{10.79a}$$

$$p_r = \tfrac{1}{2}\left[\frac{\mathbf{x}}{r}\,.\,\mathbf{p} + \mathbf{p}\,.\,\frac{\mathbf{x}}{r}\right] = \frac{1}{r}\ (\mathbf{x}.\mathbf{p} - i\hbar) \tag{10.79b}$$

we find, from Eq. 10.78 that $\boldsymbol{\alpha}.\mathbf{p} = \alpha_r\ [p_r + ir^{-1}\ \beta K]$, and hence

$$H = c\alpha_r p_r + icr^{-1}\ \alpha_r\beta K + \beta mc^2 - (Ze^2/r) \tag{10.80}$$

It is a straightforward matter to verify that K commutes with all other quantities appearing in Eq. 10.80, and hence with H itself. Therefore, we can choose a complete set of eigenstates of H which are simultaneously eigenstates of K. Considering such eigenstates, we replace K in Eq. 10.80 by its eigenvalue $\hbar k$ where $k = \pm 1, \pm 2, \ldots$ according to Eq. 10.70. The only matrices remaining in H now are α_r and β. If we adopt the representation[11]

$$\beta = \begin{pmatrix} I & 0 \\ 0 & -I \end{pmatrix},\ \boldsymbol{\alpha} = \begin{pmatrix} 0 & -i\boldsymbol{\sigma} \\ i\boldsymbol{\sigma} & 0 \end{pmatrix} \tag{10.81}$$

for $\boldsymbol{\alpha}$ and β, we have, with the notation

$$\sigma_r = r^{-1}\ (\boldsymbol{\sigma}.\mathbf{x})\ \text{and}\ \Sigma_r = \begin{pmatrix} \sigma_r & 0 \\ 0 & \sigma_r \end{pmatrix} \tag{10.82}$$

$$\alpha_r = \begin{pmatrix} 0 & -i\sigma_r \\ i\sigma_r & 0 \end{pmatrix} = \begin{pmatrix} 0 & -iI \\ iI & 0 \end{pmatrix}\Sigma_r,\quad \alpha_r\beta = \begin{pmatrix} 0 & iI \\ iI & 0 \end{pmatrix}\Sigma_r \tag{10.83}$$

It may be verified that σ_r commutes with p_r. Therefore, Σ_r commutes with all quantities in H, and by considering simultaneous eigenstates of H and Σ_r we can replace Σ_r in Eq. 10.80 by its eigenvalue $\varepsilon(= \pm 1)$ just as K was replaced by $\hbar k$. With these simplifications our eigenvalue equation $Hu = Eu$, becomes

[11] The factors of i in this representation of α serve to ensure that the equations finally obtained, Eq. 10.86, involve only real quantities.

$$\left[c\varepsilon p_r\begin{pmatrix}0 & -iI\\ iI & 0\end{pmatrix}+\frac{ic\varepsilon\hbar k}{r}\begin{pmatrix}0 & iI\\ iI & 0\end{pmatrix}+\begin{pmatrix}(mc^2-Ze^2/r)\,I & 0\\ 0 & (-mc^2-Ze^2/r)I\end{pmatrix}\right]\begin{pmatrix}v\\ w\end{pmatrix}$$
$$= E\begin{pmatrix}v\\ w\end{pmatrix} \tag{10.84}$$

where we have written the four-component wave function u in terms of the two-component parts v, w. Further, since $\mathbf{x.p} = -i\hbar\mathbf{x}.\nabla = -i\hbar(\partial/\partial r)$, Eq. 10.79b for p_r can be written as

$$p_r = -i\hbar\left(\frac{\partial}{\partial r}+\frac{1}{r}\right) \tag{10.85}$$

On carrying out the matrix multiplications in Eq. 10.84 and using Eq. 10.85, we are led to the following pair of equations for v and w

$$\left(E-mc^2+\frac{Ze^2}{r}\right)v+\hbar c\varepsilon\left(\frac{\partial}{\partial r}+\frac{1}{r}\right)w+\frac{c\varepsilon\hbar k}{r}\,w=0 \tag{10.86a}$$

$$\left(E+mc^2+\frac{Ze^2}{r}\right)w-\hbar c\varepsilon\left(\frac{\partial}{\partial r}+\frac{1}{r}\right)v+\frac{c\varepsilon\hbar k}{r}\,v=0 \tag{10.86b}$$

If these equations are solved for one sign of ε, the replacement $v\to v$, $w\to -w$ clearly gives the solution for the opposite sign of ε. We shall, therefore, confine our attention to the case $\varepsilon = +1$ hereafter.

The only operators involved in Eqs. 10.86 are derivatives with respect to r. Therefore, we can separate the orbital and spin parts of the wave functions completely and write

$$v=\frac{F(r)}{r}\,v_a,\quad w=\frac{G(r)}{r}\,w_a \tag{10.87}$$

where v_a and w_a are two-component spinors independent of r, and $F(r)$ and $G(r)$ are simple functions of r alone. Then Eqs. 10.86 reduce to coupled ordinary differential equations for F and G:

$$\frac{dF}{dr}-\frac{k}{r}\,F-\frac{1}{\hbar c}\left(E+mc^2+\frac{Ze^2}{r}\right)G=0 \tag{10.88a}$$

$$\frac{dG}{dr}+\frac{k}{r}\,G-\frac{1}{\hbar c}\left(mc^2-E-\frac{Ze^2}{r}\right)F=0 \tag{10.88b}$$

Eqs. 10.88 form the counterpart of the radial wave equation of nonrelativistic theory.

10.13 Series Solutions of the Radial Equations: Asymptotic Behaviour

As in the case of equations dealt with in Chapter 4, we first examine the asymptotic behaviour of the wave functions. As $r\to\infty$, Eqs. 10.88 become

$$\frac{dF}{dr}\approx\frac{1}{\hbar c}\,(E+mc^2)\,G,\qquad \frac{dG}{dr}\approx\frac{1}{\hbar c}\,(mc^2-E)\,F \tag{10.89}$$

so that $d^2F/dr^2 \approx [(m^2c^4-E^2)/\hbar^2c^2]\,F$. Therefore, the asymptotic behaviour of $F(r)$ [and similarly of $G(r)$] is seen to be as $e^{-\varkappa r}$, where

$$\varkappa=\frac{(m^2c^4-E^2)^{1/2}}{\hbar c}=(\varkappa_1\varkappa_2)^{1/2};\ \varkappa_1=\frac{mc^2+E}{\hbar c};\ \varkappa_2=\frac{mc^2-E}{\hbar c} \tag{10.90}$$

We now make the scale transformation $\rho = \varkappa r$ and set

$$F=f(\rho)e^{-\rho};\ G=g(\rho)e^{-\rho};\ \rho=\varkappa r \tag{10.91}$$

With the substitutions (10.90) and (10.91), Eqs. 10.88 become

$$\frac{df}{d\rho} - f - \frac{k}{\rho} f - \left(\frac{\varkappa_1}{\varkappa} + \frac{\gamma}{\rho}\right) g = 0 \tag{10.92a}$$

$$\frac{dg}{d\rho} - g + \frac{k}{\rho} g - \left(\frac{\varkappa_2}{\varkappa} - \frac{\gamma}{\rho}\right) f = 0, \quad \gamma = \frac{Ze^2}{\hbar c} = Z\alpha \tag{10.92b}$$

For any one-electron atom, $Z \ll 137$, and $\gamma \approx Z/137 \ll 1$.

Now we seek series solutions of Eqs. 10.92 as

$$f(\rho) = \rho^s(f_0 + f_1 \rho + f_2\rho^2 + \ldots),\ f_0 \neq 0$$
$$g(\rho) = \rho^s(g_0 + g_1 \rho + g_2\rho^2 + \ldots),\ g_0 \neq 0 \tag{10.93}$$

Here $f_0, f_1, \ldots, g_0, g_1, \ldots$ are constants. Substituting these expansions into Eqs. 10.92 and equating the coefficient of $\rho^{s+\nu-1}$ to zero, we obtain, for $\nu > 0$,

$$(s + \nu - k) f_\nu - f_{\nu-1} - \gamma g_\nu - (\varkappa_1/\varkappa) g_{\nu-1} = 0 \tag{10.94a}$$
$$(s + \nu + k) g_\nu - g_{\nu-1} + \gamma f_\nu - (\varkappa_2/\varkappa) f_{\nu-1} = 0 \tag{10.94b}$$

Considering the coefficients of ρ^{s-1}, we find that

$$(s - k) f_0 - \gamma g_0 = 0, \qquad (s + k) g_0 + \gamma f_0 = 0 \tag{10.95}$$

These equations have nonvanishing solutions f_0, g_0 only if $(s^2 - k^2) + \gamma^2 = 0$ so that

$$s = +\,(k^2 - \gamma^2)^{1/2} \tag{10.96}$$

(The solution $s = -\,(k^2 - \gamma^2)^{1/2}$ is not admissible since that would make f and g diverge at the origin[12]).

We observe now that on multiplying Eq. 10.94a by $\varkappa$, Eq. 10.94b by $\varkappa_1$ and subtracting, the terms in $f_{\nu-1}$ and $g_{\nu-1}$ cancel out because $\varkappa = (\varkappa_1\varkappa_2)^{1/2}$. We are then left with a relation between f_ν and g_ν (for $\nu > 0$).

$$[(s + \nu - k)\,\varkappa - \gamma\varkappa_1]\, f_\nu = [(s + \nu + k)\,\varkappa_1 + \gamma\varkappa]\; g_\nu \tag{10.97}$$

It is possible now to see how $f(\rho)$ and $g(\rho)$ behave as $\rho \to \infty$ if the series (10.93) do not terminate. This asymptotic behaviour is determined by that of the coefficients f_ν, g_ν as $\nu \to \infty$. By substituting Eq. 10.97 in Eq. 10.94 and neglecting terms not involving ν, we find that as $\nu \to \infty$

$$f_\nu \approx \frac{2}{\nu} f_{\nu-1} \text{ and } g_\nu \approx \frac{2}{\nu} g_{\nu-1} \tag{10.98}$$

This is the same behaviour as in the function $e^{2\rho}$, so that both F and G would diverge like e^ρ at infinity. We conclude, therefore, that if F and G are to be admissible wave functions, both the series in (10.93) must terminate at some ν, say $\nu = n'$.

10.14 Determination of the Energy Levels

Let us suppose then that in Eq. 10.93 $f_{n'}$, $g_{n'}$ are nonzero but $f_{n'+1}$, $g_{n'+1}$ (and all higher coefficients) vanish. If we introduce this requirement in Eqs. 10.94 with ν set equal to $n' + 1$, they reduce to

[12] When k^2 has its lowest value ($k^2 = 1$) in Eq. 10.96, s is slightly less than unity, and then v and w diverge mildly at the origin. But the volume integrals of $v^\dagger v$ and $w^\dagger w$ are still finite. So we admit this value of s. It will be recalled that a similar situation arose in the case of the Klein-Gordon equation.

$$f_{n'} = -(\varkappa_1/\varkappa)\ g_{n'} \tag{10.99}$$

This relation must be compatible with Eq. 10.97, which requires

$$[(s + n' - k)\varkappa - \gamma\varkappa_1]\ \varkappa_1 = -[(s + n' + k)\ \varkappa_1 + \gamma\varkappa]\ \varkappa \tag{10.100}$$

Since $\varkappa^2 = \varkappa_1\varkappa_2$, this reduces to

$$2\varkappa\ (s + n') = \gamma(\varkappa_1 - \varkappa_2) \tag{10.101}$$

Squaring both sides and using the expressions (10.90) for $\varkappa_1$, $\varkappa_2$, $\varkappa$, we obtain $(m^2c^4 - E^2)\ (s + n')^2 = E^2\ \gamma^2$ and hence

$$E = mc^2\left[1 + \frac{\gamma^2}{(s+n')^2}\right]^{-1/2} = mc^2\left[1 + \frac{\gamma^2}{\{(k^2-\gamma^2)^{1/2} + n'\}^2}\right]^{-1/2} \tag{10.102}$$

The derivation of this result is not strictly valid in the case $n' = 0$, since Eq. 10.100 was obtained by setting $\nu = n'$ in Eq. 10.97 which is valid only for $\nu > 0$. When $n' = 0$ the compatibility of Eq. 10.99 with Eq. 10.95 must be considered instead. This leads to the requirement

$$\frac{\gamma}{s-k} = -\frac{s+k}{\gamma} = -\frac{\varkappa_1}{\varkappa},\quad n' = 0 \tag{10.103}$$

Hence $[\gamma/(s-k)]\ [(s+k)/\gamma] = (s+k)/(s-k) = -(\varkappa_1/\varkappa)^2 = -(\varkappa_1/\varkappa_2)$, or

$$(\varkappa_1 + \varkappa_2)\ s = k\ (\varkappa_1 - \varkappa_2),\quad n' = 0 \tag{10.104}$$

On squaring this and using Eq. 10.90 we obtain $E = mc^2\ s/k$ which may be verified to be the same as Eq. 10.102 with $n' = 0$. However, unlike Eq. 10.100 for $n' > 0$ from which k cancelled off, Eq. 10.104 involves k explicitly and in fact it requires that k be *positive* since all the other factors are positive. Thus the values which n' and k in Eq. 10.102 may take are

$$\begin{aligned} n' &= 1,2,3,\ldots, & k &= \pm 1, \pm 2, \pm 3,\ldots \\ n' &= 0, & k &= 1,2,3,\ldots \end{aligned} \tag{10.105}$$

It may be noted that the denominator in the expression (10.102) deviates only slightly from

$$n \equiv n' + |k| \tag{10.106}$$

Since $\gamma \ll 1$, we can write $(k^2 - \gamma^2)^{1/2} \approx |k|\ (1 - \gamma^2/2k^2)$, so that $n' + (k^2 - \gamma^2)^{1/2} \approx n - \gamma^2/2\,|k|$. The expression for E may then be seen to be given, to order γ^4, by

$$E = mc^2\left[1 - \frac{\gamma^2}{2n^2} - \frac{\gamma^4}{2n^4}\left(\frac{n}{|k|} - \frac{3}{4}\right)\right] \tag{10.107}$$

The leading term in $(E - mc^2)$, namely $(-mc^2\gamma^2/2n^2)$ gives just the nonrelativistic formula (4.142). The term of order γ^4 represents the fine structure caused by the dependence on k. The spread in energy of the fine structure levels (i.e., the difference in E between the extreme values $|k| = 1$ and $|k| = n$ for given n) is $(mc^2\gamma^4/2n^3)\ (n-1)/2n$. This is much less than the corresponding spread, Eq. 10.20, for a spinless particle in a Coulomb field. The observed fine structure of the hydrogen atom is in fact very well reproduced by Eq. 10.102. It is a curious fact that precisely the same formula was obtained by Sommerfeld from the Old Quantum Theory — rather fortuitously, since Sommerfeld's treatment did not take into account the spin (which in fact had not been discovered at that time).

Finally we note that the energy eigenfunctions of the present problem have the property that the two upper components v are much larger than the lower ones. A quick estimate of their relative magnitudes may be got from the ratio g_0/f_0 of the leading terms in the expansion of $g(\rho)$ and $f(\rho)$, to which w and v respectively are proportional. We see from the second of Eqs. 10.95 that $g_0/f_0 = -\gamma/(s+k) \approx -\gamma/2k \ll 1$. This result is not surprising since we have already seen a similar phenomenon in the case of a (free) nonrelativistic Dirac particle: in the present problem also the particle is to be considered nonrelativistic, since $|E - mc^2| \ll mc^2$ according to Eq. 10.107. Under these circumstances one can validly ascribe to the eigenfunction u an orbital angular momentum l, namely that associated with the large components v, ignoring the different value pertaining to w (see Sec. 10.11). Of the $2(2l+1)$ states with given l, those with $j = l + \frac{1}{2}$ have a higher energy than those with $j = l - \frac{1}{2}$. This follows from the energy level formula (10.102) which shows that E increases with $|k|$ and hence with j (Remember $|k| = j + \frac{1}{2}$). In fact it is seen from this formula that E depends only on j besides n.[13]

10.15 Electron in a Magnetic Field — Spin Magnetic Moment

(*a*) *Energy Spectrum and Eigenfunctions:* When there is a magentic field $\mathcal{H}$ and no electric field, we can take the scalar potential ϕ to be zero. In such a field, Eq. 10.62 for a Dirac particle reduces to

$$(c\boldsymbol{\alpha}.\boldsymbol{\pi} + \beta mc^2)\, u = Eu \tag{10.108}$$

This differs from Eq. 10.41 of the free Dirac particle only in having $\boldsymbol{\pi} \equiv \mathbf{p} - e\mathbf{A}/c$ instead of $\mathbf{p}$ and can be reduced to a pair of equations similar to (10.44):

$$(E - mc^2)v = c(\boldsymbol{\sigma}.\boldsymbol{\pi})w; \quad (E + mc^2)w = c(\boldsymbol{\sigma}.\boldsymbol{\pi})v \tag{10.109}$$

($v(\mathbf{x})$ and $w(\mathbf{x})$ are the upper and lower halves of u). Hence $c^2(\boldsymbol{\sigma}.\boldsymbol{\pi})^2 v = (E + mc^2)\, c(\boldsymbol{\sigma}.\boldsymbol{\pi})\, w = (E^2 - m^2c^4)\, v$. This equation, which involves v only, can be expressed as

$$(E^2 - m^2c^4 - c^2\pi^2 + e\hbar c\boldsymbol{\sigma}.\mathcal{H})\; v(\mathbf{x}) = 0 \tag{10.110}$$

in view of the fact that by (8.33d) and (4.207),

$$(\boldsymbol{\sigma}.\boldsymbol{\pi})^2 = \pi^2 + i\boldsymbol{\sigma}.\boldsymbol{\pi} \times \boldsymbol{\pi} = \pi^2 - (e\hbar/c)\boldsymbol{\sigma}.\mathcal{H} \tag{10.111}$$

An identical equation obtains for $w(\mathbf{x})$ also. For a homogeneous magnetic field in the z-direction ($\boldsymbol{\sigma}.\mathcal{H} = \sigma_z \mathcal{H}$) this equation (whether for v or w) decouples into separate equations since σ_z is diagonal (with eigenvalues $+1$ and -1). Thus we have

$$c^2\pi^2\, u_i(\mathbf{x}) = (E^2 - m^2c^4 + \varepsilon_i e\hbar c\mathcal{H})\; u_i(\mathbf{x}) \tag{10.112}$$

$$\varepsilon_i = +1 \text{ (spin up) for } i = 1,3; \quad \varepsilon_i = -1 \text{ (spin down) for } i = 2,4$$

It is evident that these are just eigenvalue equations for the operator π^2. But this problem has already been solved in Sec. 4.23: From Eq. 4.214 which

[13] According to this result, the $^2s_{1/2}$ and $^2p_{1/2}$ states of the $n = 2$ level of hydrogen, should have exactly the same energy. But a small difference between the two (called the Lamb shift, after its discoverer) has been observed and precisely accounted for by quantum electrodynamics, wherein electrons (like photons) are treated as excitations of a quantum field.

gives the eigenvalue spectrum of $(\pi^2/2m)$, that of π^2 is obtained as[14]

$$(2n+1)\,|e|\,\hbar\mathcal{H}/c + P^2,\ n = 0,1,2,\ldots;\ -\infty < P < \infty.$$

On using this information in Eq. 10.112 we immediately obtain

$$E^2 = m^2c^4 + c^2P^2 + (2n_i + 1 + \varepsilon_i)\,|e|\,\hbar\mathcal{H}c \tag{10.113}$$

It should be noted that for a particular energy level E (for given momentum P in the z-direction), $(2n_i + \varepsilon_i)$ is fixed, and so the values of the quantum number n for the different components u_i of u are interrelated: $n_1 = n_2 - 1 = n_3 = n_4 - 1$. The functions $u_i(\mathbf{x})$ are then those denoted by $u_{n\perp}.(2\pi)^{-1/2}e^{iPz/\hbar}$ in Eq. 4.214, with the respective values n_i for n. Their explicit forms are determined in Sec. 4.23b. Independent constant factors in u_1 and u_2 remain arbitrary, but once these are chosen, u_3 and u_4 are determined by the second of Eqs. 10.109. It may be noted that even if the upper half of the wave function is chosen to have 'spin up' ($u_2 = 0$), the lower half will contain both 'spin up' and 'spin down' parts: u_3, u_4 are both nonzero. This is not surprising since the Hamiltonian of Eq. 10.108 does not commute with σ_z.

(*b*) *Spin Magnetic Moment:* Consider Eq. 10.110 for the two-component function $v(\mathbf{x})$ — or $w(\mathbf{x})$ — for energies E close to mc^2, i.e., $E' \equiv E - mc^2 \ll mc^2$. Then $E^2 \approx m^2c^4 - 2mc^2E'$. On making this approximation in Eq. 10.110 and dividing by $2mc^2$, it reduces to

$$\left[\frac{1}{2m}\left(\mathbf{p} - \frac{e\mathbf{A}}{c}\right)^2 - \frac{e\hbar}{2mc}\boldsymbol{\sigma}.\mathcal{H}\right]u(\mathbf{x}) = E'u(\mathbf{x}) \tag{10.114}$$

The first term within the square brackets is precisely the nonrelativistic Hamiltonian of a spinless particle in the presence of a vector potential. The presence of spin manifests itself through the second term which is just the interaction energy of a magnetic moment $\boldsymbol{\mu}_s = (e\hbar/2mc)\boldsymbol{\sigma}$ with the magnetic field. Thus the Dirac theory requires a spin magnetic moment of one Bohr magneton $(e\hbar/2mc)$ — a value which had been earlier postulated on purely empirical grounds.[15]

10.16 The Spin Orbit Energy

We shall now see that the Dirac equation also leads automatically to the spin orbit interaction (8.40) in the presence of a central potential $V(r)$. This result is also deduced from inspection of the nonrelativistic limit $|E - mc^2| \ll mc^2$. As a preliminary step we obtain an equation for the upper half of the wave function, eliminating w (without making any approximation). The starting point is Eq. 10.62 wherein we write $e\phi = V$ and introduce the representation (10.36) for α and β:

$$E\begin{pmatrix} v \\ w \end{pmatrix} = c\begin{pmatrix} 0 & \boldsymbol{\sigma}.\pi \\ \boldsymbol{\sigma}.\pi & 0 \end{pmatrix}\begin{pmatrix} v \\ w \end{pmatrix} + \begin{pmatrix} mc^2 & 0 \\ 0 & -mc^2 \end{pmatrix}\begin{pmatrix} v \\ w \end{pmatrix} + V\begin{pmatrix} v \\ w \end{pmatrix} \tag{10.115}$$

This can be separated into the pair of coupled equations

[14] In taking over (4.214), we have replaced e by $|e|$ in accordance with the remark at the end of Sec. 4.23b, since e here is negative (being the charge on the electron).

[15] The magnetic moment of the electron has been measured to a fantastic precision. It differs slightly from one magneton. The difference has been exactly accounted for by quantum electrodynamics.

$$(E - mc^2 - V)\, v = c\boldsymbol{\sigma}.\pi\, w, \; (E + mc^2 - V)\, w = c\boldsymbol{\sigma}.\pi\, v \tag{10.116}$$

Hence $w = (E + mc^2 - V)^{-1}\, c(\boldsymbol{\sigma}.\pi)v$, and substituting this into the first equation we get

$$(E - mc^2)\, v = Vv + c(\boldsymbol{\sigma}.\pi)\,(E + mc^2 - V)^{-1}c(\boldsymbol{\sigma}.\pi)v \tag{10.117}$$

It is necessary to make a number of mathematical manipulations to bring this equation to a form which permits physical interpretation. First, we use the identity

$$AB^{-1} = B^{-1}A - B^{-1}\,[A,B]\,B^{-1} \tag{10.118}$$

which follows from the obvious fact that $[A,BB^{-1}] = 0$, to write

$$(\boldsymbol{\sigma}.\pi)\,(E + mc^2 - V)^{-1}$$
$$= (E + mc^2 - V)^{-1}\,(\boldsymbol{\sigma}.\pi) - (E + mc^2 - V)^{-2}\, i\hbar(\boldsymbol{\sigma}.\text{grad } V). \tag{10.119}$$

In this step we have used the result that

$$[(\boldsymbol{\sigma}.\pi),\, (E + mc^2 - V)] = -\,[(\boldsymbol{\sigma}.\mathbf{p}),V] = i\hbar\boldsymbol{\sigma}.\text{grad } V. \tag{10.120}$$

If Eq. 10.119 is substituted in Eq. 10.117, the combinations $(\boldsymbol{\sigma}.\pi)^2$ and $(\boldsymbol{\sigma}.\text{grad } V)\,(\boldsymbol{\sigma}.\pi)$ appear therein. In view of (8.33d) they can be expressed as

$$(\boldsymbol{\alpha}.\pi)^2 = \pi^2 + i\boldsymbol{\sigma}.\pi \times \pi = \pi^2 - (e\hbar/c)\boldsymbol{\sigma}.\mathcal{H} \tag{10.121}$$
$$(\boldsymbol{\sigma}.\text{grad } V)\,(\boldsymbol{\sigma}.\pi) = (\text{grad } V.\pi) + i\boldsymbol{\sigma}.\,(\text{grad } V \times \pi)$$

The latter can be further simplified if V is spherically symmetric, as we assume:

$$\left.\begin{aligned} \text{grad } V.\mathbf{p} &= \frac{dV}{dr}\frac{\mathbf{x}}{r}\cdot(-\,i\hbar\nabla) = -\,i\hbar\frac{dV}{dr}\frac{\partial}{\partial r} \\ \text{grad } V \times \mathbf{p} &= \frac{dV}{dr}\frac{\mathbf{x}}{r} \times \mathbf{p} = \frac{1}{r}\frac{dV}{dr}\mathbf{L}. \end{aligned}\right\} \tag{10.122}$$

Introducing Eq. 10.119 in 10.117 and then using Eqs. 10.121 together with 10.122, we obtain

$$(E - mc^2)\, v = \Bigg\{ V + \frac{c^2}{E + mc^2 - V}\,\left(\pi^2 - \frac{e\hbar}{c}\,\boldsymbol{\sigma}.\mathcal{H}\right) - \frac{i\hbar c^2}{(E + mc^2 - V)^2} \times$$
$$\left[-\,i\hbar\frac{dV}{dr}\frac{\partial}{\partial r} - \text{grad } V.\frac{e\mathbf{A}}{c} + \frac{i}{r}\frac{dV}{dr}\,\boldsymbol{\sigma}.\mathbf{L} - i\boldsymbol{\sigma}.\,\left(\text{grad } V \times \frac{e\mathbf{A}}{c}\right)\right]\Bigg\}\, v \tag{10.123}$$

We shall now make the nonrelativistic approximation, which has motivated the above (exact) reduction. We write $E = E' + mc^2$, assuming $|E'| \ll mc^2$ and $|V| \ll mc^2$. Then

$$(E + mc^2 - V)^{-1} = (2mc^2 + E' - V)^{-1} \approx \frac{1}{2mc^2}\left(1 - \frac{E' - V}{2mc^2}\right) \tag{10.124}$$

where we have retained only terms up to the second order in $(mc^2)^{-1}$. In the same order of approximation, $(E + mc^2 - V)^{-2} \approx (2mc^2)^{-2}$.

To see the main features of Eq. 10.123 with these approximations, we first write it down assuming that no magnetic field is present ($\mathbf{A} = 0$, $\mathcal{H} = 0$, $\pi = \mathbf{p}$):

$$E'v = \Bigg\{\frac{1}{2m}\left(1 - \frac{E' - V}{2mc^2}\right)\,\mathbf{p}^2 + V - \frac{\hbar^2}{4m^2c^2}\frac{dV}{dr}\frac{\partial}{\partial r}$$
$$+ \frac{1}{2m^2c^2}\cdot\frac{1}{r}\frac{dV}{dr}\,\mathbf{s}.\mathbf{L}\Bigg\}\, v \tag{10.125}$$

Here $\mathbf{s}$ stands for $\frac{1}{2}\hbar\sigma$, the spin operator acting on the two-component wave function v. Grouping together the terms proportional to E' as $E'\,(1 + \mathbf{p}^2/4m^2c^2)$ on the left hand side, we multiply the equation through by $(1 + \mathbf{p}^2/4m^2c^2)^{-1}$.

Then, approximating this factor by $(1 - \mathbf{p}^2/4m^2c^2)$, we carry out the multiplication explicitly, retaining only terms up to order $(mc^2)^{-2}$. The final result is

$$E'v = \left[\frac{\mathbf{p}^2}{2m} + V - \frac{(\mathbf{p}^2)^2}{8m^3c^2} + \frac{\hbar^2}{4m^2c^2}\left\{(\nabla^2 V) + \frac{dV}{dr}\frac{\partial}{\partial r}\right\} + \frac{1}{2m^2c^2}\frac{1}{r}\frac{dV}{dr}\,\mathbf{s.L}\right] v \qquad (10.126)$$

The first two terms in Eq. 10.126 give just the nonrelativistic Schrödinger Hamiltonian. The next term is the lowest order *relativistic correction* to the kinetic energy $(\mathbf{p}^2/2m)$, as may be seen from the expansion of the relativistic expression for the energy: $(c^2\mathbf{p}^2 + m^2c^4)^{1/2} \approx mc^2 + \mathbf{p}^2/2m - (\mathbf{p}^2)^2/8m^3c^2$. The last term in Eq. 10.126 gives the *spin-orbit energy* (Sec 8.4) which thus emerges automatically from the Dirac equation. The remaining term, which is the simplified form of $(1/4m^2c^2)\,[-\mathbf{p}^2V + V\mathbf{p}^2 - \hbar^2(dV/dr)\partial/\partial r]$, does not seem to have an immediate physical interpretation in terms of classically identifiable effects.

If a magnetic field were also present, the orbital and spin magnetic moment interactions, which appear through the π^2 and $\boldsymbol{\sigma}.\mathcal{H}$ terms in Eqs. 10.123, would show up in the leading order. Other effects, not easily identifiable, occur in higher orders.

Finally, a word of caution is in order against taking the square-bracketed operator in Eqs. 10.126 too seriously as a nonrelativistic Hamiltonian. For one thing, it is not Hermitian, because the term $(dV/dr)\,(\partial/\partial r)$ is not. The reason for this peculiarity can be traced to the fact that Eq. 10.117, obtained by the elimination of w from the Dirac equation, is not in the standard form of an eigenvalue equation, since the operator on the right hand side explicitly depends on the parameter E'. A related point is that the probability density $P(\mathbf{x},t) = u^\dagger u$ when expressed entirely in terms of w, is not $w^\dagger w$ but $w^\dagger\{1 + c^2\boldsymbol{\sigma}.\boldsymbol{\pi}(E' + 2mc^2 - V)^{-2}\boldsymbol{\sigma}.\boldsymbol{\pi}\}\,w$. A more satisfactory procedure for obtaining the nonrelativistic limit has been given by Foldy and Wouthuysen,[16] but this is beyond our scope.

PROBLEMS

1. By introducing a two-component wave function ϕ with components ϕ_1, ϕ_2 defined by $\psi = \phi_1 + \phi_2$, $i\hbar\,\partial\psi/\partial t = mc^2(\phi_1 - \phi_2)$, show that the Klein - Gordon equation (10.2) can be transcribed as $i\hbar\partial\phi/\partial t = H\phi$ with $H = (\tau_3 + i\tau_2)\,\mathbf{p}^2/2m + mc^2\tau_3$, where τ_1, τ_2, τ_3 are the Pauli matrices. Express $P(\mathbf{x},t)$ in terms of ϕ.
2. Show that any 2×2 matrix is a linear combination of σ_x, σ_y, σ_z and I. Hence show that only the zero matrix anticommutes with σ_x, σ_y and σ_z.
3. States of definite helicity (in which the projection of the spin along the momentum direction has definite values) are eigenstates of $\mathbf{S.p}$. Show that the free Dirac Hamiltonian H commutes with $\mathbf{S.\,p}$. Obtain the simultaneous eigenfunctions of H and $\mathbf{S.p}$.
4. Employing the procedure of Sec. 10.4, show that the nonrelativistic limit of $P(\mathbf{x},t)$ of Eq. 10.5a is $\psi^*\,\psi$.

[16] L. L. Foldy and S. A. Wouthuysen, *Phys. Rev.*, **78,** 29, 1950.

5. Show that $K = \beta(\boldsymbol{\Sigma}.\mathbf{L} + \hbar)$ commutes with $\alpha_r = r^{-1}(\boldsymbol{\alpha}.\mathbf{x})$. Hence or otherwise, show that K commutes with the Hamiltonian for a Dirac particle in a Coulomb field.
6. Determine the continuity conditions on the Dirac wave function at a potential step. Obtain an equation for the minimum depth V_0 for a square well potential of radius a which would just bind a Dirac particle.
7. Space inversion takes the Dirac wave function $\psi(\mathbf{x},t)$ into $\psi'(\mathbf{x}, t) = \beta\,\psi(-\mathbf{x}, t)$. Verify that $\psi'(\mathbf{x},t)$ satisfies the Dirac equation.
8. Verify that the matrices γ^μ ($\mu = 0,1,2,3$) and γ_5 defined by $\gamma^k = \beta\alpha_k$ ($k = 1,2,3$), $\gamma^0 = \beta$, $\gamma_5 = i\alpha_1\alpha_2\alpha_3$ anticommute with one another and that the square of each is the unit matrix.
9. Show that γ_5 of Prob. 8 gives a constant of the motion for a massless Dirac particle.

APPENDICES

A

Classical Mechanics

Lagrangian Formulation

A system of N classical particles (with $f = 3N$ degrees of freedom) can be characterized by a *Lagrangian* $L\ (q_1, q_2, \ldots, q_f, \dot{q}_1, \dot{q}_2, \ldots, \dot{q}_f;\ t)$ which governs the dynamics of the system. The q_i are *generalized coordinates* (functions of $\mathbf{x}_1$, $\mathbf{x}_2, \ldots \mathbf{x}_N$) which specify the instantaneous positions of the particles (i.e., the configuration of the system) and $\dot{q}_i = dq_i/dt$. In the following we shall write q for the set $q_1, q_2, \ldots, q_f$ and $\dot{q}$ for $\dot{q}_1, \dot{q}_2 \ldots, \dot{q}_f$. The Lagrangian is

$$L = T - V \tag{A.1}$$

where $T\ (q, \dot{q})$ is the kinetic energy of the system and V is a potential function in terms of which the generalized force F_i pertaining to the ith degree of freedom is given as

$$F_i = -\frac{\partial V}{\partial q_i} + \frac{d}{dt}\left(\frac{\partial V}{\partial \dot{q}_i}\right) \tag{A.2}$$

The last term takes care of *velocity-dependent forces* which are present in the important case of charged particles interacting with electromagnetic fields.

Equations of motion for the system are obtained as a consequence of *Hamilton's variational principle.* It asserts that in going from some configuration $q\ (t_1)$ at time t_1, to $q\ (t_2)$ at t_2, the path $q\ (t)$ followed by the system will be such that the *action* $I = \int_{t_1}^{t_2} L\ (q(t),\ \dot{q}(t);\ t)\ dt$ is an extremum: the variation of I, on replacing the actual path $q(t)$ by a varied path $q^{(v)}\ (t) \equiv q(t) + \delta q(t)$ between the same initial and final configurations, vanishes in first order.

$$\delta \int_{t_1}^{t_2} L dt = 0, \text{ given } \delta q_i(t_1) = \delta q_i(t_2) = 0 \tag{A.3}$$

This becomes

$$0 = \int_{t_1}^{t_2} \sum_i \left[\frac{\partial L}{\partial q_i} \delta q_i + \frac{\partial L}{\partial \dot{q}_i} \delta \dot{q}_i\right] dt = \int_{t_1}^{t_2} \sum_i \left[\frac{\partial L}{\partial q_i} - \frac{d}{dt}\frac{\partial L}{\partial \dot{q}_i}\right] \delta q_i \, dt$$

with one integration by parts and use of the end conditions on $\delta q_i(t)$. Since this result is to hold for independent arbitrary variations $\delta q_i(t)$ for $i = 1, 2, \ldots, f$, one concludes that

$$\frac{\partial L}{\partial q_i} - \frac{d}{dt}\frac{\partial L}{\partial \dot{q}_i} = 0 \qquad (i = 1,2,\ldots,f) \tag{A.4}$$

With L quadratic in the $\dot{q}_i$, Eqs. A.4 give a system of $2f$ second order differential equations for the q_i. When the q_i are Cartesian coordinates, these equations reduce to the Newtonian form (1.1).

Hamiltonian Formulation

Defining the momentum conjugate to q_i as

$$p_i = \frac{\partial L}{\partial \dot{q}_i} \tag{A.5}$$

one introduces the Hamiltonian

$$H(q,p) = \Sigma p_i\dot{q}_i - L \tag{A.6}$$

As indicated by the notation, H is independent of the velocities $\dot{q}_i$: the variation of the right hand side, $\Sigma (p_i d\dot{q}_i + dp_i\dot{q}_i) - \Sigma [(\partial L/\partial q_i) dq_i + (\partial L/\partial \dot{q}_i) d\dot{q}_i]$ may be seen to be free of the $d\dot{q}_i$. In fact on using Eqs. A.5 and A.4 it reduces to $\Sigma (\dot{q}_i dp_i - \dot{p}_i dq_i)$. Equating this to the variation $\Sigma [(\partial H/\partial q_i) dq_i + (\partial H/\partial p_i) dp_i]$ of the left hand side, one obtains

$$\dot{q}_i = \frac{\partial H}{\partial p_i}, \; \dot{p}_i = -\frac{\partial H}{\partial q_i} \quad (i = 1,2,\ldots,f) \tag{A.7}$$

These are *Hamilton's canonical equations of motion*. They form a set of $2f$ first order differential equations for the quantities (q,p) which can be arbitrarily specified at any instant of time. For the time variation of any function $A(q,p;t)$ one easily obtains with the aid of Eq. A.7

$$\frac{d}{dt}A(q,p,t) = \sum_i \left(\frac{\partial A}{\partial q_i}\frac{\partial H}{\partial p_i} - \frac{\partial A}{\partial p_i}\frac{\partial H}{\partial q_i}\right) + \frac{\partial A}{\partial t} \tag{A.8}$$

Poisson Brackets

The Poisson bracket $\{A,B\}$ of any two functions of the q's and p's is defined by

$$\{A,B\} = \sum_i \left(\frac{\partial A}{\partial q_i}\frac{\partial B}{\partial p_i} - \frac{\partial A}{\partial p_i}\frac{\partial B}{\partial q_i}\right) \tag{A.9}$$

With this notation, Eq. A.8 can be succinctly written as

$$\frac{dA}{dt} = \{A,H\} + \frac{\partial A}{\partial t} \tag{A.10}$$

It is obvious that

$$\{q_i, p_j\} = \delta_{ij},\ \{q_i, q_j\} = \{p_i, p_j\} = 0 \tag{A.11}$$

Canonically conjugate coordinate-momentum pairs are thus characterized by a unit value for the Poisson bracket. Poisson brackets have the following general properties.

$$\begin{aligned} \{A_1 + A_2, B\} &= \{A_1, B\} + \{A_2, B\} \\ \{A, B\} &= -\{B, A\} \\ \{A_1\ A_2, B\} &= A_1\{A_2, B\} + \{A_1, B\}\ A_2 \\ \{\{A, B\}, C\} + \{\{B, C\}, A\} &+ \{\{C, A\}, B\} = 0 \end{aligned} \tag{A.12}$$

The last relation is known as the *Jacobi identity*.

Canonical or Contact Transformations

A set of $2f$ independent functions

$$Q_i = Q_i\,(q, p),\ P_i = P_i\,(q, p) \qquad (i = 1,2,\ldots,f) \tag{A.13}$$

can be used as new variables for the description of the system. They are canonical coordinates if there exists a function $K(Q,P;t)$ in terms of which the equations of motion have the canonical form

$$\dot{Q}_i = \frac{\partial K}{\partial P_i},\ \dot{P}_i = -\frac{\partial K}{\partial Q_i} \tag{A.14}$$

This means that Hamilton's principle must be valid whether L is $\Sigma\, p_i\dot{q}_i - H$ as in Eq. A.6 or L is $\Sigma\, P_i\dot{Q}_i - K$, i.e.,

$$\delta \int_{t_1}^{t_2} (\Sigma p_i\dot{q}_i - H)\ dt = \delta \int_{t_1}^{t_2} (\Sigma P_i\dot{Q}_i - K)\ dt = 0 \tag{A.15}$$

Now, considering an infinitesimal segment traversed in time dt along an arbitrary path, suppose the descriptions in terms of the alternative sets of variables are such that

$$\Sigma\, p_i\, dq_i - H dt = \Sigma\, P_i\, dQ_i - K dt + dF \tag{A.16}$$

where dF is the total differential of some function F. Then the difference between the values of the two integrals in Eq. A.15 is just $F(t_2) - F(t_1)$. Its variation, $\delta[F(t_2) - F(t_1)]$, vanishes if F depends on the path only through the configuration variables q, Q (and possibly t) since the variations δq_i, δQ_i vanish at the end points. So, both the equations A.15 can be satisfied with such an F in Eq. A.16. Then (A.13) is a *canonical transformation*, and F is called its *generator* because it determines Eqs. A.13 completely. In fact, if we write $F = F_1\ (q,Q;t)$, so that

$$dF = \sum \frac{\partial F_1}{\partial q_i}\, dq_i + \sum \frac{\partial F_1}{\partial Q_i}\, dQ_i + \frac{\partial F_1}{\partial t}\, dt \tag{A.17}$$

and then substitute in Eq. A.16, on equating coefficients of the independent differentials, dq_i, dQ_i, dt, we get

$$p_i = \frac{\partial F_1}{\partial q_i},\ P_i = -\frac{\partial F_1}{\partial Q_i} \tag{A.18a}$$

$$K = H + \partial F_1/\partial t \tag{A.18b}$$

It is obvious that Eqs. A.18a determine (A.13) implicitly.

A possible form for F_1 is

$$F_1\ (q,Q;t) = F_2(q,P;t) - \Sigma\ P_i Q_i \tag{A.19}$$

where F_2 is an arbitrary function of q and P only. (It is easily verified that in the differential of the right hand side, the dP_i cancel out when Eqs. A.18a are used). If in particular,

$$F_2(q,P) = \Sigma\ q_i P_i + \varepsilon G(q,P) \tag{A.20}$$

where ε is infinitesimal, one has from Eqs. A.19 and A.20,

$$dF_1 = \sum\left[\left(q_i + \varepsilon\frac{\partial G}{\partial P_i} - Q_i\right)dP_i + \left(P_i + \varepsilon\frac{\partial G}{\partial q_i}\right)dq_i - P_i dQ_i\right]$$

On substituting this for dF in Eq. A.16 and equating coefficients of the dq_i and dP_i (remembering that any two of the sets q, Q and p,P can be considered independent) we obtain

$$Q_i = q_i + \varepsilon\frac{\partial G}{\partial P_i},\ P_i = p_i - \varepsilon\frac{\partial G}{\partial q_i} \tag{A.21}$$

Thus the old and new coordinates differ by infinitesimal quantities. The function G in Eq. A.20 may, therefore, be characterized as the *generator of infinitesimal canonical transformations.* Note that to the first order in ε, G can well be written as $G\ (q,\ p)$. Then Eqs. A.21 become

$$Q_i = q_i + \varepsilon\ \{q_i, G\},\ P_i = p_i + \varepsilon\ \{p_i, G\} \tag{A.22}$$

This equation may be compared with Eqs. 7.88 and 7.89 which give the effect of certain infinitesimal unitary transformations on dynamical variables in quantum mechanics. They have the same form as Eq. A.22 except that commutator brackets appear in the place of Poisson brackets and that the generators of unitary transformations are operators.

B

Relativistic Mechanics

Fundamental Principles; Lorentz Transformations

The basic postulates of Einstein's special theory of relativity are (1) that the laws of physics *are relativistically invariant*, i.e. they have the same form in all inertial (unaccelerated) reference frames and (2) that the speed of light in vacuum has the same value c in all inertial frames.

Consistency with the second postulate in the description of physical phenomena is achieved by requiring that if S and S' are two inertial frames, the space and time coordinates ($\mathbf{x}$, t) and ($\mathbf{x}'$, t') assigned to any event (with respect to S and S' respectively) should be related by a linear transformation such that

$$c^2dt'^2 - \mathbf{dx}'^2 = c^2dt^2 - \mathbf{dx}^2 \tag{B.1}$$

S and S' may be either in uniform motion or at rest relative to each other. In the latter case, S' is related to S by a *translation* (shift of origin of space and time coordinates), described by

$$\mathbf{x}' = \mathbf{x} - \mathbf{a}, \quad t' = t - t_0, \quad (\mathbf{a},\ t_0 \text{ constants}) \tag{B.2}$$

or by a *rotation* (through θ about an axis $\mathbf{n}$):

$$\mathbf{x}' = \mathbf{x}\cos\theta + (\mathbf{x}.\mathbf{n})\,\mathbf{n}\,(1-\cos\theta) - (\mathbf{n}\times\mathbf{x})\sin\theta, \quad t' = t, \tag{B.3}$$

or a combination of both. These transformations have no essential role in relativity theory. It is when S and S' are related by a *boost*, i.e. they are in (uniform) motion relative to each other, that the essential new element, the *mixing of space and time coordinates*, comes in.

If S' moves with constant speed v in the x-direction with respect to S (and it is assumed that observers at the origins O, O' of S and S' synchronize their clocks to $t = t' = 0$ at an instant when S and S' are momentarily in coincidence), the relation between $\mathbf{x}$ and $\mathbf{x}'$ is the *Lorentz transformation*

$$x' = \gamma\ (x - \beta ct),\ y' = y,\ z' = z,\ ct' = \gamma\ (ct - \beta x) \tag{B.4}$$

where

$$\beta = v/c,\ \gamma = (1 - v^2/c^2)^{-1/2} \tag{B.5}$$

More generally, if the relative motion is along the direction $\mathbf{n}$,

$$\begin{aligned} \mathbf{x}' &= \mathbf{x} - (1 - \gamma)\ \mathbf{n}\ (\mathbf{n.x}) - \beta\gamma\ ct\mathbf{n} \\ ct' &= \gamma ct - \beta\gamma\ (\mathbf{n.x}) \end{aligned} \tag{B.6}$$

Note that these transformations make sense only if $\beta < 1$.

The fact that (B.6) mixes the space and time coordinates of any event makes it necessary to think of them as the components of a single entity, a *four-vector*. To emphasize this, one introduces the notation[1] x^μ ($\mu = 0, 1, 2, 3$) for these quantities:

$$ct = x^0,\ \mathbf{x} = (x^1,\ x^2,\ x^3) \tag{B.7}$$

Using the familiar summation convention (any index which appears twice in any term is to be summed over from 0 to 3), linear transformations of the x^μ may now be expressed succinctly as

$$x'^\mu = \Lambda^\mu{}_\nu\ x^\nu + a^\mu \tag{B.8a}$$

where the $\Lambda^\mu{}_\nu$ and a^μ are constants. In matrix form (writing x for a column with the four components x^μ, etc.) this can be written as

$$x' = \Lambda x + a \tag{B.8b}$$

The fundamental requirement of relativity (Eq. B.1), then becomes

$$\widetilde{dx'}\ G\ dx' = \widetilde{dx}\ G\ dx \quad \text{or} \quad g_{\mu\nu}\ dx'^\mu\ dx'^\nu = g_{\mu\nu}\ dx^\mu dx^\nu \tag{B.9}$$

where the tilde (~) denotes transposition, and the matrix G (with elements $g_{\mu\nu}$) is given by

$$G = \begin{pmatrix} 1 & 0 & 0 & 0 \\ 0 & -1 & 0 & 0 \\ 0 & 0 & -1 & 0 \\ 0 & 0 & 0 & -1 \end{pmatrix} \tag{B.10}$$

Since $dx' = \Lambda dx$, i.e. $dx'^\mu = \Lambda^\mu{}_\rho\ dx^\rho$, Eq. B.9 requires

$$\tilde{\Lambda} G \Lambda = G \quad \text{or} \quad g_{\mu\nu}\Lambda^\mu{}_\rho\ \Lambda^\nu{}_\sigma = g_{\rho\sigma} \tag{B.11}$$

The coefficients of the x^μ in Eqs. B.2, B.3 and B.6 do constitute matrices Λ obeying Eq. B.11. Conversely, any linear transformation (B.8) subject to (B.11) can be accomplished by a succession of boosts, rotations and translations of the reference frame. All such transformations which satisfy Eq. B.1 will generally be called Lorentz transformations.[2] The transformation is *homogeneous* if it does not include translations, i.e. $a^\mu = 0$ in (B.8a); otherwise it is *inhomogeneous*.

Under the change of reference frame (from S to S') characterized by Eqs. B.8, various physical quantities behave like four-vectors, tensors etc. A four-vector is defined as a quantity whose components (V^0, $\mathbf{V}$) and (V'^0, $\mathbf{V}'$) in S and S' are related in the same manner as dx'^μ and dx^μ:

[1] Superscripts are indices identifying the components, as in the tensor notation for contravariant vectors.

[2] In elementary works this term is restricted to Eqs. B.6 associated with boosts.

$$V'^{\mu} = \Lambda^{\mu}{}_{\nu} V^{\nu} \tag{B.12}$$

Covariant components V_{μ} of the vector are defined by

$$V_{\mu} = g_{\mu\nu} V^{\nu}; \text{ i.e. } V_0 = V^0,\ V_i = -V^i\ (i = 1,2,3) \tag{B.13a}$$

Conversely,

$$V^{\mu} = g^{\mu\nu} V_{\nu} \tag{B.13b}$$

where the $g^{\mu\nu}$ are elements of the matrix G^{-1} and are numerically equal to $g_{\mu\nu}$. The combination

$$V^{\mu} V_{\mu} \equiv g_{\mu\nu} V^{\mu} V^{\nu} = (V^0)^2 - \mathbf{V}^2 \tag{B.14}$$

is invariant under Lorentz transformations, i.e. its value is unaltered if V^{μ} is replaced by $V'^{\mu} \equiv \Lambda^{\mu}{}_{\rho} V^{\rho}$. This is a consequence of Eq. B.11.

It is found that a formulation of the mechanics of particles consistent with the first postulate of relativity is possible only if the energy E and momentum $\mathbf{P}$ of any system, or more precisely, $(E/c, \mathbf{P})$, behave as the components of a four-vector. It follows then from the counterpart of Eq. B.6 that in a frame (say S_r) which is reached by a boost characterized by $\mathbf{v} \equiv v\mathbf{n} = c^2\mathbf{P}/E$, the momentum vanishes (and the energy is E_r, say). Of course, $(E/c, \mathbf{P})$ is related to $(E_r/c, \mathbf{0})$ through the opposite boost (with relative velocity $-\mathbf{v}$, whence

$$E = \gamma E_r,\ \mathbf{P} = \gamma(\mathbf{v}/c)\,(E_r/c),\quad [\gamma = (1 - v^2/c^2)^{-1/2}] \tag{B.15}$$

In the case of a *single particle*, S_r is its rest frame and E_r, its energy at rest. To ensure that the momentum $\mathbf{P}$ has the known form $m\mathbf{v}$ in the non-relativistic limit ($v/c \ll 1$, $\gamma \approx 1$), it is necessary that $E_r = mc^2$. It follows then from Eq. B.15 that

$$E^2 - c^2\mathbf{P}^2 = E_r^2 = m^2c^4 \tag{B.16}$$

Also, from the first of Eqs. B.15, $E = \gamma mc^2$; it differs from E_r in having the 'dynamic mass' γm in the place of the rest mass m. The kinetic energy is $E - E_r = (\gamma - 1)\,mc^2$.

In the case of a *system of particles*, the frame S_r is the *centre of mass frame.* Consider, in particular, the collision of two relativistic particles of masses m_1 and m_2. If in the laboratory frame S_0, m_2 is at rest and m_1 has momentum $\mathbf{p}^{(0)}$, the total energy-momentum vector of the system is

$$[(E_1^{(0)} + m_2c^2)/c,\ \mathbf{p}^{(0)}] \tag{B.17}$$

where $E_1^{(0)} = [(c\mathbf{p}^{(0)})^2 + m_1^2c^4]^{1/2}$. Since the energy E_r in the centre of mass frame S_r is given by $E_r^2 = E^2 - c^2\mathbf{p}^2$, we have on using (B.17) for $(E/c, \mathbf{P})$,

$$E_r^2 = (E_1^{(0)} + m_2c^2)^2 - (c\mathbf{p}^{(0)})^2 = (m_1c^2)^2 + (m_2c^2)^2 + 2m_2c^2E_1^{(0)} \tag{B.18}$$

Thus in a highly relativistic situation ($E_1^{(0)} \gg m_1c^2,\ m_2c^2$), increase of the projectile energy $E_1^{(0)}$ produces only a much slower increase ($\sim \sqrt{E_1^{(0)}}$) of E_r. This result imposes severe restrictions on the possibility of increasing the energy $(E_r - m_1c^2 - m_2c^2)$ available for particle production (in inelastic collisions) by increasing the projectile energy.

For an elastic collision in which the scattering angle is θ in S_r, the momenta of particles 1 and 2 can be written respectively, as

$$(p,0,0),\ (-p,0,0) \text{ before collision} \tag{B.19}$$

$$(p\cos\theta,\ p\sin\theta,\ 0),\ (-p\cos\theta,\ -p\sin\theta,\ 0) \text{ after collision}$$

with a convenient choice of the coordinate axes in S_r. Their energies are:

$$E_1 = (c^2p^2 + m_1{}^2c^4)^{1/2},\ E_2 = (c^2p^2 + m_2{}^2c^4)^{1/2} \tag{B.20}$$

A boost with speed $v = c^2p/E_2$ in the negative x-direction takes one to the laboratory frame in which particle 2 is at rest. By applying the corresponding Lorentz transformation [the counterpart of (B.4)] to the four vector $(E_1/c,$ $p\cos\theta$, $p\sin\theta$, $0)$ one sees that the momentum $\mathbf{p}_1{}^{(0)}$ of particle 1 after collision as seen in the laboratory frame is

$$\mathbf{p}_1{}^{(0)} = [(\,p/m_2c^2)\,(E_2\cos\theta + E_1),\ \ p\sin\theta, 0) \tag{B.21}$$

Hence the scattering angle θ_0 in the laboratory frame is given by

$$\tan\theta_0 = \frac{m_2c^2\sin\theta}{E_2\cos\theta + E_1} \tag{B.22}$$

This is the relativistic generalization of Eq. 6.111a.

Relativistic Invariance of the Maxwell, Klein-Gordon and Dirac Equations

With electric and magnetic fields $\mathcal{E}$, $\mathcal{H}$, behaving as the components of an antisymmetric tensor $F^{\mu\nu}$ under Lorentz transformations $(\mathcal{E}_x = F^{10},\ \mathcal{E}_y = F^{20},$ $\mathcal{E}_z = F^{30};\ \mathcal{H}_x = F^{23},\ \mathcal{H}_y = F^{31},\ \mathcal{H}_z = F^{12})$ and charge and current densities ρ and $\mathbf{j}$ behaving like the components j^μ of a four-vector $(j^0 = c\rho, j^1 = j_x, j^2 =$ $j_y,\ j^3 = j_z)$, the relativistic invariance of Maxwell's equations becomes self evident since they can be written as the tensor equations

$$\frac{\partial}{\partial x^\mu}F^{\mu\nu} = \frac{4\pi}{c}j^\nu;\quad \frac{\partial}{\partial x^\lambda}F^{\mu\nu} + \frac{\partial}{\partial x^\mu}F^{\nu\lambda} + \frac{\partial}{\partial x^\nu}F^{\lambda\mu} = 0 \tag{B.23}$$

These two equations comprise, respectively, the first two and the last two of Eqs. 1.2. The scalar and vector potentials ϕ, $\mathbf{A}$ are the components $\mathbf{A}^\mu$ of a four-vector (with $\phi = \mathbf{A}^0$), and

$$F_{\mu\nu} = \frac{\partial A_\nu}{\partial x^\mu} - \frac{\partial A_\mu}{\partial x^\nu} \tag{B.24}$$

The *Klein-Gordon equation* can be written as

$$\left[g^{\mu\nu}\frac{\partial}{\partial x^\mu}\frac{\partial}{\partial x^\nu} + \left(\frac{mc}{\hbar}\right)^2\right]\psi = 0 \tag{B.25}$$

Its relativistic invariance is obvious.

As for the *Dirac equation* (10.26), it can be cast in the invariant-looking form

$$-\,i\hbar\,\gamma^\mu\frac{\partial\psi}{\partial x^\mu} + mc\,\psi = 0 \tag{B.26}$$

by multiplying throughout by β and introducing the matrices

$$\gamma^0 = \beta,\ \gamma^1 = \beta\alpha_x,\ \gamma^2 = \beta\alpha_y,\ \gamma^3 = \beta\alpha_z \tag{B.27}$$

The following properties of the γ^μ may be easily verified:

$$\gamma^\mu\gamma^\nu + \gamma^\nu\gamma^\mu = 2g^{\mu\nu} \tag{B.28}$$

$$\gamma^{0\dagger} = \gamma^0,\ \gamma^{i\dagger} = -\,\gamma^i\quad (i = 1, 2, 3) \tag{B.29}$$

It is to be expected that under a Lorentz transformation Λ, the Dirac wave function at a particular point will change from $\psi(\mathbf{x},t)$ to

$$\psi'\,(\mathbf{x}', t') = S(\Lambda)\,\psi(\mathbf{x}, t) \tag{B.30}$$

where S is a matrix (*cf.* the effect (7.80) of rotations on a vector wave function).

Substituting $\psi = S^{-1}\psi'$ into Eq. B.26 and multiplying on the left by S, we observe that the resulting *equation for ψ' will have the same form* as Eq. B.26 provided $S\gamma^\mu S^{-1}(\partial/\partial x^\mu) = \gamma^\nu (\partial/\partial x'^\nu)$, or

$$\gamma^\mu \frac{\partial}{\partial x^\mu} = \gamma'^\nu \frac{\partial}{\partial x'^\nu} \text{ with } \gamma'^\nu = S^{-1}\gamma^\nu S \tag{B.31}$$

Since the differential operators $(\partial/\partial x^\mu)$ form a covariant vector, the condition for (B.31) to hold is that the γ^μ behave like a contravariant vector:

$$\gamma'^\nu = \Lambda^\nu{}_\mu\gamma^\mu, \text{ i.e. } S^{-1}(\Lambda)\gamma^\nu S(\Lambda) = \Lambda^\nu{}_\mu\gamma^\mu \tag{B.32}$$

This is the *condition for relativistic invariance of the Dirac equation.* When Λ represents a boost along the x-axis with an infinitesimal value ε for v/c, this condition becomes (to first order in ε)

$$\gamma^{0\prime} = \gamma^0 - \varepsilon\gamma^1, \ \gamma'^1 = \gamma^1 - \varepsilon\gamma^0, \ \gamma'^2 = \gamma^2, \ \gamma'^3 = \gamma^3 \tag{B.33}$$

It is easily verified, with the aid of Eq. B.28, that this is indeed satisfied if $S = 1 + \frac{1}{4}\varepsilon\,(\gamma^1\gamma^0 - \gamma^0\gamma^1)$. For example,

$$\begin{aligned}\gamma^{0\prime} &= [1 - \tfrac{1}{4}\varepsilon\,(\gamma^1\gamma^0 - \gamma^0\gamma^1)]\,\gamma^0\,[1 + \tfrac{1}{4}\varepsilon\,(\gamma^1\gamma^0 - \gamma^0\gamma^1)]\\ &\approx \gamma^0 - \tfrac{1}{4}\varepsilon\,[(\gamma^1\gamma^0 - \gamma^0\gamma^1)\gamma^0 - \gamma^0(\gamma^1\gamma^0 - \gamma^0\gamma^1)] = \gamma^0 - \varepsilon\gamma^1\end{aligned}$$

since $(\gamma^0)^2 = 1$, $\gamma^0\gamma^1 = -\gamma^1\gamma^0$. Writing S as $1 + i\varepsilon G$, we see then that the generator G of boosts along the x-direction is $\frac{1}{4}i\,(\gamma^0\gamma^1 - \gamma^1\gamma^0)$. Similarly, for infinitesimal rotation about the z-axis, $(\gamma'^0 = \gamma^0,\ \gamma'^1 = \gamma^1 + \theta\gamma^2,\ \gamma'^2 = \gamma^2 - \theta\gamma^1, \gamma'^3 = \gamma^3)$, the generator is found to be $\frac{1}{4}i\,(\gamma^1\gamma^2 - \gamma^2\gamma^1)$. It reduces to $-\frac{1}{2}i\alpha_x\alpha_y = \frac{1}{2}\Sigma_z = S_z/\hbar$ where S_z, (Eq. 10.58a), is the z-component of the spin angular momentum operator. (This result is to be expected from the general theory of Sec. 7.9). Since finite Lorentz transformations can be built up from infinitesimal ones, the above results are enough to ensure that the Dirac equation is relativistically invariant.

C | The Dirac Delta Function

The delta function, which is a singular function, may be visualized as a limit of a sequence of ordinary functions. Among the various possibilities are the following:

(a) $$\delta(x) = \lim_{\varepsilon \to 0+} \delta_\varepsilon(x), \text{ where}$$

$$\delta_\varepsilon(x) = \frac{1}{2\varepsilon}, \quad |x| < \varepsilon; \quad \delta_\varepsilon(x) = 0, \quad |x| > \varepsilon \tag{C.1}$$

(b) $$\delta(x) = \lim_{\varepsilon \to 0+} \frac{1}{\sqrt{2\pi}\,\varepsilon} e^{-x^2/2\varepsilon^2} \tag{C.2}$$

(c) $$\delta(x) = \lim_{\varepsilon \to 0+} \frac{1}{\pi} \frac{\varepsilon}{x^2 + \varepsilon^2} \tag{C.3}$$

(d) $$\delta(x) = \frac{1}{2\pi} \int_{-\infty}^{\infty} e^{ikx}\, dk \tag{C.4}$$

$$= \lim_{\alpha \to \infty} \frac{1}{2\pi} \int_{-\alpha}^{\alpha} e^{ikx}\, dk = \lim_{\alpha \to \infty} \frac{\sin \alpha x}{\pi x} \tag{C.5}$$

In each of the above equations, the integral (from $-\infty$ to $+\infty$) of the function on the right side is unity, independently of the value of ε or α. So this property is possessed also by the limiting function. The latter vanishes everywhere except at $x = 0$ in cases (a), (b) and (c). In (d), $(\sin \alpha x)/\pi x$ does *not* tend to zero anywhere as $\alpha \to \infty$. But its oscillations become infinitely fast, so that in $\int f(x)\, [\sin \alpha x/\pi x]\, dx$, the contributions from adjacent regions with opposite signs of sin αx cancel each other out. The only nonvanishing contribution comes from the neighbourhood of the origin where there is a peak of height (α/π) and width $\sim (\pi/\alpha)$.

Though the sequences of functions involved in the three cases are quite different, the same symbol $\delta(x)$ is used for the limiting function in all cases because all of them satisfy the fundamental requirement that the *integral* $\int f(x)\, \delta(x)\, dx$ be equal to $f(0)$. The following equalities are also to be understood in the sense that *if both sides are multiplied by $f(x)$ and integrated, both yield the same value for the integral.*

$$x^\alpha\, \delta(x) = 0, \quad \alpha > 0 \tag{C.6}$$

$$x\delta'(x) = -\,\delta(x) \tag{C.7}$$

$$\delta\,(ax) = a^{-1}\,\delta(x), \quad a > 0 \tag{C.8}$$

$$\delta\,(x^2 - a^2) = (2a)^{-1}\,[\delta(x - a) + \delta(x + a)], \quad a > 0 \tag{C.9}$$

$$\delta(g(x)) = \sum_i \frac{1}{|g'(x_i)|}\,\delta\,(x - x_i) \tag{C.10}$$

where the x_i are the zeroes of the function $g(x)$, i.e. $g(x_i) = 0$. Eqs. C.8 and C.9 are often encountered as special cases of Eq. C.10. They may be proved by a change of the variable of integration from x to $y = g(x)$ in $\int f(x)\, \delta(g(x))\, dx$. Eq. C.7 is proved by an integration by parts in $\int f(x)\, x\, \delta'(x)\, dx$.

It may be noted that $\delta(x)$ and $\delta'(x)$ are even and odd functions respectively, that is

$$\delta(x) = \delta(-\,x), \quad \delta'(-\,x) = -\,\delta'(x) \tag{C.11}$$

D Mathematical Appendix

Definitions of the special functions referred to in the text, and a select few of their properties which are of immediate utility in the present context, are given here.

Gamma Function

$$\Gamma(z) = \int_0^\infty e^{-t}\, t^{z-1}\, dt, \text{ (Re } z > 0) \tag{D.1}$$

$$\Gamma(z+1) = z\Gamma(z), \ \Gamma(n+1) = n! \tag{D.2a}$$

$$\Gamma(1) = 1, \ \Gamma(\tfrac{1}{2}) = \sqrt{\pi}, \ \Gamma(-n) = \infty, \ (n = 0, 1, 2, \ldots) \tag{D.2b}$$

$$\Gamma(z)\,\Gamma(1-z) = \frac{\pi}{\sin \pi z} \tag{D.3}$$

$$\Gamma(2z) = \pi^{-1/2}\, 2^{2z-1}\, \Gamma(z)\, \Gamma(z+\tfrac{1}{2}) \tag{D.4}$$

The Stirling approximation:

$$n! \approx (2\pi)^{1/2}\, e^{-n}\, n^{n+1/2}, \quad (n \gg 1) \tag{D.5}$$

Hermite Polynomials: $H_n(\rho)$

They may be defined in terms of their generating function

$$G(\rho, \xi) \equiv \sum_{n=0}^{\infty} \frac{H_n(\rho)}{n!}\, \xi^n = e^{-\xi^2 + 2\rho\xi} \tag{D.6}$$

The following expressions for $H_n(\rho)$ are useful:

$$H_n(\rho) = (-)^n\, e^{\rho^2} \frac{d^n}{d\rho^n} e^{-\rho^2} = (-)^n\, e^{\frac{1}{2}\rho^2} \left(e^{\frac{1}{2}\rho^2} \frac{d}{d\rho} e^{-\frac{1}{2}\rho^2} \right)^n e^{-\frac{1}{2}\rho^2} \tag{D.7a}$$

$$= \frac{2^n}{\sqrt{\pi}} \int_{-\infty}^{\infty} (\rho + i\alpha)^n e^{-\alpha^2} d\alpha \quad \text{(D.7b)}$$

$$= (2\rho)^n - \frac{n(n-1)}{1!} (2\rho)^{n-2} + \frac{n(n-1)(n-2)(n-3)}{2!} (2\rho)^{n-4} + \dots \quad \text{(D.7c)}$$

In the second of the forms in (D.7a) and in (D.12) below, the whole quantity within brackets is an operator which acts on the function which follows it. The H_n satisfy the differential equation

$$H_n''(\rho) - 2\rho H_n'(\rho) + 2nH_n(\rho) = 0 \quad \text{(D.8)}$$

and the recurrence relations

$$2\rho H_n(\rho) = 2nH_{n-1}(\rho) + H_{n+1}(\rho) \quad \text{(D.9)}$$

$$H_n'(\rho) = 2\rho H_n(\rho) - H_{n+1}(\rho) = 2nH_{n-1}(\rho) \quad \text{(D.10)}$$

As n becomes very large,

$$H_{2n}(\rho) \to (-)^n 2^n (1.3.5\dots(2n-1))\, e^{\frac{1}{2}\rho^2} \cos(\sqrt{4n+1}\,\rho)$$

$$H_{2n+1}(\rho) \to (-)^n 2^{n+\frac{1}{2}}(1.3.5\dots(2n-1))\,(2n+1)^{\frac{1}{2}}\, e^{\frac{1}{2}\rho^2} \sin(\sqrt{4n+3}\,\rho) \quad \text{(D.11)}$$

Laguerre and Associated Laguerre Polynomials

The Laguerre polynomials $L_q(\rho)$ are given by

$$L_q(\rho) = e^{\rho} \frac{d^q}{d\rho^q} (e^{-\rho} \rho^q) = \left(e^{\rho} \frac{d}{d\rho} e^{-\rho} \right)^q \rho^q \quad \text{(D.12)}$$

The associated Laguerre polynomials $L_q^p(\rho)$ are defined by

$$L_q^p(\rho) \equiv \frac{d^p}{d\rho^p} L_q(\rho) = \frac{q!}{(q-p)!} (-\rho^{-p})\, e^{\rho} \frac{d^{q-p}}{d\rho^{q-p}} (e^{-\rho}\rho^q) \quad \text{(D.13a)}$$

$$= \frac{q!}{(q-p)!} e^{\rho} \frac{d^q}{d\rho^q} (e^{-\rho} \rho^{q-p}) \quad \text{(D.13b)}$$

They may be defined through their generating function as

$$G^{(p)}(\rho,\xi) = \sum_{q=p}^{\infty} \frac{L_q^p(\rho)\xi^q}{q!} = \left(-\frac{\xi}{1-\xi} \right)^p \frac{e^{-\rho\xi/(1-\xi)}}{1-\xi} \quad \text{(D.14)}$$

and satisfy the differential equation

$$\rho \frac{d^2 L_q^p}{d\rho^2} + (p + 1 - \rho) \frac{dL_q^p}{d\rho} + (q-p) L_q^p = 0 \quad \text{(D.15)}$$

and the recurrence relations

$$L_q^p(\rho) = q\,[L_{q-1}^{p}(\rho) - L_{q-1}^{p-1}(\rho)] \quad \text{(D.16)}$$

$$\rho\, L_q^p(\rho) = (q+1-p)\, L_q^{p-1}(\rho) - q^2\, L_{q-1}^{p-1}(\rho) \quad \text{(D.17)}$$

$$\frac{d}{d\rho} L_q^p(\rho) = L_q^{p+1}(\rho) \quad \text{(D.18)}$$

Bessel Functions

The Bessel differential equation

$$y'' + \frac{1}{\rho}y' + \left(1 - \frac{\nu^2}{\rho^2}\right)y = 0, \quad (\nu > 0) \tag{D.19}$$

has the solution

$$y = J_\nu(\rho) \equiv \sum_{s=0}^{\infty} \frac{(-)^s (\frac{1}{2}\rho)^{\nu+2s}}{s!\ \Gamma(s+\nu+1)} \tag{D.20}$$

which is regular at $\rho = 0$. The second solution, singular at $\rho = 0$, is $J_{-\nu}(\rho)$. But when ν is an integer, $J_{-\nu}$ is a constant multiple of J_ν. An alternative solution which is independent of J_ν for all ν is the Neumann function

$$N_\nu(\rho) = \frac{1}{\sin \pi\nu}[\cos \pi\nu\, J_\nu(\rho) - J_{-\nu}(\rho)] \tag{D.21}$$

The combinations

$$H_\nu^{(1)}(\rho) = J_\nu(\rho) + iN_\nu(\rho),\ H_\nu^{(2)}(\rho) = J_\nu(\rho) - iN_\nu(\rho) \tag{D.22}$$

are called the Hankel functions of the first and second kind.

The Bessel functions $J_l(\rho)$ for $l = 0, 1, 2 \ldots$ have the generating function

$$\sum_{n=-\infty}^{\infty} J_l(\rho)\, \xi^l = \exp\left[\tfrac{1}{2}\rho\left(\xi - \frac{1}{\xi}\right)\right] \tag{D.23}$$

From this, one gets the integral representation

$$J_l(\rho) = \frac{1}{2\pi}\int_{-\pi}^{\pi} e^{-i(l\theta - \rho\cos\theta)}\, d\theta \tag{D.24}$$

Recurrence relations for the Bessel functions are

$$\frac{2\nu}{\rho} J_\nu(\rho) = J_{\nu-1}(\rho) + J_{\nu+1}(\rho) \tag{D.25}$$

$$J_\nu'(\rho) = \tfrac{1}{2}[J_{\nu-1}(\rho) - J_{\nu+1}(\rho)] \tag{D.26}$$

The *spherical Bessel functions* $j_l(\rho)$ and $n_l(\rho)$ are defined by

$$j_l(\rho) = \left(\frac{\pi}{2\rho}\right)^{1/2} J_{l+1/2}(\rho) \tag{D.27}$$

$$n_l(\rho) = \left(\frac{\pi}{2\rho}\right)^{1/2} N_{l+1/2}(\rho) = (-)^{l+1}\left(\frac{\pi}{2\rho}\right)^{1/2} J_{-l+1/2}(\rho) \tag{D.28}$$

These, as well as the spherical Hankel functions defined by

$$h_l^{(1)}(\rho) = j_l(\rho) + in_l(\rho),\ h_l^{(2)}(\rho) = j_l(\rho) - in_l(\rho) \tag{D.29}$$

satisfy the equation

$$y'' + \frac{2}{\rho}y' + \left[1 - \frac{l(l+1)}{\rho^2}\right]y = 0 \tag{D.30}$$

Their behaviour near $\rho = 0$ may be deduced from their definitions, together with (D.20)

$$j_l(\rho) \approx \frac{\rho^l}{1.3.5\ldots(2l+1)}, \quad (\rho^2 < 4l + 6) \tag{D.31}$$

$$n_l(\rho) \approx -[1.3.5\ldots(2l-1)]\ \rho^{-l-1}, \ (\rho^2 < 2) \tag{D.32}$$

Their asymptotic behaviour (as $\rho \to \infty$) is given by

$$j_l(\rho) \to \rho^{-1} \sin\,(\rho - \tfrac{1}{2}l\pi), \quad n_l(\rho) \to -\rho^{-1} \cos\,(\rho - \tfrac{1}{2}\,l\pi) \tag{D.33}$$

$$h_l^{(1)}(\rho) \to i^{-(l+1)}\,\rho^{-1}\,e^{i\rho}, \quad h_l^{(2)}(\rho) \to i^{(l+1)}\,\rho^{-1}\,e^{-i\rho} \tag{D.34}$$

The Legendre Polynomials and Associated Legendre Functions

The Legendre equation

$$(1-w^2)\,\frac{d^2y}{dw^2} - 2w\,\frac{dy}{dw} + l(l+1)\,y = 0 \tag{D.35}$$

has, for $l = 0,1,2,\ldots$, the polynomial solution

$$y = P_l(w) = \frac{1}{2^l\,l!}\,\frac{d^l}{dw^l}\,(w^2-1)^l \tag{D.36}$$

$$P_l(w) = \sum_{k=0}^{[l/2]} \frac{(-)^k\,(\tfrac{1}{2})_{l-k}\,(2w)^{l-2k}}{k!\,(l-2k)!} \tag{D.37}$$

where $[l/2] = \tfrac{1}{2}l$ if l is an even integer and $\tfrac{1}{2}(l-1)$ if l is an odd integer. Also $(\tfrac{1}{2})_{l-k}$ stands for the product $\tfrac{1}{2}(\tfrac{1}{2}+1)\ldots\{\tfrac{1}{2}+(l-k)-1\}$.

$$P_l(1) = 1;\; P_l(-1) = (-)^l. \tag{D.38}$$

Unlike the *Legendre polynomial* $P_l(w)$, the second solution $Q_l(w)$ is singular at $w = \pm 1$. The generating function of the $P_l(w)$ is defined by

$$\frac{1}{(1-2\xi w+\xi^2)^{1/2}} = \sum_{l=0}^{\infty} \xi^l P_l(w), \quad |\xi| < 1 \tag{D.39}$$

(If $w = \cos\theta$, the denominator of the left hand side has a geometrical interpretation: It is the length of the third side of a triangle when the other two sides are of length 1 and ξ and include an angle θ between them). The following properties can be proved from the generating function:

$$wP_l = \frac{l+1}{2l+1}\,P_{l+1} + \frac{l}{2l+1}\,P_{l-1} \tag{D.40}$$

$$P_l' = \frac{l(wP_l - P_{l-1})}{w^2-1} = (l+1)\,\frac{(P_{l+1} - wP_l)}{w^2-1} \tag{D.41}$$

For $l \gg 1$ and $\theta \gg (1/l)$, $(\pi - \theta) \gg (1/l)$,

$$P_l(\cos\theta) \to \left(\frac{2}{l\pi\,\sin\theta}\right)^{1/2} \sin\left[(l+\tfrac{1}{2})\,\theta + \frac{\pi}{4}\right] \tag{D.42}$$

and for $l \gg 1$, $\theta \ll (1/l)$,

$$P_l(\cos\theta) \approx J_0\,(l\theta) \tag{D.43}$$

The associated Legendre functions $P_l^m(w)$ are defined by

$$P_l^m(w) = (1-w^2)^{\frac{1}{2}m}\,\frac{d^m}{dw^m}\,P_l(w), \qquad m \geqslant 0 \tag{D.44a}$$

$$P_l^m(w) = (-)^m\,\frac{(l+m)!}{(l-m)!}\,P_l^{|m|}(w), \qquad m < 0 \tag{D.44b}$$

The Spherical Harmonics

$$Y_l^m(\theta,\varphi) = \left[\frac{2l+1}{4\pi}\,\frac{(l-m)!}{(l+m)!}\right]^{1/2} (-)^m\,P_l^m\,(\cos\theta)\,e^{im\varphi} \tag{D.45}$$

$$[Y_l^m(\theta,\varphi)]^* = (-)^m\,Y_l^{-m}(\theta,\varphi) \tag{D.46}$$

$$Y_l^m(0,\varphi) = Y_l^m(0,0) = \left(\frac{2l+1}{4\pi}\right)^{1/2} \delta_{m0} \tag{D.47}$$

If the directions of two points with respect to the origin are indicated by the polar coordinates (θ,φ) and (θ',φ'), and if the angle between these directions is Θ, then

$$P_l(\cos\Theta) = \frac{4\pi}{2l+1} \sum_{m=-l}^{+l} Y_{lm}^*(\theta,\varphi)\, Y_{lm}(\theta',\varphi') \tag{D.48}$$

Expansion of the plane wave $e^{i\mathbf{k}.\mathbf{x}}$ in terms of the polar coordinates (r,θ,φ) of $\mathbf{x}$ and (k,θ',φ') of $\mathbf{k}$ can be obtained by introducing the above expression for $P_l(\cos\Theta)$ in

$$e^{i\mathbf{k}.\mathbf{x}} = \sum_{l=0}^{\infty} (2l+1)\, i^l j_l(kr)\, P_l(\cos\Theta) \tag{D.49}$$

nilarly one gets

$$\frac{1}{|\mathbf{x}-\mathbf{x}'|} = \frac{1}{r_>} \sum_{l=0}^{\infty} \left(\frac{r_<}{r_>}\right)^l \left[\frac{4\pi}{2l+1} \sum_{m=-l}^{l} Y_{lm}^*(\theta,\varphi)\, Y_{lm}(\theta',\varphi')\right] \tag{D.50}$$

where $r_>$ is the larger of the two lengths r,r', and $r_<$ is the smaller.

A very useful integral is

$$\int Y_{l'm'}^*(\theta,\varphi)\, Y_{\lambda\mu}(\theta,\varphi)\, Y_{lm}(\theta,\varphi) \sin\theta\, d\theta\, d\varphi$$
$$= \left[\frac{(2l-1)(2\lambda+1)}{4\pi(2l'+1)}\right]^{1/2} \langle\, l'm' \mid lm;\lambda\mu \,\rangle \langle\, l'0 \mid l0;\lambda 0 \,\rangle \tag{D.51}$$

where the quantities on the right hand side are Clebsch-Gordan coefficients (Sec. 8.6).

Hypergeometric Functions

The hypergeometric equation

$$\rho(\rho-1)\frac{d^2y}{d\rho^2} + [(1+a+b)\rho - c]\frac{dy}{d\rho} + aby = 0 \tag{D.52}$$

has the solution[1]

$$F(a,b;c;\rho) = 1 + \frac{a.b}{c.1!}\rho + \frac{a(a+1).b(b+1)}{c(c+1).2!}\rho^2 + \ldots, |\rho| < 1 \tag{D.53}$$

If $a = -n$ or $b = -n$, $(n = 0,1,2,\ldots.)$ then $F(a,b;c;\rho)$ reduces to a polynomial of degree n (called the *Jacobi polynomial*).

The second solution is $\rho^{1-c} F(a-c+1,\ b-c+1;\ 2-c;\rho)$. If c is a negative integer say $-n$, $F(a,b;c;\rho)$ becomes infinite, but

$$\lim_{c\to -n} \frac{F(a,b;c;\rho)}{\Gamma(c)} = \frac{\Gamma(a+n+1)\,\Gamma(b+n+1)^{2n+1}}{\Gamma(a)\Gamma(b)\ (n+1)!} \times F(a+n+1,\ b+n+1;\ n+2;\rho) \tag{D.54}$$

The identity

[1] This hypergeometric function is more precisely denoted by ${}_2F_1$ to indicate that *two* parameters (a,b) appear in the numerator and *one* (c) in the denominator.

$$F(a,b;c;\rho) = \frac{\Gamma(c)\Gamma(b-a)}{\Gamma(b)\Gamma(c-a)}(-\rho)^{-a} F\left(a,a-c+1;\ a-b+1;\frac{1}{\rho}\right)$$
$$+\frac{\Gamma(c)\Gamma(a-b)}{\Gamma(a)\Gamma(c-b)}(-\rho)^{-b} F\left(b,b-c+1;\ b-a+1;\frac{1}{\rho}\right). \quad \text{(D.55)}$$

reveals the asymptotic behaviour of $F(a,b;c;\rho)$. As $\rho \to \infty$, $(1/\rho) \to 0$, the F functions on the right side tend to unity.

The Legendre polynomial $P_l(w)$, can be written as

$$P_l(w) = F(l+1, -l;1;\rho), \quad \rho = \tfrac{1}{2}(1-w) \quad \text{(D.56)}$$

The confluent hypergeometric function $F(a;c;\rho)$ — or more precisely ${}_1F_1(a;c;\rho)$ — which is defined in Sec. 4.18, is the limit of $F(a,b;c;\rho/b)$ as $b \to \infty$. Some of the special functions considered above may be expressed as particular cases of this function:

$$H_{2s}(\rho) = (-)^n \frac{(2s)!}{s!} F(-s;\tfrac{1}{2};\rho^2) \quad \text{(D.57)}$$

$$H_{2s+1}(\rho) = (-)^n \frac{(2s+1)!}{s!} 2\rho . F(-n;\tfrac{3}{2};\rho^2) \quad \text{(D.58)}$$

$$L_n(\rho) = n!\ F(-n;1;\rho) \quad \text{(D.59)}$$

$$L_n{}^m(\rho) = (-)^m\ (n!)^2\ [m!(n-m)!]^{-1}\ F(-[n-m];\ m+1;\rho) \quad \text{(D.60)}$$

$$J_l(\rho) = \frac{(\frac{1}{2}\rho)^l}{\Gamma(l+1)} e^{-i\rho} F(l+\tfrac{1}{2};\ 2l+1;2i\rho) \quad \text{(D.61)}$$

Matrices

A rectangular array made up of numbers arranged in m rows and n columns constitutes an $m \times n$ matrix. The number standing at the intersection of the ith row and jth column is the (i,j) element of A and is denoted by A_{ij} (or $(A)_{ij}$ or a_{ij}). An $m \times 1$ matrix (a single column) is called a *column vector* and a $1 \times n$ matrix (a single row), a *row vector*. Basic operations with matrices are defined as follows ($A,B,C\ldots$ are matrices, c is a scalar, i.e. a single number):

Addition: If A, B are of the same dimensions ($m \times n$),

$$(A+B)_{ij} = A_{ij} + B_{ij} \quad \text{(D.62)}$$

Scalar multiplication:

$$(cA)_{ij} = cA_{ij} \quad \text{(D.63)}$$

Matrix multiplication: An $(m \times n)$ matrix A can be multiplied by an $(n \times k)$ matrix B on the right, yielding an $(m \times k)$ matrix AB with elements

$$(AB)_{ij} = \sum_{l=1}^{n} A_{il}B_{lj} \quad \text{(D.64)}$$

This rule is associative, that is, $(AB)\ C = A\ (BC)$. If A, B are *both* $(n \times n)$ square matrices, then AB and BA exist, but $AB \neq BA$ in general.

Unit, null and diagonal matrices: The $n \times n$ square matrix I with elements δ_{ij} has the property that for any $m \times n$ matrix A or $n \times k$ matrix B, $AI = A$, $IB = B$. So I is called the *unit matrix* (of dimension n); it is often written simply as 1. A null or zero matrix is one whose elements are all zero. A diagonal matrix A has $A_{ij} = 0$ for all $i \neq j$.

Inverse: If A,B are $n \times n$ square matrices such that $AB = 1$, then A and B are the inverses of each other and $A = B^{-1}$, $B = A^{-1}$. If A^{ij} is the matrix obtained from A by deleting the ith row and jth column,

$$(A^{-1})_{ij} = (-)^{i+j}\frac{\det A^{ji}}{\det A} \tag{D.65}$$

(If $\det A = 0$, the matrix is said to be *singular*; it has no unique inverse). The inverse of a product of square matrices is given by

$$(AB)^{-1} = B^{-1}A^{-1} \tag{D.66}$$

Transpose and Hermitian conjugate: The transpose $\tilde{A}$ or A^{T} is obtained by writing the rows of A as columns and vice versa. If this is followed by complex conjugation (*) of all elements, the result is the Hermitian conjugate of A, denoted by $A^\dagger$.

$$(A^\dagger) = \tilde{A}^*;\ \tilde{A}_{ij} = A_{ji},\ A_{ij}{}^\dagger = A_{ji}{}^* \tag{D.67}$$

$$(\widetilde{AB}) = \tilde{B}\tilde{A},\ (AB)^\dagger = B^\dagger A^\dagger \tag{D.68}$$

Note the reversal of order of factors in Eqs. D.66 and D.68.

Symmetric, Hermitian and Unitary Matrices are defined by

$$A = \tilde{A}\quad \text{(Symmetric)}\qquad A = A^\dagger\quad \text{(Hermitian)} \tag{D.69}$$

$$AA^\dagger = A^\dagger A = 1\quad \text{(Unitary)} \tag{D.70}$$

Transformations: The process $A \to SAS^{-1}$ where S is any nonsingular matrix, is a *similarity transformation*. It is called a *unitary transformation* if S is unitary ($S^{-1} = S^\dagger$). The *trace*, defined by

$$\operatorname{tr} A = \sum_i A_{ii} \tag{D.71}$$

and the determinant, are unchanged by similarity transformation. Further, the trace of a product of any number of matrices is unchanged by cyclic permutations, e.g.

$$\operatorname{tr}(ABC) = \operatorname{tr}(BCA) = \operatorname{tr}(CAB) \tag{D.72}$$

Rank: The maximum number of linearly independent rows (or columns) in a matrix A is called the rank (r) of A. The rank is also the dimension of the largest submatrix (obtained by deleting various rows and columns of A) which has a non-vanishing determinant. This means that if a square ($n \times n$) matrix has rank $r < n$ it is singular.

Solution of linear equations: If A is an ($n \times n$) matrix of rank r, the equation $Ax = 0$ for the column vector x has $(n - r)$ linearly independent nontrivial (i.e. nonzero) solutions, say $w^{(\alpha)}$, $\alpha = 1,2,\ldots, (n - r)$.

$$Ax = 0 \Rightarrow x = \sum_{\alpha=1}^{n-r} c_\alpha w^{(\alpha)},\ Aw^{(\alpha)} = 0 \tag{D.73a}$$

Similarly

$$\tilde{x}A = 0 \Rightarrow \tilde{x} = \sum_{\alpha=1}^{n-r} d_\alpha \tilde{v}^{(\alpha)},\quad \tilde{v}^{(\alpha)}A = 0 \tag{D.73b}$$

Of course, if A is nonsingular, $n - r = 0$ and Eq. D.73 has no nontrivial solutions.

The equation

$$Ax = y \tag{D.74}$$

has a unique solution $x = A^{-1}y$ if A is non-singular. If A is singular, Eq. D.74 has no solutions *unless* $\tilde{v}^{(\alpha)} y = 0$ for $\alpha = 1,2,\ldots, n-r$. If this condition is satisfied, there exists an infinity of solutions: if $x^{(P)}$ is a vector such that $Ax^{(P)} = y$ then $x^{(P)} + \Sigma\, c_\alpha w^{(\alpha)}$ is also a solution of Eq. D.74 for arbitrary coefficients c_α.

Eigenvalues and Eigenvectors: If A is an $n \times n$ square matrix and x is an $n \times 1$ column such that

$$Ax = \lambda x \tag{D.75}$$

then x is an *eigenvector* (more precisely a right-eigenvector) belonging to the *eigenvalue* λ of A. This equation which can be written as $(A - \lambda I)\, x = 0$, has non-trivial solutions only if $\det(A - \lambda I) = 0$. This is the *characteristic equation of A*. It is an nth degree algebraic equation, say $f(\lambda) = 0$, whose roots are the eigenvalues of A. The matrix A itself satisfies this equation, that is, $f(A) = 0$ (*Cayley-Hamilton Theorem*).

Suppose (for a given matrix A) the characteristic equation has only simple roots. Corresponding to any particular root λ, A has a right-eigenvector, say $w^{(\lambda)}$ (as the solution of Eq. D.75) and also a left eigenvector $\tilde{v}^{(\lambda)}$. The arbitrary normalization factors contained in these may be so chosen that $\tilde{v}^{(\lambda)} w^{(\lambda)} = 1$. Also, $\tilde{v}^{(\lambda)} w^{(\lambda')} = 0$ if $\lambda \neq \lambda'$. Thus

$$Aw^{(\lambda)} = \lambda w^{(\lambda)}; \quad \tilde{v}^{(\lambda)} A = \lambda \tilde{v}^{(\lambda)} \tag{D.76a}$$

$$\tilde{v}^{(\lambda)} w^{(\lambda')} = \delta_{\lambda\lambda'} \tag{D.76b}$$

In certain special cases v and w are related.

$$v = w \ (A \text{ symmetric}); \quad v = w^* \ (A \text{ Hermitian or Unitary}) \tag{D.77}$$

Diagonalization: The n columns $w^{(\lambda)}$, when placed side by side, make up a square matrix W; and the corresponding rows $v^{(\lambda)}$, when stacked one above the other, form a square matrix V. Eq. D.76b implies then that $VW = 1$ or $W = V^{-1}$. It follows further from Eqs. D.76 that

$$(VAW)_{\lambda\lambda'} = \tilde{v}^{(\lambda)} A w^{(\lambda')} = \lambda \delta_{\lambda\lambda'} \tag{D.78}$$

so that $VAV^{-1} = VAW$ is a diagonal matrix whose diagonal elements are the eigenvalues of A:

$$VAV^{-1} = A_D \tag{D.79}$$

A_D is called the *diagonalized* form of A. The similarity transformation $A \rightarrow VAV^{-1}$ *diagonalizes* A.

If the characteristic equation of A has multiple roots (of multiplicity m_λ with $\Sigma m_\lambda = n$), the number of linearly independent right (or left) eigenvectors belonging to the eigenvalue λ is within the limits $1 \leqslant d_\lambda \leqslant m_\lambda$. If it so happens that $d_\lambda = m_\lambda$ for all λ, so that $\Sigma d_\lambda = n$, once again square matrices V and W can be constructed as before, and used to diagonalize A.

All Hermitian matrices and unitary matrices can be diagonalized. The diagonalizing matrix V is unitary in either case, because of (D.77); and (D.79) is a unitary transformation, $A_D = VAV^\dagger$.

Non-Diagonalizable Matrices: If there is any 'defective' λ which has $d_\lambda < m_\lambda$, the matrix is *not diagonalizable*. The eigenvectors (less than n in number) have to be supplemented by other vectors to make up a nonsingular matrix V', and the simplest form obtainable for VAV'^{-1} (by suitable choice of the non-eigenvectors in V') is the *Jordan canonical form.* This form has a string of square submatrices (called *Jordan blocks*) along the diagonal, all elements outside these blocks being zero. The typical form of a Jordan block, and a numerical example illustrating the way the blocks (framed by lines) appear in the canonical form, are given in Eq. D.80.

$$\begin{pmatrix} \lambda & 1 & 0 & 0 & 0 \\ 0 & \lambda & 1 & 0 & 0 \\ 0 & 0 & \lambda & 1 & 0 \\ 0 & 0 & 0 & \lambda & 1 \\ 0 & 0 & 0 & 0 & \lambda \end{pmatrix}; \quad \left(\begin{array}{ccc|c|c|c} 3 & 1 & 0 & 0 & 0 & 0 \\ 0 & 3 & 1 & 0 & 0 & 0 \\ 0 & 0 & 3 & 0 & 0 & 0 \\ \hline 0 & 0 & 0 & 3 & 0 & 0 \\ \hline 0 & 0 & 0 & 0 & 4 & 0 \\ \hline 0 & 0 & 0 & 0 & 0 & 4 \end{array}\right) \tag{D.80}$$

Occurrence of a 'defective' eigenvalue is signalled by the appearance of at least one nontrivial Jordan block (of dimension > 1) with that eigenvalue along the diagonal ($\lambda = 3$ in the above example).

The algebraic equation of lowest degree satisfied by A is the so-called *minimal equation* which has the form

$$\prod_{\lambda} (A - \lambda)^{a_\lambda} = 0 \tag{D.81}$$

where the product is over the *distinct* eigenvalues, and a_λ is the dimension of the largest Jordan block for λ. For the numerical matrix of Eq. D.80, the minimal equation is $(A - 3)^3 (A - 4) = 0$.

E

Many-Electron Atoms

The Schrödinger equation for an atom with many electrons is quite intractable if taken in its entirety. So one begins by leaving out all but the most important aspects of the interactions among the particles as a first step, and then brings in the neglected effects one by one in subsequent steps.

Central Field Approximation

To start with, one treats the atomic electrons as if they move independently of each other. Each electron is supposed to 'see' only the averaged-out field produced by all the remaining electrons and the nucleus, this field being taken to be spherically symmetric. (Methods of determining this field are considered at the end of this Appendix). The Hamiltonian in this central field approximation evidently commutes with the angular momentum operator $\mathbf{L}_i$ of each electron. So the state of each individual electron can be characterized by the quantum numbers $n\ l\ m_l$ (Sec. 4.16). To this set must be added the quantum number m_s ($= +\frac{1}{2}$ or $-\frac{1}{2}$) specifying the spin orientation. The energy is independent of m_l and m_s at this stage because no interactions involving the orientations of the $\mathbf{L}_i$ and $\mathbf{S}_i$ (such as the spin-orbit interaction) have been taken into account. Thus, for given n and l there are $2(2l+1)$ available wave functions or orbitals, all having the same energy. Each of them can accommodate no more than one electron in view of the Pauli exclusion principle (Sec. 3.16b). This set of states or orbitals (for given n, l) is said to constitute a *shell* of the atom, and if all these orbitals are occupied, the shell is said to be *closed*. The first few shells (in the order of increasing energy) and the maximum number of electrons each can accommodate, are as follows:

n	1	2		3			4			
l	0	0	1	0	1	2	0	1	2	3
Spectroscopic Notation	1*s*	2*s*	2*p*	3*s*	3*p*	3*d*	4*s*	4*p*	4*d*	4*f*
Max. no. of electrons	2	2	6	2	6	10	2	6	10	14

In the ground state of an atom, the electrons should arrange themselves in the lowest available levels. For instance, in an atom with 11 electrons (neutral sodium) the first three shells in the above scheme would be completely filled and the last remaining electron would go into a 3*s* state. This is summarized in the notation

$$(1s)^2\,(2s)^2\,(2p)^6\,3s,$$

which specifies what is called the electronic *configuration* of the atom.

Residual Electrostatic Interaction

The picture of independent motion of electrons, which was utilized up to this point, is of course an idealization. In reality, the electrostatic interactions couple each electron to all others (as well as to the nucleus). The actual electrostatic potential energy

$$H_{\text{el}} = \sum_i \sum_{j>i} \frac{e^2}{r_{ij}} - \sum_i \frac{Ze^2}{r_i} \tag{E.1}$$

does not commute with the $\mathbf{L}_i$ of the individual electrons, nor even with $\mathbf{L}_i^2$. So, the individual angular momenta l_i (which constitute the configuration labels) are not, strictly speaking, good quantum numbers. This implies that an energy eigenfunction will not consist of a single configuration, but be a linear combination of wave functions belonging to different configurations. However, the averaged-out potential energy functions $V_i(r_i)$ employed in the central field approximation are expected to be so determined that the difference $H_{\text{el}} - \Sigma\, V_i\,(r_i) \equiv H_{\text{res}}$ (the so-called *residual* electrostatic interaction) is small. Then the admixture of configurations c' with any given configuration c will also be small. [Using first order perturbation theory (Eq. 5.19a), the amount of admixture is seen to be the ratio of the matrix element $(H_{\text{res}})_{cc'}$ to the difference $(E_c - E_{c'})$ between the energies of the two configurations in the central field approximation]. With this understanding one can continue to use the configuration labels in the identification of energy levels. Further labels distinguishing different states belonging to a given configuration are provided by the quantum numbers L, S, M_L, M_S associated with $\mathbf{L}^2$, $\mathbf{S}^2$, L_z and S_z (where $\mathbf{L} = \Sigma\, \mathbf{L}_i$ is the total orbital angular momentum operator and $\mathbf{S}$, the total spin operator for the system of electrons). They are good quantum numbers because H_{el} commutes with $\mathbf{L}$ and of course with $\mathbf{S}$. (H_{res} does not involve spin). The states with various values of L in a given configuration c, which are degenerate in the central field approximation, are separated by H_{res}. But the energy continues to be independent of M_L and M_S since there are no external forces. Thus, for given (LS), there is a multiplet of $(2L + 1) \times (2S + 1)$ states with energy E_{cL}. One can choose independent states of the multiplet to be eigenstates of $\mathbf{J}^2$ and $\mathbf{J}_z$ (instead of L_z, S_z), and identify them by the labels $cLSJM$.

Spin-Orbit Interaction; Russell Saunders and j-j Coupling Schemes

There is one more effect yet to be taken into account, namely the spin-orbit interaction

$$H_{s \cdot o} = \sum_i \xi_i (r_i)\, \mathbf{L}_i . \mathbf{S}_i \tag{E.2}$$

It commutes with the $\mathbf{J}_i \equiv \mathbf{L}_i + \mathbf{S}_i$ of the individual electrons and, of course, with $\mathbf{J} \equiv \Sigma\, \mathbf{J}_i$. It does not commute with $\mathbf{L}^2$ or $\mathbf{S}^2$, so that when $\mathbf{H}_{s \cdot o}$ is taken into account, L and S are not strictly good quantum numbers.

However, if *$H_{s \cdot o}$ is much weaker than H_{res}*, the energy difference $| E_{cL} - E_{cL'} |$ between multiplets belonging to (LS) and $(L'S')$ is much larger than the matrix elements of $H_{s \cdot o}$ between member states of the two multiplets, so that $H_{s \cdot o}$ causes only an insignificant admixture between states belonging to distinct multiplets. (It may be recalled that a similar statement was made earlier regarding mixing of configurations). Under such circumstances it is enough to concentrate attention on a single (LS) multiplet and determine the effect of $H_{s \cdot o}$ on it, using the perturbation theory appropriate to degenerate levels. Since $\mathbf{J}^2$ and J_z commute with $H_{s.o}$ as noted earlier, the secular determinant (for given cLS) is diagonal in the JM basis. So the shift in energy of the state $| cLSJM \rangle$ due to the spin-orbit interaction is directly given by $\langle cLSJM | H_{s.o} | cLSJM \rangle$. This quantity depends on J (but not on M, since no external forces which distinguish between different directions of $\mathbf{J}$ are present). Therefore, the effect of the spin-orbit interaction is to split each (LS) multiplet into submultiplets which are characterized by (LSJ) and have multiplicity $(2J + 1)$. Such a level is denoted by ${}^{2s+1}[L]_J$ where $[L]$ is the spectroscopic symbol corresponding to L. For example, for $L = 1$, $S = \frac{1}{2}$, $J = \frac{1}{2}$ the notation is ${}^2P_{1/2}$.

The above treatment of the spin-orbit effect is called the *Russell Saunders* or *L-S coupling scheme*. The latter name derives from the fact that the orbital and spin angular momenta are independently added to form L and S which are then coupled to form J. The spectra of most atoms (except very heavy ones) can be understood in terms of this scheme. The selection rules in the dipole approximation are determined by the matrix elements

$$\langle c'L'S'J'M' | e \sum_i \mathbf{x}_i | cLSJM \rangle.$$

It is evident that only the ith electron is involved in the transition due to the term $\mathbf{x}_i$, and that since $\mathbf{x}_i$ is a vector operator with respect to $\mathbf{L}_i$ and is of odd parity, it changes $\mathbf{L}_i$ by one unit. Further, since $\Sigma \mathbf{x}_i$ is independent of spin, only transitions with $S' = S$ are allowed; and since $\Sigma \mathbf{x}_i$ is a vector operator with respect to **L** or **J**, it follows that $L' - L = 0, \pm 1$; $J' - J = 0, \pm 1$ (but with $J = 0 \to J' = 0$ strictly forbidden).

In the case of very heavy atoms the spin orbit interaction becomes large (because of the factor $\xi(r) \propto dV/dr$ in $H_{s.o}$). If it becomes dominant, i.e. $H_{s.o} \gg H_{res}$, then it is the angular momenta j_i of the individual electrons that are (approximately) good quantum numbers for labelling the states (since the $\mathbf{J}_i$ commute with the dominant interaction $H_{s.o}$ while $\mathbf{L}^2$ and $\mathbf{S}^2$ do not). The resulting scheme is called the *jj coupling scheme*.

Determination of Central Field

We return now to a closer look at the central field approximation.

(*a*) *The Thomas-Fermi Statistical Method:* In this method, the atomic electrons are viewed as constituting a classical charge distribution which, together with the nuclear charge, gives rise to an electrostatic potential $V(r)/e$; and this potential in turn determines the distribution of electrons around the nucleus. If the number of electrons per unit volume at a distance r from the nucleus in $n(r)$, the charge density is $n(r)e$, and the potential is then determined by Poisson's equation as

$$\nabla^2 V(r)/e = -4\pi n(r)e \tag{E.3}$$

with the boundary conditions $V(r) \to -Ze^2/r$ as $r \to 0$ and $V(r) \to 0$ as $r \to \infty$. To determine $n(r)$ in terms of $V(r)$, the following argument is used. Consider a region of volume v about $\mathbf{x}$, wherein V is approximately constant. One views it as a 'box' within which the electrons can have various quantized values of the momentum. The maximum momentum a bound electron at $\mathbf{x}$ can have is $p_0(r) = [-2m\ V(r)]^{1/2}$. It is postulated that all the quantum states in the 'box' with momenta up to this maximum value p_0 are occupied by electrons. The number of such states divided by v then gives $n(r)$. Using the known expression $2v/(2\pi\hbar)^3$ for density of states in momentum space (wherein the factor 2 is for the two spin states for each momentum), we obtain

$$n(r) = \frac{1}{v} \cdot \frac{2v}{(2\pi\hbar)^3} \int_{|\mathbf{p}|<p_0} d\tau_{\mathbf{p}} = \frac{1}{4\pi^3\hbar^3} \cdot \frac{4\pi p_0^3}{3} = \frac{p_0^3}{3\pi^2\hbar^3} \tag{E.4}$$

with $p_0 = [-2mV]^{1/2}$ Combining the above two equations, we obtain

$$\nabla^2 V(r) = \frac{1}{r^2}\frac{d}{dr}\left(r^2 \frac{dV}{dr}\right) = \frac{-4e^2}{3\pi\hbar^3}[-2mV(r)]^{3/2} \tag{E.5}$$

This can be expressed in dimensionless form in terms of $\chi(r)$ and ρ defined by

$$\chi(r) = -\left(\frac{Ze^2}{r}\right)^{-1} V(r), \quad \rho = \frac{r}{b} \tag{E.6}$$

$$b = \frac{1}{2}\left(\frac{3\pi}{4}\right)^{2/3} \cdot \frac{\hbar^2}{me^2 Z^{1/3}} = \frac{0{\cdot}885 a_0}{Z^{1/3}} \text{ with } a_0 = \frac{\hbar^2}{me^2} \tag{E.7}$$

The result is

$$\rho^{1/2}\frac{d^2\chi}{d\rho^2} = \chi^{3/2} \tag{E.8}$$

with the boundary conditions $\chi \to 1$ as $\rho \to 0$, $\chi \to 0$ as $\rho \to \infty$. From the solution of this equation (obtained numerically) one gets $V(r)$ for a particular atom on feeding in the relevant value of the scale parameter b which is a measure of the 'size' or 'radius' of the atom. It will be noted that the radius is proportional to $Z^{-1/3}$. At this radius, $V \propto Z^{4/3}$ and the de Broglie wavelength corresponding to p_0 is $\propto V^{-1/2} \propto Z^{-2/3}$. So the fractional change in V per wavelength, $[\lambda\, dV/dr]_{r=b}$ decreases as $Z^{-1/3}$ as Z increases. This means that the larger the value of Z, the better is the 'statistical' picture—of constant-potential regions with quantized states of constant momentum, and of a degenerate Fermi gas of electrons filling up these states—employed in the derivation. So the Thomas-Fermi method is the most suitable for heavy atoms. Once $V(r)$ is obtained, the electrons (in the central field approximation) are taken to have wave functions which are eigenfunctions of the one-particle Schrödinger equation with $V(r)$ as the potential term.

(*b*) *The Hartree Self-Consistent Method:* The basic assumption of this method is that the potential energy term V_i in the Schrödinger equation for the ith electron (in the central field approximation) is given by the energy of its electrostatic interaction with the nucleus and the charge clouds associated with the remaining electrons. Taking the charge density at $\mathbf{x}_j$ due to jth electron having the wave function $u_j(\mathbf{x}_j)$ to be $e\,|\,u_j(\mathbf{x}_j)\,|^2$, one has

$$V_i(\mathbf{x}_i) = -\frac{Ze^2}{r_i} + \sum_{j \neq i} \int |\,u_j(\mathbf{x}_j)\,|^2 \frac{e^2}{r_{ij}}\, d\tau_j \tag{E.9}$$

The $u_j(\mathbf{x}_j)$ in turn are to be determined from the Schrödinger equations

$$-(\hbar^2/2m)\, \nabla_j^2 u_j(\mathbf{x}_j) + V_j\, u_j(x_j) = \varepsilon_j\, u_j(\mathbf{x}_j),\ j = 1,2,\ldots,Z \tag{E.10}$$

The problem then is to determine potentials V_j and wave functions u_j such that Eqs. E.9 and E.10 are mutually consistent. In practice, one proceeds by assuming a plausible potential function V, solving the Schrödinger equation with this potential to obtain the stationary wave functions $u_j(\mathbf{x})$, assigning these wave functions to the electrons in accordance with the exclusion principle, then constructing the V_i using these wave functions in Eq. E.9, feeding them into Eq. E.10, and repeating this cycle of operations till self-consistency is achieved, i.e. the V_i's and u_i's in successive cycles coincide. Actually it is not the V_i of Eq. E.9 which is directly employed in the Schrödinger equation at each stage, but rather a spherically symmetric approximation to it obtained by averaging $V_i(\mathbf{x}_i)$ over all directions. This central potential approximation enables the u_i to be taken as products of radial wave functions and spherical harmonics.

F

Internal Symmetry

When a quantum system has symmetry under rotations, its Hamiltonian H commutes with the rotation generators, namely the components of $\mathbf{J}$ (Sec. 7.10 and 7.12). Then, (*i*) $\hat{\mathbf{J}}$ is conserved, so that j, m are good quantum numbers for the system, and (*ii*) the $(2j + 1)$ independent states for given j belong to the same energy level. [If $\hat{H}\,|\,jm\,\rangle = E\,|\,jm\,\rangle$, then $\hat{H}\,|\,j, m+1\,\rangle = (c_{jm}^{+})^{-1}\cdot \hat{H}\hat{J}_{+}\,|\,jm\,\rangle = (c_{jm}^{+})^{-1}\,\hat{J}_{+}\hat{H}\,|\,jm\,\rangle = (c_{jm}^{+})^{-1}\,\hat{J}_{+}E\,|\,jm\,\rangle = E\,|\,j, m+1\,\rangle$, and so on]. Conversely, the occurrence of such $(2j + 1)$-fold multiplets of states indicates invariance under rotations. However, it is when the degeneracy is removed by some perturbation (not possessing rotational symmetry) that the existence of multiplicity is made manifest.

The observation that the proton and the neutron have very nearly the same mass (energy) led Heisenberg to postulate (a) that they are a doublet of states of a single entity, called the nucleon, (b) that in analogy to spin-$\frac{1}{2}$ doublets (e.g. in atomic physics), this doublet of particles can be characterized by *isospin* $I = \frac{1}{2}$, i.e. it belongs to the eigenvalue $\frac{1}{2}(\frac{1}{2} + 1)$ of $\hat{\mathbf{I}}^2 = \hat{I}_1^2 + \hat{I}_2^2 + \hat{I}_3^2$ where

$$[\hat{I}_1, \hat{I}_2] = i\hat{I}_3, \quad [\hat{I}_2, \hat{I}_3] = i\hat{I}_1, \quad [\hat{I}_3, \hat{I}_1] = i\hat{I}_2 \tag{F.1}$$

and (c) that the proton and neutron belong respectively to the eigenvalues $+\frac{1}{2}$ and $-\frac{1}{2}$ of $\hat{I}_3$. Many other multiplets of particles are now known, and appropriate isospin assignments have been made to them, e.g. the pion (π-meson) occurring in three charge states, π^+, π^0, π^- ($I = 1$; $I_3 = +1, 0, -1$).

Since (F.1) is identical in form to the commutation rules (8.1) of angular momentum, the entire theory based on the latter applies to isospin too (though physically, isospin and angular momentum are quite different). In particular, for a system consisting of nucleons, pions, etc., one can construct states of definite isospin I using the very same *C-G* coefficients as in angular momentum theory (Sec. 8.5 and 8.6). Further, the role of $\hat{\mathbf{J}}$ as the generator of rotations in ordinary space is parallelled by that of $\hat{\mathbf{I}}$ as generator of rotations in a 3-dimensional *isospace* which is an *internal* space pertaining to degrees of freedom (such as charge) which are unrelated to the ordinary space. (All operators defined in relation to the latter, such as $\hat{\mathbf{x}}$, $\hat{\mathbf{J}}$, parity, etc., commute with $\hat{\mathbf{I}}$). The equality of masses of the neutron and the proton (or of the three pions) then appears as the manifestation of an *internal symmetry*, namely the symmetry with respect to rotations in isospin space.[1] More generally, for a system made up of specified numbers of nucleons, pions, etc., isospin symmetry implies that the states forming the $(2I + 1)$-fold multiplet for given I cannot be distinguished from one another[2] (except through additional, symmetry-breaking interactions). For example, in the scattering of pions (π) by nucleons ($\mathcal{N}$), the scattering amplitude is the same in all four of the $I = \frac{3}{2}$ states of the π-$\mathcal{N}$ system, and the two $I = \frac{1}{2}$ states also have a common scattering amplitude. Consequently, the amplitudes for the *six* different scattering processes (π^+p, π^+n, π^0p, π^0n, π^-p, π^-n) are linear combinations of just *two* independent ($I = \frac{3}{2}$ and $I = \frac{1}{2}$) amplitudes. Again, in the decay of a particle with specific (I, I_3) into two particles which are members of isospin multiplets, the Clebsch-Gordan expansion

$$|II_3\rangle = \sum_{I_3', I_3''} |I'I_3'; I''I_3''\rangle \langle I'I_3'; I''I_3'' | II_3\rangle$$

implies that the final state $|I'I_3'; I''I_3''\rangle$ consisting of particles with the specific quantum numbers (I', I_3') and (I'', I_3'') appears with a probability $|\langle I'I_3'; I''I_3'' | II_3\rangle|^2$. This prediction may, however, be violated to a considerable extent by symmetry-breaking effects.

It is now known that elementary particle systems possess (approximate) symmetry with respect to a special group of 'rotations' in a *complex* 3-dimensional internal space (of which the isospace rotations form a subgroup). These 'rotations' leave the 'squared length' $z_1{}^*z_1 + z_2{}^*z_2 + z_3{}^*z_3$ of vectors in this space unchanged, and can therefore be characterized by 3×3 unitary matrices; only those rotations for which the matrices have unit determinant are to be considered. These transformations define the so-called SU(3) group, which encompasses all the isospin transformations. This symmetry has consequences qualitatively similar to, but wider in scope than, those of the isospin symmetry. Exploration of these is beyond the scope of this book.

[1] The fact that the equality is not exact is attributed to a *breaking of the symmetry* due to interaction with the electromagnetic field. This interaction singles out the third direction in isospace because it depends on the charge Q which is seen to be $I_3 + \frac{1}{2}$ for the nucleon, in view of (c) above. A similar statement holds for the pions but with the difference that $Q = I_3$ in their case. In general, $Q = I_3 + \frac{1}{2}Y$ (The Gell-Mann-Nishijima relation) where Y is the *hypercharge* of the multiplet. (Possible values of Y are $0, \pm 1, \ldots$).

[2] In the case of the nucleon-nucleon system, this result goes under the name 'charge independence of nuclear forces'.

G

Perturbation Problems—Residue-Squaring Method

Residue-Squaring Iterative Method for Perturbation Problems

The method to be presented here is an iterative one which can give the eigenvalues of a perturbed Hamiltonian $H \equiv H_0 + \lambda H_1$ correct to orders λ, λ^3, λ^7, λ^{15}, . . . in successive steps.

Schrödinger Equation as a Matrix Eigenvalue Equation

This new method makes explicit use of the fact that the eigenvalue equation for any quantum mechanical operator is equivalent to a matrix eigenvalue equation (Chapter VII). For instance, the Schrödinger equation

$$Hv(\mathbf{x}) = Wv(\mathbf{x}) \tag{G.1}$$

can be transformed into

$$\sum_n H_{mn} c_n = W c_m \tag{G.2}$$

$$H_{mn} = \int u_m^*(\mathbf{x}) H u_n(\mathbf{x}) d\tau, \quad c_m = \int u_m^*(\mathbf{x}) v(\mathbf{x}) d\tau \tag{G.3}$$

where the u_m form a complete orthonormal set. The transformation is accomplished by using the closure property $\delta(\mathbf{x} - \mathbf{x}') = \sum_n u_n(\mathbf{x})\, u_n^*(\mathbf{x}')$ to write

$$v(\mathbf{x}) = \int \sum_n u_n(\mathbf{x})\, u_n^*(\mathbf{x}')\, v(\mathbf{x}')\, d\tau' = \sum u_n(\mathbf{x})\, c_n \tag{G.4}$$

in (G.1) and then multiplying the whole equation by $u_m^*(\mathbf{x})$ and integrating. In the case of a perturbation problem[1]

$$H = H_0 + \lambda H_1 \tag{G.5}$$

[1] In this Appendix, the perturbation is denoted by H_1 because notational convenience requires the use of the symbol H' for other purposes.

the u_m are chosen to be eigenfunctions belonging to eigenvalues E_m of H_0, so that

$$H_{mn} = E_m\delta_{mn} + \lambda(H_1)_{mn} \tag{G.6}$$

The H_{mn} define a matrix, and (G.2) is its eigenvalue equation. According to Eq. G.6, the off-diagonal matrix elements, given by $\lambda(H_1)_{mn}$, $(m \neq n)$, are of the first order in the small parameter λ and hence considered small in comparison with the diagonal elements

$$d_m \equiv H_{mm} = E_m + \lambda(H_1)_{mm} \tag{G.7}$$

Our new procedure for the determination of the eigenvalues and eigenfunctions of H rests on two elementary observations. The first is that all matrices related to each other by similarity transformations have identical eigenvalues. The second is that if the off-diagonal elements of a matrix are small, the diagonal elements give the eigenvalues approximately. (The smaller the off-diagonal elements, the better the approximation). From these it follows that any similarity transformation which reduces the magnitude of the off-diagonal elements would lead to an improvement in the eigenvalues (as estimated by the diagonal elements). The construction of such a similarity transformation is the core of the present method.

The Residue-Squaring Process

Consider the transformation

$$H' = (1 + \lambda F)^{-1} H (1 + \lambda F) \tag{G.8}$$

where F is a matrix to be chosen suitably. (In the remainder of this Appendix, the symbol H is to be understood as standing for the *matrix* with elements H_{mn}). Separating H into its diagonal part D and off-diagonal part λR,

$$H = D + \lambda R,$$
$$D_{mn} = d_m\delta_{mn}, \quad R_{mn} = (H_1)_{mn}(1 - \delta_{mn}) \tag{G.9}$$

we introduce it into (G.8) and obtain

$$\begin{aligned} H' &= (1 + \lambda F)^{-1}(D + \lambda R + \lambda DF + \lambda^2 RF) \\ &= D + (1 + \lambda F)^{-1}\{\lambda(R + DF - FD) + \lambda^2 RF\} \end{aligned} \tag{G.10}$$

Let us now choose F so as to satisfy

$$R + DF - FD = 0 \tag{G.11}$$

Then the expression for H' reduces to

$$H' = D + \lambda^2(1 + \lambda F)^{-1} RF \tag{G.12}$$

It is immediately evident that the off-diagonal part of H' is[2] $O(\lambda^2)$. Denoting this part by $\lambda^2 R'$, we have

$$R' = [(1 + \lambda F)^{-1} RF]_{\text{o.d.}} \tag{G.13}$$

It should also be noted that the diagonal part D' of H' differs from that of H in the second order only:

$$D' = D + \lambda^2[(1 + \lambda F)^{-1} RF]_{\text{diag}} \tag{G.14}$$

We can now repeat the above procedure, starting from $H' \equiv D' + \lambda^2 R'$ and defining F' through $R' + D'F' - F'D' = 0$, to obtain

[2] A quantity is said to be of order λ^n, or simply $O(\lambda^n)$, if its expansion in powers of λ starts with the nth power: none of the lower powers should appear in the expansion, but powers higher than the nth may be present.

$$H'' = (1 + \lambda^2 F')^{-1} H' (1 + \lambda^2 F') = D' + \lambda^4 (1 + \lambda^2 F')^{-1} R'F' \qquad (G.15)$$

The off-diagonal part is now only $O(\lambda^4)$; so also is the difference between the diagonal elements of H'' and those of H'.

What we have thus is an iterative procedure which gives matrices H', H'', H''',...all equivalent to H, wherein the off-diagonal parts are of orders λ^2, λ^4, λ^8,... In going from $H^{(n)}$ to $H^{(n+1)}$ the diagonal elements (which provide our estimate of the eigenvalues) change only by a quantity of the order 2^{n+1} in λ. So the diagonal elements of $H^{(n)}$ give the eigenvalues of H correct to order $(2^{n+1} - 1)$ in λ. In particular, the d_m, Eq. G.7, give the eigenvalues correct to the first order, in agreement with standard perturbation theory, while

$$d_m' = d_m + \lambda^2 [(1 + \lambda F)^{-1} RF]_{mm} \qquad (G.16a)$$

$$d_m'' = d_m' + \lambda^4 [(1 + \lambda^2 F')^{-1} R'F']_{mm} \qquad (G.16b)$$

etc., are good to orders λ^3, λ^7,... respectively. Note *that the residual error at any stage of the iterative process is of the order of the square of the error at the previous stage.* This statement presupposes that $(1 + \lambda F)^{-1}$, $(1 + \lambda^2 F')^{-1}$ etc., are evaluated exactly, or at least to adequate accuracy. This point will be considered more closely later on.

Explicit Expressions for Eigenvalues

To evaluate d_m' explicitly one needs the matrix elements of F. These may be obtained from Eq. G.11 which gives

$$R_{mn} + (d_m - d_n) F_{mn} = 0 \qquad (G.17)$$

On putting $m = n$ we get $R_{mm} = 0$, i.e. R is required to be a purely off-diagonal matrix, with all diagonal elements vanishing. We have in fact assumed this from the beginning. Eq. G.17 leaves the diagonal elements of F undefined. It is simplest to take $F_{mm} = 0$ and we shall do so. We then have

$$F_{mn} = - \frac{R_{mn}}{d_m - d_n} (1 - \delta_{mn}) \qquad (G.18)$$

If there is any degeneracy (i.e. if $E_m = E_n$ for some $m \neq n$ so that $d_m - d_n$ is at most of order λ and may even be zero), a certain "preconditioning" of the matrix H is necessary before using this formula. Deferring consideration of this matter let us turn to the evaluation of the inverse matrices in Eqs. G.16.

The matrices involved in quantum mechanical problems are infinite dimensional (except when the spin degree of freedom alone is considered). Exact evaluation of $(1 + \lambda F)^{-1}$ is therefore not possible. The most obvious thing to do then is to make an expansion in powers of λ,

$$(1 + \lambda F)^{-1} = 1 - \lambda F + \lambda^2 F^2 - \lambda^3 F^3 + \cdots, \qquad (G.19)$$

and retain just the number of terms needed for the desired accuracy. If, for instance, the final results are needed only to order λ^3, it is adequate to take $(1 + \lambda F)^{-1} \approx 1 - \lambda F$ in (G.16a). We then have

$$\begin{aligned} d'_m &= d_m + \lambda^2 (RF)_{mm} - \lambda^3 (FRF)_{mm} \\ &= d_m + \lambda^2 \sum_n \frac{R_{mn} R_{nm}}{d_m - d_n} - \lambda^3 \sum_{n,k} \frac{R_{mn} R_{nk} R_{km}}{(d_n - d_m)(d_m - d_k)} \end{aligned} \qquad (G.20)$$

with $d_m = E_m + \lambda(H_1)_{mm}$ and $R_{mn} = (H_1)_{mn}$. To calculate to higher orders (anything up to order λ^7), one must determine d''_m, taking d'_m, R' and F' correct to the requisite order. For instance, if the full accuracy — $O(\lambda^7)$ — is desired, one must clearly take R' and F' in (G.16b) to $O(\lambda^3)$ and d'_m to $O(\lambda^7)$, while approximating $(1 + \lambda^2 F')^{-1}$ as $1 - \lambda^2 F'$. This requires in turn that the expansion of $(1 + \lambda F)^{-1}$ in (G.13) be taken up to the cubic term, and in (G.14) up to the $(\lambda F)^5$ term. It is to be particularly noted that it is not enough to take (G.20) for d'_m while making higher order calculations.

An important feature of the expression (G.20) for d'_m is that it is not a simple expansion in powers of λ. The denominator also contains λ through d_m. (A similar statement can of course be made also about d''_m etc.) The effect of this can be appreciated if one considers a case where the $(H_1)_{mm}$ are all nonzero and unequal, and λ is very large instead of being small as was originally supposed. If λ is large enough that $\lambda(H_1)_{mm} \gg E_m$ for all m then the F_{mn} become of order $(1/\lambda)$. It is easy to see then that d'_m increases only linearly with λ for large λ, and the same is true of d''_m etc. This is precisely how the exact eigenvalues would behave too since $H \approx \lambda H_1$ for large λ. (In contrast, the behaviour of the formulae of conventional perturbation theory is quite unrealistic for large λ). It appears therefore that despite the truncation of binomial expansions, the validity of the present method is not restricted to very small λ. In fact if $(1 + \lambda F)^{-1}$, $(1 + \lambda^2 F')^{-1}$, etc., can be calculated exactly (as for example when the matrix to be diagonalized is of finite dimension) the results can be obtained with any desired accuracy even for large λ by performing the iterations a sufficient number of times.

The Degenerate Case

If it so happens that some of the unperturbed eigenvalues are equal, i.e. $E_m = E_n$ for some $m \neq n$, and the corresponding R_{mn} is nonzero, then F_{mn} becomes $-R_{mn}/\lambda(H_{mm} - H_{nn})$ which is very large, being of order λ^{-1}. In such a case it would no longer be possible to treat λF as small in $(1 + \lambda F)^{-1}$, and the binomial expansion would not be justified. The way out of this difficulty is to "precondition" the matrix H to make R_{mn} zero whenever $d_m - d_n$ is $O(\lambda)$ or smaller. When this is done we have from (G.17) that $(d_m - d_n) F_{mn} = 0$ so that F_{mn} can be taken to be zero even though $d_m - d_n$ is $O(\lambda)$. The diagonalization process given above then proceeds without difficulty.

The "preconditioning" is done as follows: Suppose there is an r-fold degeneracy, i.e. r of the d_m's differ from each other by $O(\lambda)$. The elements belonging to the r rows and columns containing these particular d_m's form an $r \times r$ submatrix of H. A similarity transformation of H which diagonalizes this $r \times r$ submatrix will accomplish our objective, since it makes all the off-diagonal elements (R_{mn} of the transformed matrix) *within this submatrix* zero. If $A^{(r)}$ is an r-dimensional matrix which diagonalizes the $r \times r$ submatrix by itself, the required similarity transformation of H is $A^{-1}HA$ where the elements of A are given by δ_{mn} *except* within an $r \times r$ block pertaining to those rows and columns to which the degenerate d_m's belong. This $r \times r$ block of A is to be taken as $A^{(r)}$.

Whenever there is degeneracy, the above preconditioning is to be carried out for all degenerate eigenvalues before starting the residue-squaring process. It may be noted that this step is similar to the treatment of degenerate levels in standard perturbation theory.

The Eigenfunctions

Eq. G.2 defines the elements c_m of a column eigenvector c belonging to the eigenvalue W of the matrix H. In matrix notation this equation reads $Hc = Wc$. Multiplying by $(1 + \lambda F)^{-1}$ and using (G.8) we can rewrite this as $H'(1 + \lambda F)^{-1} c = W(1 + \lambda F)^{-1}c$, bringing out the elementary fact that $(1 + \lambda F)^{-1}c$ is an eigenvector of H'. If the latter is known, we can find c by premultiplication by $(1 + \lambda F)$.

Now, when we make the approximation of taking the diagonal elements d'_m of H' as the eigenvalues of H and hence of H', in effect we take $H' \approx D'$. So the eigenvectors of H' should be approximated by those of the diagonal matrix D'. Any such eigenvector has all elements zero except one (say in the kth position) which may be taken as unity for normalization.

Let this (approximate) eigenvector of H' be called e.

$$H'e \approx D'e = We; \; e_n = \delta_{nk} \tag{G.21}$$

Let the corresponding eigenvector of H, to this approximation, be denoted by c'. Then $c' = (1 + \lambda F)\, e$, and its elements are given by

$$c'_m = \Sigma\, (1 + \lambda F)_{mn}\, e_n = (1 + \lambda F)_{mk} = \delta_{mk} - \frac{\lambda R_{mk}}{d_m - d_k} \tag{G.22}$$

For various values of k (the position in which the unit element of e occurs) one gets various eigenvectors of H' and corresponding eigenvectors c' of H.

If in the determination of the eigenvalues we had carried the iterative process one step further and approximated H'' (and not H') by its diagonal part, then we would have e as an eigenvector of H'' (rather than of H'). It is easily verified that the corresponding eigenvector of H to this approximation (say c'') will be given by $c'' = (1 + \lambda^2 F')\,(1 + \lambda F)\, e$. We would thus get

$$c''_m = [(1 + \lambda^2 F')(1 + \lambda F)]_{mk} = (1 + \lambda F + \lambda^2 F' + \lambda^3 F'F)_{mk} \tag{G.23}$$

as an improved approximation to c_m. Note that (G.23) differs from (G.22) through terms of $O(\lambda^2)$. So c_m' is correct to order λ only. Similarly c_m'' may be seen to be correct to order λ^3 since the next approximation, $c_m''' = [(1 + \lambda^4 F'')(1 + \lambda^2 F')\,(1 + \lambda F)]_{mk}$, differs from it in terms of $O(\lambda^4)$. It is interesting to note that at any given stage of the iterative process, the accuracy of the eigenvector is limited to that of the eigenvalues at the *previous stage:* $c_m', c_m'', c_m''' \ldots$ are correct to the same order as $d_m, d_m', d_m'' \ldots$ respectively.

Having obtained the eigenvectors c of the matrix H to any desired approximation, we can return to the Schrödinger language and obtain the eigenfunctions $v(\mathbf{x})$ by substituting c_m' or $c_m'' \ldots$ for c_m in (G.4).

Physical Constants*

	Symbol	*Value*	*Error*	*Units*
Velocity of light	c	2·9979250	(10)	10^{10} cm/sec
Planck's constant	h	6·626196	(50)	10^{-27} erg sec
	$\hbar$	1·0545919	(50)	10^{-27} erg sec
Electron charge	e	4·803250	(21)	10^{-10} esu
Fine structure constant ($e^2/\hbar c$)	α	7·297351	(11)	10^{-3}
	α^{-1}	137·03602	(21)	
Magnetic flux quantum ($hc/2e$)	Φ_0	2·0678538	(69)	gauss cm^2
Electron mass	m_e	9·109558	(54)	10^{-28} gm
		5·485930	(34)	10^{-4} amu
Electron charge/mass ratio (e/m_e)		5·272759	(16)	10^{17} esu/gm
Classical electron radius (e^2/m_ec^2)	r_0	2·817939	(13)	10^{-13} cm
Bohr radius ($\hbar^2/m_ee^2$)	a_0	5·2917715	(81)	10^{-9} cm
Compton wavelength of electron (h/m_ec)	λ_c	2·4263096	(74)	10^{-10} cm
Rydberg's constant ($m_ee^4/4\pi\hbar^3c$)	R_∞	1·09737312	(11)	10^5 cm^{-1}
Bohr magneton ($e\hbar/2m_ec$)	μ_B	9·274096	(65)	10^{-21} erg/gauss
Electron magnetic moment	μ_e	1·0011596389	(31)	Bohr magnetons
		9·284851	(65)	10^{-21} erg/gauss
Proton mass	M_p	1·00727661	(8)	amu
		1·672614	(11)	10^{-24} gm
		1836·109	(11)	m_e
Neutron mass	M_n	1·00866520	(10)	amu
		1·674920	(11)	10^{-24} gm
Proton magnetic moment	μ_p	1·52103264	(46)	10^{-3} μ_B
		1·4106203	(99)	10^{-23} erg/gauss
Nuclear magneton	μ_N	5·050951	(50)	10^{-24} erg/gauss
Avogadro's number	N	6·022169	(40)	10^{23}/mole
Faraday constant (Ne)	F	2·892599	(16)	10^{14} esu/mole
Gas constant	R	8·31434	(35)	10^7 erg/mole/deg
Boltzmann's constant	k	1·380622	(59)	10^{-16} erg/deg
Stefan-Boltzmann constant	σ	5·66961	(96)	10^{-5}erg/sec/cm^2/deg^4
Gravitational constant	G	6·6732	(31)	10^{-8} dyne.cm^2/gm^2

1 Mev =	1·6021917	(70)	10^{-6} erg
1 amu =	931·4812	(52)	Mev
=	1·660531	(11)	10^{-24} gm.

*These constants are taken from the compilation by B. N. Taylor, W. H. Parker and D. N. Langenberg, *Rev. Mod. Phys.*, Vol. 41, No. 2, July 1969. The last digits of the values are uncertain to the extent shown under "errors".

INDEX